Wireless Sensor Networks in Smart Environments

IEEE Press
445 Hoes Lane
Piscataway, NJ 08854

Wireless Sensor Networks in Smart Environments

Enabling Digitalization from Fundamentals to Advanced Solutions

Edited by

Domenico Ciuonzo
University of Naples Federico II
Italy

Pierluigi Salvo Rossi
Norwegian University of Science and Technology
Trondheim, Norway

IEEE Press Series on Sensors
Vladimir Lumelsky, Series Editor

IEEE*Press*

WILEY

Published by John Wiley & Sons, Inc., Hoboken, New Jersey.
Published simultaneously in Canada.

Library of Congress Cataloging-in-Publication Data Applied for:

Hardback ISBN: 9781394249824

Cover Design: Wiley
Cover Image: © metamorworks/Shutterstock

Set in 9.5/12.5pt STIXTwoText by Straive, Chennai, India

To Rosaria, Lucia, Giorgia, and my family

—Domenico Ciuonzo

My MSc thesis was dedicated to my parents (Claudio and Teresa)
My PhD thesis was dedicated to my wife (Francesca)
My first book was dedicated to my children (Federico, Daniele, and Claudio)
Now it is my turn: this book is dedicated to myself

—Pierluigi Salvo Rossi

Contents

About the Editors

Domenico Ciuonzo received his PhD in 2013 from the University of Campania. He has held the following visiting appointments: NATO CMRE (IT), UCONN (US), NTNU (NOR), and CTTC (ES). He has been a Track-Chair for IEEE WCNC 2024, a publication chair for IEEE TMA 2023, an elected member of IEEE SPS SPCOM Technical Committee, and a member of the "Conferences and Events Committee" of the IEEE IoT Technical Community. He has served as Associate EiC for the *IEEE Communications Letters*, Technical Editor for the *IEEE Transactions on Aerospace and Electronic Systems*, and Lead Guest Editor for the *IEEE IoT Magazine*. His reviewing and editorial activities were recognized by the *IEEE Communications Letters*, *IEEE Transactions on Communications*, *IEEE Transactions on Wireless Communications*, and *MDPI*, which nominated him as Exemplary Reviewer and Editor for 14 times. He is the recipient of two Best Paper awards (IEEE ICCCS 2019 and Elsevier Computer Networks 2020), the 2019 Exceptional Service award from IEEE AESS, the 2020 Early-Career Technical Achievement award from IEEE SENSORS COUNCIL for sensor networks/systems, and the 2021 Early-Career Award from IEEE AESS. His research interests fall within the areas of data fusion, network traffic analysis, statistical signal processing, IoT and wireless sensor networks, and AI. He has co-authored over 150 journal and conference publications in top-notch venues. Since 2016, he has been an IEEE Senior Member.

Pierluigi Salvo Rossi (Senior Member, IEEE) was born in Naples, Italy, in 1977. He received his DrEng degree (summa cum laude) in telecommunications engineering and PhD in computer engineering from the University of Naples "Federico II," Italy, in 2002 and 2005, respectively. He is currently a Full Professor and the Deputy Head of the Department of Electronic Systems, Norwegian University of Science and Technology (NTNU), Trondheim, Norway. He is also a (part-time) Senior Research Scientist of the Department of Gas Technology, SINTEF Energy Research, Norway. Previously, he worked with the University of Naples "Federico II," Italy; with the Second University of Naples, Italy; with NTNU, Norway; and with Kongsberg Digital AS, Norway. He held visiting appointments with Drexel University (USA), Lund University (Sweden), NTNU (Norway), and Uppsala University (Sweden). His research interests fall within the areas of communication theory, data fusion, machine learning, and signal processing. Prof Salvo Rossi was awarded the Exemplary Senior Editor Award of the *IEEE Communications Letters* in 2018. He is (or has been) in the Editorial Board of the *IEEE Sensors Journal, IEEE Transactions on Signal and Information Processing over Networks, IEEE Open Journal of the Communications Society, IEEE Communications Letters*, and *IEEE Transactions on Wireless Communications.*

List of Contributors

Paolo Casari
Department of Information Engineering and Computer Science
University of Trento
Trento
Italy

and

CNIT
Parma
Italy

Charles C. Cavalcante
Department of Teleinformatics Engineering
Federal University of Ceara
Fortaleza
Ceara
Brazil

Themistoklis Charalambous
University of Cyprus
Nicosia
Cyprus

Apoorva Chawla
Department of Electronic Systems
Norwegian University of Science and Technology
Trondheim
Norway

Domenico Ciuonzo
Department of Electrical Engineering and Information Technologies (DIETI)
University of Naples Federico II
Naples
Italy

Emma Green
Department of Engineering Cybernetics
Norwegian University of Science and Technology
Trondheim
Norway

José M. B. da Silva Jr
Department of Information Technology
Uppsala University
Uppsala
Sweden

Lital Dabush
School of ECE
Ben-Gurion University of the Negev
Beer-Sheva
Israel

Waltenegus Dargie
Faculty of Computer Science
Institute of Systems Architecture
Chair of Distributed and Networked Systems
TU Dresden
Dresden
Germany

Subhrakanti Dey
Department of Electrical Engineering
Signals and Systems Division
Uppsala University
Uppsala
Sweden

Jun Fang
National Key Laboratory of Wireless Communications
University of Electronic Science and Technology of China
Chengdu
China

Carlo Fischione
School of Electrical Engineering and Computer Science and Digital Futures Research Center
KTH Royal Institute of Technology
Stockholm
Sweden

Morad Halihal
School of ECE
Ben-Gurion University of the Negev
Beer-Sheva
Israel

Song Han
School of Computing
University of Connecticut
Storrs
CT
USA

Aditya K. Jagannatham
Electrical Engineering
Indian Institute of Technology Kanpur
Kanpur
India

Usman A. Khan
Tufts University
Medford
MA
USA

Vikram Krishnamurthy
Department of Electrical and Computer Engineering
Cornell University
Ithaca
NY
USA

Mehmet N. Kurt
Department of Electrical Engineering
Columbia University
New York
NY
USA

Hongbin Li
Department of Electrical and
Computer Engineering
Stevens Institute of Technology
Hoboken
NJ
USA

Natong Lin
School of Computing
University of Connecticut
Storrs
CT
USA

Ashkan Moradi
Department of Electronic Systems
NTNU
Trondheim
Norway

Gal Morgenstern
School of ECE
Ben-Gurion University of the Negev
Beer-Sheva
Israel

Subhas C. Mukhopadhyay
School of Engineering
Macquarie University
Sydney
New South Wales
Australia

Anindya Nag
Faculty of Electrical and Computer
Engineering
Technische Universität Dresden
Dresden
Germany

and

Centre for Tactile Internet with
Human-in-the-Loop (CeTI)
Technische Universität Dresden
Dresden
Germany

Chen Quan
Faculty of Electrical Engineering
Mathematics, and Computer Science
Delft University of Technology
Delft
Netherlands

Muhammad I. Qureshi
Tufts University
Medford
MA
USA

Marco Di Renzo
Laboratoire des Signaux et Systèmes
CNRS
CentraleSupélec
Université Paris-Saclay
Gif-sur-Yvette
France

and

Department of Engineering
Centre for Telecommunications
Research
King's College London
London
United Kingdom

Apostolos I. Rikos
Boston University
Boston
MA
USA

Michel Rottleuthner
Faculty of Computer Science
HAW Hamburg
Hamburg
Germany

Tirza Routtenberg
School of ECE
Ben-Gurion University of the Negev
Beer-Sheva
Israel

and

ECE Department
Princeton University
Princeton
NJ
USA

Pierluigi Salvo Rossi
Department of Electronic Systems
Norwegian University of Science and Technology
Trondheim
Norway

Thomas C. Schmidt
Faculty of Computer Science
HAW Hamburg
Hamburg
Germany

Luke Snow
Department of Electrical and Computer Engineering
Cornell University
Ithaca
NY
USA

Diego P. Sousa
Department of Teleinformatics Engineering
Federal University of Ceara
Fortaleza
Ceara
Brazil

Nagender K. Suryadevara
School of Computer and Information Sciences
University of Hyderabad
Hyderabad
India

Luca Turchet
Department of Information Engineering and Computer Science
University of Trento
Trento
Italy

Naveen K. D. Venkategowda
Department of Science and Technology
Linköping University
Norrköping
Sweden

Pramod K. Varshney
Department of Electrical Engineering and Computer Science
Syracuse University
Syracuse
NY
USA

H. Vincent Poor
ECE Department
Princeton University
Princeton
NJ
USA

Matthias Wählisch
Faculty of Computer Science
Institute of Systems Architecture
Chair of Distributed and Networked Systems
TU Dresden
Dresden
Germany

Hongwei Wang
National Key Laboratory of Wireless Communications
University of Electronic Science and Technology of China
Chengdu
China

Xiaodong Wang
Department of Electrical Engineering
Columbia University
New York
NY
USA

Stefan Werner
Department of Electronic Systems
NTNU
Trondheim
Norway

Yasin Yilmaz
Department of Electrical Engineering
University of South Florida
Tampa
FL
USA

Zelin Yun
School of Computing
University of Connecticut
Storrs
CT
USA

Alessio Zappone
Department of Electrical and Information Engineering "Maurizio Scarano"
University of Cassino and Southern Lazio
Cassino
Italy

Shengli Zhou
Department of Electrical and Computer Engineering
University of Connecticut
Storrs
CT
USA

Preface

The rapid evolution of **Wireless Sensor Networks (WSNs)** and their integration into **smart environments** are redefining the digital landscape. From **intelligent cities** to **precision agriculture**, from **predictive maintenance** to **digital healthcare**, WSNs serve as the sensing backbone of the **Internet of Things (IoT)**—an ecosystem that continues to grow at an exponential pace. As both industry and research communities push the boundaries of WSN capabilities, a comprehensive and forward-thinking reference has become indispensable.

This book, *Wireless Sensor Networks in Smart Environments: Enabling Digitalization from Fundamentals to Advanced Solutions*, is a much-needed contribution to the field. By bringing together leading experts across different domains, it offers a well-structured, interdisciplinary approach to the challenges and innovations shaping the future of WSNs. The book's four-part organization—**Signal Processing, Communication Technologies, Cybersecurity, and Applications in Smart Environments**—ensures a **coherent and integrated exploration** of key topics. Readers will gain **deep theoretical insights** while also benefiting from **real-world case studies** that illustrate the impact of WSNs in diverse application areas.

One of the book's most valuable contributions lies in its ability to bridge **fundamental concepts** with **emerging technologies**. The integration of **AI, statistical signal processing, advanced wireless communication techniques, and cybersecurity tools** into WSNs is reshaping how data is collected, processed, and utilized in **smart environments**.

This book provides not only **methodological foundations** but also a glimpse into **future trends**, making it a **timely and enduring resource** for researchers, graduate students, and industry professionals alike.

In the coming years, WSNs will play an even more central role in the **digital transformation** of industries. This book serves as both a foundational guide and an advanced reference, equipping readers with the **knowledge and skills** needed to contribute to this rapidly evolving field.

Whether the reader is an academic, an engineer, or a technology strategist, this book serves as an **indispensable and authoritative resource,** offering essential insights into the evolving landscape of **intelligent wireless sensing**.

June 5th 2025

Domenico Ciuonzo, Naples (IT)
Pierluigi Salvo Rossi, Trondheim (Nor)

Acknowledgments

The editors would like to thank Michelle Dunckley and Nandhini Karuppiah for their precious help and administrative support during the editorial preparation of the book. Also, they would like to recognize the following collaborators who helped with the reviewing phase of book chapters: Mohammed Ayalew Belay (NTNU), Omkar Vilas Bhoite (NTNU), Muhammad Asaad Cheema (NTNU), Venkata Sateeshkrishna Dhuli (NTNU), Amirshayan Haghipour (NTNU), Manju James (NTNU), Amanda Ledell (NTNU), Johan Nicolas Suarez Lojan (NTNU), Peter Keenan Morris (NTNU), Vitor Rosa Meireles Elias (NTNU), Dawit Kiros Redie (NTNU), Khadija Shaheen (NTNU), and Gianluca Tabella (SINTEF Energy Research). Last but not least, the editors acknowledge the huge effort from contributing authors in delivering high-quality products, which led to the present book outcome.

Domenico Ciuonzo
Pierluigi Salvo Rossi

Introduction

The Internet-of-Things (IoT) revolution is no longer a distant dream. Today, billions of connected devices are seamlessly and pervasively integrated into our daily lives, driving innovation in industries including surveillance, healthcare, urban planning, and environmental and industrial monitoring. These devices, primarily sensors and actuators, form the backbone of Wireless Sensor Networks (WSNs), however, unique opportunities and challenges exist in the adoption of WSNs for any specific vertical area. For the mentioned reason, the global WSN market is projected to grow exponentially, with estimates valuing it at over USD 148.67 billion by 2026 (Source: Fortune Business Insights), while IoT-related investments are expected to exceed USD 1 trillion globally in the same projected timeframe (Source: International Data Corporation). WSNs usually consists of a number of small, inexpensive, heterogeneous, and geographically dispersed nodes, which reflects in limited computational and storage capabilities, and limited energy availability. Based on these premises, WSNs are in charge of (i) monitoring a physical asset and gathering vast amounts of data, (ii) disseminating (to other nodes) or report (to a collective unit) such data over the wireless medium, (iii) elaborating high-level analytics for the scenario at hand, (iv) (possibly) controlling the environment, and (v) preserving the security of the whole monitored/controlled physical asset while implementing all these functionalities.

Accordingly, this edited book deals with the design and deployment of WSNs, offering a comprehensive journey from fundamental to advanced solutions. The text is organized into **four parts**, each addressing a critical aspect of WSN technology. **Part I** (Chapters 1–3) delves into the core of signal processing techniques crucial for efficient data handling in WSNs. **Part II** (Chapters 4–6) shifts the focus to the communication technologies allowing these networks to operate reliably and efficiently, even under demanding conditions (e.g. low-energy budget). **Part III** (Chapters 7–9) tackles cybersecurity, an essential area given the vulnerabilities these networks face in a hyper-connected world. Finally,

Part IV (Chapters 10–15) showcases practical applications of WSNs, highlighting their transformative potential in smart environments, from urban monitoring to industrial automation.

This structure provides a cohesive understanding of the evolving role that WSNs play in enabling IoT, ensuring readers are well-equipped to harness the full potential of this rapidly expanding technological landscape.

Part I – Signal Processing in Wireless Sensor Networks

WSNs gather complex, high-dimensional data, representing a serious challenge for efficient processing. Signal processing, specifically adapted, is crucial to manage effectively WSN data. The first part of this book focuses on signal processing for WSNs, offering strategies to handle and analyze the data they generate. It includes *three chapters*, each addressing different but connected aspects. It will make readers gain both theoretical and practical tools to master WSN-data processing and maximize WSN potential.

In **Chapter 1** (by *Morgenstern, Dabush, Huleihel, Routtenberg, and Poor*) introduces the foundational concepts of graph signal processing (GSP), a cutting-edge approach tailored to handle signals over irregular domains such as those encountered in WSNs. GSP provides a robust framework for analyzing the intricate structures inherent in WSN data. Essential GSP tools are explored, including the graph Fourier transform and Laplacian-based regularization. Additionally, recent advancements in GSP methodologies are covered, such as smoothness validation, signal recovery, anomaly detection, and topology identification, all within the context of WSN applications. This chapter sets the basis for understanding how GSP can enhance the interpretation and utility of WSN data.

In **Chapter 2** (by *Qureshi, Rikos, Charalambous, and Khan*) shifts focus to distributed learning methods, essential for harnessing the full potential of WSNs in real-world applications. Here, the significance of distributed learning is discussed with related practical challenges and with limitations of current methodologies. The chapter aims to equip the readers with the knowledge to develop innovative approaches building on existing work. Practical applications, such as fine-tuning vision transformers in WSN environments, illustrate how distributed learning can be applied to enhance performance and efficiency in diverse scenarios.

In **Chapter 3** (by *Cuthbert and Dey*) addresses the critical task of decentralized and distributed non-Bayesian quickest change detection in energy-harvesting WSNs. This chapter delves into the mechanisms of how sensors operate within energy constraints, periodically sampling and computing log-likelihood ratios (LLRs) to detect changes. Both decentralized and distributed scenarios are examined, highlighting the balance between quantization rates and energy consumption for accurate decision-making. Additionally, an optimal sensing and quantization rate allocation problem is presented, providing analytical solutions and asymptotic expressions for detection delays and false alarm times at the fusion center.

Part II – Communications Technologies in Wireless Sensor Networks

In the fast-paced world of WSNs, efficient communication technologies are key to their success, which should be adaptable to diversified application scenarios. Part II of this book explores the latest innovations that boost WSN performance, reliability, and efficiency. It includes *three* chapters, each tackling unique challenges and presenting cutting-edge solutions. From integrating reconfigurable intelligent surfaces (RIS) for better decision fusion to developing low-complexity rules for massive multi-input multi-output (MIMO) systems and real-time WiFi protocols, these chapters offer insights into advanced communication methods that enhance WSNs across diverse applications.

In **Chapter 4** (by *Ciuonzo, Zappone, Salvo Rossi, and Di Renzo*) focuses on the distributed detection of phenomena of interest (POI) through decision fusion techniques in WSNs. It examines how decisions from multiple sensors, collected by a fusion center over a shared flat-fading channel with multiple antennas, can be integrated to make more accurate global decisions. The chapter introduces channel-aware fusion techniques supported by smart wireless environments, emphasizing the role of RIS. RIS aids in conveying the state of the POI to the fusion center efficiently, promoting energy-efficient data analytics aligned with the IoT paradigm. The chapter progresses from presenting an optimal decision fusion rule to deriving a suboptimal joint fusion rule and RIS design. This approach balances performance with reduced complexity and system knowledge requirements. Simulation-based evaluations underscore the benefits of incorporating RIS, even with suboptimal designs.

In **Chapter 5** (by *Chawla, Ciuonzo, Jagannatham, and Salvo Rossi*), the focus shifts to the development of low-complexity fusion rules for detecting unknown parameters in millimeter-wave (mmWave) massive MIMO WSNs. The chapter explores both centralized and distributed MIMO antenna topologies, evaluating system performance based on false alarm and detection probabilities. It delves into the optimization of sensor gains to enhance detection performance and examines power scaling laws for extended sensor battery life without sacrificing performance. Additionally, this chapter addresses the challenges of channel state information uncertainty, leveraging sparse Bayesian learning for mmWave massive MIMO channel estimation. Extensive simulations validate the effectiveness of the proposed detectors, highlighting their practical applicability and performance under various conditions.

In **Chapter 6** (by *Yun, Lin, Zhou, and Han*) reviews three innovative 802.11-based WiFi solutions tailored to meet the urgent need for real-time, high-speed wireless communication protocols in time- and mission-critical wireless sensing and control systems. The first solution, an RT-WiFi protocol, employs a time division multiple access (TDMA)-based data link layer scheduler to guarantee deterministic packet delivery timings using commercial off-the-shelf (COTS) devices. The second solution, SRT-WiFi, is a software-defined radio (SDR)-based implementation on an FPGA platform, offering full-stack configurability in line with evolving IEEE 801.11 standards. Lastly, the chapter explores the implementation of 802.11a/g/n/ac physical layers on GNU Radio-based SDR platforms, supporting both single-user and multiuser MIMO transmissions. The efficacy of these solutions is demonstrated through real-world testbed deployments and extensive simulations, confirming their suitability for high-speed, reliable communication in WSNs.

Part III – Cyber Security in Wireless Sensor Networks

Despite WSNs have transformed how we engage with our environment, the distributed and resource-limited nature of WSNs makes them highly susceptible to cyberthreats. This part explores the key challenges and related innovative solutions for securing these networks. The *three chapters* offer a deep dive into WSN security, covering topics from enhancing privacy in distributed filters to unified frameworks for anomaly detection, while balancing energy efficiency and security. As reliance on WSNs grows, the insights presented here are vital for protecting our connected world.

In **Chapter 7** (by *Venkategowda, Moradi, and Werner*) explores the application of distributed Kalman filters (DKFs) in multi-agent networks, a crucial technique for enhancing tracking and information sharing among agents. While DKFs significantly improve tracking accuracy, they also introduce vulnerabilities by exposing shared information to potential adversaries. This chapter addresses these privacy and security challenges head-on. A privacy-preserving DKF (PP-DKF) is introduced that enhances privacy through techniques such as noise injection and state decomposition, providing theoretical bounds on privacy leakage even in the presence of an honest-but-curious adversary. Furthermore, a robust DKF is presented, with the aim of designing to counteract Byzantine adversaries who intentionally falsify data. This robust approach transforms the problem into a distributed optimization task, utilizing the total variation penalty term and the distributed subgradient method for resilient state estimation. Through rigorous numerical simulations and theoretical analysis, the chapter demonstrates the effectiveness of these innovative algorithms in protecting the integrity and privacy of multi-agent networks.

In **Chapter 8** (by *Yilmaz, Kurt, and Wang*) addresses the critical issue of anomaly detection in the rapidly evolving landscape of complex networks such as the IoT. These networks are characterized by their high dimensionality, heterogeneity, and the dynamic nature of both systems and threats. The chapter focuses on the application of the quickest detection theory, which is essential for the timely identification of anomalies and intrusions in such environments. The novelty of this chapter lies in its unified framework approach, integrating various challenges and constraints including resource limitations and privacy concerns. By developing comprehensive quickest detection algorithms, the chapter provides a robust methodology for early detection of sudden anomalies, ensuring the security and reliability of critical network infrastructures. This holistic approach contrasts with existing methods that often address these challenges in isolation, thereby paving the way for more efficient and secure network operations.

In **Chapter 9** (by *Quan and Varshney*), the interplay between energy efficiency and security within WSNs is examined, focusing on the impact of Byzantine attacks. These attacks represent a significant threat to the decision-making processes in WSNs, particularly in schemes designed for energy efficiency. The chapter specifically investigates ordered transmission-based (OT-based) schemes, which enhance energy efficiency by transmitting only the most informative data while omitting less critical information. The challenge arises from the limited ability of the fusion center to fully comprehend the behavior of all sensors due to intermittent data reception. This makes OT-based schemes vulnerable to Byzantine attacks, where adversaries could compromise some sensors and disrupt the network's decision-making process. The chapter provides an in-depth analysis of how these attacks affect OT-based schemes and discusses potential countermeasures to mitigate their impact, thereby ensuring both energy efficiency and security in WSNs.

Part IV – Applications in Smart Environments

As we enter an age of unprecedented connectivity, smart environments have become a key innovation. These spaces use advanced technologies to interact with and adapt to their users. Part IV of this book explores various applications in smart environments, showing how IoT enhances functionality, efficiency, and user experience across different fields. This section highlights the diverse ways smart technologies are transforming urban life, from cultural engagement and resource management to security and strategy. Each of the *six chapters* offers insights into the potential and challenges of creating intelligent environments in today's world.

The Internet of Musical Things (IoMusT) represents a novel convergence of IoT and music technology. **Chapter 10** (by *Casari and Turchet*) envisions the transformative potential of IoMusT in smart cities, introducing the concept of smart musical cities. By integrating IoMusT devices with 5G and beyond, we explore a range of connected applications such as networked music performances, interactive audience experiences, and virtual agent interactions. This paradigm shift promises to enrich cultural heritage and pedagogy, making urban environments more engaging and immersive.

WSNs play a critical role in target tracking due to their enhanced reliability and precision. However, real-world applications often encounter measurement outliers caused by sensor faults or disturbances. **Chapter 11** (by *Wang, Li, and Fang*) introduces centralized and decentralized robust tracking schemes designed to detect and remove outliers, thereby improving tracking accuracy. Utilizing variational Bayesian inference, these schemes enhance the reliability of target tracking in smart environments.

Efficient water management is crucial for sustainable urban living. **Chapter 12** (by *Perdigão Sousa, da Silva Júnior, Cavalcante, and Fischione*) presents a cutting-edge approach to leakage detection in water distribution networks (WDNs) using machine learning techniques. Focusing on a residential area in Stockholm, the proposed federated learning approach preserves data privacy while improving detection rates. The chapter highlights significant improvements in detection purity rates, demonstrating the effectiveness of the proposed method in maintaining the integrity of WDNs.

Understanding the coordination and objectives of adversarial UAV networks is vital for security and defense applications. **Chapter 13** (by *Krishnamurthy and Snow*) offers an abstract interpretation of such interactions, framing coordination as a multi-objective optimization problem. By applying tools from microeconomic theory, the chapter presents methods to detect coordination and infer UAV objectives from radar signals. This innovative approach bridges the gap between physical-layer radar technology and strategic UAV behavior analysis.

In **Chapter 14** (by *Mukhopadhyay, Nag, and Suryadevara*) explores cutting-edge sensor systems and immersive technologies used to create responsive, data-driven smart environments. It discusses advancements in MEMS, nanomaterials, and wireless protocols, highlighting their roles in enhancing real-time monitoring and decision-making. By integrating immersive technologies such as AR and VR with IoT systems, this work demonstrates innovative approaches to improving human interaction, system efficiency, and resource management in smart environments. The chapter provides essential insights for optimizing smart technology applications across various domains.

The IoT is integral to the realization of smart environments, enabling extensive monitoring and management of resources through interconnected devices. **Chapter 15** (by *Dargie, Rottleuthner, Schmidt, and Wählisch*) addresses the challenges posed by harsh deployment environments and the vast data generated by IoT systems. Through two complementary use cases, it outlines the requirements and recommendations for implementing low-cost, low-power networks. The chapter provides valuable insights into overcoming obstacles and maximizing the benefits of IoT in smart environments.

Part I

Signal Processing in Wireless Sensor Networks

1

Graph Signal Processing in Wireless Sensor Networks

Gal Morgenstern[1], Lital Dabush[1], Morad Halihal[1], Tirza Routtenberg[1,2], and H. Vincent Poor[2]

[1] *School of ECE, Ben-Gurion University of the Negev, Beer-Sheva, Israel*
[2] *ECE Department, Princeton University, Princeton, NJ, USA*

1.1 Introduction

Wireless Sensor Networks (WSNs) have become central to modern data acquisition tasks, facilitating data gathering through interconnected sensor nodes [Akyildiz et al., 2002; Kandris et al., 2020]. Data originating from WSNs are usually distributed nonuniformly in space and time, which differs from regular-domain signals such as digital images and discrete-time sequences. Furthermore, WSNs are often implemented in applications that are characterized by complex and nonlinear models [He et al., 2004; Werner-Allen et al., 2006; Kim et al., 2019]. The processing of signals in WSNs is hence often intractable, especially for large networks, so it is crucial to develop advanced tools to model, process, and analyze them.

The field of graph signal processing (GSP) has gained considerable interest in the last decade due to the growing importance of networked data in various settings such as social, energy, transportation, sensor, and neural networks [Sandryhaila and Moura, 2013; Shuman et al., 2013; Ortega, 2022]. GSP theory expands concepts and techniques from traditional digital signal processing (DSP) to data indexed by graphs. GSP concepts include the graph Fourier transform (GFT), graph signal smoothness, graph filter design, and sampling and recovery of graph signals. Various GSP tools have been recruited to solve many fundamental engineering problems, such as signal denoising [Shuman et al., 2011], data reconstruction [Feng et al., 2021; Morgenstern and Routtenberg, 2024], node clustering [Sahai et al., 2012], consensus algorithms [Sandryhaila et al., 2014], and anomaly detection [Egilmez and Ortega, 2014]. Since these problems frequently

Wireless Sensor Networks in Smart Environments: Enabling Digitalization from Fundamentals to Advanced Solutions, First Edition. Edited by Domenico Ciuonzo and Pierluigi Salvo Rossi.

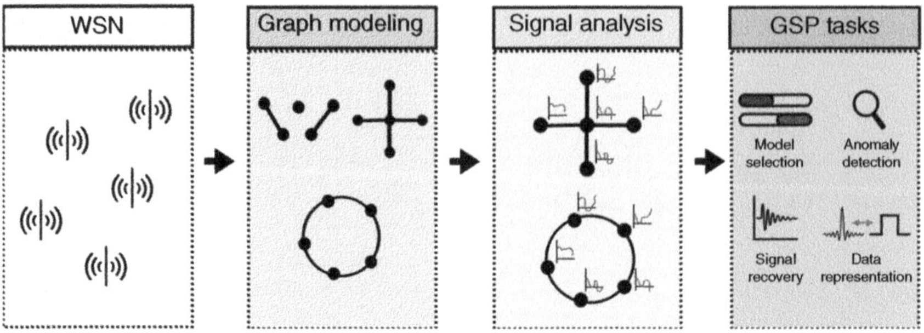

Figure 1.1 Overview of Chapter 1: end-to-end approach for processing networked signals using GSP tools in WSN applications. The approach comprises of sensor readings, graph modeling and signal analysis options, and the application of GSP tools to the acquired data.

feature in WSN data-processing tasks, the utilization of GSP tools in this context is particularly efficacious and could mark a significant shift in the methodologies and approaches employed in this sphere.

This chapter presents an end-to-end approach for processing WSN signals with GSP tools (see Figure 1.1). First, we introduce GSP-based models for WSNs as undirected weighted graphs in Section 1.2. Fundamental GSP concepts, including the graph Laplacian matrix, graph signal smoothness, and the graph spectrum, are then introduced in Section 1.3. Additionally, the chapter discusses the concept of graph signal smoothness validation in Section 1.4, which enables analyzing the signal with respect to (w.r.t.) the underlying graph. Next, based on the smoothness assumption, methods for graph signal recovery and anomaly detection are proposed in Sections 1.5 and 1.6, respectively. Utilizing the graph signal properties, we then present approaches to discover the underlying network structure. Finally, concluding remarks and future directions are given in Section 1.8.

1.2 Graph Models for WSNs

Graph theory enables the creation of intricate models that can effectively represent various types of relationships. In this context, we model WSNs as undirected weighted graphs. This approach is inherently intuitive, given that WSN deployments typically feature sensor nodes interconnected by communication links that are easily represented by graph structures. Consequently, this modeling approach enables the application of GSP tools in WSNs.

The graphical representation should capture the inherent spatial and functional relationships among sensor nodes and facilitate the translation of complex WSN data into an analyzable graph format. Different models can be considered to this end. This section outlines the general elements of the WSN in graph terms and introduces three optional graph models [Egilmez and Ortega, 2014]: the distance-based model, the correlation-based model, and the hybrid distance and correlation-based model. These perspectives provide a nuanced understanding of the diverse ways in which graph theory can be applied to WSN data.

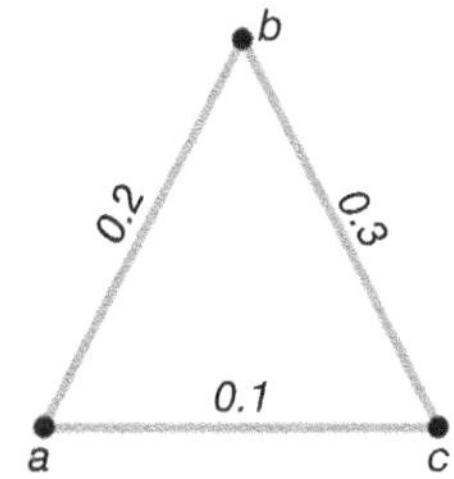

Figure 1.2 Illustration of an undirected graph with 3 nodes, $\mathcal{M} = \{a, b, c\}$, the edge set, $\xi = \{(a, b), (a, c), (b, c)\}$, and the edge weights, $w_{a,b} = 0.2$, $w_{a,c} = 0.1$, and $w_{b,c} = 0.3$.

We consider a WSN with $|\mathcal{M}|$ sensors, where each sensor measures a specific attribute. The underlying relation between the measured entities can be modeled by an undirected and weighted graph $\mathcal{G} = \{\mathcal{M}, \xi\}$, which consists of a set of nodes $\mathcal{M} = \{1, \dots, |\mathcal{M}|\}$ and a set of edges ξ. A positive edge weight between nodes k and m, $w_{k,m} > 0$, indicates that the nodes are connected, that is, $(k, m) \in \xi$, and quantifies the similarity between these nodes. Conversely, we define $w_{k,m} = 0$ if the nodes are not connected, that is, $(k, m) \notin \xi$. We assume that the graph is connected, the edge weights are positive, and no more than one edge can connect any pair of nodes, as illustrated in Figure 1.2.

In all the models below, the node set $\mathcal{M}$ is considered to be fixed. Thus, the graph model is equivalently determined by the graph adjacency matrix $\boldsymbol{W}$, defined by

$$W_{k,m} = \begin{cases} w_{k,m} & (k, m) \in \xi \\ 0 & \text{otherwise} \end{cases} \qquad k, m = 1, \dots, |\mathcal{M}|. \tag{1.1}$$

Hence, the adjacency matrix enables a matrix representation of the graph, which is often used for the design and formulation of graph models. In the following, we introduce different graph-based models appropriate for WSNs by defining their adjacency matrices.

1.2.1 Distance-Based Model

In this model, we consider the case where the locations of the sensors are known. This is often the case when the system sensors obtain a Global Positioning System (GPS) tag, or, in static systems. The underlying assumption in this model is that the distance between the sensors also defines the relation between the entities. For example, in environmental monitoring systems, it is a fair assumption that measurements gathered from closely positioned sensors will exhibit similar values [Sandryhaila and Moura, 2013]. By denoting $D(k, m)$ as the 2D-Euclidean distance between the locations of the sensors at nodes k and m, we define the distance-based adjacency matrix, $\boldsymbol{W}^{(d)}$ elementwisely by [Shuman et al., 2013]:

$$W^{(d)}_{k,m} = \begin{cases} e^{\frac{-D(k,m)^2}{\Delta_d^2}} & D(k,m) \leq \gamma_d \\ 0 & \text{otherwise,} \end{cases} \tag{1.2}$$

where Δ_d determines the exponential decay rate, and γ_d is the threshold determining the graph connectivity. It can be seen that selecting a low value for γ_d results in a low number of edges, i.e. with a sparse graph.

The distance-based modeling proposed in (1.2) is illustrated in Figure 1.3. It is important to highlight that (1.2) can be readily generalized for 3D-Euclidean distance and for alternative distance metrics, such as the Manhattan distance [Cardei et al., 2008].

1.2.2 Correlation-Based Model

In this model, we consider the case where the underlying relation between the system nodes can be characterized due to their spatial and/or temporal correlations [Pradhan et al., 2002; Vuran and Akyildiz, 2006]. By defining $\rho(k,m)$ as the correlation coefficient between the entities measured by the sensors at nodes k and m, we define the correlation-based adjacency matrix, $\boldsymbol{W}^{(c)}$ elementwisely by

$$W^{(c)}_{k,m} = \begin{cases} |\rho(k,m)| & |\rho(k,m)| \geq \gamma_c \\ 0 & \text{otherwise,} \end{cases} \tag{1.3}$$

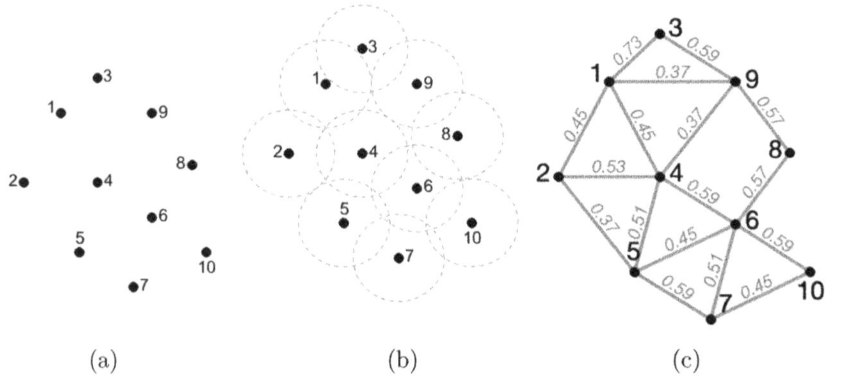

Figure 1.3 Illustration of the distance-based model: (a) system sensors: sensor network depicted in a 2D-Euclidean space. (b) Sensor-centered circles of radius $\gamma_d = 2.2$, each sensor is associated with a circle of radius $\gamma_d/2$. (c) The resulting distance-based graph, where nodes are connected by an edge if the spheres corresponding to two (or more) circles intersect, indicating that the distance between vertices is smaller than γ_d. The parameter Δ_d has been set to $\Delta_d = \gamma_d = 2.2$.

where γ_c is the threshold determining the graph connectivity. In a similar manner to in (1.2), selecting a high value of γ_c in (1.3) will result in a sparse graph.

Hybrid models that take into account both the correlation between the application entities and the geometric positions of the sensors can also be considered. For instance, here we define the hybrid-distance- and correlation-based adjacency matrix, $\mathbf{W}^{(b)}$ elementwisely by

$$W^{(b)}_{k,m} = \begin{cases} e^{-\frac{D(k,m)^2}{\Delta_d^2}} e^{-\frac{(1-|\rho(k,m)|)^2}{\Delta_c^2}} & |\rho(k,m)| \geq \gamma_c \text{ and } D(k,m) \leq \gamma_d \\ 0 & \text{otherwise,} \end{cases} \quad (1.4)$$

where $D(k,m)$ and $\rho(k,m)$ are the 2D-Euclidean distance and the correlation coefficient between nodes k and m, respectively. The parameters Δ_c and Δ_d determine the exponential decay rate for each of the exponents, and the thresholds γ_d and γ_c determine the graph connectivity. It can be seen that in order for two nodes to be connected, they must be sufficiently correlated, i.e. $|\rho(k,m)| \geq \gamma_c$, and obtain a short enough distance, i.e. $D(k,m) \leq \gamma_d$. Additionally, the parameters Δ_c and Δ_d also determine which of the properties is more dominant in the relation quantified by the edge weight.

1.2.3 Alternative Models

It should be noted that the modeling of the underlying graph in WSN applications is not limited to the models described above and can be specified depending on the application. For example, in WSNs deployed in power systems [Tariq and Poor, 2016], the underlying graph may naturally emerge from the physical interactions of the system. Moreover, additional elements in WSNs, such as sink nodes and communication links, can also be considered as part of the modeling [Schizas et al., 2008]. Thus, certain scenarios may favor a communication-based graph over a distance-based one, emphasizing communication paths and accommodating the representation of communication losses due to obstacles. Some additional graph models may also be considered: (i) a grid-based model for structured layouts, where nodes are connected if they share an edge in the grid [Servetto and Barrenechea, 2002] (this model is suitable for applications like agricultural monitoring [Díaz et al., 2011]); (ii) a random geometric graph model (this is realistic for scenarios where sensor nodes are deployed randomly and communication is confined to nearby nodes [Ramamoorthy et al., 2005]); (iii) a hierarchical/tree topology (this is beneficial for efficient data aggregation and transmission to a central point [Hasheminejad and Barati, 2021]); and (iv) a mesh topology, where each sensor node serves as a router, enabling multi-hop communication and offering redundancy and multiple paths for enhanced network reliability (this is particularly suitable in applications where robustness and fault tolerance are critical [Nurlan et al., 2021]).

1.3 Concepts in GSP

In WSNs, sensor measurements can be represented as graph signals defined as

$$\mathbf{x} : \mathcal{M} \rightarrow \mathbb{R}^{|\mathcal{M}|}. \tag{1.5}$$

Each signal element in (1.5) is a real-valued parameter that is associated with a single node of the graph. This definition can be extended to more complicated scenarios, including multidimensional vectors at each node and signals incorporating missing measurements.

A graph shift operator (GSO), **S**, operates on graph signals, similar to how time shift operates on time series in DSP. However, while time shift adjusts the position of signal values along the time axis, a graph shift redistributes signal values based on the structure of the underlying graph. The GSO, **S**, is an $|\mathcal{M}| \times |\mathcal{M}|$ matrix with entries that satisfy

$$S_{k,m} = 0, \quad \text{if } k \neq m \quad \text{and} \quad (k,m) \notin \xi, \ \forall k,m \in \mathcal{M}. \tag{1.6}$$

Consequently, the GSO defines a local operator that, when applied on a graph signal, **x**, results in

$$[\mathbf{Sx}]_k = S_{k,k}x_k + \sum_{m:(k,m)\in\xi} S_{k,m}x_m. \tag{1.7}$$

That is, the signal value x_k at node k is replaced with a linear combination of values at the node itself and the neighbors of node k.

GSP tools can be developed for various GSOs such as the adjacency matrix in (1.1). For the sake of simplicity, the tools presented in this chapter are based on the specific GSO of the graph Laplacian matrix, **L**. The elements of the matrix **L** are defined as

$$L_{k,m} = \begin{cases} \sum\limits_{(k,j)\in\xi} w_{k,j} & k = m \\ -w_{k,m} & (k,m) \in \xi \\ 0 & \text{otherwise} \end{cases} \quad k,m = 1, \dots, |\mathcal{M}|. \tag{1.8}$$

Similar to the adjacency matrix in (1.1), the graph Laplacian matrix fully captures the graph structure. The relation between those matrices is given by

$$\mathbf{L} = \text{diag}(\mathbf{1}^T\mathbf{W}) - \mathbf{W}, \tag{1.9}$$

where $\text{diag}(\mathbf{1}^T\mathbf{W})$ is a diagonal matrix whose (k,k)th entry is $\sum_{m=1}^{|\mathcal{M}|} W_{k,m}$, and $\mathbf{1}$ is the all-one vector.

1.3.1 Graph Spectrum

The graph Laplacian matrix of the graph $\mathcal{G}$, which is defined in (1.8), is a real, symmetric, and positive semidefinite matrix. Thus, its singular value decomposition (SVD) is given by

$$\mathbf{L} = \mathbf{V}\text{diag}(\boldsymbol{\lambda})\mathbf{V}^T, \tag{1.10}$$

where the columns of $\mathbf{V}$, $\{\mathbf{V}_i\}_{i\in\mathcal{M}}$, are the eigenvectors of $\mathbf{L}$ and thus satisfy $\mathbf{V}^T = \mathbf{V}^{-1}$. The diagonal matrix $\text{diag}(\boldsymbol{\lambda}) \in \mathbb{R}^{|\mathcal{M}|\times|\mathcal{M}|}$ consists of the eigenvalues of $\mathbf{L}$, $\lambda_1, \dots, \lambda_{|\mathcal{M}|}$, which satisfy $0 = \lambda_1 \leq \lambda_2 \leq \cdots \leq \lambda_{|\mathcal{M}|}$. Additionally, under the assumption that the graph is connected, it can be verified that the eigenvalues satisfy $\lambda_k > 0$, $k = 2, \dots, |\mathcal{M}|$.

The SVD of the graph Laplacian matrix enables a definition for the graph spectrum of $\mathcal{G}$. Specifically, the eigenvalues, $\lambda_1, \dots, \lambda_{|\mathcal{M}|}$, are interpreted as graph frequencies with the eigenvectors in $\mathbf{V}$ as their corresponding graph frequency components. Using this interpretation, we can represent the graph signal in (1.5) in the graph frequency domain by

$$(a)\ \tilde{\mathbf{x}} = \mathbf{V}^T\mathbf{x}, \ (b)\ \mathbf{x} = \mathbf{V}\tilde{\mathbf{x}}, \tag{1.11}$$

where (a) represents the GFT of the vector $\mathbf{x}$ and (b) represents the inverse GFT of $\tilde{\mathbf{x}}$.

1.3.2 Graph Signal Properties

A central focus of GSP is analyzing the graph signal defined in (1.5) and identifying its unique properties w.r.t. the underlying graph. In this context, an important property in GSP is graph signal smoothness, where a graph signal is considered to be smooth when its values exhibit moderate variations across the graph, i.e. signal elements have similar values at neighboring nodes. The graph total variation (GTV) is a key measure of smoothness [Shuman et al., 2013], which is defined in the node domain by

$$TV^{\mathcal{G}}(\mathbf{x}) \triangleq \mathbf{x}^T\mathbf{L}\mathbf{x} = \frac{1}{2}\sum_{k=1}^{|\mathcal{M}|}\sum_{m=1}^{|\mathcal{M}|} w_{k,m}\left(x_k - x_m\right)^2. \tag{1.12}$$

By substituting (1.10) and (1.11) in (1.12), we obtain the GTV definition in the graph frequency domain:

$$TV^{\mathcal{G}}(\mathbf{x}) = \mathbf{x}^T\mathbf{V}\text{diag}(\boldsymbol{\lambda})\mathbf{V}^T\mathbf{x} = \tilde{\mathbf{x}}^T\ \text{diag}(\boldsymbol{\lambda})\tilde{\mathbf{x}} = \sum_{k=1}^{|\mathcal{M}|}\lambda_k\tilde{\mathbf{x}}_k^2. \tag{1.13}$$

According to (1.12), a graph signal $\mathbf{x}$ is smooth if its GTV, $TV^{\mathcal{G}}(\mathbf{x})$, is small in terms of the specific application. It can be seen that in order for a graph signal to be considered smooth, its elements in connected nodes need to have similar values (according to the right-hand side (r.h.s.) of (1.12)), and its graph spectrum needs to be concentrated in the small eigenvalues region (according to the r.h.s. of (1.13)). The assessment of graph variation can also be formulated using alternative vector norms, such as other ℓ_p norms [Chen et al., 2015a].

An example of a smooth graph signal is a low-frequency bandlimited graph signal, defined as follows.

Definition 1.1 ***(Bandlimited graph signal)*** A graph signal, $\mathbf{x}$, is ideal β-bandlimited in the graph frequency domain w.r.t. the GFT basis $\mathbf{V}$ if

$$\tilde{x}_k = 0, \quad k = \beta + 1, \dots, |\mathcal{M}|, \tag{1.14}$$

where the parameter β is referred to as the cutoff graph frequency.

Definition 1.1 implies sparsity in the signal's representation in the spectral, graph frequency domain. Intuitively, similar to bandlimited DSP signals, which are characterized by a low variation over consecutive time slots, a graph-bandlimited signal is expected to obtain a low GTV.

In Figure 1.4, we present the first, third, fifth, and eight eigenvectors of the graph Laplacian matrix associated with the graph from Figure 1.3c, to illustrate the concept of graph signal smoothness. It can be seen that the eigenvectors display an increasing variation w.r.t. the graph as the eigenvalue (graph frequency) increases.

1.3.3 Graph Filters

Filtering constitutes a fundamental concept in GSP applications, similar to in DSP. A graph filter is a function $h(\cdot)$ applied to a GSO, which is associated with the underlying graph. By selecting the GSO as the graph Laplacian matrix, $\mathbf{L}$, we present the following definition for graph filters based on the SVD in (1.10).

Definition 1.2 ***(Graph filters)*** For a given graph associated with the graph Laplacian matrix $\mathbf{L}$, a graph filter $h(\mathbf{L})$ is an $|\mathcal{M}| \times |\mathcal{M}|$ matrix given by

$$h(\mathbf{L}) = \mathbf{V}\text{diag}(h(\boldsymbol{\lambda}))\mathbf{V}^T, \tag{1.15}$$

where $h(\boldsymbol{\lambda}) \triangleq (h(\lambda_1), \dots, h(\lambda_{|\mathcal{M}|}))$ is the graph frequency response and corresponds to the graph frequencies λ_k, $k = 1, \dots, |\mathcal{M}|$ defined in (1.10). Moreover, if the eigenvalues are not distinct, then $h(\lambda_k) = h(\lambda_m)$ for any k and m that satisfy $\lambda_m = \lambda_k$ [Ortega, 2022].

The output of the graph filter, provided with the input signal $\mathbf{x}$, i.e.

$$\mathbf{y} = h(\mathbf{L})\mathbf{x}, \tag{1.16}$$

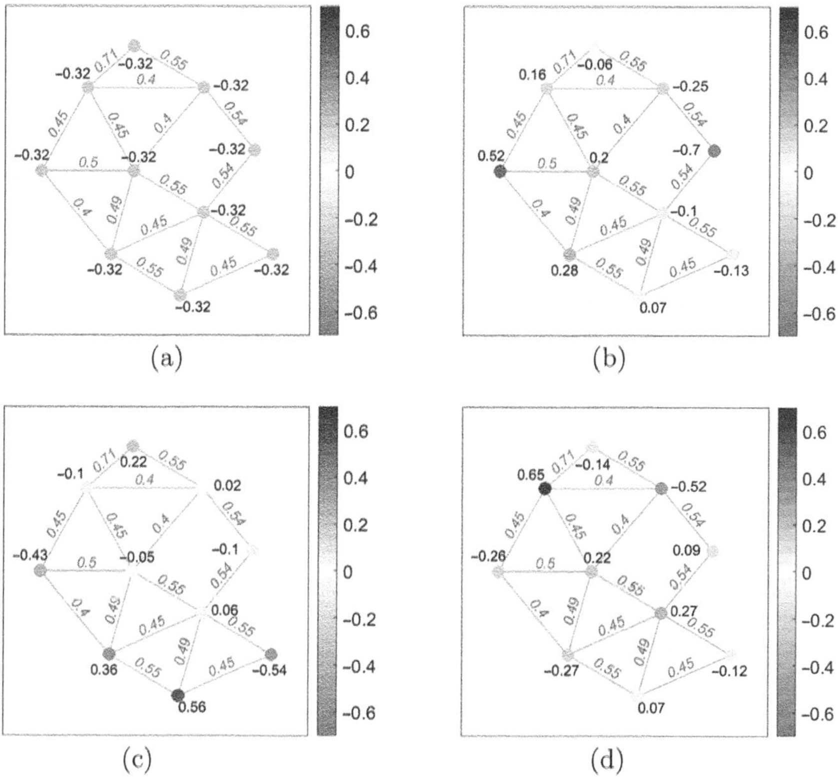

Figure 1.4 Example of Four eigenvectors of the graph Laplacian matrix associated with the graph in Figure 1.3. Each eigenvector is a graph signal, and the shading (and the number) at each node represents the signal value at the node. (a) $\lambda_1 = 0$, (b) $\lambda_3 = 0.81$, (c) $\lambda_5 = 1.52$, and (d) $\lambda_8 = 2.51$.

is also a graph signal, as defined in (1.5). Moreover, as shown in the following theorem, in the graph spectral domain, the output signal is the result of a Hadamart product (elementwise product) between the input signal and the graph frequency response. A block diagram of this filtering process is presented in Figure 1.5.

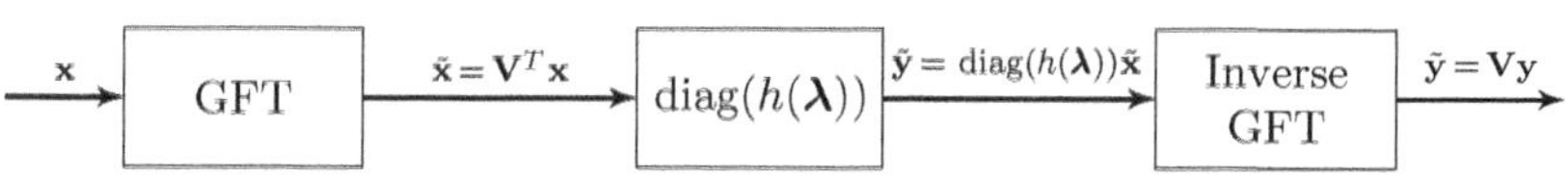

Figure 1.5 Block diagram illustrating the filtering of a graph signal **x** by the graph filter in (1.15).

Theorem 1.1 The graph spectral representation of the filtered signal satisfies

$$\tilde{\mathbf{y}} = \mathrm{diag}(h(\boldsymbol{\lambda}))\tilde{\mathbf{x}} \tag{1.17}$$

if and only if $h(\mathbf{L})$ is defined by (1.15).

Proof: ($\rightarrow$) By multiplying $\mathbf{V}$ on both sides of (1.17) and using the definitions of the GFT and the inverse GFT from (1.11), we obtain

$$\mathbf{y} = \mathbf{V}\tilde{\mathbf{y}} = \mathbf{V}\mathrm{diag}(h(\boldsymbol{\lambda}))\tilde{\mathbf{x}} = \mathbf{V}\mathrm{diag}(h(\boldsymbol{\lambda}))\mathbf{V}^T\mathbf{x}.$$

Hence, the graph filter in this case is defined by (1.15).

($\leftarrow$) By substituting (1.15) in the definition of the inverse GFT from (1.11), we obtain

$$\tilde{\mathbf{y}} = \mathbf{V}^T\mathbf{y} = \mathbf{V}^T\mathbf{V}\mathrm{diag}(h(\boldsymbol{\lambda}))\mathbf{V}^T\mathbf{x} = \mathrm{diag}(h(\boldsymbol{\lambda}))\tilde{\mathbf{x}}. \tag{1.18}$$

From Theorem 1.1, we observe that filtering a signal with the graph filter in (1.15) is equivalent in the graph frequency domain to multiplying the signal's spectrum by the frequency response of the filter. Thus, (1.15) can be perceived as an extension of the *convolution theorem* from DSP to graphs [Shuman et al., 2013]. Consequently, in a similar manner to DSP, the graph filter in (1.15) can be categorized based on its filter frequency response, e.g. as low-pass, band-pass, high-pass, or all-pass [Sandryhaila and Moura, 2014].

The following definition and theorem present an alternative representation of the graph filter in (1.15). In this representation, the graph filter is defined as a polynomial of the graph Laplacian matrix.

Definition 1.3 ***(Graph filters – alternative representation)*** For a given graph associated with the graph Laplacian matrix $\mathbf{L}$, a shift-invariant graph filter $h(\mathbf{L})$ is an $N \times N$ matrix that can be written as a polynomial of $\mathbf{L}$:

$$h(\mathbf{L}) = p(\mathbf{L}) = \sum_{j=0}^{J} a_j \mathbf{L}^j, \tag{1.19}$$

where $\mathbf{L}^0 = \mathbf{I}$ and the scalars $\{a_j\}_j$ are the polynomial coefficients.

Theorem 1.2 Any shift-invariant graph filter, defined as in (1.19), can be defined as the graph filter in (1.15).

Proof: The proof can be found in Page 86 in Ortega [2022].

The shift-invariant graph filter in (1.19) is a local operator. That is, the filter output at node k, y_k, is a linear combination of the input signal at nodes with a geodesic distance smaller than or equal to J from node k. This filter is a generalization of the conventional DSP shift-invariant filter, applied to graph signals. Moreover, analogously to DSP shift-invariant filters, matrix multiplication between shift-invariant graph filters is commutative.

It is noted that the graph filters presented in this chapter can be alternatively defined using other GSOs instead of the graph Laplacian matrix [Sandryhaila et al., 2014].

1.4 GSP-Based Smoothness Validation for WSN Signals

Graph signals obtained from WSN data exhibit smoothness, i.e. a low GTV, as defined in (1.12). This observation seems intuitive when considering the correlation-based model in Section 1.2.2 and may also hold true for the distance-based model in Section 1.2.1, given that WSN data often exhibit spatial similarity features [Pattem et al., 2008; Kong et al., 2013]. Graph signal smoothness is the basis for a variety of GSP approaches that can be utilized for WSN applications, including signal recovery (see Section 1.5) and anomaly detection (see Section 1.6). Therefore, in these applications, assessing the smoothness level of system signals w.r.t. the underlying graph is expected to provide insights that can enhance the application performance.

1.4.1 Smooth Graph Filters

A smooth graph filter is a special case of the graph filter defined in (1.15), in which the output signal $\mathbf{y}$ from (1.16) has a low GTV as defined in (1.12). The following definition gives a mathematical expression for this concept [Shaked and Routtenberg, 2021; Dabush and Routtenberg, 2024].

Definition 1.4 ***(Smooth graph filter)*** Let the elements of the input graph signal, $\mathbf{x}$, be independent and identically distributed (i.i.d.) zero-mean random variables. Additionally, denote $\mathbf{y}$ as the output of the graph filter. Then, $h(\cdot)$ in (1.15) is a smooth graph filter if

$$r \triangleq \frac{\mathrm{E}[||\mathbf{x}||^2]}{\mathrm{E}[||\mathbf{y}||^2]} \times \frac{\mathrm{E}[\mathbf{y}^T\mathbf{L}\mathbf{y}]}{\mathrm{E}[\mathbf{x}^T\mathbf{L}\mathbf{x}]} < 1. \tag{1.20}$$

It can be seen that Definition 1.4 is based on the GTV. In [Dabush and Routtenberg, 2023], it is shown that r can be written in the graph frequency domain as

$$r = \lambda_{avg}^{-1} \times \frac{\sum_{k=1}^{|\mathcal{M}|} \lambda_k h^2(\lambda_k)}{\sum_{k=1}^{|\mathcal{M}|} h^2(\lambda_k)} < 1, \quad \text{where} \quad \lambda_{avg} \triangleq \frac{1}{|\mathcal{M}|}\sum_{k=1}^{|\mathcal{M}|} \lambda_k. \tag{1.21}$$

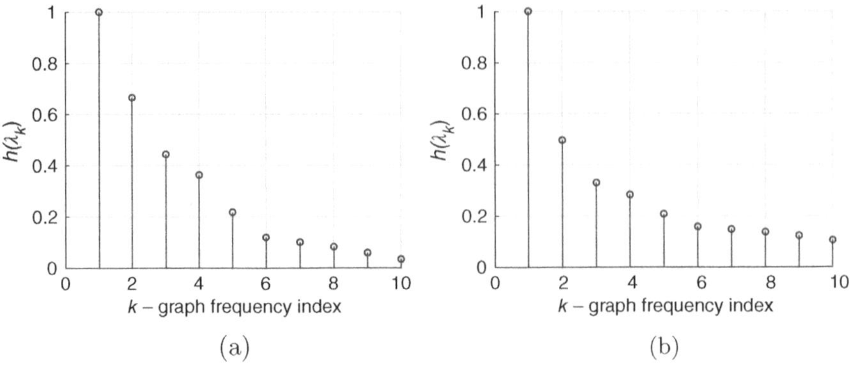

Figure 1.6 Examples of smooth graph filters. (a) Heat diffusion Kernel filter and (b) Laplacian (Tikhonov) filter.

Thus, if the energy of the frequency response is uniformly distributed across all graph frequencies, the ratio will be 1, indicating that the graph filter is not smooth. If the energy is biased toward low graph frequencies, the ratio will be lower than 1, indicating that the graph filter is smooth.

In Figure 1.6, we present two smooth graph filters [Isufi et al., 2024]: (i) the heat diffusion kernel filter, $h(\lambda_k) = \exp\{-\lambda_k\}$, $k = 1, \dots, |\mathcal{M}|$ and (ii) the Laplacian (Tikhonov) filter, $h(\lambda_k) = 1/(1 + 2.5\lambda_k)$, $k = 1, \dots, |\mathcal{M}|$. It can be seen that both filters are graph low-pass filters (GLPFs) that preserve the energy in the lower graph frequencies of the input signal, while reducing the energy of the signal at the higher graph frequency regime. Thus, they are smooth graph filters (1.21).

In Figure 1.7 we compare the input signal $\mathbf{x}$, which is drawn from the Gaussian distribution $\mathcal{N}(\mathbf{0}, \mathbf{I})$, with the output $\mathbf{y}$, which is filtered by the heat-diffusion-kernel GLPF. As expected, the output signal exhibits smoother variation over the graph in the vertex domain. In addition, it can be seen that the higher graph frequencies of the input signal have been attenuated.

We conclude our discussion on smooth graph filters with the following two remarks.

Remark 1.1 An alternative definition that evaluates whether a graph filter can be considered as a graph low-pass (GLP) filter is described in Definition 1 in Ramakrishna et al. [2020].

Remark 1.2 Modeling graph signals as outputs of graph filters is common practice for signal analysis widely used in GSP and graph neural networks (GNNs) [Schultz et al., 2021; He and Wai, 2022; Kroizer et al., 2022]. Specifically, smooth graph signals are often modeled as the output of smooth graph filters, where the input is a white Gaussian noise vector [Kalofolias, 2016; Dong et al., 2020].

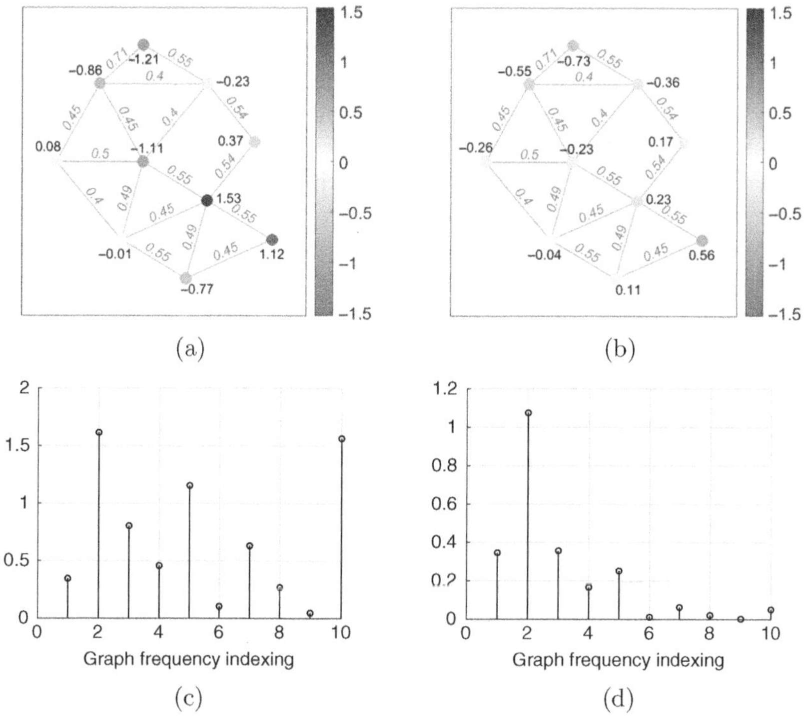

Figure 1.7 Example: comparison between the input signal $\mathbf{x}$, which is drawn from the Gaussian distribution $\mathcal{N}(\mathbf{0}, \mathbf{I})$, and the output $\mathbf{y}$, which is filtered by the heat-diffusion-kernel GLPF $h(\lambda_k) = \exp\{-\lambda_k\}$, $k = 1, \ldots, |\mathcal{M}|$. (a) Input: $\mathbf{x}$, $TV^{\mathcal{G}}(\mathbf{x}) = 13.2942$, (b) Output: $\mathbf{y}$, $TV^{\mathcal{G}}(\mathbf{y}) = 0.7179$, (c) $\tilde{\mathbf{x}} = \mathbf{V}^T\mathbf{x}$, and (d) $\tilde{\mathbf{y}} = \mathbf{V}^T(\mathbf{y})$.

1.4.2 Semi-parametric Graph Signal Smoothness Detector

In this section, our goal is to determine whether a sequence of signals $\{\mathbf{y}[n]\}_{n=1}^N$ are smooth graph signals. Based on Remark 1.2, the signals $\{\mathbf{y}[n]\}_{n=1}^N$ are modeled as the outputs of a linear graph filter, as defined in (1.16):

$$\mathbf{y}[n] = h(\mathbf{L})\mathbf{x}[n], \qquad n = 1, \ldots, N, \tag{1.22}$$

where $\{\mathbf{x}[n]\}_{n=1}^N$ are i.i.d. Gaussian random vectors, $\mathbf{x}[n] \sim \mathcal{N}(\mathbf{0}, \sigma^2\mathbf{I})$ and $\sigma^2 = 1$. The smoothness validation problem is formulated as the following composite hypothesis testing problem:

$$\begin{cases} \mathcal{H}_0: & h(\mathbf{L}) \text{ is a smooth graph filter} \\ \mathcal{H}_1: & h(\mathbf{L}) \text{ is a non-smooth graph filter.} \end{cases} \tag{1.23}$$

It should be noted that since we are dealing with real data from WSNs, we cannot presume knowledge of the graph filter, $h(\mathbf{L})$. Thus, as an integral part of the detection approach, we employ a data-driven approach to the estimation of the graph filter. For the sake of simplicity, we assume that $h(\mathbf{L})$ is a non-singular matrix, and all the eigenvalues of $\mathbf{L}$ are distinct. These assumptions are chosen for simplicity, and the proposed detector can be developed without them, as shown in [Dabush and Routtenberg, 2023].

In order to solve the composite hypothesis testing problem in (1.23), we estimate the graph filter frequency response, i.e. $\{h^2(\lambda_k), k = 1, \dots, |\mathcal{M}|\}$, from the measurements, and use the condition in (1.21) in order to determine whether the graph filter is smooth.

The log-likelihood function of the measurement model from (1.22) parametrized by $\{h^2(\lambda_k) \mid k = 1, \dots, |\mathcal{M}|\}$ after removing constant terms is

$$\log f(\mathbf{y}; h^2(\mathbf{L})) \propto \frac{N}{2}\log(|h^2(\mathbf{L})|) - \frac{1}{2}\sum_{n=1}^{N}\mathbf{y}^T[n](h^2(\mathbf{L}))^{-1}\mathbf{y}[n], \tag{1.24}$$

where $|\cdot|$ denotes the determinant of its argument matrix.

Based on (1.15), we replace $h(\mathbf{L})$ with its SVD, $\mathbf{V}\text{diag}(h(\lambda))\mathbf{V}^T$, in the log-likelihood in (1.24). Accordingly, the Maximum Likelihood (ML) estimator (see Chapter 7 in [Kay, 1993a]) is reduced to

$$\begin{aligned}\hat{h}^2(\lambda) &= \arg\max_{h^2(\lambda)\in\mathbb{R}^N} \log f(\mathbf{y}; h^2(\lambda)) \\ &= \arg\max_{h^2(\lambda)\in\mathbb{R}^N} \frac{N}{2}\log\left(|\mathbf{V}\left(\text{diag}(h^2(\lambda))\right)^{-1}\mathbf{V}^T|\right) \\ &\quad - \frac{1}{2}\sum_{n=1}^{N}\mathbf{y}^T[n]\mathbf{V}\left(\text{diag}(h^2(\lambda))^{-1}\right)\mathbf{V}^T\mathbf{y}[n],\end{aligned} \tag{1.25}$$

where $h(\lambda) = (h(\lambda_1), \dots, h(\lambda_{|\mathcal{M}|}))$. By using the GFT definition in (1.11) and the fact that $\mathbf{V}$ is a unitary matrix, we can write (1.25) in the graph frequency domain as

$$\hat{h}^2(\lambda) = \arg\max_{h^2(\lambda)\in\mathbb{R}^N} \frac{N}{2}\sum_{k=1}^{|\mathcal{M}|}\log\left(\left(h^2(\lambda_k)\right)^{-1}\right) - \frac{1}{2}\sum_{k=1}^{|\mathcal{M}|}\left(h^2(\lambda_k)\right)^{-1}\sum_{n=1}^{N}\tilde{y}_k^2[n]. \tag{1.26}$$

By equating the derivative of the log-likelihood function from (1.26), w.r.t. each of the graph filter frequencies, $h^2(\lambda_k)$, $k = 1, \dots, |\mathcal{M}|$, to zero, one obtains

$$\hat{h}^2(\lambda_k) = \frac{1}{N}\sum_{n=1}^{N}\tilde{y}_k^2[n], \quad k = 1, \dots, |\mathcal{M}|. \tag{1.27}$$

By substituting (1.27) in the condition for smooth graph filters in (1.21), one obtains

$$\hat{r} = \lambda_{\text{avg}}^{-1} \frac{\sum_{k=1}^{|\mathcal{M}|} \sum_{n=1}^{N} \lambda_k \tilde{y}_k^2[n]}{\sum_{k=1}^{|\mathcal{M}|} \sum_{n=1}^{N} \tilde{y}_k^2[n]} < 1. \tag{1.28}$$

By replacing the order of summation in both the numerator and the denominator, substituting (1.13) in the numerator and (1.11) in the denominator, and using the unitary matrix property $||\mathbf{V}\mathbf{y}||^2 = ||\mathbf{y}||^2$, one obtains

$$\hat{r} = \lambda_{avg}^{-1} \frac{\sum_{n=1}^{N} \mathbf{y}^T[n]\mathbf{L}\mathbf{y}[n]}{\sum_{n=1}^{N} ||\mathbf{y}[n]||^2} < 1. \tag{1.29}$$

The detector in (1.29) is the sample mean of the GTV of the filtered signal $\mathbf{y}$, w.r.t. to the underlying graph, normalized by its sample variance and the average graph frequency λ_{avg}. Thus, the proposed detector can be interpreted as an empirical evaluation of the GTV of the output graph signal, $\mathbf{y}$.

1.5 GSP-Based Signal Recovery in WSN Models with Missing Data

Data loss is a frequent problem in WSNs that may be caused by a variety of factors such as noise, collisions, unreliable links, and damage. Consequently, many WSN applications operate under partial observation models [Kong et al., 2013]. Several signal reconstruction techniques have been developed to address this issue, including compressive-sensing-based methods [Kong et al., 2013], K-nearest neighbors-based methods [Pan and Li, 2010], and spatial–temporal imputation-based methods [Li and Parker, 2008]. In general, signal recovery from inaccessible and/or corrupted measurements requires additional knowledge of signal properties. To compensate for missing data over the graph, one can leverage the properties of graph signals that are often bandlimited or smooth graph signals [Chen et al., 2015a; Marques et al., 2015; Romero et al., 2016]. In this section, we consider the latter approach and utilize the graph signal smoothness of signals in WSN applications. It should be noted that the efficiency of the proposed methods is influenced by the graph signal smoothness level of the application's signals.

We consider the following observation model:

$$\mathbf{y} = \mathbf{\Phi}\mathbf{x} + \mathbf{n}, \tag{1.30}$$

where $\mathbf{\Phi} \in \mathbb{R}^{Q\times|\mathcal{M}|}$ represents a linear operation on the graph signal $\mathbf{x}$, and $\mathbf{n}$ is modeled by the Gaussian vector $\mathbf{n} \sim \mathcal{N}(\mathbf{0}, \mathbf{R}) \in \mathbb{R}^Q$. Furthermore, it is assumed that the measurements in the set $\{Q \backslash S\}$ are the measurements missing. Thus, the task is to recover the signal, $\mathbf{x}$, based on the missing data model

$$\mathbf{y}_{\mathcal{S}} = \mathbf{\Phi}_{\mathcal{S},\mathcal{M}}\mathbf{x} + \mathbf{n}_{\mathcal{S}}, \tag{1.31}$$

where $\mathbf{\Phi}_{\mathcal{S},\mathcal{M}}$ is the submatrix of $\mathbf{\Phi}$ that contains only the rows associated with the index set $\mathcal{S}$.

1.5.1 Signal Recovery Approaches

Approach 1. Weighted Least Squares (WLS): A common approach for estimating $\mathbf{x}$ based on the measurement model in (1.31) is by the WLS optimization problem:

$$\begin{aligned} \hat{\mathbf{x}}^{\text{WLS}} &= \arg\min_{\mathbf{x}\in\mathbb{R}^{|\mathcal{M}|}} (\mathbf{y}_{\mathcal{S}} - \mathbf{\Phi}_{\mathcal{S},\mathcal{M}}(\mathbf{L})\mathbf{x})^T\mathbf{R}^{-1}(\mathbf{y}_{\mathcal{S}} - \mathbf{\Phi}_{\mathcal{S},\mathcal{M}}(\mathbf{L})\mathbf{x}) \\ &= \left(\mathbf{\Phi}_{\mathcal{S},\mathcal{M}}(\mathbf{L})\right)^{\dagger}\mathbf{y}_{\mathcal{S}}, \end{aligned} \tag{1.32}$$

where $(\cdot)^{\dagger}$ denotes the pseudo-inverse operator. However, this approach may not be suitable for the partial observation model in (1.31). Specifically, when the columns of $\mathbf{\Phi}_{\mathcal{S},\mathcal{M}}$ become linearly dependent, it is required to incorporate additional properties beyond the measurement model in (1.31) to achieve a unique estimator of $\mathbf{x}$.

Approach 2. WLS with GTV-based regularization: The recovery of smooth graph signals by incorporating regularization terms has been well-studied in the literature [Ortega et al., 2018; Puy and Pérez, 2018]. In particular, recovery with a regularization using the Laplacian quadratic form has been used in various applications including image processing, data classification, and supervised learning on graphs [Belkin et al., 2004; Wang and Zhang, 2006; Elmoataz et al., 2008; Cai et al., 2010; Zheng et al., 2010]. This approach involves incorporating a constraint on the GTV of the graph signal in the WLS problem in (1.32), which is formulated by

$$\begin{aligned} \hat{\mathbf{x}}^{\text{GSP-WLS}} &= \arg\min_{\mathbf{x}\in\mathbb{R}^{|\mathcal{M}|}} \left((\mathbf{y}_{\mathcal{S}} - \mathbf{\Phi}_{\mathcal{S},\mathcal{M}}(\mathbf{L})\mathbf{x})^T\mathbf{R}^{-1}(\mathbf{y}_{\mathcal{S}} - \mathbf{\Phi}_{\mathcal{S},\mathcal{M}}(\mathbf{L})\mathbf{x})\right) \\ &\qquad \text{such that } \mathbf{x}^T\mathbf{L}\mathbf{x} \leq \varepsilon. \end{aligned} \tag{1.33}$$

The parameter ε and the efficiency of the GSP-WLS estimation depend on the smoothness properties of the signal that can be validated as discussed in Section 1.4.

By using the Karush–Kuhn–Tucker (KKT) conditions, the minimization problem in (1.33) can be replaced by the following regularized optimization problem:

$$\hat{\mathbf{x}}^{\text{GSP-WLS}} = \arg\min_{\mathbf{x}\in\mathbb{R}^{|\mathcal{M}|}} \left((\mathbf{y}_{\mathcal{S}} - \mathbf{\Phi}_{\mathcal{S},\mathcal{M}}\mathbf{x})^T\mathbf{R}^{-1}(\mathbf{y}_{\mathcal{S}} - \mathbf{\Phi}_{\mathcal{S},\mathcal{M}}\mathbf{x}) + \mu\mathbf{x}^T\mathbf{L}\mathbf{x}\right). \tag{1.34}$$

The term $\mathbf{x}^T\mathbf{L}\mathbf{x}$ is a regularization term that is based on the smoothness constraint from (1.33). The parameter $\mu \geq 0$ is a Lagrange multiplier, which is a tuning parameter that replaces ε. The GSP-WLS estimator from (1.34) is obtained by equating the derivative of (1.34) w.r.t. $\mathbf{x}$ to zero, which results in [Wieringen, 2015]

$$\hat{\mathbf{x}}^{\text{GSP-WLS}} = \left(\mathbf{\Phi}_{S,\mathcal{M}}^T \mathbf{R}^{-1} \mathbf{\Phi}_{S,\mathcal{M}} + \mu \mathbf{L}\right)^{-1} \mathbf{\Phi}_{S,\mathcal{M}}^T \mathbf{R}^{-1} \mathbf{y}_S. \tag{1.35}$$

For an insufficiently measured system, the matrix $\mathbf{\Phi}_{S,\mathcal{M}}^T \mathbf{R}^{-1} \mathbf{\Phi}_{S,\mathcal{M}}$ is a singular matrix. Thus, the addition of the term $\mu\mathbf{L}$ is essential for the numerical stability of the proposed GSP-WLS estimator.

Approach 3. WLS with graph bandlimitness-based regularization: Another GSP approach involves using the bandlimitness assumption in Definition 1.1. Thus, in this case, we formulate the GLP-WLS estimator by incorporating the graph-bandlimitness property in (1.14) as the constraint on the WLS problem in (1.32) as follows:

$$\begin{aligned} \hat{\mathbf{x}}^{GLP\text{-}WLS} &= \arg\min_{\mathbf{x}\in\mathbb{R}^{|\mathcal{M}|}} (\mathbf{y}_S - \mathbf{\Phi}_{S,\mathcal{M}}(\mathbf{L})\mathbf{x})^T \mathbf{R}^{-1} (\mathbf{y}_S - \mathbf{\Phi}_{S,\mathcal{M}}(\mathbf{L})\mathbf{x}) \\ &\quad \text{such that } [\mathbf{V}^T\mathbf{x}]_k = 0, \quad k = \beta+1, \dots, |\mathcal{M}|, \end{aligned} \tag{1.36}$$

where $\mathbf{V}^T\mathbf{x}$ is the graph spectral representation of $\mathbf{x}$, as defined in (1.11). Due to the constraint in (1.36), the estimated signal, $\hat{\mathbf{x}}^{GLP\text{-}WLS}$, obtains nonzero elements only in the graph frequencies $\{\lambda_1, \dots, \lambda_\beta\}$ that are located in the lower regime of the graph spectrum.

By substituting $\mathbf{x} = \mathbf{I}\mathbf{x} = \mathbf{V}\mathbf{V}^T\mathbf{x}$ in (1.36), denoting $\mathbf{\Theta} \triangleq \mathbf{\Phi}_{S,\mathcal{M}}\mathbf{V}$, and then placing the constraint in the cost function, we obtain the following WLS problem:

$$\begin{aligned} \hat{\tilde{\mathbf{x}}}_{1:\beta}^{GLP\text{-}WLS} &= \arg\min_{\tilde{\mathbf{x}}_{1:\beta}\in\mathbb{R}^{\beta}} (\mathbf{y}_S - \mathbf{\Theta}_{1:\beta}\tilde{\mathbf{x}}_{1:\beta})^T \mathbf{R}^{-1} (\mathbf{y}_S - \mathbf{\Theta}_{1:\beta}\tilde{\mathbf{x}}_{1:\beta}) \\ &= (\mathbf{\Theta}_{1:\beta}^T \mathbf{R}^{-1} \mathbf{\Theta}_{1:\beta})^{-1} \mathbf{\Theta}_{1:\beta} \mathbf{R}^{-1} \mathbf{y}_S, \end{aligned} \tag{1.37}$$

where the submatrix $\mathbf{\Theta}_{1:\beta}$ includes the rows in $\mathbf{\Theta}$ associated with the indices $1, \dots, \beta$, and $\tilde{\mathbf{x}}_{1:\beta}$ includes the elements of $\tilde{\mathbf{x}}$ in the positions $1, \dots, \beta$. Now, by setting $\hat{\tilde{\mathbf{x}}}_{\beta+1:|\mathcal{M}|} = \mathbf{0}$, we solve (1.36) by $\hat{\mathbf{x}}^{GLP\text{-}WLS} = \mathbf{V}\hat{\tilde{\mathbf{x}}}^{GLP\text{-}WLS}$.

1.5.2 GSP-Based Sampling Policies

Managing energy consumption in WSNs is crucial, especially concerning sensing and data forwarding tasks, as they significantly influence node lifespan and network efficiency [Chiumento et al., 2019]. While improving sensor energy efficiency or devising specialized radio protocols can help conserve energy, an equally potent solution involves selective sensing in which sensors are activated only when and where needed. In traditional DSP, downsampling involves reducing the number of samples of a time series. Similarly, in GSP, downsampling refers to sampling a graph signal across a subset of nodes. By incorporating graph topology, additional information on how a signal propagates across vertices can be considered in WSNs. This raises the question: What constitutes a good sampling subset, given limitations on bandwidth, power, and the number of sensors for the sampled graph signal?

We consider a situation where the WSN operates with constrained sensing resources, possibly due to limitations in energy and communication budgets. In such instances, optimizing sensor placements becomes crucial, and various criteria can guide this optimization process. Existing sampling policies include:

- Task-based sampling in which a sample allocation rule is designed for the sensing model in (1.31) with the goal of minimizing the mean-squared-error (MSE).
- Experimentally designed (E-design) sampling [Chen et al., 2015b] aims to minimize the worst-case errors by maximizing the smallest singular value of the matrix $\mathbf{V}_{\mathcal{S},\mathcal{M}}^T \mathbf{V}_{\mathcal{S},\mathcal{M}}$.
- A-optimal design (A-design) sampling [Chen et al., 2015b] aims to minimize the average errors by seeking $\mathcal{S}$, which minimizes the trace of the matrix inverse $\mathrm{Tr}((\mathbf{V}_{\mathcal{S},\mathcal{M}}^T \mathbf{V}_{\mathcal{S},\mathcal{M}})^{-1})$.
- Cramér–Rao bound (CRB) minimization-based methods have been designed for the general model discussed in (1.31). These methods are based on minimizing the CRBs on the MSE performance (see, e.g. [Dabush et al., 2023; Routtenberg, 2021]).

These sampling policies address the challenge of optimizing sensor placements or selecting a subset of activated sensors under resource constraints in WSNs. The choice among these strategies depends on the specific objectives and criteria relevant to the application scenario, providing flexibility in adapting to different constraints and requirements.

1.6 GSP-Based Anomaly Detection for WSN

Detecting anomalies in WSNs is a critical task. These anomalies often emanate from sensor malfunctions or disruptions in communication links, which may potentially damage hardware and/or affect application performance [Rajasegarar et al., 2008; Xie et al., 2011; Erhan et al., 2021]. Detecting these anomalies is challenging, particularly when dealing with a large number of sensors in the network and/or when anomalies are deliberately concealed.

In this section, we leverage smoothness and GLP signal properties in order to detect anomalies. This approach has been presented in the context of temperature sensors in Sandryhaila and Moura [2014], WSNs in Egilmez and Ortega [2014], and detection of false data injection (FDI) attacks in power systems in Drayer and Routtenberg [2019], Dabush and Routtenberg [2022], and Morgenstern et al. [2024].

1.6.1 Hypothesis Testing Problem

We consider the following hypothesis testing problem:

$$\begin{cases} \mathcal{H}_0: & \mathbf{z} = \mathbf{x} + \mathbf{n} \\ \mathcal{H}_1: & \mathbf{z} = \mathbf{x} + \mathbf{a} + \mathbf{n}, \end{cases} \tag{1.38}$$

where the null hypothesis $\mathcal{H}_0$, indicates regular system operations, and $\mathcal{H}_1$, indicates disruptive interference in the system operation. Here, the vector $\mathbf{z}$ represents either sensor readings or their estimates. The system signal, $\mathbf{x}$, is assumed to satisfy local properties w.r.t. the graph. For example, the signal could be smooth, i.e. with a small GTV as defined in (1.12), to be a GLP signal, or even bandlimited graph signal as defined in Definition 1.1. The noise vector $\mathbf{n}$ is considered to be random. The anomaly is modeled by the deterministic vector $\mathbf{a}$, which is an arbitrary vector. Hence, it is not considered smooth, low-pass, or bandlimited w.r.t. the graph.

1.6.2 Graph High-Pass Filter (GHPF)-Based Detection

As mentioned below (1.13), the smooth graph signal, $\mathbf{x}$, can be considered as a GLP signal. Thus, under the assumption that the influence of noise is limited, we can use the following general detector:

$$\left\| h(\mathbf{L})\mathbf{z} \right\|^2 \underset{\mathcal{H}_0}{\overset{\mathcal{H}_1}{\gtrless}} \gamma, \tag{1.39}$$

where γ is the detection threshold determined based on the tested WSN application. Here, $h(\mathbf{L})$ is a graph high-pass filter (GHPF) that preserves the energy at the higher graph frequencies of its input, while reducing the content of the signal at the lower graph frequency regime. This detector is based on the assumption that the anomaly, $\mathbf{a}$, is neither smooth nor small enough to be neglected, and thus, it is expected to obtain energy in the higher graph frequencies. Consequently, it is expected that under $\mathcal{H}_1$, the measurement signal, $\mathbf{z}$, will obtain energy in higher graph frequencies. Thus, as a result, it is expected that the l.h.s. of (1.39) will exceed the threshold under hypothesis $\mathcal{H}_1$.

Examples of GHPFs include the ideal GHPF, which is defined by the graph frequency response

$$h^{id}(\lambda_k) = \begin{cases} 0 \leq \lambda_{cut} \\ 1 \;\; \lambda_k > \lambda_{cut} \end{cases} \quad k = 1, \dots, |\mathcal{M}|, \tag{1.40}$$

where λ_{cut} is the cutoff frequency. Another example is the GTV graph filter, which is defined by $h^{TV}(\mathbf{L}) = \mathbf{L}^{0.5}$. By substituting this graph filter in (1.39), we obtain that for this case the detector is reduced to

$$||h^{TV}(\mathbf{L})\mathbf{z}||^2 = \mathbf{z}^T\mathbf{L}^{0.5}\mathbf{L}^{0.5}\mathbf{z} = \mathbf{z}^T\mathbf{L}\mathbf{z}. \tag{1.41}$$

Thus, the GHPF detector in (1.39) is a generalization of the smoothness detector that has been used in Sandryhaila and Moura [2013] and Drayer and Routtenberg [2019].

1.6.3 Illustrative Example

In this example, our goal is to demonstrate the influence of a malfunction in one sensor of a WSN on the measurements of a smooth graph signal. We consider the graph in Figure 1.3b and define the following GLP signal in the graph frequency domain: $\tilde{\mathbf{x}} = \{0.32, 0.28, 0.15, 0.01, 0.01, 0.01, 0, 0, 0, 0\}$. Thus, $\tilde{\mathbf{x}}$ is a GLP signal and a β-bandlimited graph signal, as defined in (1.14), with a cutoff frequency of $\beta = 6$. The signal in the node domain is computed by the inverse GFT in (1.11). We model the anomaly by the additive vector $\mathbf{a} = \{0, 0, 0, 0, -0.3, 0, 0, 0, 0, 0\}$, where an anomaly is inserted at the fifth sensor.

In Figure 1.8, we present the signals $\mathbf{x}$ and $\mathbf{x} + \mathbf{a}$ in both domains. First, it can be seen that the differences between the signals are not clearly evident in the graph

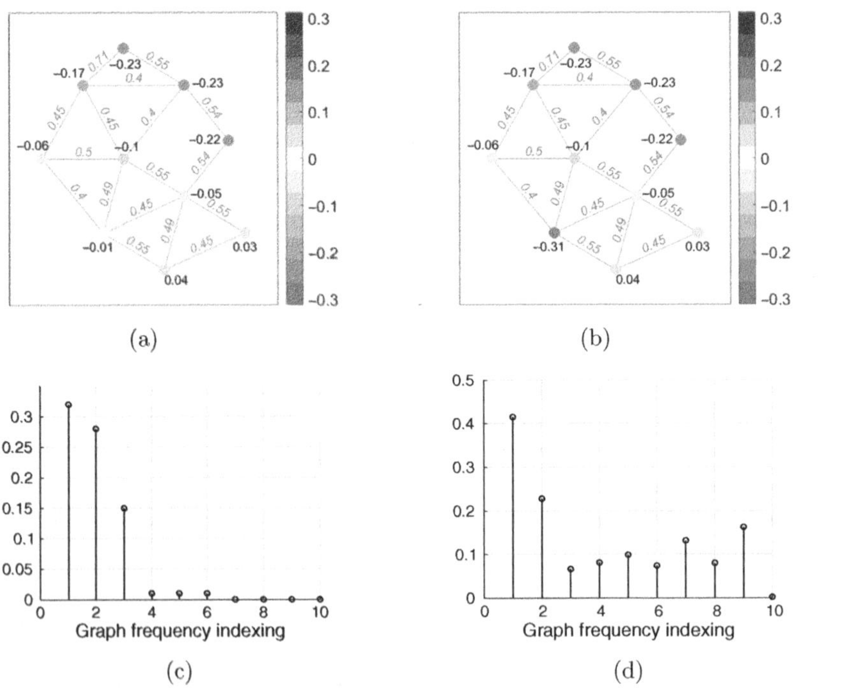

Figure 1.8 Results from the illustrative example in Section 1.6.3 showing the influence of an additive anomaly, $\mathbf{a}$, on a graph signal, $\mathbf{x}$, presented in both the graph node and the graph frequency domains. (a) $\mathbf{x}$, $TV^{\mathcal{G}}(\mathbf{x}) = 0.051$, (b) $\mathbf{x} + \mathbf{a}$, $TV^{\mathcal{G}}(\mathbf{x} + \mathbf{a}) = 0.189$, (c) $\tilde{\mathbf{x}} = \mathbf{V}^T\mathbf{x}$, and (d) $\tilde{\mathbf{x}} + \tilde{\mathbf{a}} = \mathbf{V}^T(\mathbf{x} + \mathbf{a})$.

node domain while the GTVs $TV^{\mathcal{G}}(\mathbf{x}) = 0.0212$ and $TV^{\mathcal{G}}(\mathbf{x} + \mathbf{a}) = 0.2114$ are significantly different. Moreover, the addition of the anomaly results in abnormal energy in the higher graph frequencies. Consequently, the detector in (1.39) is suitable for detecting this anomaly.

1.7 GSP-Based Graph Topology Identification for Modeling WSNs

In Section 1.2, several approaches based on GSP for modeling WSNs are presented. These models rely on prior knowledge of certain factors such as the location of the sensor nodes (i.e. the distance-based model in (1.2)), the correlation between the sensor nodes (i.e. the correlation-based model in (1.3)), or structural data (e.g. transportation networks and power systems). Unfortunately, this prior information may be unavailable or unreliable in some cases. For instance, the exact locations of some of the sensor nodes in a WSN may be unknown [Boukerche et al., 2007]. Additionally, in most cases, the correlation between the sensor nodes in the WSNs is unknown and must be estimated based on data samples [Vuran et al., 2004]. Furthermore, in infrastructure networks, such as power systems, the topology is subject to edge disconnections due to line outages [Shaked and Routtenberg, 2021]. Therefore, there is a need for methods to validate or estimate the underlying interactions in WSNs.

Graph topology inference approaches rely on algebraic and statistical methods. Classic examples include correlation-based methods [Kolaczyk and Csárdi, 2014], Graphical Lasso [Friedman et al., 2008], and GSP-based models [Kalofolias, 2016; Egilmez et al., 2017; Segarra et al., 2017; Medvedovsky et al., 2024]. In this context, GSP-based topology identification is vital for understanding and managing complex systems. Topology identification has applications in diverse fields such as gene regulatory, brain, power, and social networks [Giannakis et al., 2018]. Consequently, using GSP for topology identification in WSNs holds significant promise for enhancing WSN applications.

A key method for topology identification is based on the estimation of the graph Laplacian matrix of the graph, which captures the graph structure and its fundamental qualities. As presented in Sections 1.3–1.6, the graph Laplacian matrix is used in the spectral analysis of graph signals, graph filters, anomaly detection, and signal sampling and recovery, and thus, its accurate estimation is of great importance.

1.7.1 ML Estimation of the Graph Laplacian Matrix

In this section, we derive the general ML estimator under graph Laplacian constraints. These constraints can be implemented by requiring the ML estimator to belong to the set of Laplacian matrices for connected graphs, which can be defined as [Ying et al., 2020b]

$$\mathcal{L} = \left\{ \mathbf{L} \in \mathcal{S}_+^p \mid \mathrm{L}_{k,m} \leq 0, \forall\, k \neq m,\ \mathbf{L}\mathbf{1} = \mathbf{0},\ \mathrm{rank}(\mathbf{L}) = |\mathcal{M}| \right\}. \tag{1.42}$$

where $\mathcal{S}_+^p$ is the set of $p \times p$ symmetric positive semi-definite matrices. The graph estimation problem is approached by formulating it as a Laplacian learning problem based on a probabilistic graphical model [Banerjee et al., 2008; Koller and Friedman, 2009]. Thus, we assume that we have i.i.d. data samples, $\mathbf{x}_1, \ldots, \mathbf{x}_n$, drawn from a zero-mean Gaussian distribution parametrized by a positive semi-definite precision matrix, $\mathbf{L}$, i.e. $\mathbf{x} \sim \mathcal{N}(\mathbf{0}, \mathbf{L}^\dagger)$. This defines an improper Laplacian-constrained Gaussian Markov Random Field (LGMRF) model. The ML estimator of $\mathbf{L}$ under this model can be obtained by solving the following constrained minimization of the negative log-likelihood:

$$\hat{\mathbf{L}}^{ML} = \arg \min_{\mathbf{L} \in \mathcal{L}} \left\{ Tr(\mathbf{L}\mathbf{S}) - \log |\mathbf{L}|_+ \right\}, \tag{1.43}$$

where $\mathbf{S} \triangleq \frac{1}{n} \sum_{i=1}^n \mathbf{x}_i \mathbf{x}_i^T$ is the sample covariance matrix, $Tr(\cdot)$ denotes the trace operator, and $(\cdot)_+$ denotes the pseudo-determinant. It should be noted that various objective functions can be considered for topology identification under different assumed models within the constrained setting of $\mathbf{L} \in \mathcal{L}$. For example, in [Grotas et al, 2019] and [Halihal and Routtenberg, 2022], the topology of a power system is identified by solving the ML estimator of the Laplacian-constrained setting, where the samples are modeled using power flow equations.

Sparsity, which plays an important role in high-dimensional learning, can also be incorporated into the problem in (1.43). A sparse graph estimation problem under the LGMRF model can be formulated by adding a sparse penalty function to the estimator in (1.43). Specifically, selecting the penalty function as $|\mathbf{L}|_{1,\text{off}} = \sum_{k \neq m} |L_{k,m}|$ yields the same objective function as the well-known Graphical Lasso problem. However, due to the Laplacian constraints in (1.43), this penalty function is ineffective [Ying et al., 2020a]. Alternative approaches utilizing nonconvex penalties have been proposed to address this issue, as discussed in Medvedovsky et al. [2024]. This result highlights the fact that Laplacian-based GSP approaches are not merely straightforward extensions of conventional methods, but require careful consideration.

1.7.2 Topology Change Identification

Dynamic topology estimation, especially in WSNs with limited cooperation, presents unique challenges. Advances such as those already in 5G technology and those expected in 6G technology [Xia et al., 2020; Yeh et al., 2023] are making WSNs increasingly crucial in many aspects, and understanding their structure has become a significant concern. A critical application is in the security context, where comprehending the structure of adversary communication networks is of paramount importance. Traditional methods, typically designed for static, unchanging network structures, fall short in addressing the complexities of these modern dynamic environments. In this section, we discuss the application of detecting changes in the topology within dynamic settings.

Specifically, we consider the problem of identifying the underlying graph associated with a set of smooth graph signals $\{\mathbf{y}[n]\}_{n=1}^{N}$, obtained as outputs of a smooth graph filter as defined in (1.22). The underlying graph can be either the original graph, denoted as $\mathcal{G}^0 = (\mathcal{M}, \xi^0)$, or any graph from the set $\{\mathcal{G}^d = (\mathcal{M}, \xi^d)\}_{d=1}^{D}$ that is obtained by disconnecting a set C^d of edges from the original graph. This problem is formulated in the following multiple hypothesis testing problem:

$$\begin{cases} \mathcal{H}_0 : \mathbf{y}[n] = h(\mathbf{L}^{(0)})\mathbf{x}[n] \\ \mathcal{H}_d : \mathbf{y}[n] = h(\mathbf{L}^{(d)})\mathbf{x}[n], \quad n = 1, \dots, N, \end{cases} \tag{1.44}$$

for $d = 1, \dots, D$, where under each hypothesis $\mathcal{H}_d$, $d = 0, \dots, D$, the graph filter $h(\mathbf{L}^{(d)})$ is a smooth graph filter as in (1.20). Additionally, in each of the hypothesis $\mathcal{H}_d, d = 1, \dots, D$, the graph Laplacian matrix is given by

$$\mathbf{L}^{(d)} = \mathbf{L}^{(0)} + \mathbf{E}^d, \qquad \mathbf{E}^d = \sum_{(i,j)\in C^{(d)}} \mathbf{E}^{(i,j)}, \tag{1.45}$$

where the addition of $\mathbf{E}^{(i,j)}$ to the graph Laplacian matrix models the removal of the edge (i,j) from the graph. The signal-edge disconnection matrix $\mathbf{E}^{(i,j)}$ is defined elementwisely by

$$E_{k,m}^{(i,j)} = L_{k,m}^{(0)} \times \begin{cases} -1 & \{k = m = i\} \cup \{k = m = j\} \\ 1 & \{k = i, m = j\} \cup \{k = j, m = i\} \\ 0 & \text{otherwise.} \end{cases} \tag{1.46}$$

For the sake of simplicity, we assume that under each alternative hypothesis $\mathcal{H}_d$, the graph filter that is used to generate the measurements is a non-singular matrix, and the eigenvalues of the graph Laplacian matrix are distinct. In addition, we assume that $\mathbf{x}[n] \overset{\text{i.i.d.}}{\sim} \mathcal{N}(\mathbf{0}, \mathbf{I})$. While these assumptions are chosen for simplicity, the problem in (1.44) can be solved without these assumptions, as shown in [Shaked and Routtenberg, 2021].

We solve the multiple hypothesis testing problem in (1.44) with the ML decision rule [Kay, 1993b]:

$$\begin{aligned} \hat{d} &= \arg\max_{0\le d\le D} \log f(\mathbf{y}; \mathbf{L}^{(d)}) \\ &= \arg\max_{0\le d\le D} -\frac{1}{2}\sum_{n=1}^{N} \mathbf{y}^T[n](h^2(\mathbf{L}^{(d)}))^{-1}\mathbf{y}[n] + \log(|h^2(\mathbf{L}^{(d)})|), \end{aligned} \tag{1.47}$$

where $\log(f(\mathbf{y};\mathbf{L}^{(d)}))$ are the log-likelihoods under each hypothesis $\mathcal{H}_d$, $d = 0,\dots,D$. The last equality results from substituting the distribution of the smooth output graph signal $\mathbf{y}[n] \overset{i.i.d.}{\sim} \mathcal{N}(\mathbf{0}, h^2(\mathbf{L}^{(d)}))$ in the log-likelihood function, and then removing constant terms.

In order to analyze the result in (1.47) in the graph frequency domain, based on (1.10) and (1.15), we use the notation $\{\lambda_1^{(d)} \dots \lambda_{|\mathcal{M}|}^{(d)}\}$, $\mathbf{V}^{(d)}$, and $\{h(\lambda_1^{(d)}),\dots,h(\lambda_{|\mathcal{M}|}^{(d)})\}$ for the graph frequencies, eigenvectors, and graph filter response, associated with the graph Laplacian matrix $\mathbf{L}^{(d)}$. In [Shaked and Routtenberg, 2021], it was shown that (1.47) can be written as

$$\hat{d} = \arg\max_{0\le d\le D} -\frac{1}{2}\sum_{k=1}^{|\mathcal{M}|}(h^2(\lambda_k^{(d)}))^{-1}\sum_{n=1}^{N}(\tilde{y}_k^{(d)})^2[n] + \sum_{k=1}^{|\mathcal{M}|}\log\ h^2(\lambda_k^{(d)}), \tag{1.48}$$

where the kth element of the mean-squared GFT of the output graph signal is defined as

$$\psi_k^{(d)} \triangleq \frac{1}{|\mathcal{M}|}\sum_{n=1}^{N}(\tilde{y}_k^{(d)})^2[n]. \tag{1.49}$$

By substituting (1.49) with (1.48), we obtain

$$\hat{d} = \arg\max_{0\le d\le D} -\sum_{k=1}^{|\mathcal{M}|}\frac{\psi_k^{(d)}}{h^2(\lambda_k^{(d)})} + \sum_{k=1}^{|\mathcal{M}|}\log\ h^2(\lambda_k^{(d)}). \tag{1.50}$$

It can be seen from (1.50) that sufficient statistics for the ML decision rule are the graph frequency energy levels $\{\psi_k^{(d)}\}_{k=1}^{|\mathcal{M}|}$, $d = 0,\dots,D$. Additionally, for smooth graph filters, as defined in Definition 1.4, the weights of the graph frequency levels in (1.50), $(1/h^2(\lambda_k^{(d)}))$, amplify the influence of low graph-frequencies. Therefore, the ML decision rule is governed by the low-graph frequencies, which can be associated with the graph signal smoothness property.

1.8 Conclusions and Future Directions

In this chapter, we have outlined the fundamentals of GSP and showcased an end-to-end GSP-based approach for signal processing in WSN applications. In Section 1.2, we discussed the modeling of WSNs as undirected weighted graphs and presented several models that consider features such as the distance and correlation between the sensor nodes. In Section 1.3, we outlined fundamental concepts in GSP, including graph signals, graph filters, graph signal properties (e.g. graph signal smoothness), and graph spectral analysis. We then expanded upon graph signal smoothness in Section 1.4. Specifically, we introduced the concept of smooth graph filters, formulated a composite hypothesis testing problem differentiating between outputs of smooth or non-smooth graph filters, and derived a semi-parametric detector that solves this composite testing problem. Then, we utilized the graph signal smoothness property for practical applications in WSNs. In Section 1.5, we derived an ML-based estimator for signal recovery in models with missing data. In Section 1.6, we used graph signal smoothness for anomaly detection. Finally, in Section 1.7, we presented approaches for identifying the topology of the WSN underlying graph.

The exploration of GSP within the context of WSNs has unveiled significant insights into the effectiveness of GSP-based tools for the analysis and manipulation of WSN data. As we chart the course for future directions, several promising avenues merit attention. First, in addition to GSP, the advent of GNNs is expected to provide useful tools for modeling intricate graph-structured data. Future research should delve into synergies between GSP and GNNs to enhance the understanding and processing capabilities of WSN data. Additionally, investigating the synergy between GSP and the deployment of WSNs could lead to optimized network architectures that leverage the inherent strengths of both. Their considerations should incorporate practical aspects such as network layers and physical attributes, including capacity and coverage. Specifically, utilizing GSP for clustering within WSNs holds the potential to reveal hidden patterns and improve overall network efficiency. Furthermore, to address inherent limitations on energy resources, processing power, communication constraints, and computational costs [Egilmez and Ortega, 2014], it is crucial to develop distributed GSP techniques to minimize communication costs, enhance energy and processing efficiency, and adhere to the physical constraints of the system. Moreover, as the field of GSP grows, it is becoming clear that methods for identifying topologies need to be flexible in order to develop robust and efficient communication systems. Thus, incorporating GSP tools for this task may enable a better understanding of the underlying network structure promoting better decision-making in WSN applications. Finally, the analysis of specific structures inherent in typical WSN networks, such as the star topology, hierarchical/tree topology, and mesh topology (see Figure 1.9), may lay the basis to develop tailored GSP techniques for optimized performance for diverse WSN architectures. In essence, the integration of GSP into WSN research opens up a spectrum of possibilities,

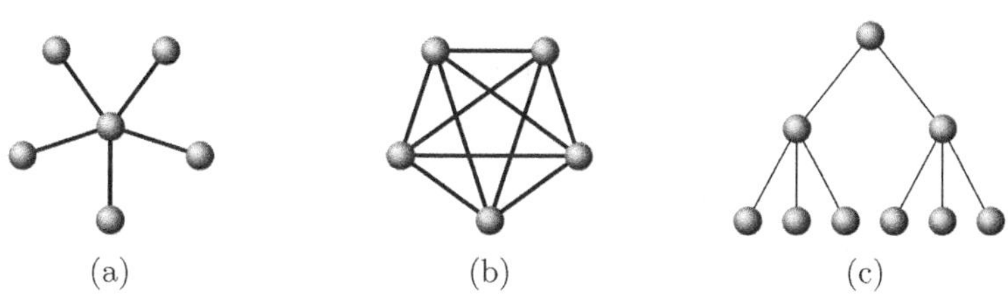

Figure 1.9 Network topologies. (a) Star, (b) Mesh, and (c) Tree.

and these future directions can further enhance the synergy between GSP and WSNs, thereby advancing the capabilities and applications of this dynamic field.

Acknowledgments

This work is partially supported by the Israel Science Foundation (Grant no. 1148/22), the Jabotinsky Scholarship from the Israel Ministry of Technology and Science, the Israel Ministry of National Infrastructure and Energy, and the US National Science Foundation under Grants CNS-2128448 and ECCS-2335876.

Bibliography

I. F. Akyildiz, W. Su, V. Sankarasubramaniam, and E. Cayirci. Wireless sensor networks: A survey. *Computer Networks*, 38(4):393–422, 2002.

O. Banerjee, L. El Ghaoui, and A. d'Aspremont. Model selection through sparse maximum likelihood estimation for multivariate Gaussian or binary data. *The Journal of Machine Learning Research*, 9:485–516, 2008.

M. Belkin, I. Matveeva, and P. Niyogi. Regularization and semi-supervised learning on large graphs. In *Proceedings of the 17th Annual Conference on Learning Theory*, pages 624–638. Springer, 2004.

A. Boukerche, H. A. Oliveira, E. F. Nakamura, and A. Loureiro. Localization systems for wireless sensor networks. *IEEE Wireless Communications*, 14(6):6–12, 2007.

D. Cai, X. He, J. Han, and T. S. Huang. Graph regularized nonnegative matrix factorization for data representation. *IEEE Transactions on Pattern Analysis and Machine Intelligence*, 33(8):1548–1560, 2010.

M. Cardei, Y. Yang, and J. Wu. Non-uniform sensor deployment in mobile wireless sensor networks. In *Proceedings of the IEEE International Symposium on a World of Wireless, Mobile and Multimedia Networks*, pages 1–8, 2008.

S. Chen, A. Sandryhaila, J. M. F. Moura, and J. Kovačević. Signal recovery on graphs: Variation minimization. *IEEE Transactions on Signal Processing*, 63(17):4609–4624, 2015a.

S. Chen, R. Varma, A. Sandryhaila, and J. Kovačević. Discrete signal processing on graphs: Sampling theory. *IEEE Transactions on Signal Processing*, 63(24):6510–6523, 2015b.

A. Chiumento, N. Marchetti, and I. Macaluso. Energy efficient WSN: A cross-layer graph signal processing solution to information redundancy. In *Proceeding of the International Symposium on Wireless Communication Systems (ISWCS)*, pages 645–650. IEEE, 2019.

L. Dabush and T. Routtenberg. Detection of false data injection attacks in unobservable power systems by Laplacian regularization. In *Proceedings of the Sensor Array and Multichannel Signal Processing Workshop (SAM)*, pages 415–419, 2022.

L. Dabush and T. Routtenberg. "Verifying the Smoothness of Graph Signals: A Graph Signal Processing Approach," in *IEEE Transactions on Signal Processing*, vol. 72, pp. 4349–4365, 2024, doi: 10.1109/TSP.2024.3439554.

L. Dabush, A. Kroizer, and T. Routtenberg. State estimation in partially observable power systems via graph signal processing tools. *Sensors*, 23(3):1387, 2023.

S. E. Díaz, J. C. Pérez, A. C. Mateos, M. C. Marinescu, and B. B. Guerra. A novel methodology for the monitoring of the agricultural production process based on wireless sensor networks. *Computers and Electronics in Agriculture*, 76(2):252–265, 2011.

X. Dong, D. Thanou, L. Toni, M. Bronstein, and P. Frossard. Graph signal processing for machine learning: A review and new perspectives. *IEEE Signal Processing Magazine*, 37(6):117–127, 2020.

E. Drayer and T. Routtenberg. Detection of false data injection attacks in smart grids based on graph signal processing. *IEEE Systems Journal*, 14(2):1886–1896, 2019.

H. E. Egilmez and A. Ortega. Spectral anomaly detection using graph-based filtering for wireless sensor networks. In *Proceedings of the IEEE International Conference on Acoustics, Speech and Signal Processing (ICASSP)*, pages 1085–1089, 2014.

H. E. Egilmez, E. Pavez, and A. Ortega. Graph learning from data under Laplacian and structural constraints. *IEEE Journal on Selected Topics in Signal Processing*, 11(6):825–841, 2017.

A. Elmoataz, O. Lezoray, and S. Bougleux. Nonlocal discrete regularization on weighted graphs: A framework for image and manifold processing. *IEEE Transactions on Image Processing*, 17(7):1047–1060, 2008.

L. Erhan, M. Ndubuaku, M. D. Mauro, W. Song, M. Chen, G. Fortino, O. Bagdasar, and A. Liotta. Smart anomaly detection in sensor systems: A multi-perspective review. *Information Fusion*, 67:64–79, 2021.

J. Feng, F. Chen, and H. Chen. Data reconstruction coverage based on graph signal processing for wireless sensor networks. *IEEE Wireless Communications Letters*, 11(1):48–52, 2021.

J. Friedman, T. Hastie, and R. Tibshirani. Sparse inverse covariance estimation with the graphical lasso. *Biostatistics*, 9(3):432–441, 2008.

G. B. Giannakis, Y. Shen, and G. V. Karanikolas. Topology identification and learning over graphs: Accounting for nonlinearities and dynamics. *Proceedings of the IEEE*, 106(5):787–807, 2018.

S. Grotas, Y. Yakoby, I. Gera, and T. Routtenberg. Power systems topology and state estimation by graph blind source separation. *IEEE Transactions on Signal Processing*, 67(8):2036–2051, 2019.

M. Halihal and T. Routtenberg. Estimation of the admittance matrix in power systems under Laplacian and physical constraints. In *Proceeding of the IEEE International Conference on Acoustics, Speech and Signal Processing (ICASSP)*, pages 5972–5976, 2022.

E. Hasheminejad and H. Barati. A reliable tree-based data aggregation method in wireless sensor networks. *Peer-to-Peer Networking and Applications*, 14(2):873–887, 2021.

Y. He and H. T. Wai. Detecting central nodes from low-rank excited graph signals via structured factor analysis. *IEEE Transactions on Signal Processing*, 70:2416–2430, 2022.

T. He, S. Krishnamurthy, J. A. Stankovic, T. Abdelzaher, L. Luo, R. Stoleru, T. Yan, L. Gu, J. Hui, and B. Krogh. Energy-efficient surveillance system using wireless sensor networks. In *Proceedings of the International Conference on Mobile Systems, Applications, and Services*, pages 270–283, 2004.

E. Isufi, F. Gama, D. I. Shuman and S. Segarra. “Graph Filters for Signal Processing and Machine Learning on Graphs,” in *IEEE Transactions on Signal Processing*, vol. 72, pp. 4745–4781, 2024. doi: 10.1109/TSP.2024.3349788.

V. Kalofolias. How to learn a graph from smooth signals. In *Artificial Intelligence and Statistics (PMLR)*, pages 920–929, 2016.

D. Kandris, C. Nakas, D. Vomvas, and G. Koulouras. Applications of wireless sensor networks: An up-to-date survey. *Applied System Innovation*, 3(1):14, 2020.

S. M. Kay. *Fundamentals of Statistical Signal Processing: Estimation Theory*, volume 1. Prentice Hall PTR, New Jersey, NJ, USA, 1993a.

S. M. Kay. *Fundamentals of Statistical Signal Processing, Volume II: Detection Theory*. Prentice Hall PTR, New Jersey, NJ, USA, 1993b.

B. S. Kim, K. I. Kim, B. Shah, F. Chow, and K. H. Kim. Wireless sensor networks for big data systems. *Sensors*, 19(7):1565, 2019.

E. D. Kolaczyk and G. Csárdi. *Statistical Analysis of Network Data with R*, volume 65. Springer, Cham, Switzerland, 2014.

D. Koller and N. Friedman. *Probabilistic Graphical Models: Principles and Techniques*. MIT Press, Massachusetts, MA, USA, 2009.

L. Kong, M. Xia, X. Y. Liu, G. Chen, Y. Gu, M. Y. Wu, and X. Liu. Data loss and reconstruction in wireless sensor networks. *IEEE Transactions on Parallel and Distributed Systems*, 25(11):2818–2828, 2013.

A. Kroizer, T. Routtenberg, and Y. C. Eldar. Bayesian estimation of graph signals. *IEEE Transactions on Signal Processing*, 70:2207–2223, 2022.

Y. Li and L. E. Parker. A spatial-temporal imputation technique for classification with missing data in a wireless sensor network. In *Proceedings of the International Conference on Intelligent Robots and Systems*, pages 3272–3279, 2008.

A. G. Marques, S. Segarra, G. Leus, and A. Ribeiro. Sampling of graph signals with successive local aggregations. *IEEE Transactions on Signal Processing*, 64(7):1832–1843, 2015.

Y. Medvedovsky, E. Treister, and S. T. Routtenberg. Efficient Graph Laplacian Estimation by Proximal Newton. Proceedings of The 27th International Conference on Artificial Intelligence and Statistics, *Proceedings of Machine Learning Research*, 238:1171–1179, 2024. Available from https://proceedings.mlr.press/v238/medvedovsky24a.html.

G. Morgenstern and T. Routtenberg (2024). "Efficient Recovery of Sparse Graph Signals From Graph Filter Outputs," in *IEEE Transactions on Signal Processing*, vol. 72, pp. 5550–5566. doi: 10.1109/TSP.2024.3495225.

G. Morgenstern, J. Kim, J. Anderson, G. Zussman, and T. Routtenberg. Protection against graph-based false data injection attacks on power systems. *IEEE Transactions on Control of Network Systems*, 11(4):1924–1936, 2024.

Z. Nurlan, T. Zhukabayeva, M. Othman, A. Adamova, and N. Zhakiyev. Wireless sensor network as a mesh: Vision and challenges. *IEEE Access*, 10:46–67, 2021.

A. Ortega. *Introduction to Graph Signal Processing*. Cambridge University Press, 2022.

A. Ortega, P. Frossard, J. Kovačević, J. M. F. Moura, and P. Vandergheynst. Graph signal processing: Overview, challenges, and applications. *Proceedings of the IEEE*, 106(5):808–828, 2018.

L. Pan and J. Li. K-nearest neighbor based missing data estimation algorithm in wireless sensor networks. *Wireless Sensor Network*, 2(02):115, 2010.

S. Pattem, B. Krishnamachari, and R. Govindan. The impact of spatial correlation on routing with compression in wireless sensor networks. *ACM Transactions on Sensor Networks*, 4(4):1–33, 2008.

S. S. Pradhan, J. Kusuma, and K. Ramchandran. Distributed compression in a dense microsensor network. *IEEE Signal Processing Magazine*, 19(2):51–60, 2002.

G. Puy and P. Pérez. Structured sampling and fast reconstruction of smooth graph signals. *Information and Inference: A Journal of the IMA*, 7(4):657–688, 2018.

S. Rajasegarar, C. Leckie, and M. Palaniswami. Anomaly detection in wireless sensor networks. *IEEE Wireless Communications*, 15(4):34–40, 2008.

R. Ramakrishna, H. T. Wai, and A. Scaglione. A user guide to low-pass graph signal processing and its applications: Tools and applications. *IEEE Signal Processing Magazine*, 37(6):74–85, 2020.

A. Ramamoorthy, J. Shi, and R. D. Wesel. On the capacity of network coding for random networks. *IEEE Transactions on Information Theory*, 51(8):2878–2885, 2005.

D. Romero, M. Ma, and G. B. Giannakis. Kernel-based reconstruction of graph signals. *IEEE Transactions on Signal Processing*, 65(3):764–778, 2016.

T. Routtenberg. Non-Bayesian estimation framework for signal recovery on graphs. *IEEE Transactions on Signal Processing*, 69:1169–1184, 2021.

T. Sahai, A. Speranzon, and A. Banaszuk. Hearing the clusters of a graph: A distributed algorithm. *Automatica*, 48(1):15–24, 2012.

A. Sandryhaila and J. M. F. Moura. Discrete signal processing on graphs. *IEEE Transactions on Signal Processing*, 61(7):1644–1656, 2013.

A. Sandryhaila and J. M. F. Moura. Discrete signal processing on graphs: Frequency analysis. *IEEE Transactions on Signal Processing*, 62(12):3042–3054, 2014.

A. Sandryhaila, S. Kar, and J. M. F. Moura. Finite-time distributed consensus through graph filters. In *Proceedings of the IEEE International Conference on Acoustics, Speech and Signal Processing (ICASSP)*, pages 1080–1084, 2014.

I. D. Schizas, A. Ribeiro, and G. B. Giannakis. Consensus in Ad Hoc WSNs with noisy links—Part I: Distributed estimation of deterministic signals. *IEEE Transactions on Signal Processing*, 56(1):350–364, 2008.

K. Schultz, A. Saksena, E. P. Reilly, R. Hingorani, and M. Villafae-Delgado. Detecting anomalous swarming agents with graph signal processing. In *Proceedings of the IEEE International Conference on Autonomous Systems (ICAS)*, pages 1–5, 2021.

S. Segarra, A. G. Marques, G. Mateos, and A. Ribeiro. Network topology inference from spectral templates. *IEEE Transactions on Signal and Information Processing Over Networks*, 3(3):467–483, 2017.

S. D. Servetto and G. Barrenechea. Constrained random walks on random graphs: Routing algorithms for large scale wireless sensor networks. In *Proceedings of the ACM International Workshop on Wireless Sensor Networks and Applications*, pages 12–21, 2002.

S. Shaked and T. Routtenberg. Identification of edge disconnections in networks based on graph filter outputs. *IEEE Transactions on Signal and Information Processing Over Networks*, 7:578–594, 2021.

D. I. Shuman, P. Vandergheynst, and P. Frossard. Chebyshev polynomial approximation for distributed signal processing. In *Proceedings of the IEEE International Conference on Distributed Computing in Sensor Systems and Workshops (DCOSS)*, pages 1–8, 2011.

D. I. Shuman, S. K. Narang, P. Frossard, A. Ortega, and P. Vandergheynst. The emerging field of signal processing on graphs: Extending high-dimensional data analysis to networks and other irregular domains. *IEEE Signal Processing Magazine*, 30(3):83–98, 2013.

M. Tariq and H. V. Poor. Real time electricity theft detection in microgrids through wireless sensor networks. 2016 *IEEE Sensors*, pages 1–3, Orlando, FL, USA, 2016. doi: 10.1109/ICSENS.2016.7808729.

M. C. Vuran and I. F. Akyildiz. Spatial correlation-based collaborative medium access control in wireless sensor networks. *IEEE/ACM Transactions On Networking*, 14(2):316–329, 2006.

M. C. Vuran, O. B. Akan, and I. F. Akyildiz. Spatio-temporal correlation: Theory and applications for wireless sensor networks. *Computer Networks*, 45(3):245–259, 2004.

F. Wang and C. Zhang. Label propagation through linear neighborhoods. In *Proceedings of the International Conference on Machine Learning*, pages 985–992, 2006.

G. Werner-Allen, K. Lorincz, J. Johnson, J. Lees, and M. Welsh. Fidelity and yield in a volcano monitoring sensor network. In *Proceedings of the 7th Symposium on Operating Systems Design and Implementation*, pages 381–396, 2006.

W. N. V. Wieringen. Lecture notes on ridge regression. *arXiv preprint: 1509.09169*, 2015.

T. Xia, M. M. Wang, J. Zhang, and L. Wang. Maritime internet of things: Challenges and solutions. *IEEE Wireless Communications*, 27(2):188–196, 2020.

M. Xie, S. Han, B. Tian, and S. Parvin. Anomaly detection in wireless sensor networks: A survey. *Journal of Network and Computer Applications*, 34(4):1302–1325, 2011.

C. Yeh, G. Do Jo, Y. J. Ko, and H. K. Chung. Perspectives on 6G wireless communications. *ICT Express*, 9(1):82–91, 2023.

J. Ying, J. V. D M. Cardoso, and D. P. Palomar. Does the ℓ_1-norm learn a sparse graph under Laplacian constrained graphical models? *arXiv preprint: 2006.14925*, 2020a.

J. Ying, J. V. D. M. Cardoso, and D. Palomar. Nonconvex sparse graph learning under Laplacian constrained graphical model. *Advances in Neural Information Processing Systems*, 33:7101–7113, 2020b.

M. Zheng, J. Bu, C. Chen, C. Wang, L. Zhang, G. Qiu, and D. Cai. Graph regularized sparse coding for image representation. *IEEE Transactions on Image Processing*, 20(5):1327–1336, 2010.

2

Learning and Optimization in Wireless Sensor Networks

Muhammad I. Qureshi[1], Apostolos I. Rikos[2], Themistoklis Charalambous[3], and Usman A. Khan[1]

[1] *Tufts University, Medford, MA, USA*
[2] *The Hong Kong University of Science and Technology (Guangzhou), Guangzhou, China*
[3] *University of Cyprus, Nicosia, Cyprus*

2.1 Introduction

In many science, engineering, and social domains, the synergy between learning and optimization stands as a fundamental pillar. Learning is a process where a model or a system is trained using historical data to execute particular tasks. At the core of learning methods (including machine learning and deep learning), optimization techniques play a pivotal role where a certain loss or an objective function is minimized in order to improve the performance of an underlying model through iterative refinements. Most learning problems can be divided into two main categories:

- **Classification problems:** These types of problems require learning model parameters to distinguish between predefined categories (or classes). This is typically achieved by training a model to learn some parameters that minimize a specific loss function while considering realistic constraints; see, e.g., Forero et al. [2010], Raja and Bajwa [2016], Safavi et al. [2018], and Reisizadeh et al. [2020].
- **Generative problems:** This class of problems aims to learn the underlying distribution from a set of input data in order to generate (realistic) samples from that distribution. For instance, in some scenarios, min–max optimization techniques are applied to identify a saddle point while optimizing the loss functions of generative adversarial networks (GANs): see, e.g. Goodfellow et al. [2014] and Lin et al. [2020a,2020b].

Wireless Sensor Networks in Smart Environments: Enabling Digitalization from Fundamentals to Advanced Solutions, First Edition. Edited by Domenico Ciuonzo and Pierluigi Salvo Rossi.

Both classification and generative problems demand a significant amount of data for effective model training. The widespread use of computational devices (such as cellphones, Internet of Things (IoT) devices like sensors, and security cameras) has given rise to a substantial surge in the amount of information that is generated, leading to a notable rise in the applications that are built on this information with the help of learning and optimization, Lee and Zavlanos [2018], Safavi et al. [2018], Forero et al. [2010], Raja and Bajwa [2016], Yang et al. [2019], Bottou et al. [2018], Benzi et al. [2005], Goodfellow et al. [2014], Sinha et al. [2018], Lin et al. [2020a,2020b], and Liang and Stokes [2019].

For many such large-scale machine learning applications, data is often available at several geographically distributed devices, which makes centralized computation and processing infeasible, e.g. due to privacy concerns, computational limitations, and/or communication constraints. Consequently, there is an increasing interest in developing distributed learning and optimization methods that guarantee the same performance as centralized methods however by only utilizing the local datasets, [Xin et al., 2020]. This interest has led to the development of various distributed optimization techniques that are designed to address these challenges by enabling collaborative computation and communication among the geographically distributed, computational nodes. The related distributed optimization methods mainly follow two architectures based on the network topology:

- **Server-client networks (federated settings)**: In this architecture, the data is divided among a network of clients, and the server coordinates the model updates. Figure 2.1 shows a general example of the server–client network architecture.
- **Peer-to-peer/mesh networks (distributed settings)**: In this architecture, the nodes communicate only with their neighbors to exchange information and update their local models. Figure 2.2 shows a general example of a peer-to-peer network architecture.

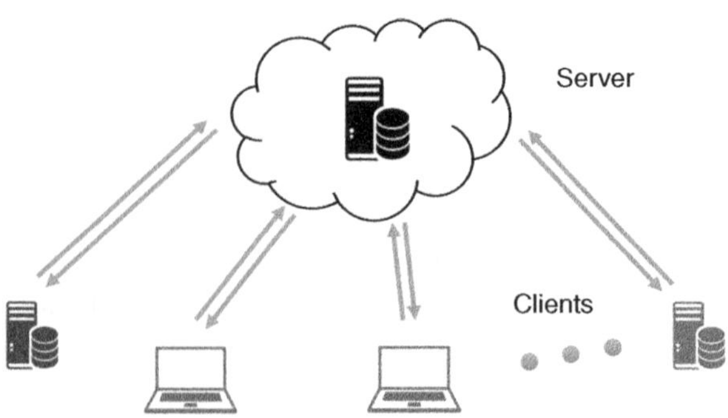

Figure 2.1 Server–client network used for federated learning problems.

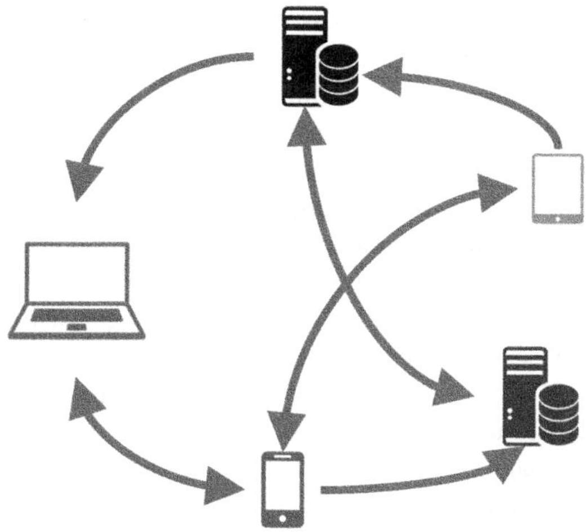

Figure 2.2 Peer-to-peer network of computational nodes over a strongly connected directed graph.

Server–client networks or other well-connected topologies typically model highly structured scenarios, such as data centers. In contrast, weakly connected peer-to-peer/mesh networks are good candidates to model ad hoc WSNs, where communication is more restricted and it is not possible to have a central coordinator. Recently, learning over a peer-to-peer network architecture has been described as near-shot learning, Qureshi [2024b]. In this chapter, we will focus on such near-shot learning methods, i.e. distributed optimization methods over strongly connected peer-to-peer networks.

In the remaining of this chapter, we will describe some foundational work on distributed optimization methods in WSNs. In particular, Section 2.2 provides some useful definitions and notations. Section 2.3 introduces the problem formulation and Section 2.4 discusses the key optimization methods and their extensions to restricted communication scenarios. Finally, Section 2.7 concludes the chapter and discusses potential future directions.

2.1.1 Related Work

Distributed optimization has gained huge traction in recent years, Ram et al. [2010], Lian et al. [2017], Chen and Sayed [2012], Xu et al. [2015], Zhu and Martínez [2010], Lorenzo and Scutari [2016], Qu and Li [2017], Nedić et al. [2017], Xin and Khan [2020], and Qureshi and Khan [2023b]. For smooth objective functions, several gradient-based methods were proposed in Tsitsiklis et al. [1986], Nedić and Ozdaglar [2009], Ram et al. [2010], Lian et al. [2017], and Xin et al. [2020]. These methods require each node to compute the gradient of its local loss function using its local data, update its local state estimates using gradient descent, and then exchange information with the neighboring nodes.

Notable works are Distributed Gradient Descent (DGD), Nedić and Ozdaglar [2009], and Distributed Stochastic Gradient Descent (DSGD) Ram et al. [2010]. DGD employs a deterministic offline model assuming each node has access to the entire local dataset. Every node computes the gradient for each update based on the complete local dataset, which is computationally expensive and not feasible in many practical applications. In contrast, DSGD is a stochastic variant that is particularly beneficial for online and streaming data applications. DSGD exhibits the same operation as DGD with the difference being that each node executes local updates utilizing the *stochastic gradient* calculated by processing only a subset of the local dataset.

The above-mentioned distributed optimization methods perform well for distributed data settings but do not converge to the optimal solution of the global problem. This, as we will explain later, is due to the difference in gradient evaluations caused by the heterogeneous data distribution. In general, the gradient evaluated at each node directs the state estimation vectors to move towards a different direction as compared to the direction of the global gradient (gradient with respect to the data possessed in centralized settings). For smooth and strongly convex objective functions, this causes inexact linear convergence of DGD and DSGD using a constant stepsize. Nedić and Ozdaglar [2009] propose a method for exact convergence to the optimal solution using a decaying stepsize but at a sublinear rate. Recent work proposed in, Di Lorenzo and Scutari [2015] and Xin and Khan [2018] uses the gradient tracking technique to overcome the problems caused by data heterogeneity. Each node possesses an additional gradient tracking state vector, which asymptotically converges to the global gradient. Hence, each node updates its local state moving in the opposite direction of the gradient tracking term, see Xin et al. [2020] for a detailed review. Among other first-order gradient-based methods are Qureshi and Khan [2022], Xin et al. [2021], Saadatniaki et al. [2020], and Assran et al. [2019]. Qureshi and Khan [2022] use different variance reduction techniques to eliminate the variance caused by stochastic gradients. Saadatniaki et al. [2020] and Assran et al. [2019] quantify the performance of the above-mentioned methods under constrained scenarios, i.e. when the communication between nodes is time varying, Saadatniaki et al. [2020], when the nodes have different computational power, or when there are communication bottlenecks, Assran et al. [2019]. To expand more on wireless networks, we now explain some useful definitions.

2.2 Notations and Definitions

The sets of real numbers, rational numbers, and integers are denoted by $\mathbb{R}$, $\mathbb{Q}$, and $\mathbb{Z}$, respectively. The symbols $\mathbb{Z}_{\geq 0}$ ($\mathbb{Z}_{>0}$) represent the set of nonnegative (positive) integers. Similarly, $\mathbb{Z}_{\leq 0}$ ($\mathbb{Z}_{<0}$) denote the sets of nonpositive (negative) integers. Vectors are represented by lowercase letters, and matrices are denoted by uppercase letters. For a matrix $A \in \mathbb{R}^{n \times n}$, the element in the ith row and jth column is denoted by A_{ij}. The transpose of A is indicated by $A^\top$. The all-ones vector of size n is denoted by $\mathbf{1}_n$, and the identity matrix is denoted by I (with dimensions implied by the context).

2.2.1 Graph-Theoretic Notions

We now introduce fundamental graph theory concepts to analyze how information flows among nodes in wireless networks. Let us consider a wireless network consisting of n nodes (where $n \geq 2$) that communicate exclusively with their immediate neighbors. We assume that each node can directly transmit information to some or all neighboring nodes, without necessarily being able to receive information from them. Our network is modeled as a directed graph (or digraph), denoted as $\mathcal{G}_d = (\mathcal{V}, \mathcal{E})$. Within this directed graph $\mathcal{G}_d$, the set of nodes is defined as $\mathcal{V} = \{v_1, v_2, \dots, v_n\}$, and its size is indicated as $|\mathcal{V}| = n$. The set of edges is specified as $\mathcal{E} \subseteq \mathcal{V} \times \mathcal{V} - \{(v_j, v_j) \mid v_j \in \mathcal{V}\}$ (excluding self-edges), and its size is represented by $m = |\mathcal{E}|$. A directed edge from node v_i to node v_j is denoted as $m_{ji} \triangleq (v_j, v_i) \in \mathcal{E}$, indicating that node v_j can receive information from node v_i (but not the other way around). The set of nodes that can directly transmit information to node v_j is known as the in-neighbors of v_j, denoted by $\mathcal{N}_j^- = \{v_i \in \mathcal{V} \mid (v_j, v_i) \in \mathcal{E}\}$. The count of nodes in $\mathcal{N}_j^-$ is referred to as the in-degree of v_j, denoted as $\mathcal{D}_j^- = |\mathcal{N}_j^-|$. Similarly, the set of nodes that can directly receive information from node v_j is termed the out-neighbors of v_j, represented by $\mathcal{N}_j^+ = \{v_l \in \mathcal{V} \mid (v_l, v_j) \in \mathcal{E}\}$. The number of nodes in $\mathcal{N}_j^+$ is the out-degree of v_j, denoted by $\mathcal{D}_j^+ = |\mathcal{N}_j^+|$. We assume that the given directed graph $\mathcal{G}_d = (\mathcal{V}, \mathcal{E})$ is strongly connected, ensuring the existence of a directed path from any node v_i to v_j for all distinct nodes $v_j, v_i \in \mathcal{V}$. An undirected graph $\mathcal{G}$ implies bidirectional links between each node and its neighbors, leading to a doubly stochastic weight matrix representing the communication graph. Even with possible unidirectional links among nodes of $\mathcal{G}$, the weight matrix could still be doubly stochastic, classifying it as a weight-balanced graph. Additionally, the diameter D of a directed graph $\mathcal{G}_d$ is defined as the length of the longest shortest path between any pair of nodes $v_i, v_j \in \mathcal{V}$ in the network.

2.2.2 Summary of Variables

In this section, we provide a summary of useful variables essential for the development of our results. Table 2.1 describes some important terms that are used throughout the rest of the chapter.

Table 2.1 Description of important variables.

Variables	Description
l	Loss function
f_i	Local cost function of node v_i
∇f_i	Derivative of local cost function of node v_i
F	Global objective function
∇F	Derivative of global objective function
$x^{[k]}$	State vector estimate at kth iteration
$x_i^{[k]}$	State vector estimate at kth iteration at node v_i
Δ	Quantization step size
D	Digraph diameter
α	Optimization step size
y_i, z_i	Communication variables (or mass variables) of node v_i
M_i, m_i	Voting variables of node v_i
ε_s	Optimization error bound
c_r	Refinement constant of quantization level
γ_β	Optimization convergence time step
$f_{i,j}$	Decomposable local cost function of node v_i
$\nabla f_{i,j}$	Derivative of decomposable local cost function of node v_i

2.3 Problem Formulation

In this section, we formally describe the setup for distributed optimization problems over peer-to-peer networks. However, we first elaborate on the centralized settings, where a single computational node possesses all the data. In most learning and optimization problems, the goal is to find a model $h(m, x)$ parameterized by $x \in \mathbb{R}^p$, which maps an input $m \in \mathbb{R}^q$ to an output $y \in \mathbb{R}^r$. A loss function $l(h(m, x), y)$ is used to evaluate the performance of the optimizer in estimating the model parameters x. Additionally, in practice, the (m, y) pairs belong to some joint probability distribution $\mathcal{P}(m, y)$. Thus, the optimization problem is expressed as follows:

$$\min_x \left\{F(x) := \mathbb{E}_{(m,x)\sim\mathcal{P}(m,x)} l(h(m, x), y)\right\}.$$

In most practical scenarios, we do not know the joint probability distribution $\mathcal{P}(m, y)$ in which the (m, y) pairs belong. We assume that each pair (m, y) is sampled from a large dataset $\{(m_i, y_i)\}_{i=1}^n$ under i.i.d. conditions (independent and identically distributed). Therefore, we solve the empirical risk minimization problem:

$$\min_{x} \left\{ F(x) := \frac{1}{n} \sum_{i=1}^{n} l_i(h_i(m_i, x), y_i) \right\}. \tag{2.1}$$

This finite-sum formulation encapsulates a wide range of learning problems. For very large-scale problems, we often consider each l_i to be distributed over a network of n computational nodes communicating over a strongly connected directed graph with the global problem being to minimize the average of local costs distributed over the network. Thus, in the distributed setting, the optimization problem has the following form:

$$\min_{x} \left\{ F(x) := \frac{1}{n} \sum_{i=1}^{n} l_i(h_i(m_i, x), y_i) := \frac{1}{n} \sum_{i=1}^{n} f_i(x) \right\}, \tag{2.2}$$

where $f_i(x) : \mathbb{R}^p \to \mathbb{R}$ is the cost function private to node v_i. Additionally, every node can only communicate with its neighbors. Therefore, unlike the centralized settings, each node cannot directly evaluate the gradient of the global cost (i.e. $\nabla F(x) := \nabla \left(\frac{1}{n} \sum_{i=1}^{n} f_i(x) \right)$, but it can only access its first-order oracle (i.e. local cost $f_i(\cdot)$ and local gradient $\nabla f_i(\cdot)$). Considering this distributed setting, in Sections 2.4 and 2.5, we describe various methods to solve distributed optimization problems.

2.4 Distributed Optimization Methods

In this section, we will focus on first-order distributed optimization methods. Before delving into the methods, we present some assumptions that are required for analyzing their convergence.

Assumption 2.1 Communication among nodes occurs through a strongly connected directed graph.

Assumption 2.2 Communication among nodes takes place over a weight-balanced digraph. Let $W \in \mathbb{R}^{n \times n}$ represent the weight matrix of the underlying communication graph. Then W is doubly stochastic, i.e. $W\mathbf{1}_n = \mathbf{1}_n$ and $W^\top \mathbf{1}_n = \mathbf{1}_n$.

Assumption 2.3 For each node v_i, the local cost function f_i is L_i smooth and the global cost function F is μ-strongly convex. This implies that for every $x, y \in \mathbb{R}^p$

- there exists a positive constant L_i such that $\forall i$,

$$\|\nabla f_i(y) - \nabla f_i(x)\|_2 \leq L_i \|y - x\|_2, \tag{2.3}$$

- there exists positive a constant μ such that

$$F(y) \geq F(x) + \nabla F(x)^\top (y - x) + \frac{\mu}{2}\|y - x\|_2^2. \tag{2.4}$$

The assumptions mentioned above are common in the literature of distributed optimization, [Qu and Li, 2017; Xin and Khan, 2018]. Assumption 2.1 is related to the connectivity of network topology. It ensures that there are no disconnects in the communication graph (i.e. each node is linked with every other node through a maximum of n-hops and there are no isolated nodes). Assumption 2.3 governs the behavior of the cost functions. Lipschitz-continuity guarantees that the function is smooth with a quadratic upper bound (2.3). Strong convexity, on the other hand, ensures that the function is lower bounded by a quadratic and has a unique minimum x^* (2.4). It is noteworthy that both strong connectivity of the underlying network and Assumption 2.3 are necessary for distributed optimization methods to attain a linear convergence rate.[1] A linear rate is often used in the literature on optimization, [Nedić and Ozdaglar, 2009; Xin et al., 2020] to describe an exponentially decaying term. Now, we describe the distributed implementation of gradient descent.

2.4.1 Distributed Gradient Descent

Minimizing a cost function through gradient-based methods is well-established in research, with gradient descent being extensively explored in the literature, see Bottou et al. [2018] for a detailed report on optimization methods. In centralized settings, the algorithm commences by initializing with a random state vector $x^{[0]} \in \mathbb{R}^p$. Subsequently, for all $k \geq 0$, it updates the state vector according to the following rule:

$$x^{[k+1]} = x^{[k]} - \nabla F(x^{[k]}).$$

In distributed settings, the local cost functions are private to each node. Therefore, the nodes can only compute their local gradients $\nabla f_i(x)$ and use them to evaluate their local state vectors $x_i \in \mathbb{R}^p$. The DGD is formally defined in Algorithm 2.1.

During the operation of Algorithm 2.1, nodes communicate over a network modeled as a strongly connected and weight-balanced digraph $W = \{w_{ir}\}_{r=1}^n$. Each node i initializes with a random state vector $x_i^{[0]} \in \mathbb{R}^p$. At iteration k, every

1 Note that a linear rate is essentially an exponential decay of the error, which is *linear* on the log-scale.

Algorithm 2.1 Distributed Gradient Descent (DGD)

Input: $x_i^{[0]} \in \mathbb{R}^p, \alpha > 0, \{w_{ir}\}_{r=1}^n$
Iteration: For $k = 0, 1, 2, \dots, K$, each node $v_i \in \mathcal{V}$ does:

1. $x_i^{[k+1]} = \sum_{i=1}^n w_{ir} \left(x_r^{[k]} - \alpha_k \nabla f_r(x_r^{[k]}) \right)$

Output: Each node $v_i \in \mathcal{V}$ estimates x^* by $x_i^{[K]}$ to solve (2.2)

node computes its local gradient $\nabla f_i(x_i^{[k]})$ and then takes a step in the negative direction of that gradient with a step-size $\alpha_k > 0$. Then each node updates its next state estimate $x_i^{[k+1]}$ by sharing these updated states $\left(x_i^{[k]} - \alpha_k \nabla f_i(x_i^{[k]}) \right)$ with its neighbors and summing them up according to the weights $\{w_{ir}\}_{r=1}^n$.

In the following theorem, we describe the main convergence result for Algorithm 2.1 under certain assumptions. The formal proof can be found in Nedić and Ozdaglar [2009].

Theorem 2.1 Nedić and Ozdaglar [2009]. Consider the problem in (2.2) under Assumptions 2.2 and 2.3. For a small enough constant step-size $\alpha_k = \alpha > 0$, Algorithm 2.1 (DGD) linearly converges to an error-ball around the optimal solution x^*.

In Theorem 2.1, it is shown that for a constant step-size $\alpha > 0$, Algorithm 2.1 converges in a linear fashion but the convergence is inexact (i.e. it does not evaluate the exact optimal solution). It is important to note that Algorithm 2.1 does not converge to the optimal solution because of local versus global cost gaps (i.e. $\|\nabla F - \nabla f_i\| \neq 0$ in general). Therefore, at each step k, the local gradient ∇f_i directs the state estimate $x_i^{[k]}$ toward its local minimum, which, in general, is not the global minimum (except if all the local cost functions are the same). These cost gaps arise due to the heterogeneous data distribution between different nodes, a common characteristic in WSNs. Section 2.5 describes a gradient tracking scheme, Di Lorenzo and Scutari [2015] and Xin and Khan [2018], that can be used to overcome this dissimilarity. Next, we provide a useful theorem to establish the conditions for convergence of DGD to the optimal solution.

Theorem 2.2 Nedić and Ozdaglar [2009]. Consider the problem in (2.2) under Assumptions 2.2 and 2.3. For a decaying step-size $\alpha_k \sim \mathcal{O}\left(\frac{1}{k}\right)$, Algorithm 2.1 (DGD) convergences to the optimal solution at a sublinear rate.

The proof of Theorem 2.2 can be found in Nedić and Ozdaglar [2009]. Another limitation of Algorithm 2.1 (DGD) is that it requires the network topology to be weight-balanced. Therefore, we need to ensure that the communication between the nodes is either bidirectional or the corresponding weight matrix (that models the communication network) is doubly stochastic. This condition is practically hard to meet due to bandwidth limitations or the nature of links between sensors. Thus, for a more general class of networks (directed), DGD is not applicable. In Section 2.5, we provide some useful extensions of DGD designed to remove various practical limitations.

2.5 Extensions of DGD

In this section, we discuss several practical aspects of interest encountered by real-life applications and provide some useful extensions of DGD.

2.5.1 Extension to Directed Communication

Let us now consider the case where nodes communicate over a network modeled as a strongly connected directed graph (e.g. a wireless sensor network, WSN). The weight matrix describing the communication network is either row or column stochastic, [Nedić and Olshevsky, 2016; Xin et al., 2019]. We consider the case when $B = \{b_{ij}\} \in \mathbb{R}^{n\times n}$ is primitive and column stochastic. In this case, the extension of DGD can be written as:

$$x_i^{[k+1]} = \sum_{r=1}^{n} b_{ir} \left(x_r^{[k]} - \alpha \nabla f_r(x_r^{[k]}) \right).$$

For $\alpha = 0$, the above iterations do not converge to the average $\frac{1}{n}\sum_{i=1}^{n} x_r^{[k]}$ because the right eigenvector (let us call it $\boldsymbol{\pi}_B$) corresponding to the eigenvalue that is equal to 1 is not $\mathbf{1}_n$. It can be verified that $\mathbf{1}_n^\top B = \mathbf{1}_n$, $B\boldsymbol{\pi}_B = \boldsymbol{\pi}_B$, and $\lim_{k\to\infty} B^k = \boldsymbol{\pi}_B \mathbf{1}_n^\top$. Therefore, each node converges to

$$x_i^{[k]} \to [\boldsymbol{\pi}_B]_i \sum_{r=1}^{n} x_r^{[k]},$$

where $[\boldsymbol{\pi}_B]_i$ is the ith element of vector $\boldsymbol{\pi}_B$. To cater this weighted sum, we would like to divide each $x_i^{[k]}$ with corresponding $[\boldsymbol{\pi}_B]_i$ so we can get

$$\frac{x_i^{[k]}}{n[\boldsymbol{\pi}_B]_i} \to \frac{[\boldsymbol{\pi}_B]_i}{n[\boldsymbol{\pi}_B]_i} \sum_{r=1}^{n} x_r^{[k]}.$$

However, nodes do not have the knowledge of $\boldsymbol{\pi}_B$ vector. Push-sum [Hadjicostis and Charalambous, 2014 and Nedić and Olshevsky, 2016] is a method that can be used to allow each node to iteratively evaluate $[\boldsymbol{\pi}_B]_i$ locally.

The second error stems from the local versus global cost gaps for heterogeneous data distributions. We note that unless data distribution is homogeneous, the state estimate evaluated at each node is directed toward an inexact solution because $\nabla f_i \neq \nabla F$. This can be dealt with, using a gradient tracking scheme described below:

$$t_i^{[k+1]} = \sum_{i=1}^{n} b_{ir} \left(t_i^{[k]} + \nabla f_r(x_r^{[k+1]}) - \nabla f_r(x_r^{[k]}) \right), \qquad \forall k > 0.$$

It can be verified that when $t_i^{[0]} = \nabla f_i(x_i^{[k]})$, $t_i^{[0]} \rightarrow \nabla F(x_i^{[k]})$, see Di Lorenzo and Scutari [2015] and Xin and Khan [2018] for more details. Hence, the local versus global cost dissimilarity is removed and the state estimate evaluated at each node converges to the optimal solution x^*.

A distributed optimization method that relies on push-sum consensus to deal with the asymmetry caused by directed communication and a global gradient tracking scheme is called accelerated distributed directed optimization method (ADD-OPT) and is formally defined in Algorithm 2.2.

Algorithm 2.2 Accelerated Distributed Directed Optimization (ADD-OPT)

Input: $x_i^{[0]} = z_i^{[0]} \in \mathbb{R}^p, y_i^{[0]} = 1, t_i^{[0]} = \nabla f_i(z_i^{[0]}), \alpha > 0, \{b_{ir}\}_{r=1}^n$

Iteration: For $k = 0, 1, 2, \ldots, K$, each node $v_i \in \mathcal{V}$ does:

1. $x_i^{[k+1]} = \sum_{i=1}^n b_{ir} \left(x_r^{[k]} - \alpha t_r^{[k]} \right)$
2. $y_i^{[k+1]} = \sum_{i=1}^n b_{ir} y_r^{[k]}$
3. $z_i^{[k+1]} = x_i^{[k]} / y_i^{[k]}$
4. $t_i^{[k+1]} = \sum_{i=1}^n b_{ir} \left(t_i^{[k]} + \nabla f_r(z_r^{[k+1]}) - \nabla f_r(z_r^{[k]}) \right)$

Output: Each node $v_i \in \mathcal{V}$ estimates x^* by $z_i^{[K]}$ to solve (2.2)

The parameter estimate at each node $x_i^{[0]}$ is initialized randomly. An additional variable $y_i^{[0]}$ is used to help estimate the left eigenvector of the weight matrix. It can be verified that $y_i^{[k+1]} = \sum_{i=1}^n b_{ir} y_r^{[k]} \rightarrow n[\boldsymbol{\pi}_B]_i$ at each node because $\lim_{k\rightarrow\infty} B^k = \boldsymbol{\pi}_B \mathbf{1}_n^\top$. Therefore, the effect of directed communication is balanced out using these extra iterations. Furthermore, the gradient tracking term converges to the global gradient, i.e. $t_i^{[k]} \rightarrow \nabla F(z_i^{[k]})$ at each node. This ensures convergence of each state estimate to the exact solution, i.e. $z_i \rightarrow x^*$. We now describe the main convergence result for Algorithm 2.2.

Theorem 2.3 Xi et al. [2017]. Consider the problem in (2.2) under Assumptions 2.1 and 2.3. For a small enough constant step-size $\alpha > 0$, Algorithm 2.2 (ADD-OPT) linearly converges to the optimal solution x^* of F.

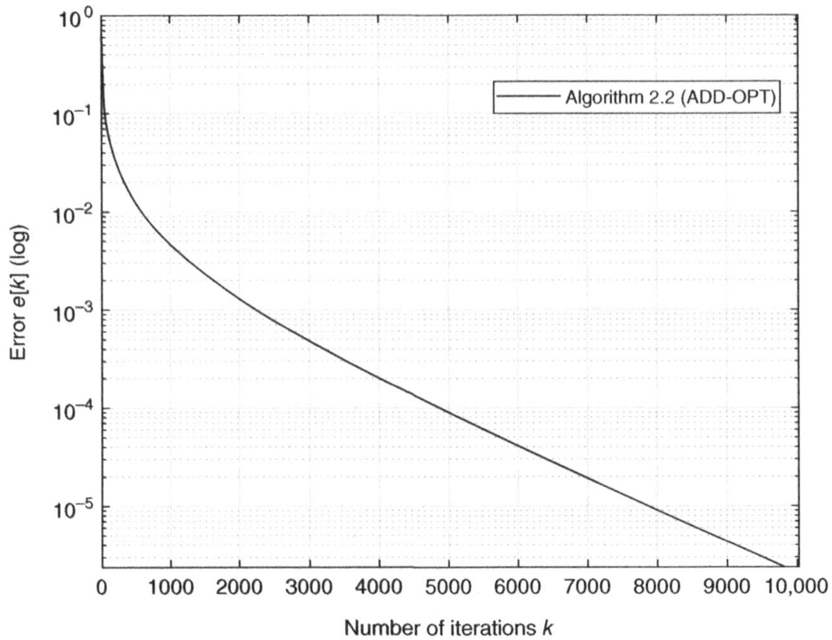

Figure 2.3 Performance results of ADD-OPT (Algorithm 2.2). Source: Adapted from [Xi et al., 2017].

Theorem 2.3 highlights the convergence properties of ADD-OPT method. It states the conditions under which linear convergence is achieved while the underlying communication graph is directed (can have unidirectional links) and strongly connected. The formal proof can be found in Xi et al. [2017].

In Figure 2.3, we present the performance results for ADD-OPT (Algorithm 2.2) over a randomly generated directed graph with 16 nodes. The network aims to solve a logistic regression problem for classifying images from MNIST dataset by minimizing the global cost F. Each node v_i possesses a local cost f_i and private dataset. To evaluate the performance of ADD-OPT, we evaluate the error $e[k] = \sum_{i=1}^{n} \|z_i^{[k]} - x^*\|$. Figure 2.3 shows the linear convergence of ADD-OPT to the optimal solution x^* at each node, which aligns with Theorem 2.3, see Qureshi [2020] for complete code.

2.5.2 Operation Over Wireless Networks

Wireless communication networks play a crucial role in distributed network control systems. By their very nature, these networks lack a physical infrastructure.

This allows for their deployment in scenarios where fixed infrastructure is unavailable, making them highly versatile and attractive for various applications, including military, civil, industrial, and environmental monitoring in challenging environments. However, it is essential to acknowledge that wireless networks come with inherent limitations. These limitations rely on various factors such as e.g. packet-dropping communication links, synchronicity between the network components, bandwidth-constrained communication channels, and noisy communication channels. In Section 2.5.2.1, we focus on the specific challenges posed by bandwidth-constrained channels and noisy communication channels. The topics of packet-dropping communication links and synchronicity between network components will be discussed toward the conclusion of this section, providing a comprehensive understanding of the inherent limitations that they impose in wireless communication networks for distributed control systems.

2.5.2.1 Quantized Communication

A popular technique in wireless networks to address the challenges of limited bandwidth and noisy channels is quantization. Quantization is a technique to reduce the amount of information transmitted by encoding a continuous signal into a discrete signal, [Rabbat and Nowak, 2005]. This technique significantly reduces the required bandwidth for communication among agents. Some of the most common quantization techniques are uniform, asymmetric, and logarithmic quantization, [Wei et al., 2019]. For developing our results in this chapter, we rely on asymmetric quantization. Asymmetric quantizers are defined as

$$q^a_\Delta(\xi) = \left\lfloor \frac{\xi}{\Delta} \right\rfloor, \tag{2.5}$$

where $\xi \in \mathbb{R}$ represents the input value for quantization, $q^a_\Delta(\xi) \in \mathbb{Q}$ denotes the quantized form of ξ, and $\Delta \in \mathbb{Q}$ indicates the quantization step size. Note that even though for the development of our results, we will utilize the asymmetric quantization technique, our results can also be extended to different quantization techniques (e.g. uniform or logarithmic).

2.5.2.2 Distributed Gradient Descent with Quantized Communication

We now present a distributed optimization algorithm for the case where nodes exchange quantized valued messages. Before presenting our algorithm, we make the following assumption.

Assumption 2.4 Each node $v_j \in \mathcal{V}$ possesses knowledge of the network's diameter D or an upper bound D' (where $D' \geq D$), as well as a common quantization level Δ.

Assumption 2.4 facilitates node coordination. In particular, with awareness of the network diameter D, each node can determine when every other node in the network has improved its estimation of the optimal solution. Moreover, the shared understanding of a common quantization level Δ enables nodes to transmit quantized messages to their neighbors, ensuring efficient communication with consistent precision.

Algorithm 2.3 Distributed gradient descent with quantized communication

---

Input: A strongly connected digraph $\mathcal{G}$ with $n = |\mathcal{V}|$ nodes and $m = |\mathcal{E}|$ edges. Static step-size $\alpha \in \mathbb{R}$, digraph diameter D, initial value $x_j^{[0]}$, local cost function f_j, quantization level $\Delta \in \mathbb{Q}$, for every node $v_j \in \mathcal{V}$

Iteration: For $k = 0, 1, 2, \dots, K$, each node $v_j \in \mathcal{V}$ does:

1. $x_j^{[k+\frac{1}{2}]} = x_j^{[k]} - \alpha \nabla f_j(x_j^{[k]})$
2. $x_j^{[k+1]} =$ Algorithm 2.3a$(x_j^{[k+\frac{1}{2}]}, D, \Delta)$

Output: Each node $v_j \in \mathcal{V}$ estimates x^* by $x_i^{[K]}$ to solve (2.2)

The intuition behind Algorithm 2.3 (QuAGD) is as follows. Initially, each node holds an approximation of the optimal solution and the specified quantization level. At each time step k, every node updates its estimate of the optimal solution through gradient descent towards the direction opposite to its gradient. Subsequently, each node updates its solution estimate using Algorithm 2.3a. More specifically, Algorithm 2.3a enables nodes to compute the quantized average of each node's estimate in finite time by exchanging and processing quantized messages, with precision determined by the quantization level. The intuition behind FAQuA is elaborated later.

The intuition behind Algorithm 2.3a (FAQuA) is as follows. Initially, each node v_j quantizes its state. Subsequently, at each time step λ, node v_j updates its state variables to match its mass variables and divides $y_j[\lambda]$ into $z_j[\lambda]$ equal parts (some parts may have slightly higher values than others). It selects the part with the minimum value and transmits it to itself. Then sends the remaining parts to randomly chosen out-neighbors or to itself. After receiving messages from its neighbors, it combines these messages with the stored ones. For every D time steps, a max and min consensus operation is carried out. If the termination condition is met, the solution is adjusted based on the quantization level.

The convergence of Algorithm 2.3 is analyzed via the following theorem.

Algorithm 2.3a Finite time quantized coordination

Input: $x_i^{[k+\frac{1}{2}]}, D, \Delta$

Initialization: Each node $v_i \in \mathcal{V}$ does the following:

1. Assigns probability b_{li} to each out-neighbor $v_l \in \mathcal{N}_i^+ \cup \{v_i\}$, as follows:

$$b_{li} = \begin{cases} \frac{1}{1+\mathcal{D}_i^+}, & \text{if } l = i \text{ or } v_l \in \mathcal{N}_i^+, \\ 0, & \text{if } l \neq i \text{ and } v_l \notin \mathcal{N}_i^+ \end{cases}$$

2. sets $z_i = 2, y_i = 2\, q_\Delta^a(x_i^{[k+\frac{1}{2}]})$

Iteration: For $\lambda = 1, 2, \ldots$, each node $v_i \in \mathcal{V}$, does:

1. **if** $\lambda \mod (D) = 1$ **then** $M_i = \lceil y_i/z_i \rceil, m_i = \lfloor y_i/z_i \rfloor$
2. broadcasts M_i, m_i to every $v_l \in \mathcal{N}_i^+$; receives M_j, m_j from every $v_j \in \mathcal{N}_i^-$; sets $M_i = \max_{v_j \in \mathcal{N}_i^- \cup \{v_i\}} M_j$,
   $m_i = \min_{v_j \in \mathcal{N}_i^- \cup \{v_i\}} m_j$
3. sets $c_i^z = z_i$;
4. **while** $c_i^z > 1$ **do**
   4.1. $c_i^y = \lfloor y_i \,/\, z_i \rfloor$
   4.2. sets $y_i = y_i - c_i^y, z_i = z_i - 1$, and $c_i^z = c_i^z - 1$
   4.3. transmits c_i^y to randomly chosen out-neighbor $v_l \in \mathcal{N}_i^+ \cup \{v_i\}$ according to b_{li}
   4.4. receives c_j^y from $v_j \in \mathcal{N}_i^-$ and sets

$$y_i = y_i + \sum_{j=1}^{n} w_{\lambda,ij}^{[r]} c_j^y, \tag{2.6}$$

$$z_i = z_i + \sum_{j=1}^{n} w_{\lambda,ij}^{[r]}, \tag{2.7}$$

   where $w_{\lambda,ij}^{[r]} = 1$ when node v_i receives c_i^y, 1 from v_j at time step λ (otherwise $w_{\lambda,ij}^{[r]} = 0$ and v_i receives no message at time step λ from v_j)
5. **if** $\lambda \mod D = 0$ **and** $M_i - m_i \leq 1$ **then** sets $x_i^{[k+1]} = m_i\Delta$ and stops operation

Output: $x_i^{[k+1]}$

Theorem 2.4 When the step-size α satisfies $\alpha \in (\frac{n(\mu+L)}{4\mu L}, \frac{2n}{\mu+L})$ and $\delta \in (0, \frac{n[4\alpha\mu L - n(\mu+L)]}{2\alpha[n(\mu+L)-2\alpha\mu L]})$, where $L = \sum_{i=1}^{n} L_i, \mu = \sum_{i=1}^{n} \mu_i$, Algorithm 2.3 generates a sequence of points $\{x^{[k]}\}$ (i.e. the variable $x_i^{[k]}$ of each node $v_i \in \mathcal{V}$) which satisfy

$$||\hat{x}^{[k+1]} - x^*||^2 < \vartheta||\hat{x}^{[k]} - x^*||^2 + \mathcal{O}(\Delta^2), \tag{2.8}$$

where Δ is the quantizer and

$$\vartheta := 2(1 + \frac{\alpha\delta}{n})(1 - \frac{2\alpha\mu L}{n(\mu+L)}) \in (0, 1), \tag{2.9a}$$

$$\mathcal{O}(\Delta^2) = (8 + 32\hat{\alpha}^2 L^2 + \frac{32\hat{\alpha}L^2}{\delta})\Delta^2. \tag{2.9b}$$

Proof: Full proof is provided in Rikos et al. [2023b, Theorem 1]. □

In Figure 2.4, we present a performance comparison of Algorithm 2.3 at various quantization levels with the works by Jiang and Charalambous [2022] and Khatana et al. [2020]. The logarithmic plot shows the error $e^{[k]}$ plotted against the number of iterations. This error, defined as

$$e^{[k]} = \sqrt{\sum_{j=1}^{n} \frac{(x_j^{[k]} - x^*)^2}{(x_j^{[0]} - x^*)^2}}, \tag{2.10}$$

is calculated with respect to the optimal solution x^* of the optimization problem (2.2). Our findings indicate that Algorithm 2.3 performs similarly to Khatana et al. [2020], especially when the quantization level matches the predefined tolerance value ρ (refer to Khatana et al. [2020]). Our proposed algorithm utilizes quantized values, making it suitable for channels with limited or finite capacity. Moreover, Algorithm 2.3 demonstrates comparable performance to Jiang and Charalambous [2022], even in cases where the results by Jiang and Charalambous [2022] do not consist of an error floor. It is important to note that Algorithm 2.3 can approximate the optimal solution with accuracy based on the chosen quantization level. Refining the quantization level enables nodes to achieve a more precise approximation of the optimal solution. In contrast, the approach in Jiang and Charalambous [2022] involves constructing the Hankel matrix and performing additional computations when the matrix loses rank, requiring exact values from each node and implying nodes exchanging messages of infinite capacity. Therefore, the key advantage of Algorithm 2.3 over Jiang and Charalambous [2022] is that nodes operate with quantized values, while in Jiang and Charalambous [2022], nodes exchange values of infinite precision.

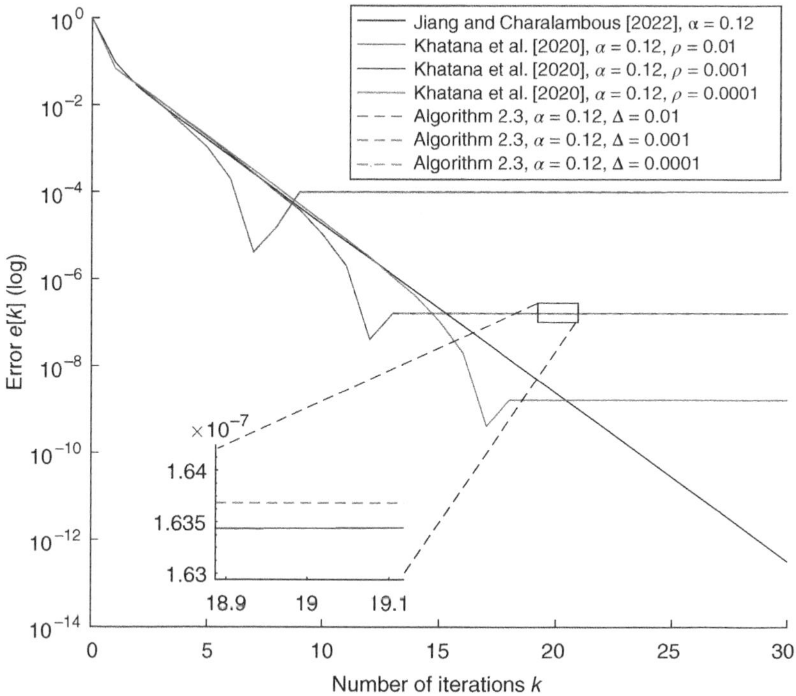

Figure 2.4 Comparison of Algorithm 2.3 for different quantization levels. Source: Jiang and Charalambous [2022] and Khatana et al. [2020].

2.5.2.3 Enhancing Accuracy of Optimal Solution

During the execution of Algorithm 2.3, the term $\mathcal{O}(\Delta^2)$ in (2.11) arises due to quantized communication between nodes, impacting the precision of the optimal solution. Determining the appropriate quantization level poses a challenge: if too coarse, the optimization solution may have an unacceptably large error floor; if too fine, it may lead to increased communication delays and packet losses. Since the exact solution is unknown a priori, selecting an optimal quantization level is challenging. Our idea is to build on Algorithm 2.3 and refine the quantization interval based on the error floors resulting from different quantization intervals. Specifically, by comparing solutions across various quantization intervals, we iteratively adjust the interval. If the solutions surpass a specified threshold, we further refine the quantization interval; otherwise, we terminate the operation. While exact accuracy cannot be guaranteed, we are able to reach a desired level of accuracy by selecting an appropriate threshold.

The idea behind Algorithm 2.4 is as follows. At each iteration k, nodes update their solution estimate through a gradient descent step. Then, they implement a

Algorithm 2.4 Gradient descent with zoomed quantized communication

Input: Strongly connected directed graph $\mathcal{G}$ with $n = |\mathcal{V}|$ nodes and $m = |\mathcal{E}|$ edges. Static step-size $\alpha \in \mathbb{R}$, digraph diameter D, initial value $x_i^{[0]}$, local cost function f_i, error bound ε_s, quantization level $\Delta \in \mathbb{Q}$, refinement constant $c_r \in \mathbb{N}$, for every node $v_j \in \mathcal{V}$. Assumptions 2.3, 2.4 hold.

Initialization: Each node $v_i \in \mathcal{V}$ sets $\text{ind}_i = 0$, $\beta = \text{ind}_i$, $S_i = \{0\}$.

Iteration: For $k = 0, 1, 2, \ldots, K$, each node $v_i \in \mathcal{V}$ does the following:

1. $x_i^{[k+\frac{1}{2}]} = x_i^{[k]} - \alpha \nabla f_i(x_i^{[k]})$
2. $x_i^{[k+1]} =$ Algorithm 2.3a($x_i^{[k+\frac{1}{2}]}, D, \Delta$)
3. **if** $x_i^{[k+1]} = x_i^{[k]}$, **then**
   3a. set $\text{ind}_i = \text{ind}_i + 1$, $\beta = \text{ind}_i$, $\gamma_\beta = k$
   3b. set $S_i = S_i \cup \{\gamma_\beta\}$
   3c. **if** $\|f_i(x_i^{[\gamma_{\beta-1}]}) - f_i(x_i^{[\gamma_\beta]})\| \leq \varepsilon_s$, **then** set $\text{vot}_i = 0$
   - **else** set $\text{vot}_i = 1$

   3d. $\text{flag}_i = \max$ - Consensus (vot_i)
   3e. **if** $\text{flag}_i = 0$ **then** terminate operation
   - **else** set $\Delta = \Delta / c_r$ and go to Step 1

Output: Each node $v_i \in \mathcal{V}$ estimates x^* by $x_i^{[K]}$ to solve (2.2)

finite-time quantized coordination algorithm (Algorithm 2.3a). When the nodes converge to a vicinity of the optimal solution (indicating convergence at the current quantization level), they store the time step. Subsequently, they assess if the difference in computed optimal solution values between the current and previous convergence instances falls below a threshold ε_s. Based on this difference, nodes determine a voting variable (either 0 or 1). A max-consensus is then performed to collectively decide on whether to proceed or halt the execution of Algorithm 2.4. If a unanimous agreement is reached among all nodes that the optimal solution values from the current and previous convergence instances are sufficiently close, the operation ceases. However, if there is discord, nodes adjust the quantization level and repeat the process.

Theorem 2.5 When the step-size α satisfies $\alpha \in (\frac{n(\mu+L)}{4\mu L}, \frac{2n}{\mu+L})$ and $\delta \in (0, \frac{n[4\alpha\mu L - n(\mu+L)]}{2\alpha[n(\mu+L) - 2\alpha\mu L]})$, where $L = \max\{L_i\}, \mu = \min\{\mu_i\}$, Algorithm 2.4 generates a sequence of points $\{x^{[k]}\}$ (i.e. the variable $x_i^{[k]}$ of each node $v_i \in \mathcal{V}$), which satisfies

$$\|\hat{x}^{[k+1]} - x^*\|^2 < \vartheta \|\hat{x}^{[k]} - x^*\|^2 + \mathcal{O}(\Delta^2), \tag{2.11}$$

where Δ is the quantizer and

$$\vartheta := 2\left(1+\frac{\alpha\delta}{n}\right)\left(1-\frac{2\alpha\mu L}{n(\mu+L)}\right) \in (0,1), \tag{2.12a}$$

$$\mathcal{O}(\Delta^2) = \left(8+32n^2\hat{\alpha}^2L^2+\frac{32n^2\hat{\alpha}L^2}{\delta}\right)\Delta^2. \tag{2.12b}$$

Proof: The proof is a direct extension of the proof for Theorem 2.4, with the key distinction being the restart of the process as outlined in iteration step (3*e*) of Algorithm 2.4. □

In Figure 2.5, we display the performance of our algorithm on a randomly generated directed graph with 20 nodes. Each node v_i is characterized by parameters such as $\alpha = 0.12$, initial values $x_i^{[0]} \in [1,5]$, threshold $\varepsilon_s = 0.003$, step size $\Delta = 0.001$, and a constant $c_r = 10$. The error $e^{[k]}$ is depicted on a logarithmic scale against the number of iterations in the same figure. This error, defined by the equation

$$e^{[k]} = \sqrt{\sum_{j=1}^{n}\frac{(x_j^{[k]}-x^*)^2}{(x_j^{[0]}-x^*)^2}}, \tag{2.13}$$

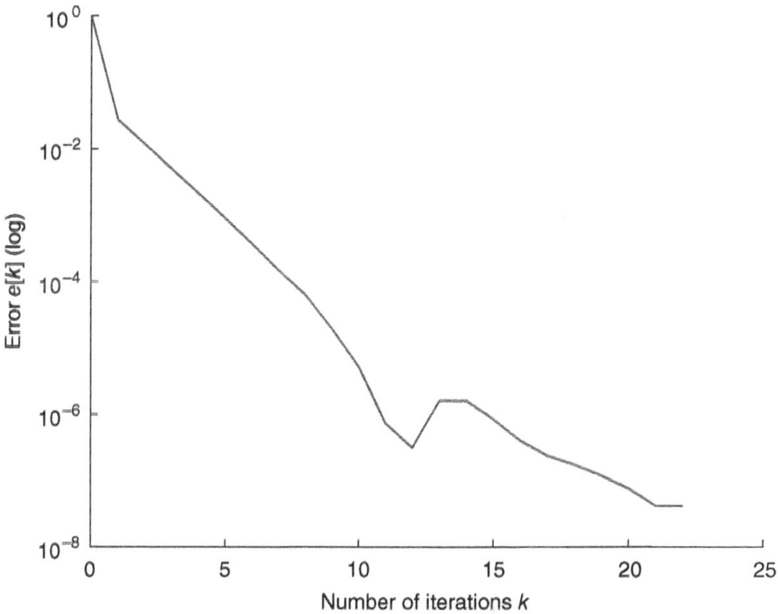

Figure 2.5 Execution of Algorithm 2.4 on a random directed graph consisting of 20 nodes.

is measured concerning the optimal solution x^* of the optimization problem (2.2). The outcomes showcase the algorithm successfully converging to the optimal solution. Specifically, let us focus at time steps $k = 13, 14$, and $k = 21, 22$. During $k = 13, 14$, the condition in Iteration Step 3 holds (i.e. $x_i^{[13]} = x_i^{[14]}$ for every node $v_i \in \mathcal{V}$), and $e^{[13]} = e^{[14]}$. Consequently, by time Step 14, nodes evaluate the collective improvement of their local cost functions (Iteration Step 3*c*). Since this criterion is not met for at least one node, they choose to adjust the quantization level by setting $\Delta = \Delta/10 = 0.0001$ and continue Algorithm 2.4. Between time steps $k = 14, \ldots, 21$, nodes achieve a more accurate approximation of the optimal solution, with the precision dependent on the quantization level as per Theorem 2.5. At time steps $k = 21, 22$, the condition in Iteration Step 3 is met again. However, by time Step 22, the overall improvement in each node's local cost function is below the specified threshold ε_s (i.e. $\|f_i(x_i^{[14]}) - f_i(x_i^{[22]})\| \leq \varepsilon_s$) for all nodes $v_i \in \mathcal{V}$ (see Iteration Step 3*c*). Consequently, nodes decide to conclude the operation at time step $k = 22$ (Iteration Step 3*e*). It should be noted that choosing a smaller ε_s may prompt nodes to further refine the quantization level, allowing for an even more precise approximation of the optimal solution.

2.5.3 Stochastic Implementation

We now describe a stochastic extension of DGD, which is very useful for large local datasets. In general, DGD requires every node to evaluate the full batch gradient at each iteration to update the local state estimates. However, this process becomes computationally demanding with large local datasets. Therefore, when the local cost f_i at node v_i can be further decomposable into m_i component cost functions $f_i = \sum_{j=1}^{m_i} f_{i,j}$, stochastic gradients $\nabla f_{i,s_i^{[k]}}$ are often used (where $s_i^{[k]}$ is chosen uniformly at random from the set $\{1, \ldots, m_i\}$ at each iteration k). We note that stochastic gradients are assumed to have bounded variance, i.e. $\forall i, k$,

$$\mathbb{E}_{s_i^{[k]}}\left[\|\nabla f_{i,s_i^{[k]}}(\mathbf{x}_i^{[k]}) - \nabla f_i(\mathbf{x}_i^{[k]})\|_2^2 \mid \mathbf{x}_i^{[k]}\right] \leq \sigma^2,$$

which leads to inexact convergence. However, SAGA-based variance reduction technique can be used to eliminate this error and estimate the exact local gradient while evaluating only the stochastic local gradients $\nabla f_{i,s_i^{[k]}}$ at each iteration.

Algorithm 2.5 describes a distributed optimization method, which integrates node-level variance reduction to eliminate the uncertainty introduced by stochastic gradients, network-level gradient tracking to tackle the distributed nature of the data, and push-sum consensus to address directed information exchange. Similar to ADD-OPT (Algorithm 2.2), the state variables are randomly initialized. However, Push-SAGA maintains the gradient table $\{\nabla f_{i,j}\}_{j=1}^{m_i}$ at each node v_i to store the component gradients.

Algorithm 2.5 Push-SAGA

Input: $x_i^{[0]} = z_i^{[0]} \in \mathbb{R}^p, y_i^{[0]} = 1, t_i^{[0]} = g_i^{[0]} = \nabla f_i(z_r^{[0]}), \alpha > 0,$
$\{b_{ir}\}_{r=1}^n, \{\nabla f_{i,j}(z_i^{[0]})\}_{j=1}^{m_i}, \{h_{i,j}^{[1]} = z_i^{[0]}\}_{j=1}^{m_i}.$

Iteration: For $k = 0, 1, 2, \dots, K$, each node $v_i \in \mathcal{V}$ does the following:

1. $x_i^{[k+1]} = \sum_{i=1}^n b_{ir}\left(x_r^{[k]} - \alpha t_i^{[k]}\right)$
2. $y_i^{[k+1]} = \sum_{i=1}^n b_{ir} y_r^{[k]}$
3. $z_i^{[k+1]} = x_i^{[k+1]} / y_i^{[k+1]}$
4. **Select** $s_i^{[k+1]}$ uniformly at random from the set $\{1, \cdots, m_i\}$,
5. $g_i^{[k+1]} = \nabla f_{i,s_i^{[k+1]}}(z_i^{[k+1]}) - \nabla f_{i,s_i^{[k+1]}}(h_{i,s_i^{[k+1]}}^{[k+1]}) + \frac{1}{m_i}\sum_{j=1}^{m_i} \nabla f_{i,j}(h_{i,j}^{[k+1]})$
6. $\nabla f_{i,s_i^{[k+1]}}(z_i^{[k+1]}) = \nabla f_{i,s_i^{[k+1]}}(h_{i,s_i^{[k+1]}}^{[k+1]})$
7. $t_i^{[k+1]} = \sum_{i=1}^n b_{ir}\left(t_i^{[k]} + g_r^{[k+1]} - g_r^{[k]}\right)$
8. **If** $j = s_i^{[k+1]}$, **then** $h_{i,j}^{[k+2]} = z_i^{[k+1]}$, **else** $h_{i,j}^{[k+2]} = h_{i,j}^{[k+1]}$
9. **end if**

Output: Each node $v_i \in \mathcal{V}$ estimates x^* by $z_i^{[K]}$ to solve (2.2)

At every iteration k, each node v_i first computes an ADD-OPT-type iterate $z_i^{[k]}$ using push-sum correction. We note that the descent direction in the $\mathbf{x}_i^{[k]}$-update (and thus in the $z_i^{[k]}$-update) is $t_i^{[k]}$, which is the (stochastic) global gradient tracker. Subsequently, node v_i generates a random index $s_i^{[k]}$ and evaluates the SAGA-based gradient estimator $g_i^{[k]}$ using the current iterate iterate $z_i^{[k]}$ and the elements from the gradient table. The gradient table is updated only at the $s_i^{[k]}$-th element, while the other entries remain unchanged. Finally, these estimators $g_i^{[k]}$ are aggregated over the network using dynamic average consensus to calculate $t_i^{[k]}$ that track the global gradient.

Next, we describe the main convergence result of Push-SAGA.

Theorem 2.6 Qureshi et al. [2022]. Consider the problem in (2.2) and let $M := \max_i m_i$, $m := \min_i m_i$, and each $f_{i,j}$ be L-smooth and each f_i to be μ-strongly convex. For the stepsize $\alpha \in (0, \overline{\alpha})$, for some $\overline{\alpha} > 0$, Push-SAGA linearly converges, at each node, to the global minimum x^* of F. In particular, for $\alpha = \overline{\alpha}$, Push-SAGA achieves an ϵ-optimal solution in

$$\mathcal{O}\left(\max\left\{M, \frac{M}{m}\frac{\kappa^2\psi}{(1-\lambda)^2}\right\}\log\frac{1}{\epsilon}\right),$$

component gradient computations (in parallel) at each node, where $\kappa := L/\mu$, $(1-\lambda)$ is the spectral gap of the weight matrix, and ψ is a directivity constant.

Theorem 2.6 provides the conditions under which PushSAGA converges linearly to the global solution. The exact convergence of this stochastic method is only possible because of SAGA-based variance-reduction technique. The result also characterizes the directivity constant ψ, which encapsulates the effects of asymmetric communication over directed network. The formal analysis and the complete proof of Theorem 2.6 can be found in Qureshi et al. [2022].

Figure 2.6 shows the performance comparison of Push-SAGA and ADD-OPT (Algorithm 2.2) over a randomly generated directed graph of 16 nodes. We consider the classification problem to distinguish images belonging to classes sampled from MNIST dataset. Each node v_i possesses a local cost f_i and private dataset and we compare the error computed at each epoch $e[k] = \sum_{i=1}^{n} \|z_i^{[k]} - x^*\|$, see Qureshi [2023] for complete code. Figure 2.6 shows the linear convergence of Push-SAGA and ADD-OPT to the optimal soluton x^* at each node. However, Push-SAGA converges much faster than ADD-OPT in terms of *epochs*. It is noteworthy that if node v_i holds m_i data samples, one epoch of ADD-OPT is one iteration, whereas one epoch of Push-SAGA is m_i iterations, as described in Algorithm 2.5.

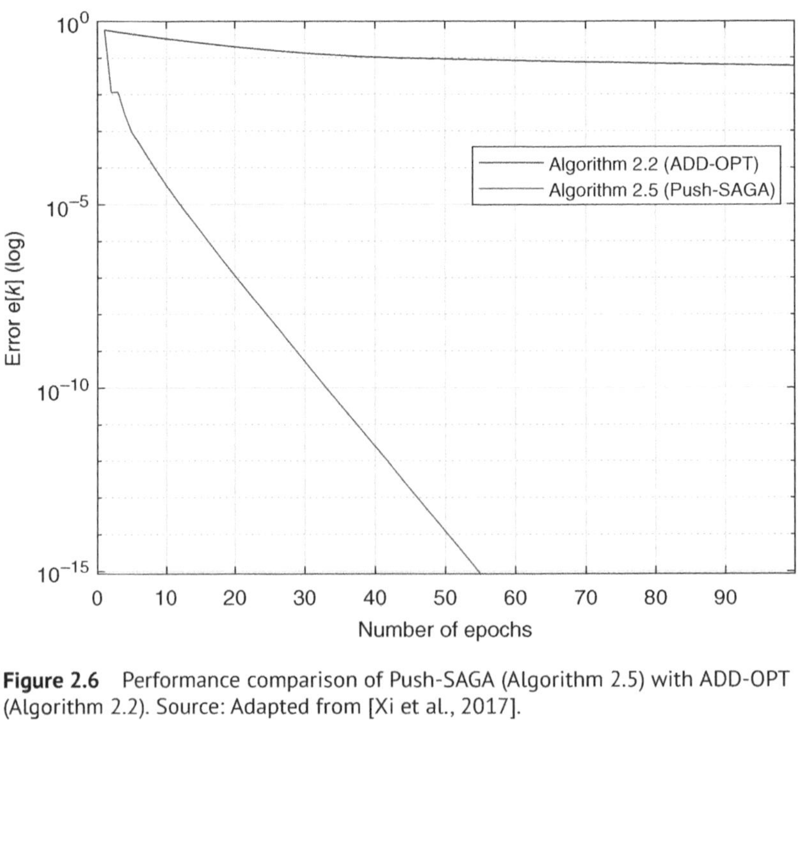

Figure 2.6 Performance comparison of Push-SAGA (Algorithm 2.5) with ADD-OPT (Algorithm 2.2). Source: Adapted from [Xi et al., 2017].

2.6 Distributed Fine-Tuning of Vision Transformers

The methods discussed in this chapter play a fundamental role in several modern applications. In this section, we consider the training process of large language models (LLMs) and demonstrate the performance results of the stochastic variant of ADD-OPT (Algorithm 2.2) [Qureshi et al., 2021] for fine-tuning vision transformers (ViTs).

LLMs have gained a lot of attention recently due to their wide range of applications in natural language understanding, text completion, content generation, image recognition, and sentiment analysis, Luong et al. [2015], Vaswani et al. [2017], Dosovitskiy et al. [2021], Devlin et al. [2019], and Radford and Narasimhan [2018]. These models use positional information and input features to apply the "*attention mechanism*," enabling LLMs to understand the underlying task. The primary objective of an ideal LLM is to cultivate a general-purpose model from which more targeted tasks can be derived, exhibiting intelligence across a wide range of tasks. During the training process, developing a general understanding is often referred to as *pretraining*, and developing a new intricate skill is called *fine-tuning*.

Training LLMs is exceptionally resource-intensive since these models are trained on large datasets and involves billions of tuning parameters, which results in extended training times to achieve optimal performance. Moreover, as discussed earlier, it is often not practical to bring all data to a single computational node. Conversely, when nodes are trained on local datasets, they often struggle to generalize well. Since training an entire model from scratch is practically infeasible at distributed nodes, leveraging an existing pre-trained model and fine-tuning for new downstream tasks becomes very significant.

In the sequel, we examine pretrained vision transformer models (ViT), Dosovitskiy et al. [2021], distributedly fine-tuned using ADD-OPT (Algorithm 2.2), over an unforeseen image dataset distributed across multiple nodes. A key challenge in this scenario is the error caused by the local versus global cost gap, which arises due to the inherent limitation that the local dataset (at each node) may not accurately represent the global data. As a consequence, Figure 2.7 illustrates the attention maps learned at a node when specific classes (flowers and dogs) of images are absent from the local fine-tuning dataset. It is evident that independent fine-tuning (Local-FT) struggles to apply attention to the correct objects.

The stochastic variant of ADD-OPT addresses this challenge using gradient tracking (eliminating the aforementioned cost gap) and learns attention maps that are comparable to the centralized fine-tuning results. Table 2.2 shows the classification accuracy for a randomly selected ith node during the fine-tuning of ViT under heterogeneous data distribution for different datasets, [Parkhi et al., 2012; Nilsback and Zisserman, 2008; Krizhevsky et al., 2010a,2010b]. The complete

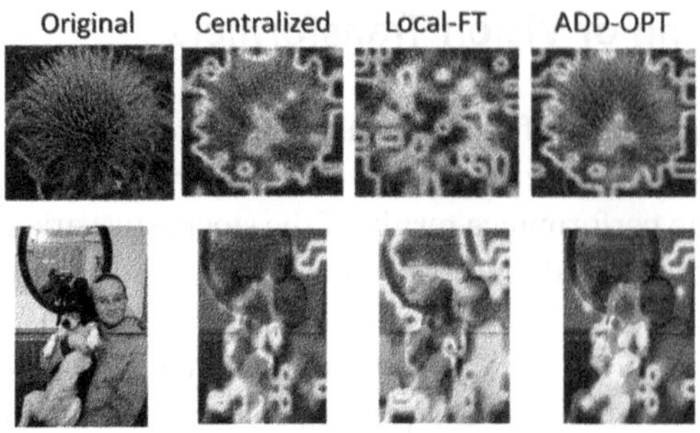

Figure 2.7 Visualizing attention maps of a pretrained ViT locally fine-tuned on downstream tasks, highlighting the impact when classes "flowers" and "dogs" are absent from the local dataset.

Table 2.2 Accuracy after fine-tuning ViT model for 100 epochs.

ViT	**Local-FT**	**ADD-OPT**
Datasets	**Accuracy at Node v_i (%)**	**Accuracy at Node v_i (%)**
Pets	13.80	87.21
Flowers	13.36	99.80
CIFAR-10	20.07	97.44
CIFAR-100	9.86	87.40

code can be found in Qureshi [2024a], where the authors fine-tune distributed ViT, DeiT, and Swin-Transformer models over a peer-to-peer network of nodes.

2.7 Discussion and Future Directions

The exploration of distributed algorithms extends beyond our proposed approaches. For instance, when dealing with bandwidth limitations and possibly non-differentiable local cost functions, a noteworthy method integrates the alternating direction method of multipliers (ADMM) with finite-time quantized coordination algorithms (similar to Algorithm 2.3a), [Rikos et al., 2023a]. In this scenario, nodes engage in the exchange of quantized valued messages, while exhibiting asynchronous operation. Another approach involves the problem of online distributed learning in the presence of communication bandwidth limitations. In this concept, learning models undergo training on diverse, distributed data sources. Stochastic gradients drive local training, while a finite-time quantized coordination protocol facilitates the aggregation of locally trained models, [Bastianello et al., 2023]. Another approach involves tackling the problem of time-varying networks, [Saadatniaki et al., 2020]. This constraint refers to the limitation of computational nodes that may not always be accessible to contribute to solving the global problem due to limitations in resources. More generally, realistic constraints can be included on parameters and cost functions and the new problem can be solved as a min–max optimization problem using DGD ascent methods, [Qureshi and Khan, 2023a,2023c].

Looking ahead, there are several open problems that require attention. Some examples include addressing bandwidth limitations in the presence of packet-dropping links, handling asynchronicities in the context of bandwidth limitations, and calculating the exact optimal solution in the face of bandwidth limitations. In general, these (along with other) open problems highlight the intricate nature of optimizing distributed systems, especially in scenarios where communication bandwidth constraints play a pivotal role. As we delve into open problems, it becomes evident that the evolution of distributed algorithms demands a holistic understanding of the interplay between various factors (such as communication constraints, system dynamics, and the nature of local cost functions). By tackling these challenges, we aim to pave the way for advancements that not only optimize distributed learning and control but also contribute to the broader landscape of wireless communication networks.

Acknowledgments

The work of U. A. Khan and M. I. Qureshi has been supported by NSF under award PIRE-2230630. The work of T. Charalambous has been partially supported by the project MINERVA, a project funded by the European Research Council (ERC) under the European Union's Horizon 2022 research and innovation programme (Grant Agreement 101044629). Usman A. Khan holds concurrent appointments as a professor at Tufts University and as an Amazon Scholar with Amazon Robotics. This chapter describes work performed at Tufts University and is not associated with Amazon.

Bibliography

M. Assran, N. Loizou, N. Ballas, and M. G. Rabbat. Stochastic gradient-push for distributed deep learning. In *Proceedings of International Conference on Machine Learning*, volume 97, pages 344–353, 2019.

N. Bastianello, A. I. Rikos, and K. H. Johansson. Online distributed learning with quantized finite-time coordination. In *Proceedings of 62nd IEEE Conference on Decision and Control*, pages 5026–5032, 2023.

M. Benzi, G. H. Golub, and J. Liesen. Numerical solution of saddle point problems. *Acta Numerica*, 14:1–137, 2005.

L. Bottou, F. E. Curtis, and J. Nocedal. Optimization methods for large-scale machine learning. *SIAM Review*, 60(2):223–311, 2018.

J. Chen and A. H. Sayed. Diffusion adaptation strategies for distributed optimization and learning over networks. *IEEE Transactions on Signal Processing*, 60(8):4289–4305, 2012.

J. Devlin, M.-W. Chang, K. Lee, and K. Toutanova. BERT: Pre-training of deep bidirectional transformers for language understanding. In *Proceedings of the Conference of the North American Chapter of the Association for Computational Linguistics: Human Language Technologies, NAACL-HLT 2019, Minneapolis, MN, USA, June 2-7, 2019, Volume 1 (Long and Short Papers)*, pages 4171–4186, 2019.

P. Di Lorenzo and G. Scutari. Distributed nonconvex optimization over networks. In *6th International Workshop on Computational Advances in Multi-Sensor Adaptive Processing*, pages 229–232, 2015.

A. Dosovitskiy, L. Beyer, A. Kolesnikov, D. Weissenborn, X. Zhai, T. Unterthiner, M. Dehghani, M. Minderer, G. Heigold, S. Gelly, J. Uszkoreit, and N. Houlsby. An image is worth 16×16 words: Transformers for image recognition at scale. In *International Conference on Learning Representations*, 2021. URL https://openreview.net/forum?id=YicbFdNTTy.

P. A. Forero, A. Cano, and G. B. Giannakis. Consensus-based distributed support vector machines. *Journal of Machine Learning Research*, 11:1663–1707, 2010.

I. Goodfellow, J. Pouget-Abadie, M. Mirza, B. Xu, D. Warde-Farley, S. Ozair, A. Courville, and Y. Bengio. Generative adversarial nets. In Z. Ghahramani, M. Welling, C. Cortes, N. Lawrence, and K. Q. Weinberger, editors, *Advances in Neural Information Processing Systems*, volume 27. Curran Associates, Inc., 2014. URL https://proceedings.neurips.cc/paper/2014/file/5ca3e9b122f61f8f06494c97b1afccf3-Paper.pdf.

C. N. Hadjicostis and T. Charalambous. Average consensus in the presence of delays in directed graph topologies. *IEEE Transactions on Automatic Control*, 59(3):763–768, 2014.

W. Jiang and T. Charalambous. A fast finite-time consensus based gradient method for distributed optimization over digraphs. In *Proceedings of IEEE Conference on Decision and Control*, pages 6848–6854, 2022.

V. Khatana, G. Saraswat, S. Patel, and M. V. Salapaka. Gradient-consensus method for distributed optimization in directed multi-agent networks. In *2020 American Control Conference*, pages 4689–4694, 2020.

A. Krizhevsky, V. Nair, and G. Hinton. CIFAR-10 (Canadian Institute for Advanced Research). 5(4):1, 2010a. URL http://www.cs.toronto.edu/kriz/cifar.html.

A. Krizhevsky, V. Nair, and G. Hinton. CIFAR-100 (Canadian Institute for Advanced Research), 2010b. URL http://www.cs.toronto.edu/kriz/cifar.html.

S. Lee and M. M. Zavlanos. Approximate projection methods for decentralized optimization with functional constraints. *IEEE Transactions on Automatic Control*, 63(10):3248–3260, 2018.

X. Lian, C. Zhang, H. Zhang, C. Hsieh, W. Zhang, and J. Liu. Can decentralized algorithms outperform centralized algorithms? A case study for decentralized parallel stochastic gradient descent. In *Advances in Neural Information Processing Systems*, volume 30, pages 5330–5340, 2017.

T. Liang and J. Stokes. Interaction matters: A note on non-asymptotic local convergence of generative adversarial networks. In *Proceedings of the 22nd International Conference on Artificial Intelligence and Statistics*, volume 89, pages 907–915. PMLR, 2019.

T. Lin, C. Jin, and M. Jordan. On gradient descent ascent for nonconvex-concave minimax problems. In *Proceedings of the 37th International Conference on Machine Learning*, volume 119, pages 6083–6093. PMLR, 2020a.

T. Lin, C. Jin, and M. I. Jordan. Near-optimal algorithms for minimax optimization. In *Proceedings of 33rd Conference on Learning Theory*, volume 125, pages 2738–2779. PMLR, 2020b.

P. D. Lorenzo and G. Scutari. NEXT: In-network nonconvex optimization. *IEEE Transactions on Signal and Information Processing over Networks*, 2(2):120–136, 2016.

T. Luong, H. Pham, and C. D. Manning. Effective approaches to attention-based neural machine translation. In *Proceedings of the Conference on Empirical Methods in Natural Language Processing*, pages 1412–1421, 2015.

A. Nedić and A. Olshevsky. Stochastic gradient-push for strongly convex functions on time-varying directed graphs. *IEEE Transactions on Automatic Control*, 61(12):3936–3947, 2016.

A. Nedić and A. Ozdaglar. Distributed subgradient methods for multi-agent optimization. *IEEE Transactions on Automatic Control*, 54(1):48, 2009.

A. Nedić, A. Olshevsky, and W. Shi. Achieving geometric convergence for distributed optimization over time-varying graphs. *SIAM Journal on Optimization*, 27(4):2597–2633, 2017.

M.-E. Nilsback and A. Zisserman. Automated flower classification over a large number of classes. In *Indian Conference on Computer Vision, Graphics and Image Processing*, Dec 2008.

O. M. Parkhi, A. Vedaldi, A. Zisserman, and C. V. Jawahar. Cats and dogs. In *IEEE Conference on Computer Vision and Pattern Recognition*, 2012.

G. Qu and N. Li. Harnessing smoothness to accelerate distributed optimization. *IEEE Transactions on Control of Network Systems*, 5(3):1245–1260, 2017.

M. I. Qureshi. S-ADDOPT. https://github.com/qureshi-mi/S-ADDOPT, 2020.

M. I. Qureshi. PushSAGA. https://github.com/qureshi-mi/PushSAGA, 2023.

M. I. Qureshi. Distributed transformers. https://github.com/qureshi-mi/Distributed-Transformers, 2024a.

M. I. Qureshi. Near-shot learning. https://github.com/qureshi-mi/NearShotLearning, 2024b.

M. I. Qureshi and U. A. Khan. Stochastic first-order methods over distributed data. In *Proceedings of the 12th Sensor Array and Multichannel Signal Processing Workshop (SAM)*, pages 405–409, 2022.

M. I. Qureshi and U. A. Khan. Distributed saddle point problems for strongly concave-convex functions. *IEEE Transactions on Signal and Information Processing over Networks*, 9:679–690, 2023a.

M. I. Qureshi and U. A. Khan. A distributed first-order optimization method for strongly concave-convex saddle point problems. In *Proceedings of the 9thIEEE International Workshop on Computational Advances in Multi-Sensor Adaptive Processing (CAMSAP)*, 2023b.

M. I. Qureshi and U. A. Khan. A distributed stochastic first-order method for strongly concave-convex saddle point problems. In 62nd *IEEE Conference of Decision and Control*, pages 4170–4175, 2023c.

M. I. Qureshi, R. Xin, S. Kar, and U. A. Khan. S-ADDOPT: Decentralized stochastic first-order optimization over directed graphs. *IEEE Control Systems Letters*, 5(3):953–958, 2021.

M. I. Qureshi, R. Xin, S. Kar, and U. A. Khan. Push-SAGA: A decentralized stochastic algorithm with variance reduction over directed graphs. *IEEE Control Systems Letters*, 6:1202–1207, 2022.

M. G. Rabbat and R. D. Nowak. Quantized incremental algorithms for distributed optimization. *IEEE Journal on Selected Areas in Communications*, 23(4):798–808, 2005.

A. Radford and K. Narasimhan. Improving language understanding by generative pre-training. 2018. URL https://api.semanticscholar.org/CorpusID:49313245.

H. Raja and W. U. Bajwa. Cloud *K*-SVD: A collaborative dictionary learning algorithm for big, distributed data. *IEEE Transactions on Signal Processing*, 64(1):173–188, 2016.

S. S. Ram, A. Nedić, and V. V. Veeravalli. Distributed stochastic subgradient projection algorithms for convex optimization. *Journal of Optimization Theory and Applications*, 147(3):516–545, 2010.

A. Reisizadeh, A. Mokhtari, H. Hassani, A. Jadbabaie, and R. Pedarsani. FedPAQ: A communication-efficient federated learning method with periodic averaging and quantization. In *Proceedings of the 23rd International Conference on Artificial Intelligence and Statistics*, pages 2021–2031, 2020.

A. I. Rikos, W. Jiang, T. Charalambous, and K. H. Johansson. Asynchronous distributed optimization via ADMM with efficient communication. In *Proceedings of the 62ndIEEE Conference on Decision and Control*, pages 7002–7008, 2023a.

A. I. Rikos, W. Jiang, T. Charalambous, and K. H. Johansson. Distributed optimization with gradient descent and quantized communication. *IFAC-PapersOnLine*, 56(2):5900–5906, 2023b.

F. Saadatniaki, R. Xin, and U. A. Khan. Decentralized optimization over time-varying directed graphs with row and column-stochastic matrices. *IEEE Transactions on Automatic Control*, 65(11):4769–4780, 2020.

S. Safavi, U. A. Khan, S. Kar, and J. M. F. Moura. Distributed localization: A linear theory. *Proceedings of the IEEE*, 106(7):1204–1223, Jul 2018.

A. Sinha, H. Namkoong, and J. Duchi. Certifiable distributional robustness with principled adversarial training. In *International Conference on Learning Representations*, 2018. URL https://openreview.net/forum?id=Hk6kPgZA-.

J. Tsitsiklis, D. Bertsekas, and M. Athans. Distributed asynchronous deterministic and stochastic gradient optimization algorithms. *IEEE Transactions on Automatic Control*, 31(9):803–812, 1986.

A. Vaswani, N. Shazeer, N. Parmar, J. Uszkoreit, L. Jones, A. N. Gomez, L. Kaiser, and I. Polosukhin. Attention is all you need. In *Advances in Neural Information Processing Systems*, volume 30, 2017.

J. Wei, X. Yi, H. Sandberg, and K. H. Johansson. Nonlinear consensus protocols with applications to quantized communication and actuation. *IEEE Transactions on Control of Network Systems*, 6(2):598–608, 2019.

C. Xi, R. Xin, and U. A. Khan. ADD-OPT: Accelerated distributed directed optimization. *IEEE Transactions on Automatic Control*, 63(5):1329–1339, 2017.

R. Xin and U. A. Khan. A linear algorithm for optimization over directed graphs with geometric convergence. *IEEE Control Systems Letters*, 2(3):315–320, 2018.

R. Xin and U. A. Khan. Distributed Heavy-Ball: A generalization and acceleration of first-order methods with gradient tracking. *IEEE Transactions on Automatic Control*, 65(6):2627–2633, 2020.

R. Xin, C. Xi, and U. A. Khan. FROST—Fast row-stochastic optimization with uncoordinated step-sizes. *EURASIP Journal on Advances in Signal Processing*, 2019:1–14, 2019.

R. Xin, S. Pu, A. Nedić, and U. A. Khan. A general framework for decentralized optimization with first-order methods. *Proceedings of the IEEE*, 108(11):1869–1889, 2020.

R. Xin, U. Khan, and S. Kar. A hybrid variance-reduced method for decentralized stochastic non-convex optimization. In *Proceedings of the*38th*International Conference on Machine Learning*, pages 11459–11469, 2021.

J. Xu, S. Zhu, Y. C. Soh, and L. Xie. Augmented distributed gradient methods for multi-agent optimization under uncoordinated constant stepsizes. In *Proceedings of the 54thIEEE Conference on Decision and Control*, pages 2055–2060, 2015.

T. Yang, X. Yi, J. Wu, Y. Yuan, D. Wu, Z. Meng, Y. Hong, H. Wang, Z. Lin, and K. H. Johansson. A survey of distributed optimization. *Annual Reviews in Control*, 47:278–305, 2019.

M. Zhu and S. Martínez. Discrete-time dynamic average consensus. *Automatica*, 46(2):322–329, 2010.

3

Distributed Non-Bayesian Quickest Change Detection with Energy Harvesting Sensors

Emma Green[1] and Subhrakanti Dey[2]

[1]*Department of Engineering Cybernetics, Norwegian University of Science and Technology, Trondheim, Norway*
[2]*Department of Electrical Engineering, Signals and Systems Division, Uppsala University, Uppsala, Sweden*

3.1 Introduction

Wireless sensor networks (WSNs) have recently been applied in various application domains such as smart grid monitoring [Gungor et al., 2010], industrial process monitoring [Gungor and Hancke, 2009], and mobile robots and autonomous vehicles [Chong and Kumar, 2003]. In a significant number of applications, the sensors are usually located in remote places. Periodically replacing batteries for such a scenario can be cumbersome and expensive. For mitigating this issue, renewable energy harvesting from the surrounding environment by individual sensors has been explored in the literature. For such a system, the sensors are outfitted with a rechargeable battery of limited capacity which is capable of energy harvesting from ambient sources. The drawback of such a system is the inherent uncertain nature of the process of harvesting energy. Additionally, due to each sensor having limited energy storage capacity, the problem of finding the optimal energy allocation strategy for sensing, information processing, and transmission is quite challenging in practice.

One important task of WSNs is to detect changes in the observation signal distribution. In a parametric setting, this can be achieved using either classical methods [Kay, 1998] or sequential detection techniques, such as quickest change detection [Poor and Hadjiliadis, 2008]. The quickest detection techniques are either applied locally at the individual sensors or at the fusion center (FC) after collating information from the sensors [Tartakovsky and Veeravalli, 2008]. These detection problems focus on detecting sudden changes in the probability

Wireless Sensor Networks in Smart Environments: Enabling Digitalization from Fundamentals to Advanced Solutions, First Edition. Edited by Domenico Ciuonzo and Pierluigi Salvo Rossi.

distribution of a stochastic process by finding the stopping time for minimizing the detection delay, subject to false alarm rate constraint. In classical literature, there are two approaches for analyzing such problems. The first one is the Bayesian framework, where the unknown change point is to be assumed to be drawn from a specific probability distribution [Banerjee and Veeravalli, 2011]. The other framework is non-Bayesian, where the change point is considered unknown but deterministic in nature [Lorden, 1971; Pollak, 1985].

In standard settings over multiple time slots, the sensors in a WSN can sense the observation signal during every time slot in the quickest change detection framework. However, this can't be ensured in the harvesting-based scenario. The quickest change detection problem with energy constraints has been studied for both the non-Bayesian [Geng and Lai, 2013] and the Bayesian frameworks [Geng et al., 2014] in centralized settings. Decentralized quickest change detection has been studied without energy constraints in Moustakides [2006] and Veeravalli [2001]. Researchers have also studied the distributed detection problems for the energy harvesting WSN but for the non-sequential hypothesis testing framework [Kalus et al., 2015; Li et al., 2018; Ciuonzo et al., 2019]. We have focused on the non-Bayesian quickest change detection with energy harvesting in both *decentralized* and *distributed* settings in this chapter. In the decentralized scenario, finding an optimal quantization policy at each individual sensor is essential because the detection performance is sensitive to the accuracy of the information collated at the FC. For the distributed settings, the asymptotic analysis of first passage times is of high research significance.

Thus, this chapter focuses on the following two topics:

1) The average detection delay minimization for the non-Bayesian decentralized quickest change detection in a WSN.
2) Asymptotic analysis for the non-Bayesian quickest change detection problem for the distributed scenarios when the average harvested energy at all individual sensors is greater or equal to the energy required for sensing.

3.2 System Model

3.2.1 Decentralized Detection Scenario

As shown in Figure 3.1, the system has N wireless sensors capable of energy harvesting and an FC. The time interval is slotted and the measurement signal is observed by each sensor over M time slots. During the kth time slot, the ith sensor decides whether to sense or refrain from sensing depending on the energy available in its battery. Corresponding sensing decision is represented by

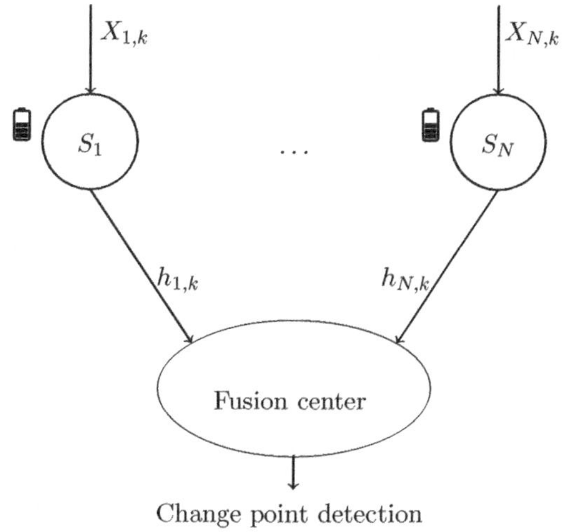

Figure 3.1 Quickest change detection with multiple sensors.

$v_{i,k}$ $1 \leq i \leq N,\ 1 \leq k \leq M$, and $v_{i,k} \in \{0, 1\}$, and 1 (or 0) denotes the decision to sense or not, respectively.

We assume that the observation signal is generated from one of the two probability distribution functions f_0 or f_1, corresponding to the periods before and after the change point λ, respectively. The observations signals $\{X_{i,k}\}$ are assumed to be independent and identically distributed (*i.i.d.*) across both time and all sensors, before and after the change. The corresponding hypothesis testing problem is then formulated as follows:

$$\mathcal{H}_0 : X_{i,k} \sim f_0(x), \quad \text{if } k < \lambda,$$
$$\mathcal{H}_1 : X_{i,k} \sim f_1(x), \quad \text{if } k \geq \lambda.$$

After receiving the observation signal $X_{i,k}$, during the kth time slot, the ith sensor S_i computes the log-likelihood ratio (LLR), $Z_{i,k} = \log \frac{f_1(X_{i,k})}{f_0(X_{i,k})}$. Furthermore, it is quantized to $q_{i,k}$ bits by comparing the LLR with the $2^{q_{i,k}} - 1$ thresholds resulting in the quantized information $U_{i,k}$, which is forwarded to the FC. Without loss of generality, the quantized information $U_{i,k}$ is assumed to be limited to the set $\{0, 1, \dots, 2^{q_{i,k}} - 1\}$, representing the $2^{q_{i,k}}$ quantization bins. During each time slot, these quantized messages are transmitted to the FC from the sensors with sufficient energy for transmission. The FC then applies the cumulative sum (CUSUM) test to detect a change in the distribution of the observation signal using the CUSUM-based sequential detection algorithm as described in Tartakovsky and Veeravalli [2008].

During every time slot, the fading channel gains between the sensors and the FC, denoted by $\{h_{i,k}\}$, $1 \leq i \leq N,\ 1 \leq k \leq M$, are assumed to remain fixed, but it is assumed to change from slot to slot. In this chapter, we consider the scenario when a sensor chooses to sense the observation signal, only if it is feasible, based on the available energy in the sensor's battery. Thus, $v_{i,k}$ is determined by the following rule:

$$v_{i,k} = \begin{cases} 1, \text{ if } B_{i,k} \geq E_s + q_{i,k}E^b_{i,k}, \\ 0, \text{ otherwise,} \end{cases} \tag{3.1}$$

where E_s represents the energy required for sensing during each time slot. The battery state, the energy needed to transmit each quantized bit, and the number of quantized bits at the ith sensor during the kth time slot are represented by $B_{i,k}$, $E^b_{i,k}$, and $q_{i,k}$, respectively. The total energy spent by the ith sensor during that time slot, denoted by $E_{i,k}$, is then calculated as:

$$E_{i,k} = v_{i,k}(E_s + q_{i,k}E^b_{i,k}). \tag{3.2}$$

The maximum battery capacity for all sensors is denoted as B_{max}. Furthermore, we assume that $H_{i,k}$ denotes the harvested energy by the ith sensor in the kth time slot. Then, the battery state is computed by the following recursive expression:

$$B_{i,k+1} = \min\left\{B_{max}, B_{i,k} + H_{i,k} - E_{i,k}\right\}. \tag{3.3}$$

3.2.2 Distributed Detection Scenario

For the fully distributed case, when the local sensors perform the CUSUM test, they only need sufficient energy required for sensing and processing a sample, denoted by E_s. However, when the CUSUM test statistic surpasses the threshold, corresponding local decision is transmitted to the FC. So for that scenario, the sensor needs to have at a minimum $E_s + E_b$ amount of energy, where E_b represents the transmission energy. Assuming $B_{max} = \infty$, we can also express the battery state recursively as $B_{k+1} = B_k + H_k - 1_{(B_k > E_s)}E_s$.

Thus, the modified version of the CUSUM test statistics for the energy harvesting sensor can be expressed as follows:

$$\overline{W}_k = \max\left\{0, \overline{W}_{k-1} + \xi_k \log\frac{f_1(x_k)}{f_0(x_k)}\right\}, \quad \overline{W}_0 = 0, \tag{3.4}$$

where $\xi_k = 1_{(B_k > E_s)}$. Obviously, $\xi_k = 1$ happens with probability $P(B_k > E_s)$ and $\xi_k = 0$, with probability $1 - P(B_k > E_s)$. We assume that the energy harvesting process H_k is independent and identically distributed (i.i.d) with mean value of $E(H_k) = \overline{H}$. It is also assumed to be independent of the observation sequence $\{X_k\}$. It should be noted that the energy harvesting processes across the different sensors are assumed to be independent, though not necessarily identically distributed. To summarize, average harvested energy $\overline{H}_i$ for individual sensors can be different. The change point is detected at a sensor when this modified version of the CUSUM test statistics surpasses the threshold h. Consequently, the battery state can be further expressed using the following recursive equation:

$$B_{k+1} = \begin{cases} B_k + H_k - 1_{(B_k > E_s)} E_s, & \text{if } \overline{W}_k \leq h, \\ B_k + H_k - 1_{(B_k > E_s + E_b)} (E_s + E_b), & \text{otherwise.} \end{cases} \tag{3.5}$$

In this chapter, we only study the random process ξ_k for the scenario, where the average harvested energy $\overline{H}$ exceeds the amount of energy used E_k, i.e. $\overline{H} \geq E_s$.

3.3 Quickest Change Detection at the FC

For the decentralized scenario, the discrete valued quantized messages $U_{i,k}$ are assumed to be distributed according to the probability mass function (pmf) g_i^j, if the observations are drawn from the hypothesis $\mathcal{H}_j$. For quantizing $Z_{i,k}$ to $q_{i,k}$ bits, we have to compute $2^{q_{i,k}} - 1$ number of thresholds. If we denote the lth quantization threshold for the ith sensor as t_l^i, the corresponding pmfs can then be expressed as follows:

$$g_i^1(l) = \mathbb{F}_1(t_{l+1}^i) - \mathbb{F}_1(t_l^i), \tag{3.6}$$

$$g_i^0(l) = \mathbb{F}_0(t_{l+1}^i) - \mathbb{F}_0(t_l^i), \tag{3.7}$$

where $\mathbb{F}_1$ and $\mathbb{F}_0$ are the corresponding cumulative distribution functions to the probability distribution functions f_1 and f_0, respectively. They can be computed as follows:

$$\mathbb{F}_1(x) = \int_{-\infty}^{x} f_1(x)dx,$$

$$\mathbb{F}_0(x) = \int_{-\infty}^{x} f_0(x)dx.$$

After obtaining the quantized information $U_{i,k}$ from the individual sensors, the FC calculates the quantized LLR between hypotheses $\mathcal{H}_1$ and $\mathcal{H}_0$ as follows:

$$Z^q(k) = \sum_{i=1}^{N} \log \frac{g_i^1(U_{i,k})}{g_i^0(U_{i,k})}. \tag{3.8}$$

We denote T as the stopping time, which is the time instant when the quickest change detection algorithm identifies a change in the distribution of the observation signal. The sensing strategy is defined as $\boldsymbol{\nu} = \{\nu_{i,k};\ i = 1, \dots, N;\ k = 1, \dots, M\}$, and the quantization function is represented as $\mathbf{q} = \{q_{i,k};\ i = 1, \dots, N;\ k = 1, \dots, M\}$. Together, the stopping time T and these parameters form the policy $\phi = (\boldsymbol{\nu}, \mathbf{q}, T)$.

The non-Bayesian quickest change detection algorithm is designed to detect the change point as rapidly as possible after it occurs. Therefore, our objective is to identify the joint sensing and quantization policy ϕ, that minimizes the average worst-case detection delay (supremum average detection delay [SADD]) [Pollak, 1985], which is expressed as follows:

$$SADD(\phi) = \sup_{1 \le \lambda \le \infty} \mathbb{E}_{\lambda}(T - \lambda | T \ge \lambda), \tag{3.9}$$

where $\mathbb{E}_{\lambda}$ denotes the expectation for the change point λ. We want to determine the optimal sensing decision $\mathbf{v}^{\star}$ and the optimal quantization function $\mathbf{q}^{\star}$, and the corresponding policy tuple $\tilde{\phi} = (\mathbf{v}^{\star}, \mathbf{q}^{\star}, T)$. The optimal stopping time is then computed by the minimax change point detection procedure as follows:

$$T^{\star} = \min_{T} SADD(\tilde{\phi}), \text{ s.t. } \mathbb{E}_{\infty}[T] > \gamma; \ \gamma > 1. \tag{3.10}$$

Here, $\mathbb{E}_{\infty}[T]$ denotes the expected stopping time when the change does not occur, i.e. $\lambda = \infty$.

In the decentralized scenario, it is asymptotically optimal for the sensors to quantize their individual LLRs in a way that maximizes the individual Kullback–Leibler (KL) divergence measures between the distributions after and before the change [Tartakovsky and Veeravalli, 2008]. Consequently, the corresponding CUSUM test statistic at the FC is defined by the following recursive equation:

$$W^{q}(k) = \max\left\{0, W^{q}(k-1) + Z^{q}(k)\right\}, \quad W^{q}(0) = 0. \tag{3.11}$$

The optimal stopping time for the CUSUM test can be computed as follows:

$$T^{\star} = \min\left\{k \ge 1 : W^{q}(k) \ge r\right\}, \text{ where } r = \log \gamma. \tag{3.12}$$

3.4 Optimization Problem Formulation

The asymptotic worst-case detection delay (as $\gamma \to \infty$) of the optimal decentralized detection scheme can be expressed as Tartakovsky and Veeravalli [2008]:

$$SADD(T) \sim \frac{\log \gamma}{\mathbf{I}_{tot}^{q}} \quad \text{as } \gamma \to \infty, \tag{3.13}$$

where $\mathbf{I}_{tot}^{q}$ represents the total Kullback–Leibler (KL) information number between the non-null and null hypothesis respectively, calculated by using the quantized information obtained from the local sensors. The number of active users in the kth time slot is denoted by n_k. It is a random variable which depends on the channel state information (CSI) and energy state information (ESI), for a given sensing and quantization strategy. In that case, $\mathbf{I}_{tot}^{q}$ (for the kth time slot) can be computed as:

$$\mathbf{I}_{tot}^{q} = \sum_{i=1}^{n_k} \mathbf{I}(g_i^1, g_i^0) = \sum_{i=1}^{n_k} \sum_{l=0}^{2^{q_{i,k}}-1} g_i^1(l) \log \frac{g_i^1(l)}{g_i^0(l)}, \tag{3.14}$$

where $\mathbf{I}(g_i^1, g_i^0)$ denotes the KL divergence between the probability mass functions g_i^1 and g_i^0 of the sensor S_i.

Thus, the optimization problem is formulated as follows:

$$\max_{\nu_{i,k}, q_{i,k}} \sum_{k=1}^{M} \mathbb{E}_{n_k} \left\{ \sum_{i=1}^{n_k} \left\{ \nu_{i,k} \sum_{l=0}^{2^{q_{i,k}}-1} g_i^1(l) \log \frac{g_i^1(l)}{g_i^0(l)} \right\} \right\}, \tag{3.15}$$

$$\text{s.t. } \nu_{i,k} \in \{0,1\};\ \forall i,k, \tag{3.16}$$

$$q_{i,k} \in \left\{1, \dots, Q_{max}\right\};\ \forall i,k, \tag{3.17}$$

$$E_{i,k} \leq B_{i,k};\ \forall i,k. \tag{3.18}$$

3.4.1 Optimal Threshold Quantization

The KL divergence for the kth slot can be computed by utilizing the quantized LLR from n_k active sensor. It can be written as follows:

$$\mathcal{F}(\{t_l^i : l \in \{0, \dots, 2^{q_{i,k}} - 1\}\}) = \sum_{i=1}^{n_k} \sum_{l=0}^{2^{q_{i,k}}-1} g_i^1(l) \log \frac{g_i^1(l)}{g_i^0(l)}, \tag{3.19}$$

while assuming that $\nu_{i,k}$ and $q_{i,k}$ satisfy (3.16), (3.17), and (3.18).

The Kullback–Leibler divergence (KLD) contribution from individual sensors depends on their own quantization thresholds. Thus, we maximize the KLD of individual sensors separately with respect to their own quantization thresholds. Therefore, the optimal thresholds for the ith sensor can be computed by solving for $\frac{\partial \mathcal{F}_i}{\partial t_l^i} = 0$, where $\mathcal{F}_i = \sum_{l=0}^{2^{q_{i,k}}-1} g_i^1(l) \log \frac{g_i^1(l)}{g_i^0(l)}$. To simplify the threshold notation t_l^i, we decide to drop the sensor index i from it. It is noticeable that only two consecutive terms in the sum of the above expression are functions of t_l, i.e. for a specific sensor only $g^1(l), g^0(l), g^1(l-1)$, and $g^0(l-1)$ depend on t_l. Hence, the corresponding gradient expression can be expressed as:

$$\frac{\partial \mathcal{F}_i}{\partial t_l} = \frac{\partial}{\partial t_l} \{\mathcal{F}_i^1 + \mathcal{F}_i^2\}, \tag{3.20}$$

where $\mathcal{F}_i^1 = g^1(l-1) \log \frac{g^1(l-1)}{g^0(l-1)}$ and $\mathcal{F}_i^2 = g^1(l) \log \frac{g^1(l)}{g^0(l)}$, and the individual gradients is simplified as follows:

$$\frac{\partial \mathcal{F}_i^1}{\partial t_l} = \frac{\partial g^1(l-1)}{\partial t_l} \log \frac{g^1(l-1)}{g^0(l-1)} + g^0(l-1) \frac{\partial}{\partial t_l} \left\{ \frac{g^1(l-1)}{g^0(l-1)} \right\}, \tag{3.21}$$

$$\frac{\partial \mathcal{F}_i^2}{\partial t_l} = \frac{\partial g^1(l)}{\partial t_l} \log \frac{g^1(l)}{g^0(l)} + g^0(l) \frac{\partial}{\partial t_l} \left\{ \frac{g^1(l)}{g^0(l)} \right\}. \tag{3.22}$$

Thus, the optimal thresholds are computed by solving the following equations:

$$\frac{\partial \mathcal{F}_i^1}{\partial t_l} + \frac{\partial \mathcal{F}_i^2}{\partial t_l} = 0,\ l = 0, 1, \dots, 2^{q_{i,k}} - 1. \tag{3.23}$$

With some algebraic manipulations, (3.23) is simplified to:

$$\frac{\partial g^1(l-1)}{\partial t_l}\left\{\log\left\{\frac{\frac{g^1(l-1)}{g^0(l-1)}}{\frac{g^1(l)}{g^0(l)}}\right\}\right\}$$
$$=\frac{\partial g^0(l-1)}{\partial t_l}\left\{\frac{g^1(l-1)}{g^0(l-1)}-\frac{g^1(l)}{g^0(l)}\right\}, \tag{3.24}$$

which can be further simplified to:

$$\frac{\frac{\partial g^1(l-1)}{\partial t_l}}{\frac{\partial g^0(l-1)}{\partial t_l}}=\left\{\frac{\frac{g^1(l-1)}{g^0(l-1)}-\frac{g^1(l)}{g^0(l)}}{\log\left\{\frac{\frac{g^1(l-1)}{g^0(l-1)}}{\frac{g^1(l)}{g^0(l)}}\right\}}\right\},\ l=1,2,\ldots,2^{q_{i,k}}. \tag{3.25}$$

As previously stated, the nonadaptive optimal thresholds are calculated by solving the above-mentioned equation using a nonlinear solver, once the optimal number of quantization bits is determined through the dynamic programming (DP) algorithm.

3.5 Detection Delay Analysis When $\overline{H} \geq E_s$ for the Distributed Scenario

In this section, we study the case when $\overline{H} \geq E_s$ and we prove that $P(\xi_k = 1) = 1$ for $k > N$ and sufficiently large enough N. We use the strong law of large number (SLLN) for the proof. It is utilized on the i.i.d. sequence $\{H_k\}$. The analysis concerns the asymptotic case, when $\gamma \to \infty$ or $h \to \infty$. For that case, we can write (provided that the CUSUM statistic doesn't exceed h):

$$B_k = B_0 + \sum_{m=0}^{k-1} H_m - E_s \sum_{m=0}^{k-1} 1_{(B_m > E_s)}. \tag{3.26}$$

Therefore, the constraint $B_k > E_s + E_b$ holds iff

$$B_0 + \sum_{m=0}^{k-1} H_m - E_s \sum_{m=0}^{k-1} 1_{(B_m > E_s)} > E_s + E_b. \tag{3.27}$$

Equivalently, $B_k > E_s + E_b$ is satisfied if and only if

$$\sum_{m=0}^{k-1} H_m > E_s \sum_{m=0}^{k-1} 1_{(B_m > E_s)} - (B_0 - E_s - E_b). \tag{3.28}$$

If we divide both sides by k, we can write:

$$\frac{\sum_{m=0}^{k-1} H_m}{k} > E_s \frac{\sum_{m=0}^{k-1} 1_{(B_m > E_s)}}{k} - \frac{B_0 - E_s - E_b}{k}. \tag{3.29}$$

Using SLLN, we can infer that there is a sufficiently large $N(\epsilon)$, such that for $k > N(\epsilon)$, the average harvested energy across k slots meets the following constraint:

$$\left| \frac{\sum_{m=0}^{k-1} H_m}{k} - \overline{H} \right| < \epsilon \tag{3.30}$$

for any $\epsilon > 0$. Therefore, we examine the following two distinct scenarios.

- For the first case, assume $\frac{\sum_{m=0}^{k-1} H_m}{k} < \overline{H}$. For this case, (3.30) simplifies to $\frac{\sum_{m=0}^{k-1} H_m}{k} > \overline{H} - \epsilon$.
- For the second case, we assume $\frac{\sum_{m=0}^{k-1} H_m}{k} > \overline{H}$. Thus, (3.30) simplifies to $\frac{\sum_{m=0}^{k-1} H_m}{k} < \overline{H} + \epsilon$.

Now, we let the initial battery state be chosen arbitrarily, with the assumption that $B_0 > E_s + E_b$. We begin by examining the first case. In this case, the right-hand side of Eq. (3.28) can be upper bounded as follows:

$$\begin{aligned} & E_s \frac{\sum_{m=0}^{k-1} 1_{(B_m > E_s)}}{k} - \frac{B_0 - E_s - E_b}{k} \\ & \leq E_s - \frac{B_0 - E_s - E_b}{k} \leq \overline{H} - \frac{B_0 - E_s - E_b}{k} \\ & < \overline{H} - \epsilon < \frac{\sum_{m=0}^{k-1} H_m}{k}. \end{aligned} \tag{3.31}$$

The penultimate step in (3.31) can be justified by the condition $B_0 > E_s + E_b$. Consequently, $\frac{B_0 - E_s - E_b}{k}$ can be lower bounded by a positive constant ϵ, where $\epsilon < \frac{B_0 - E_s - E_b}{k}$.

Similarly, we analyze the second case, where $\frac{\sum_{m=0}^{k-1} H_m}{k} > \overline{H}$. For this case, we again concentrate on the right-hand side of Eq. (3.28), which can be upper bounded as follows:

$$\begin{aligned} & E_s \frac{\sum_{m=0}^{k-1} 1_{(B_m > E_s)}}{k} - \frac{B_0 - E_s - E_b}{k} \\ & \leq E_s - \frac{B_0 - E_s - E_b}{k} \leq \overline{H} - \frac{B_0 - E_s - E_b}{k} \\ & < \overline{H} < \frac{\sum_{m=0}^{k-1} H_m}{k}. \end{aligned} \tag{3.32}$$

In this case, the penultimate step in (3.32) can be justified by the condition $B_0 > E_s + E_b$. Therefore, $\frac{B_0 - E_s - E_b}{k}$ is always positive, leading to $\overline{H} - \frac{B_0 - E_s - E_b}{k} < \overline{H}$. The final step follows directly from the assumptions related to this case. This indicates that the constraint in (3.28) is satisfied for both cases.

Thus, since $P(\xi_k = 1) = 1$ for significantly large $k > N$, in the asymptotic case where $\gamma \to \infty$, the modified CUSUM test reduces to the standard CUSUM test. Consequently, the results of the standard CUSUM test are applicable in this scenario.

In Sections 3.5.1 and 3.5.2, we summarize the results for the average detection delay and the distribution of the first passage time to false alarm for this scenario.

3.5.1 Average Detection Delay

To derive the average detection delay expression, we define the random walk as $S_n = \sum_{k=0}^{n} Z_k$, where $S_0 = 0$. The expectation and probability measure under the distribution f_1 are denoted as $\mathbb{E}_1$ and P_1 respectively. Therefore, the mean and variance of Z_k can be represented as $\mathbb{E}_1(Z_k) = \mathcal{I}_{KL}$, $\mathbb{E}_1\left\{(Z_k - \mathcal{I}_{KL})^2\right\} = \sigma_1^2 < \infty$. The running minimum of the random walk S_n is denoted by η_n which is defined as $\eta_n = -\min_{0 \le k \le n} S_k$. The perturbed version of the random walk S_n, with the additional perturbation term η_n, i.e. W_n can be expressed as $W_n = S_n - \min_{0 \le k \le n} S_k = S_n + \eta_n$.

Most results related to the first passage time for the random walk crossing a threshold were investigated in the literature for the original random walk S_n. However one can extend these results to the perturbed random walk W_n using nonlinear renewal theory, if the "slowly varying" conditions [Tartakovsky et al., 2014] hold for the corresponding perturbation term η_n. Starting from the first passage time which is defined as $\tau_h = \inf\left\{n \ge 1 : W_n > h\right\}$, we express the corresponding overshoot as $\kappa(h) = W_{\tau_h} - h$. Furthermore, we define the first ladder epoch T_+, which is the first time instant when the random walk S_n takes a positive value, as $T_+ = \inf\left\{n \ge 1 : S_n > 0\right\}$. The corresponding ladder height at the first ladder epoch is denoted by S_{T_+}, which is the value of the random walk at time T_+.

The author of Dey [2020] found that for an energy harvesting sensor whose average harvested energy $\overline{H} \ge E_s$ and implementing the CUSUM test algorithm, both the average detection delay and the distribution of the detection delay under the non-null hypothesis (f_1) don't depend on $\overline{H}$. These parameters can be obtained from the following first-order asymptotic expressions:

$$\mathbb{E}_1\left\{\tau_h\right\} = \frac{1}{\mathcal{I}_{KL}}\left\{h + \frac{\mathbb{E}_1\left\{S_{T_+}^2\right\}}{\mathbb{E}_1\left\{S_{T_+}\right\}} - \frac{\mathbb{E}_1\left\{Z_1^2\right\}}{2\mathcal{I}_{KL}}\right\} + o(1), \tag{3.33}$$

$$P_1(\overline{\tau}_h \leq x, \kappa(h) \leq y) = \Phi(x)H(y),$$
$$\text{as } h \to \infty, \quad \forall x \in \{-\infty, \infty\}, \; y \geq 0, \tag{3.34}$$

where $\overline{\tau}_h = \frac{\tau_h - \frac{h}{I_{KL}}}{\frac{h\sigma_1^2}{I_{KL}^3}}$ and $\lim_{n\to\infty} P(\kappa(h) \leq y) = H(y)$. Furthermore, $\Phi(x)$ represents the cumulative distribution function of the standard normal distribution $\mathcal{N}(0, 1)$.

3.5.1.1 Average Detection Delay for Distributed Change Detection with Local Detection at the Sensors

In this section, we will analyze the average detection delay in the distributed detection case. Specifically, we focus on three decision fusion rules: OR, AND, and r out of N rule. The corresponding first passage times are simply the minimum, maximum and the rth-order statistics of all the first passage times obtained from the individual sensors, respectively. Let the normalized average first passage times be denoted as $\overline{\tau}_1, \overline{\tau}_2, \ldots, \overline{\tau}_N$ and let their arrangement in increasing order be given by $\overline{\tau}_{(1)} \leq \overline{\tau}_{(2)} \leq \cdots \leq \overline{\tau}_{(N)}$. Then, the normalized average first passage times for the OR, AND, and r out of N rule (denoted by T_{min}, T_{max}, and T_r respectively) can be expressed as follows:

$$T_{min} = \overline{\tau}_{(1)} = \min\{\overline{\tau}_1, \overline{\tau}_2, \ldots, \overline{\tau}_N\},$$
$$T_{max} = \overline{\tau}_{(N)} = \max\{\overline{\tau}_1, \overline{\tau}_2, \ldots, \overline{\tau}_N\},$$
$$T_r = \overline{\tau}_{(r)}. \tag{3.35}$$

To compute the normalized average first passage times for OR, AND, and r out of N rule, we use the moments of different order statistics of a standard normal variate. According to Nadarajah [2008], if $\overline{\tau}_1, \overline{\tau}_2, \ldots, \overline{\tau}_N$ follow a standard normal distribution ($\mathcal{N}(0, 1)$), then the jth-order moment of the sth-order statistic can be computed using the following expression:

$$\mathbb{E}(\overline{\tau}_{s:N}^j) = \frac{N! 2^{j/2+1-s}}{\pi^{1/2}(s-1)!(N-s)!} \sum_{l=0}^{N-s} \binom{N-s}{l} \left(-\frac{1}{2}\right)^l$$
$$\times \sum_{p=0, p+j \text{ even}}^{s+l-1} \binom{s+l-1}{p} \pi^{-p/2} 2^p \Gamma\left(\frac{p+j+1}{2}\right)$$
$$\times F_A^{(p)}\left(\frac{p+j+1}{2}; \frac{1}{2}, \ldots, \frac{1}{2}; \frac{3}{2}, \ldots, \frac{3}{2}; -1, \ldots, -1\right), \tag{3.36}$$

where F_A^p is the Lauricella function of type A with parameter p.

This result can be used to prove that the normalized average detection delay for OR, AND, and r out of N rule can be expressed as:

$$\mathbb{E}\left\{T_{min}\right\} = \frac{N2^{1/2}}{\pi^{1/2}}\sum_{l=0}^{N-1}\binom{N-1}{l}\left(-\frac{1}{2}\right)^{l}$$
$$\times \sum_{p=0,p\ odd}^{l}\binom{l}{p}\pi^{-p/2}2^{p}\Gamma\left(\frac{p+2}{2}\right)$$
$$\times F_{A}^{(p)}\left(\frac{p+2}{2};\frac{1}{2},\dots,\frac{1}{2};\frac{3}{2},\dots,\frac{3}{2};-1,\dots,-1\right). \tag{3.37}$$

$$\mathbb{E}\left\{T_{max}\right\} = \frac{N2^{3/2-N}}{\pi^{1/2}}\sum_{p=0,p\ odd}^{N-1}\binom{N-1}{p}\pi^{-p/2}2^{p}\Gamma\left(\frac{p+2}{2}\right)$$
$$\times F_{A}^{(p)}\left(\frac{p+2}{2};\frac{1}{2},\dots,\frac{1}{2};\frac{3}{2},\dots,\frac{3}{2};-1,\dots,-1\right). \tag{3.38}$$

$$\mathbb{E}\left\{T_{r}\right\} = \frac{N!2^{3/2-r}}{\pi^{1/2}(r-1)!(N-r)!}\sum_{l=0}^{N-r}\binom{N-r}{l}\left(-\frac{1}{2}\right)^{l}$$
$$\times \sum_{p=0,p\ odd}^{r+l-1}\binom{r+l-1}{p}\pi^{-p/2}2^{p}\Gamma\left(\frac{p+2}{2}\right)$$
$$\times F_{A}^{(p)}\left(\frac{p+2}{2};\frac{1}{2},\dots,\frac{1}{2};\frac{3}{2},\dots,\frac{3}{2};-1,\dots,-1\right). \tag{3.39}$$

3.5.2 Asymptotic Distribution of the First Passage Time to a False Alarm

The author of Dey [2020] also found that for an energy-harvesting sensor with average harvested energy $\overline{H} \geq E_s$ implementing the CUSUM test, the asymptotic tail distribution of the (normalized) first passage time to a false alarm which is independent of $\overline{H}$, can be expressed as:

$$P_{\infty}\left(e^{-h}\tau_{\infty}(h) > x\right) = e^{-\overline{\beta}x}, \quad h \to \infty, \tag{3.40}$$

where $\overline{\beta} = \mathcal{I}_{KL}\overline{\delta}^{2}$, $\overline{\delta} = \lim_{h\to\infty}\mathbb{E}_1\{\exp\{-(S_{\tau_{\infty}(h)} - h)\}\}$ and $\mathbb{E}_{\infty}[\tau_{\infty}(h)] = \frac{e^{h}}{\mathcal{I}_{KL}\overline{\delta}^{2}}(1+o(1))$.

3.5.2.1 Asymptotic Distribution of First-Passage Time to False Alarm for Distributed Change Detection with Local Detection at the Sensors

This section focuses on the asymptotic analysis of the tail distribution of the first passage time to a false alarm for the distributed detection case. We examine the tail distribution for OR, AND, and r out of N fusion rule. From (3.40), it can be noted that the first-passage times at each sensor asymptotically follows an exponential distribution with mean $\frac{1}{\lambda} = \frac{e^h}{\mathcal{I}_{KL}\overline{\delta}^2}$ with cumulative distribution function $F(x) = 1 - e^{-\lambda x}$. Utilizing the theory of order statistics [David and Nagaraja, 2004], we derive the asymptotic cumulative distribution functions for the first-passage time to a false alarm for OR, AND, and r out of N fusion rule. These cumulative distribution functions are computed using the following expressions:

- The cumulative distribution function of the first-passage time for the OR fusion rule is:

$$F_{min}(x) = 1 - (1 - F(x))^N = 1 - e^{-N\lambda x}, \tag{3.41}$$

  which indicates that it follows an exponential distribution with mean $N\lambda$.
- The cumulative distribution function of the first-passage time for the AND fusion rule is:

$$F_{max}(x) = F(x)^N = \left(1 - e^{-\lambda x}\right)^N, \tag{3.42}$$

  and for the asymptotic case as $N \to \infty$, the distribution approaches a Gumbel distribution with the following cumulative distribution function:

$$F_{max}^{asym}(x) = e^{-e^{(\log(N)-\lambda x)}}. \tag{3.43}$$

- The cumulative distribution function of the first-passage time for r out of N rule is:

$$\begin{aligned} F_r(x) &= \sum_{i=r}^{N} \binom{N}{i} F^i(x)(1 - F(x))^{N-i} \\ &= \sum_{i=r}^{N} \binom{N}{i} (1 - e^{-\lambda x})^i e^{-(N-i)\lambda x}. \end{aligned} \tag{3.44}$$

3.5.2.2 Average First-Passage Time to False Alarm for Distributed Change Detection with Local Detection at the Sensors

This section focuses on the asymptotic analysis of the average first passage time to a false alarm under the null hypothesis for the OR, AND, and r out of N rules. We study the 1st, Nth, and rth-order moments of the first-passage time distribution, which are essentially the average first-passage time for OR, AND, and r out of N rule, respectively. In order to accomplish this, we utilize the following theorem from Nagaraja [2006]:

Theorem 3.1 If $\tau_1, \tau_2, \dots, \tau_N$ follow exponential distribution with parameter λ, then the ith-order statistics $\tau_{(i)}$ follows the distribution:

$$(\tau_{(i)}, i = 1, \dots, N) \overset{d}{=} \frac{1}{\lambda}\left(\sum_{j=1}^{i} \frac{Y_j}{N - j + 1}\right) \tag{3.45}$$

where $\stackrel{d}{=}$ represents the distributional equality and Y_j are i.i.d standard exponential random variables. Additionally, the first-order moment of the ith-order statistics can be derived by the following expression:

$$\mathbb{E}(\tau_{(i)}) = \sum_{j=1}^{i} \frac{1}{\lambda(N-j+1)}. \tag{3.46}$$

Using Theorem 3.1 and the order statistics theory, we derive the asymptotic average first passage time to a false alarm for the OR, AND, and r out of N rules. They are calculated by the following expressions:

$$\mathbb{E}_{\infty}(T_{min}) = \frac{1}{N\lambda}, \tag{3.47}$$

$$\mathbb{E}_{\infty}(T_{max}) = \sum_{j=1}^{N} \frac{1}{\lambda(N-j+1)}, \tag{3.48}$$

$$\mathbb{E}_{\infty}(T_r) = \sum_{j=1}^{r} \frac{1}{\lambda(N-j+1)}. \tag{3.49}$$

3.6 Simulation Results

3.6.1 Decentralized Detection Results

In this section, the simulation results are presented for both causal and noncausal information with finite battery capacity and for both optimal and uniform quantization strategies for the decentralized detection case. The power gain for the channel between the ith sensor S_i and the FC for the kth time slot, $h_{i,k}$, is modeled as a random variable exponential distribution with unity mean. The amount of energy being harvested, $H_{i,k}$, for the sensor S_i during the kth time slot is also assumed to be a random variable with exponentially distribution with a mean of 1 μJ. The observations are assumed to be sampled from a Gaussian distribution. The mean and the variance of the Gaussian distribution under hypothesis $\mathcal{H}_1$ are assumed to be $\mu = 1.5$ and $\sigma^2 = 1$, respectively. We also assumed that the probability of bit error for transmitting the quantized observation to the FC is $P_e = 0.005$. The noise power spectral density is taken to be $N_0 = 0.02$ μW/Hz. The sensing energy is taken to be $E_s = 0.1$ μJ. The number of sensors is assumed to be $N = 2$. The maximum quantization bits for the simulation is assumed to be $Q_{max} = 5$. This choice of Q_{max} is motivated by Figure 3.2, which demonstrates that the Kullback–Leibler (KL) divergence measure for both the optimal and uniform quantization policies, as well as the unquantized case, converges and becomes

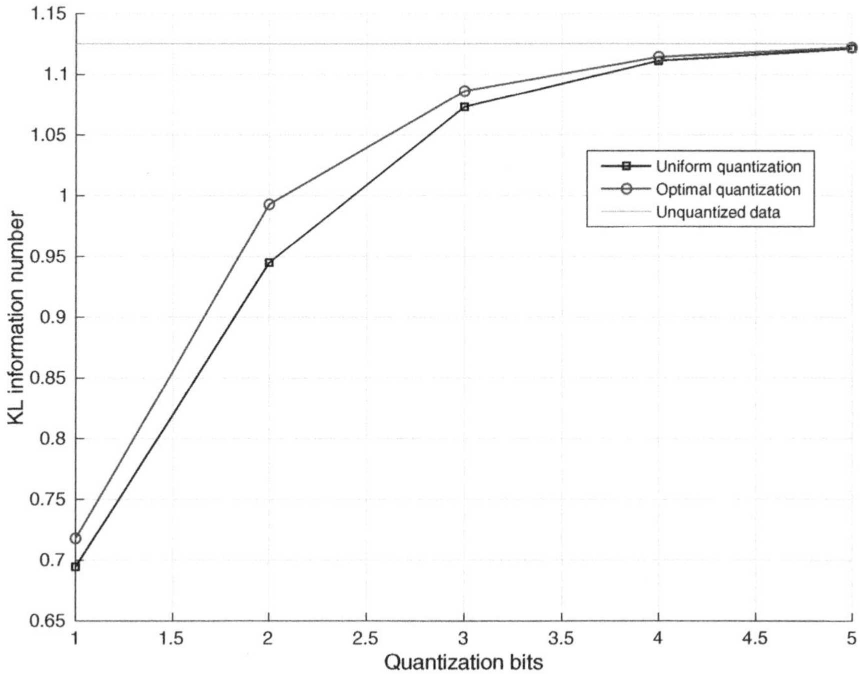

Figure 3.2 KL information number for a single-slot, single-user scenario under optimal quantization, uniform quantization, and no quantization, as a function of the number of quantization bits $q_{i,k}$.

nearly identical when $q_{i,k} \geq 5$. The initial battery level for all sensors is assumed to be 0.4 μJ. In the DP algorithm implementation, the channel power gain $h_{i,k}$ and the battery state $B_{i,k}$ are both quantized to four discrete levels. The average detection delay of the change point is calculated based on 10^4 Monte Carlo iterations.

Figure 3.2 compares the performance of the KL information number evaluated for the single sensor and one-time slot scenario. We have plotted the KL information number for the unquantized observation, optimal quantization, and uniform quantization policies while varying the number of quantization bits. The KL information number for the unquantized observation can be computed as $\frac{\mu^2}{2\sigma^2}$. $\frac{\mu^2}{2\sigma^2}$. This represents an upper bound for the quantized observation cases. As expected, Figure 3.2 illustrates that the optimal quantization policy outperforms the uniform quantization policy. Additionally, we observe that the KL information number for all three policies converges and becomes comparable as the number of quantization bits increases.

For the remaining plots, we set the probability of false alarm to $P_{fa} = 0.01$. In Figure 3.3, we plot the average detection delay as a function of the average

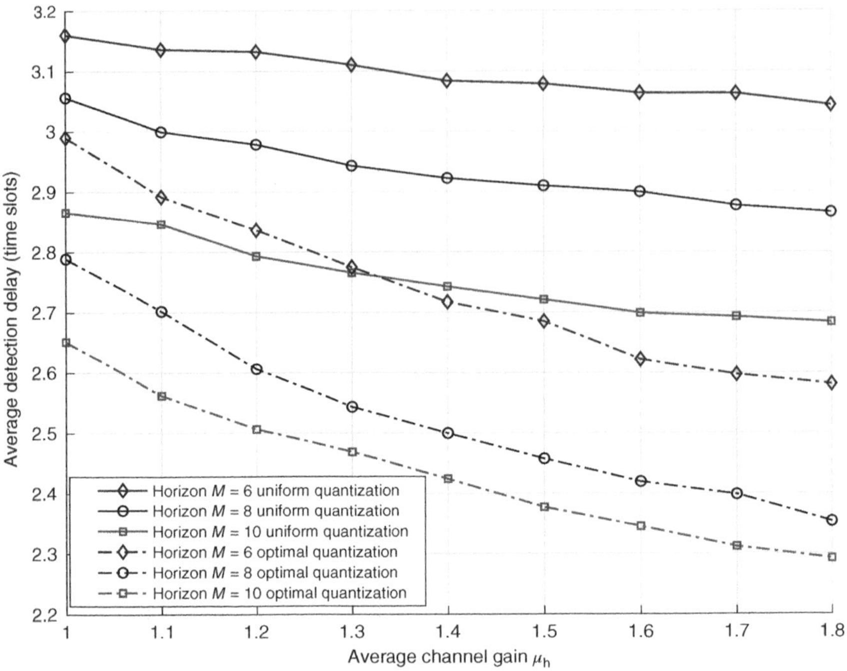

Figure 3.3 Average detection delay (in time slots) versus mean channel gain μ_h for noncausal CSI, comparing optimal and uniform quantization policies.

channel power gain parameter μ_h, with the battery capacity fixed at $B_{max} = 0.8\,\mu J$ for the noncausal CSI scenario. It is observed that the average detection delay for the optimal quantization policy decreases more rapidly compared to the uniform quantization policy. Additionally, the average detection delay reduces as the horizon length increases. This is intuitive for the non-causal scenario, as a longer horizon length provides more information before the transmission process, allowing sensors to better plan their quantization, sensing strategies, and energy usage. Figure 3.4 presents a comparative plot of the average detection delay with varying battery capacities B_{max} while keeping the mean channel gain $\mu_h = 1$. This comparison is made for both optimal and uniform quantization policies under noncausal and causal CSI conditions, with the horizon length set to $M = 8$. The numerical comparison indicates that, for $B_{max} = 0.7\,\mu J$, the average detection delay with the optimal quantization policy for the causal CSI scenario is 15.3% higher than for the noncausal scenario.

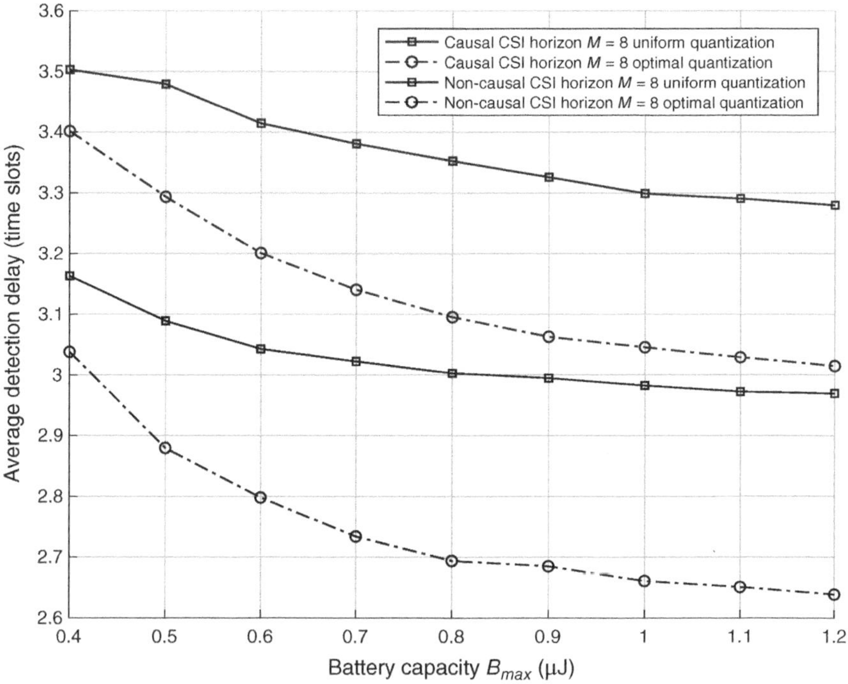

Figure 3.4 Average detection delay (in time slots) versus battery capacity B_{max} for noncausal and causal CSI, comparing optimal and uniform quantization policies.

3.6.2 Distributed Detection Results

We include the simulation results for the distributed detection case in this subsection. For this scenario, we assume that the observation signal is modeled as being drawn from a Gaussian distribution, $\mathcal{N}(0, \sigma^2)$ under the null hypothesis and $\mathcal{N}(\mu, \sigma^2)$ under the alternative hypothesis where $\mu = 0.5$ and $\sigma^2 = 1$. The sensing energy E_s is assumed to be 0.5 mJ. We perform Monte-Carlo simulations using 10^5 samples with results averaged over 35,000 iterations to obtain insights into the expected detection delay and the tail distribution to false alarm when $\overline{H} > E_s$. The detection threshold is set to $h = \log(200)$. For this simulation, the KL divergence $\mathcal{I}_{KL}$ is calculated as $\frac{\mu^2}{2\sigma^2}$. Additionally, under the alternative hypothesis $\mathcal{H}_1$, we assume the change happens at $\nu = 1$.

Figure 3.5 shows the expression $1 - F_\infty(x)$ with respect to x, where x represents the random variable for the detection delay and $F_\infty(x)$ represents the asymptotic

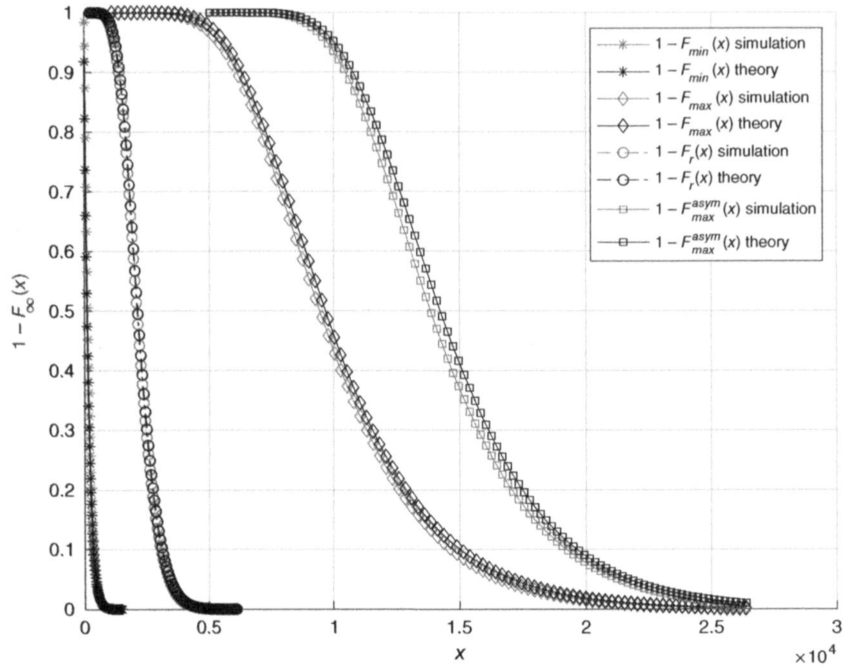

Figure 3.5 Simulated and theoretical cumulative distribution function comparison for the null hypothesis $\mathcal{H}_0$.

cumulative distribution function of the first passage time to a false alarm. It shows that the plot corresponding to the OR rule (T_{min}) is characteristically distinct from the AND rule (T_{max}) and r out of N rule (T_r) as for the OR rule, the corresponding cumulative distribution function is exponentially distributed. Figure 3.5 also shows that the performance of the r out of N rule falls between the OR and the AND rule. For these simulations, we set $N = 20$ and $r = 11$ for the r out of N rule, which essentially represents the majority logic fusion rule. Additionally, we also add the asymptotic result for the AND rule when $N \to \infty$ in Figure 3.5. As a numerical comparison, in Figure 3.5, the simulated value of $1 - F_{max}^{asym}(x)$ for $x = 1.5 \times 10^4$ is approximately 5.11% lower than the theoretical value.

In Figure 3.6, we plot the expression $1 - F_\nu(x)$ with respect to x, where x represents the random variable for the detection delay and $F_\nu(x)$ represents the asymptotic cumulative distribution function of the normalized detection delay evaluated under the assumption that change happens at point ν. Interpretations similar to Figure 3.5 can be made for these plots as well.

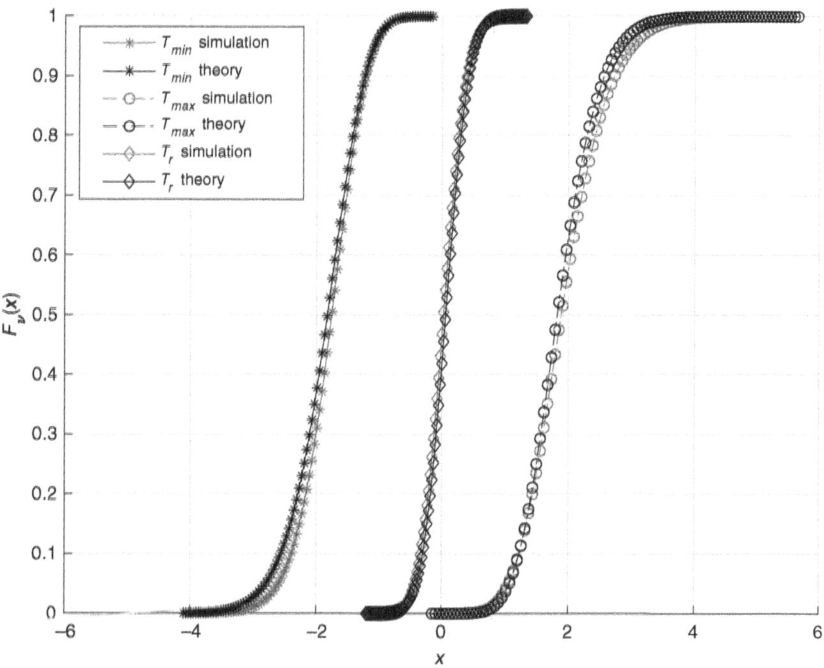

Figure 3.6 Simulated and theoretical cumulative distribution function comparison for the non-null hypothesis $\mathcal{H}_1$.

3.7 Conclusions and Future Work

In this chapter, we study the minimization of detection delay in the context of the quickest change detection framework for both the decentralized and distributed multiple sensor scenarios, where individual sensors harvest energies from their surroundings. We propose an optimal sensing and quantization strategy for such an average delay minimization problem for the decentralized case with quantized information over a finite number of time slots, by utilizing DP algorithm for both causal and noncausal CSI cases. The noncausal CSI case provides a performance benchmark for the causal counterpart. We derive the analytical expression for the optimal thresholds when the number of quantization bits is kept fixed for each time slot. Furthermore, we put forward a uniform quantization-based heuristic policy. Simulation results indicate the optimal quantizer performs significantly better than its uniform counterpart for small number of quantization bits. This performance difference becomes negligible as the number of quantization bits becomes larger. For the distributed scenario, we shift our attention from the average detection delay and the tail distribution of the run length to a false alarm and their relevant asymptotic expressions. This study concentrates on the case, when the amount of average energy harvested by individual sensors, i.e. $\overline{H}$ exceeds E_s. We demonstrate that the standard asymptotic result for a single sensor without any energy constraints holds in this case. We extend this result to the distributed case by using the order statistics theory.

In the future, we will further expand on our findings to the situation when $\overline{H} < E_s$ through the application of the Markov random walk theory and non-linear renewal theory. For real-world applications, in a WSN, the average harvested energy for some sensors will exceed E_s and the rest will not. Therefore, examining the average detection delay and the tail distribution of the run length leading to a false alarm for such a case would extend this research. The asymptotic evaluation of these parameters in a consensus network utilizing energy-harvesting sensors is another possible research direction. Other possible extensions of both distributed and decentralized detection problems could be into the domain of Bayesian framework, when the change point is considered unknown and is assumed to be generated by a random process with a known probability distribution, or to investigate the quickest change detection of the generalized-likelihood ratio test, when the distributional parameters after the change are unknown. Extensions in the domain of nonparametric change detection can also be studied with window-based sampling techniques. However, there are nontrivial issues related to an appropriate optimization problem formulation in the absence of closed-form expressions or bounds for the worst-case average detection delay for such nonparametric algorithms.

Bibliography

T. Banerjee and V. V. Veeravalli. Bayesian quickest change detection under energy constraints. In *2011 Information Theory and Applications Workshop*, pages 1–10, Feb 2011.

C.-Y. Chong and S. P. Kumar. Sensor networks: Evolution, opportunities, and challenges. *Proceedings of the IEEE*, 91(8):1247–1256, Aug 2003.

D. Ciuonzo, G. Gelli, A. Pescapé, and F. Verde. Decision fusion rules in ambient backscatter wireless sensor networks. In *2019 IEEE 30th Annual International Symposium on Personal, Indoor and Mobile Radio Communications (PIMRC)*, pages 1–6. IEEE, 2019.

H. A. David and H. N. Nagaraja. *Order Statistics*. Wiley Series in Probability and Statistics. Wiley, 2004.

S. Dey. Asymptotic performance analysis of non-Bayesian quickest change detection with an energy harvesting sensor. *arXiv preprint arXiv:2001.04752*, 2020.

J. Geng and L. Lai. Non-Bayesian quickest change detection with stochastic sample right constraints. *IEEE Transactions on Signal Processing*, 61(20):5090–5102, Oct 2013.

J. Geng, E. Bayraktar, and L. Lai. Bayesian quickest change-point detection with sampling right constraints. *IEEE Transactions on Information Theory*, 60(10):6474–6490, Oct 2014.

V. C. Gungor and G. P. Hancke. Industrial wireless sensor networks: Challenges, design principles, and technical approaches. *IEEE Transactions on Industrial Electronics*, 56(10):4258–4265, Oct 2009.

V. C. Gungor, B. Lu, and G. P. Hancke. Opportunities and challenges of wireless sensor networks in smart grid. *IEEE Transactions on Industrial Electronics*, 57(10):3557–3564, Oct 2010.

D. Kalus, M. Muma, and A. M. Zoubir. Distributed robust change point detection for autoregressive processes with an application to distributed voice activity detection. In *2015 IEEE International Conference on Acoustics, Speech and Signal Processing (ICASSP)*, pages 3906–3910. IEEE, 2015.

S. M. Kay. *Fundamentals of Statistical Signal Processing: Detection theory*. Prentice Hall Signal Processing Series. Prentice-Hall PTR, 1998.

D. Li, S. Kar, and S. Cui. Distributed quickest detection in sensor networks via two-layer large deviation analysis. *IEEE Internet of Things Journal*, 5(2):930–942, 2018.

G. Lorden. Procedures for reacting to a change in distribution. *Annals of Mathematical Statistics*, 42(6):1897–1908, Dec 1971.

G. V Moustakides. Decentralized CUSUM change detection. In *2006 9th International Conference on Information Fusion*, pages 1–6. IEEE, 2006.

S. Nadarajah. Explicit expressions for moments of order statistics. *Statistics & Probability Letters*, 78(2):196–205, 2008.

H. N. Nagaraja. Order statistics from independent exponential random variables and the sum of the top order statistics. In N. Balakrishnan, J. M. Sarabia, and E. Castillo, editors, *Advances in Distribution Theory, Order Statistics, and Inference*, pages 173–185. Springer, 2006.

M. Pollak. Optimal detection of a change in distribution. *The Annals of Statistics*, 13(1):206–227, Mar 1985.

H. V. Poor and O. Hadjiliadis. *Quickest Detection*. Cambridge University Press, 2008.

A. G. Tartakovsky and V. V. Veeravalli. Asymptotically optimal quickest change detection in distributed sensor systems. *Sequential Analysis*, 27(4):441–475, 2008.

A. Tartakovsky, I. Nikiforov, and M. Basseville. *Sequential Analysis: Hypothesis Testing and Changepoint Detection*. CRC Press, 2014.

V. V. Veeravalli. Decentralized quickest change detection. *IEEE Transactions on Information Theory*, 47(4):1657–1665, May 2001.

Part II

Communications Technologies in Wireless Sensor Networks

4

RIS-Assisted Channel-Aware Decision Fusion

Domenico Ciuonzo[1], *Alessio Zappone*[2], *Pierluigi Salvo Rossi*[3], *and Marco Di Renzo*[4,5]

[1]*Department of Electrical Engineering and Information Technologies (DIETI), University of Naples Federico II, Naples, Italy*
[2]*Department of Electrical and Information Engineering Maurizio Scarano, University of Cassino and Southern Lazio, Cassino, Italy*
[3]*Department of Electronic Systems, Norwegian University of Science and Technology, Trondheim, Norway*
[4]*Laboratoire des Signaux et Systèmes, CNRS, CentraleSupélec, Université Paris-Saclay, Gif-sur-Yvette, France*
[5]*Department of Engineering, Centre for Telecommunications Research, King's College London, London, United Kingdom*

4.1 Introduction

The Internet of Things (IoT) pushes for broad deployment of small devices with sensing, processing and communication abilities. This technological paradigm stands as a transformative force in the wireless communications and sensing sector. The "sensing arm" of the IoT, embodied by wireless sensor networks (WSNs), casts a multifaceted shadow over our daily lives. Within this expansive landscape, distributed detection (DD) emerges as a mature research topic, its applications spanning from cognitive radio systems [Chen and Zhang, 2019] to industrial contexts [Tabella et al., 2021].

A retrospective analysis highlights the evolution of DD literature through three distinct "waves." The first wave, marking its inception, can be traced back to the foundational work of Tenney and Sandell [1981]. This phase saw significant contributions from Chair and Varshney [1986], Reibman and Nolte [1987], and Warren and Willett [1989]. In this early stage, the literature focused on quantizing (via one or multiple bits) measurements and likelihoods, exploring the complexities of local detectors and fusion rule design. Implicit in these studies was the assumption of decoupled (or noise-free) reporting channels.

Wireless Sensor Networks in Smart Environments: Enabling Digitalization from Fundamentals to Advanced Solutions, First Edition. Edited by Domenico Ciuonzo and Pierluigi Salvo Rossi.

This evolved in the early 2000s (second wave) to include channel-aware fusion strategies as WSNs became more common [Chen et al., 2004]. Exploration extended to diverse reporting protocols, including type/time/frequency/code-based schemes [Mergen et al., 2007; Yiu and Schober, 2008] and interfering access [Berger et al., 2009]. Further endeavors for performance improvement led to investigations into power allocation [Zhang et al., 2008], censoring schemes [Appadwedula et al., 2005], and sensor subset selection [Ahmadi and Vosoughi, 2009].

In recent years, the third wave of DD has emerged, harnessing innovative technologies like backscattering [Ciuonzo et al., 2019] and energy harvesting [Tarighati et al., 2017] toward the concept of green (namely, near zero-energy) sensors. The integration of mmWave sensors [Chawla et al., 2021] and massive MIMO [Ciuonzo et al., 2015; Chawla et al., 2019] has been explored to curtail the energy expenditure of WSNs while achieving optimal performance. Intriguingly, the untapped potential of flexible reconfigurable intelligent surfaces (RISs) [Huang et al., 2019; Di Renzo et al., 2020] in the realm of DD beckons further exploration, which is the objective of this chapter. The main benefits of RISs in the IoT ecosystem are showcased in Zappone et al. [2022].

This chapter explores DD involving sensors that make local decisions and communicate them to a fusion center (FC) via a multiple-access channel. The FC features a receive array designed to counteract fading attenuation and reduce interference from sensors reporting simultaneously [Zhang et al., 2008; Ciuonzo et al., 2012]. The above assumption configures a distributed MIMO setup [Zhang et al., 2008; Ciuonzo et al., 2012]. In this context, information alignment of the sensors' contributions is facilitated through the deployment and design of an appropriate RIS [Huang et al., 2019; Di Renzo et al., 2020].

The aim of this chapter is then to provide a discussion on the design and optimization in DD when capitalizing on the concept of smart environments [Mudkey et al., 2022]. Specifically, a joint fusion rule and RIS design solution is discussed. Regrettably, pursuing an optimal joint approach is impractical because of the considerable complexity associated with the log-likelihood ratio (LLR). This complexity complicates the process of deriving a suitable RIS design, primarily due to the lack of theoretical performance metrics for the LLR in this context. Accordingly, this design solution leverages a simplification originating from the "Ideal Sensors" (IS) assumption [Chen et al., 2004; Ciuonzo et al., 2012, 2015], namely, it is assumed perfect sensing at the design stage. The resulting simpler design provides an effective joint fusion rule and RIS design which is agnostic of the sensor performance. The associated optimization problem is solved by means of the alternating optimization (AO) and minorization–maximization [Sun et al., 2016].

The rest of the chapter is organized as follows. Section 4.2 describes the system model considered, whereas Section 4.3 develops the IS-based joint fusion and

RIS design. The benefits of such approach are evaluated in a relevant numerical setup in Section 4.4. Wrap-up considerations and suggested further reading in Section 4.5 end the chapter.

4.2 System Model

In this chapter, a distributed binary hypothesis testing framework is considered, employing K sensors to discriminate between hypotheses in the binary set $\mathcal{H} \triangleq \{\mathcal{H}_0, \mathcal{H}_1\}$. This is visually illustrated in Figure 4.1. The pair of hypotheses may represent the absence or presence of a particular anomalous phenomenon. Each sensor, indexed as $k \in \mathcal{K} \triangleq \{1, 2, \dots, K\}$, takes a binary local decision $\xi_k \in \mathcal{H}$ based on its individual measurements. This decision ξ_k is encoded into $x_k \in \mathcal{X} = \{-1, +1\}$, which corresponds to a binary phase-shift keying (BPSK) modulation. For simplicity, it is assumed that the hypothesis $\mathcal{H}_1$ is represented by $x_k = +1$, while the decision corresponding to $\mathcal{H}_0$ maps to $x_k = -1$.

The BPSK-mapped decisions from all the sensors are collected in the $K \times 1$ transmitted signal vector $\boldsymbol{x}$ for compactness. Accordingly, the WSN sensing quality is completely defined by the conditional joint probability mass functions

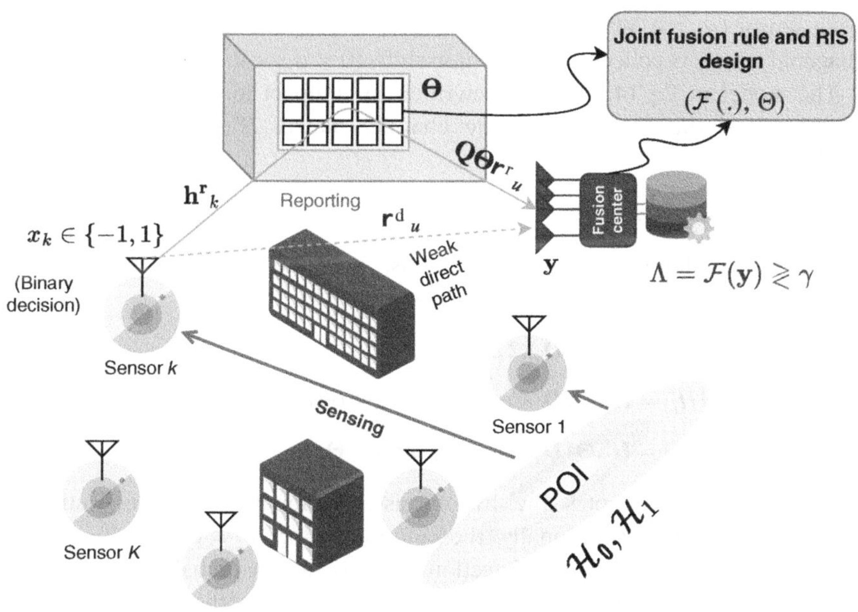

Figure 4.1 The RIS-assisted DF system model considered.

$\Pr(\boldsymbol{x}|\mathcal{H}_i)$, $i \in \{0,1\}$. The content of this chapter does not rely on any conditional mutual independence assumption on x_k, given the specific hypothesis $\mathcal{H}_i \in \mathcal{H}$. Furthermore, $P_{D,k} \triangleq \Pr\left(x_k = 1|\mathcal{H}_1\right)$ and $P_{F,k} \triangleq \Pr\left(x_k = 1|\mathcal{H}_0\right)$ are used to denote the probability of detection and false alarm of the kth sensor, respectively. In what follows, it is assumed that $P_{D,k} \geq P_{F,k}$, i.e. each sensor decision policy is reasonably above the chance line.

Sensors communicate via a flat-fading multiple access channel [Ciuonzo et al., 2012] with a Fusion Center (FC) equipped with N antennas, aided by an RIS comprising M elements, see Figure 4.1. The equivalent channels connecting the WSN to the FC ($\boldsymbol{H}^d$), the WSN to the RIS ($\boldsymbol{H}^r$), and the RIS to the FC ($\boldsymbol{G}$) represented as complex-valued matrices of dimensions $N \times K$, $M \times K$, and $N \times M$, respectively.

The $N \times 1$ received signal vector $\boldsymbol{y}$, collecting all the contributions from the receive antennas at the FC, is described by the equation:

$$\boldsymbol{y} = \left(\underbrace{\boldsymbol{G}\boldsymbol{\Theta}\boldsymbol{H}^r}_{\text{WSN–RIS–FC path}} + \underbrace{\boldsymbol{H}^d}_{\text{WSN–FC direct path}} \right) \boldsymbol{D}_\alpha \boldsymbol{x} + \boldsymbol{w}. \tag{4.1}$$

In contrast, $\boldsymbol{w} \sim \mathcal{N}_{\mathbb{C}}(\boldsymbol{0}_N, \sigma_w^2 \boldsymbol{I}_N)$ denotes a Gaussian noise vector characterized by a zero mean and a scaled identity covariance matrix. Furthermore, the diagonal matrix $\boldsymbol{D}_\alpha$ collects on its main diagonal the possibly unequal energies spent from each sensor ($\alpha_k \geq 0$, $\forall k \in \mathcal{K}$) for reporting its decision. Finally, in Eq. (4.1), the diagonal matrix $\boldsymbol{\Theta}$ collects the RIS phase shifts ($0 \leq \varphi_m < 2\pi, \forall m = 1, \ldots, M$).

The model in Eq. (4.1) can be rewritten in compact form by leveraging the definition of the $N \times K$ composite channel matrix $\boldsymbol{H}^e(\boldsymbol{\Theta}) \triangleq \left(\boldsymbol{G}\boldsymbol{\Theta}\boldsymbol{H}^r + \boldsymbol{H}^d\right)$, leading to:

$$\boldsymbol{y} = \boldsymbol{H}^e(\boldsymbol{\Theta})\, \boldsymbol{D}_\alpha \boldsymbol{x} + \boldsymbol{w}. \tag{4.2}$$

Based on the aforementioned assumptions, the vector $\boldsymbol{y}|\mathcal{H}_i$ has the following statistical characterization in the second order:

$$\begin{aligned}
\mathbb{E}\{\boldsymbol{y}|\mathcal{H}_i\} &= \boldsymbol{H}^e(\boldsymbol{\Theta})\, \boldsymbol{D}_\alpha (2\,\boldsymbol{\rho}_i - \boldsymbol{1}_K), \\
\mathrm{Cov}(\boldsymbol{y}|\mathcal{H}_i) &= \boldsymbol{H}^e(\boldsymbol{\Theta})\, \boldsymbol{D}_\alpha \,\mathrm{Cov}\left(\boldsymbol{x}|\mathcal{H}_i\right)\, \boldsymbol{D}_\alpha \boldsymbol{H}^e(\boldsymbol{\Theta})^\dagger + \sigma_w^2 \boldsymbol{I}_N, \\
\mathrm{PCov}(\boldsymbol{y}|\mathcal{H}_i) &= \boldsymbol{H}^e(\boldsymbol{\Theta})\, \boldsymbol{D}_\alpha \,\mathrm{Cov}(\boldsymbol{x}|\mathcal{H}_i)\, \boldsymbol{D}_\alpha \boldsymbol{H}^e(\boldsymbol{\Theta})^T,
\end{aligned} \tag{4.3}$$

where $\mathrm{Cov}\left(\boldsymbol{x}|\mathcal{H}_i\right)$ denotes the hypothesis-conditional covariance matrix of the decision vector. Additionally, the column vectors $\boldsymbol{\rho}_1 \triangleq [P_{D,1} \cdots P_{D,K}]^T$ and $\boldsymbol{\rho}_0 \triangleq [P_{F,1} \cdots P_{F,K}]^T$ collect the detection and false-alarm probabilities of all the sensors, respectively. It is worth underlining that, given the improper nature of the vector $\boldsymbol{y}|\mathcal{H}_i$, a complete second-order characterization must include the pseudo-covariance $\mathrm{PCov}(\boldsymbol{y}|\mathcal{H}_i)$.

The following of the chapter describes a design strategy which is intended to achieve high (goal-oriented) performance, consisting in the joint design of a fusion rule statistic $\Lambda = \mathcal{F}(\mathbf{y})$ and of the RIS shifts $\boldsymbol{\Theta}$. This is to fully leverage the collective sensing capabilities of the WSN.

4.3 Combined Design of Fusion Rule and RIS

The formulation of the optimal fusion rule for the current problem, as outlined in [Kay, 1998], is given by:

$$\left\{\Lambda_{opt} \triangleq \ln\left[\frac{p(\mathbf{y}|\mathcal{H}_1)}{p(\mathbf{y}|\mathcal{H}_0)}\right]\right\} \underset{\hat{\mathcal{H}}=\mathcal{H}_0}{\overset{\hat{\mathcal{H}}=\mathcal{H}_1}{\gtrless}} \gamma, \tag{4.4}$$

where $\hat{\mathcal{H}}$, Λ_{opt}, and γ represent the estimated hypothesis, the LLR, and the threshold for LLR comparison, respectively. The threshold γ can be adjusted to maintain a fixed system false-alarm rate or to minimize the probability of error [Kay, 1998]. It is worth noticing that the above result holds conditioned to any value of the RIS coefficients in $\boldsymbol{\Theta}$.

An explicit expression for the LLR in Eq. (4.4) is obtained as

$$\begin{aligned}\Lambda_{opt} &= \ln\left[\frac{\sum_{\mathbf{x}\in\mathcal{X}^K} p(\mathbf{y}|\mathbf{x})\Pr(\mathbf{x}|\mathcal{H}_1)}{\sum_{\mathbf{x}\in\mathcal{X}^K} p(\mathbf{y}|\mathbf{x})\Pr(\mathbf{x}|\mathcal{H}_0)}\right] \\ &= \ln\left[\frac{\sum_{\mathbf{x}\in\mathcal{X}^K}\exp\left(-\frac{\|\mathbf{y}-\mathbf{H}^e(\boldsymbol{\Theta})\mathbf{D}_\alpha\mathbf{x}\|^2}{\sigma_w^2}\right)\Pr(\mathbf{x}|\mathcal{H}_1)}{\sum_{\mathbf{x}\in\mathcal{X}^K}\exp\left(-\frac{\|\mathbf{y}-\mathbf{H}^e(\boldsymbol{\Theta})\mathbf{D}_\alpha\mathbf{x}\|^2}{\sigma_w^2}\right)\Pr(\mathbf{x}|\mathcal{H}_0)}\right].\end{aligned} \tag{4.5}$$

The above expression leverages the independence of $\mathbf{y}$ from $\mathcal{H}_i$, when conditioning on the transmitted vector $\mathbf{x}$. Direct inspection of Eq. (4.5) highlights that LLR-based fusion requires a computational complexity growing as $\mathcal{O}(2^K)$, thus becoming prohibitive in the case of a large number of sensors. Equally important, the LLR is not amenable to a tractable closed-form performance analysis. This implies that explicit expressions for system detection and false-alarm probabilities are not available when LLR is adopted. Accordingly, RIS design leveraging LLR-based fusion is not a viable option.

This kind of limitation for the LLR persists even when simpler evaluation metrics are considered, such as the deflection measure:

$$D_i(\Lambda) \triangleq \frac{\left(\mathbb{E}\{\Lambda|\mathcal{H}_1\} - \mathbb{E}\{\Lambda|\mathcal{H}_0\}\right)^2}{\mathrm{var}\{\Lambda|\mathcal{H}_i\}}, \tag{4.6}$$

where $D_0(\cdot)$ and $D_1(\cdot)$ correspond to the normal [Picinbono, 1995] and modified [Quan et al., 2008] deflections, respectively. Accordingly, a simplified approach for joint fusion rule and RIS design is developed in what follows.

Specifically, by constraining the fusion rule to be Widely-Linear (WL), the deflection becomes tractable, enabling an efficient joint design for both the fusion rule and RIS phase shifts, namely:

$$\Lambda_{\text{wl}} = \underline{\boldsymbol{a}}^{\dagger}\underline{\boldsymbol{y}}. \tag{4.7}$$

In the above equation, $\underline{\boldsymbol{a}}$ denotes the augmented vector of $\boldsymbol{a}$, that is $\underline{\boldsymbol{a}} \triangleq \left[\, \boldsymbol{a}^T \;\; \boldsymbol{a}^{\dagger} \,\right]^T$. Also, the same considerations apply to $\underline{\boldsymbol{y}}$. WL fusion rules have demonstrated appealing performance in comparable WSN scenarios, particularly in distributed MIMO configurations without RIS support [Ciuonzo et al., 2015].

By leveraging WL fusion statistic, the deflection measure specializes into the following form:

$$D_i(\Lambda_{\text{wl}}) \triangleq \frac{\left(\underline{\boldsymbol{a}}^{\dagger}\left(\mathbb{E}\{\underline{\boldsymbol{y}}|\mathcal{H}_i\} - \mathbb{E}\{\underline{\boldsymbol{y}}|\mathcal{H}_0\}\right)\right)^2}{\underline{\boldsymbol{a}}^{\dagger}\,\text{Cov}(\underline{\boldsymbol{y}}|\mathcal{H}_i)\,\underline{\boldsymbol{a}}}. \tag{4.8}$$

Additionally, it is evident that the design of the fusion rule can be streamlined even further when employing the IS assumption [Lei and Schober, 2010; Ciuonzo et al., 2015], i.e. $\Pr(\boldsymbol{x} = \mathbf{1}_K|\mathcal{H}_1) = \Pr(\boldsymbol{x} = -\mathbf{1}_K|\mathcal{H}_0) = 1$. Indeed, when compared to Eq. (4.3), the statistical characterization of $\boldsymbol{y}|\mathcal{H}_i$ at second order simplifies as:

$$\begin{aligned} \mathbb{E}\{\boldsymbol{y}|\mathcal{H}_i\} &= \boldsymbol{H}^e(\boldsymbol{\Theta})\,\boldsymbol{D}_{\alpha}\,(2i-1)\mathbf{1}_K. \\ \text{Cov}(\boldsymbol{y}|\mathcal{H}_i) &= \sigma_w^2\,\boldsymbol{I}_N. \\ \text{PCov}(\boldsymbol{y}|\mathcal{H}_i) &= \boldsymbol{O}_{N\times N}. \end{aligned} \tag{4.9}$$

The last row highlights a null pseudo-covariance term when the IS assumption is made. As a result, the proposed design demands less system knowledge since the performance of the WSN's local decisions is not necessary. Utilizing the IS assumption, both deflection measures converge into a single, unified metric [Ciuonzo et al., 2015]:

$$D\left(\underline{\boldsymbol{a}}, \boldsymbol{\Theta}\right) = \frac{4}{\sigma_w^2}\,\frac{\left(\underline{\boldsymbol{a}}^{\dagger}\left[\underline{\boldsymbol{H}}^e(\boldsymbol{\Theta})\,\boldsymbol{D}_{\alpha}\,\mathbf{1}_k\right]\right)^2}{\underline{\boldsymbol{a}}^{\dagger}\underline{\boldsymbol{a}}}, \tag{4.10}$$

where $\underline{\boldsymbol{H}}^e(\boldsymbol{\Theta})$ denotes the augmented matrix built as $[\boldsymbol{H}^e(\boldsymbol{\Theta})^T \;\; \boldsymbol{H}^e(\boldsymbol{\Theta})^{\dagger}\;]^T$. Consequently, the IS-simplified deflection presented in Eq. (4.10) can be maximized by collaboratively optimizing both the WL vector $\underline{\boldsymbol{a}}$ (which establishes the fusion rule) and the phase shift matrix $\boldsymbol{\Theta}$ (which defines the RIS configuration).

Hence, the resulting optimization is formulated as:

$$\mathcal{P}_1 : \begin{array}{ll} \underset{\underline{\boldsymbol{a}},\boldsymbol{\Theta}}{\text{maximize}} & \frac{(\underline{\boldsymbol{a}}^\dagger [\underline{\boldsymbol{H}}^e(\boldsymbol{\Theta})\,\boldsymbol{D}_\alpha\,\boldsymbol{1}_k])^2}{\underline{\boldsymbol{a}}^\dagger \underline{\boldsymbol{a}}} \\ \text{subject to} & \begin{array}{l} \|\underline{\boldsymbol{a}}\| = 1 \\ \boldsymbol{\Theta} = \text{diag}(e^{j\varphi_1}, \ldots, e^{j\varphi_M}) \end{array} \end{array} . \tag{4.11}$$

Note that each diagonal element in the phase shift matrix has unit modulus, as specified via the second constraint of $\mathcal{P}_1$. The combination of this non-convex constraint with the non-convex objective function categorizes $\mathcal{P}_1$ as a non-convex problem.

In this context, problem $\mathcal{P}_1$ is solved efficiently resorting to the AO approach. Specifically, the optimization method alternates between maximizing $\underline{\boldsymbol{a}}$ (with RIS coefficients fixed to $\boldsymbol{\Theta}_{\text{fix}}$) and $\boldsymbol{\Theta}$ (with WL fusion statistic fixed to $\underline{\boldsymbol{a}}_{\text{fix}}$). Despite leading to a suboptimal solution for nonconvex problems, AO has been shown to be successful empirically and relevant in different applications. In the following, the **two update steps** constituting the proposed AO approach are detailed.

Fusion rule update step (A): The optimization of the WL vector $\underline{\boldsymbol{a}}$ for a fixed phase-shift matrix $\boldsymbol{\Theta}_{\text{fix}}$ can be expressed in closed form. In particular, the design problem for the WL vector is articulated as follows:

$$\mathcal{P}_2 : \underset{\|\underline{\boldsymbol{a}}\|=1}{\text{maximize}} \frac{\left(\underline{\boldsymbol{a}}^\dagger \left[\underline{\boldsymbol{H}}^e(\boldsymbol{\Theta}_{\text{fix}})\,\boldsymbol{D}_\alpha\,\boldsymbol{1}_k\right]\right)^2}{\underline{\boldsymbol{a}}^\dagger \underline{\boldsymbol{a}}}. \tag{4.12}$$

$\mathcal{P}_2$ can be recognized as a Cauchy–Schwarz problem [Ciuonzo et al., 2015], whose optimal value of $\underline{\boldsymbol{a}}$ is a vector aligned to $\underline{\boldsymbol{H}}^e(\boldsymbol{\Theta}_{\text{fix}})\,\boldsymbol{D}_\alpha\,\boldsymbol{1}_k$, namely:

$$\underline{\boldsymbol{a}}^\star(\boldsymbol{\Theta}_{\text{fix}}) = \frac{\underline{\boldsymbol{H}}^e(\boldsymbol{\Theta}_{\text{fix}})\,\boldsymbol{D}_\alpha\,\boldsymbol{1}_k}{\|\underline{\boldsymbol{H}}^e(\boldsymbol{\Theta}_{\text{fix}})\,\boldsymbol{D}_\alpha\,\boldsymbol{1}_k\|}. \tag{4.13}$$

RIS update step (B): Optimizing the phase-shift matrix $\boldsymbol{\Theta}$ with a fixed WL vector $\underline{\boldsymbol{a}}_{\text{fix}}$ presents a greater challenge. The related optimization problem is defined as follows:

$$\mathcal{P}_3 : \begin{array}{l} \underset{\boldsymbol{\Theta}}{\text{maximize}} \frac{\left(\underline{\boldsymbol{a}}_{\text{fix}}^\dagger [\underline{\boldsymbol{H}}^e(\boldsymbol{\Theta})\,\boldsymbol{D}_\alpha\,\boldsymbol{1}_k]\right)^2}{\underline{\boldsymbol{a}}_{\text{fix}}^\dagger \underline{\boldsymbol{a}}_{\text{fix}}} \\ \text{subject to } \boldsymbol{\Theta} = \text{diag}(e^{j\varphi_1}, \ldots, e^{j\varphi_M}) \end{array} . \tag{4.14}$$

Upon separating the influence of the RIS phase shifts, the IS-based deflection presented in (4.10) can be reformulated in the following manner:

$$D\left(\underline{\boldsymbol{a}}_{\text{fix}}, \boldsymbol{\Theta}\right) = \frac{4}{\sigma_w^2}\,\underline{\tilde{\boldsymbol{\theta}}}^\dagger\,\boldsymbol{\Xi}\left(\underline{\boldsymbol{a}}_{\text{fix}}\right)\underline{\tilde{\boldsymbol{\theta}}}, \tag{4.15}$$

where $\tilde{\boldsymbol{\theta}} \triangleq \left[e^{j\varphi_1} \;\cdots\; e^{j\varphi_M} \;1\right]^T$ and $\boldsymbol{\Xi}(\underline{\boldsymbol{a}}_{\text{fix}}) \triangleq \frac{(\boldsymbol{N}^\dagger \underline{\boldsymbol{a}}_{\text{fix}})(\boldsymbol{N}^\dagger \underline{\boldsymbol{a}}_{\text{fix}})^\dagger}{\|\underline{\boldsymbol{a}}_{\text{fix}}\|^2}$. In the latter term, the complex-valued matrix $\boldsymbol{N}$, sized $2N \times 2(M+1)$, is explicitly defined as follows:

$$\boldsymbol{N} \triangleq \begin{bmatrix} (\boldsymbol{N}_r \;\boldsymbol{n}_d) & \boldsymbol{O}_{N\times(M+1)} \\ \boldsymbol{O}_{N\times(M+1)} & (\boldsymbol{N}_r \;\boldsymbol{n}_d)^* \end{bmatrix}, \tag{4.16}$$

where $\boldsymbol{N}_r \triangleq \boldsymbol{G}\,\text{diag}(\boldsymbol{H}^r \boldsymbol{D}_\alpha \boldsymbol{1}_K)$ and $\boldsymbol{n}_d \triangleq \left(\boldsymbol{H}^d \boldsymbol{D}_\alpha \boldsymbol{1}_K\right)$. Hence, $\boldsymbol{N}_r$ is an $N \times M$ complex-valued matrix, whereas $\boldsymbol{n}_d$ is an $N \times 1$ complex-valued vector. Therefore, optimization problem $\mathcal{P}_3$ can be recast in an equivalent form as:

$$\mathcal{P}_4 : \quad \begin{array}{ll} \underset{\tilde{\boldsymbol{\theta}}}{\text{maximize}} & g(\tilde{\boldsymbol{\theta}}) = \underline{\tilde{\boldsymbol{\theta}}}^\dagger \boldsymbol{\Xi}\left(\underline{\boldsymbol{a}}_{\text{fix}}\right) \underline{\tilde{\boldsymbol{\theta}}} \\ \text{subject to} & \left\{|\theta_m| = 1\right\}_{m=1}^{M}, \; \theta_{M+1} = 1 \end{array}. \tag{4.17}$$

Since the matrix $\boldsymbol{\Xi}(\underline{\boldsymbol{a}}_{\text{fix}})$ has rank-one, an unconstrained optimization would imply the alignment of the vector $\underline{\tilde{\boldsymbol{\theta}}}$ towards the direction of the vector $\boldsymbol{N}^\dagger \underline{\boldsymbol{a}}_{\text{fix}}$. Unfortunately, the resulting problem is non-convex (and has no closed form) because of the unit-modulus constraint of all the RIS elements. Accordingly, to enhance the computational efficiency at the FC, the problem $\mathcal{P}_4$ can be solved via the minorization–maximization technique [Sun et al., 2016].

Specifically, referring to ℓth iteration of AO and denoting with $\tilde{\boldsymbol{\theta}}^\star_{(\ell)}$ the current optimized value for $\tilde{\boldsymbol{\theta}}$, a lower bound on the objective function $g(\tilde{\boldsymbol{\theta}})$ is constructed. This lower bound, denoted as $f(\tilde{\boldsymbol{\theta}}|\tilde{\boldsymbol{\theta}}^\star_{(\ell)})$, is designed to touch the objective function at the point $\tilde{\boldsymbol{\theta}}$. Subsequently, this lower bound is utilized as a surrogate objective function, and the maximizer obtained from it is taken as the updated value of $\tilde{\boldsymbol{\theta}}$ for the next AO iteration, represented as $\tilde{\boldsymbol{\theta}}^\star_{(\ell+1)}$. This approach guarantees a monotonic increase in the objective value across iterations, ensuring that $g(\tilde{\boldsymbol{\theta}}^\star_{(\ell+1)}) \geq g(\tilde{\boldsymbol{\theta}}^\star_{(\ell)})$. Thus first-order optimality is attained by resorting to the aforementioned strategy. The success of the minorization–maximization technique relies on the careful construction of the surrogate objective function $f(\tilde{\boldsymbol{\theta}}|\tilde{\boldsymbol{\theta}}^\star_{(\ell)})$, allowing for a straightforward determination of the maximizer $\tilde{\boldsymbol{\theta}}^\star_{(\ell+1)}$.

For the phase-shift matrix optimization problem $\mathcal{P}_4$, a surrogate objective function is formulated by leveraging the convexity of $g(\tilde{\boldsymbol{\theta}})$ with respect to $\tilde{\boldsymbol{\theta}}$. Specifically, $g(\tilde{\boldsymbol{\theta}})$ is lower-bounded by its first-order approximation:

$$\underbrace{\underline{\tilde{\boldsymbol{\theta}}}^\dagger \boldsymbol{\Xi}\left(\underline{\boldsymbol{a}}_{\text{fix}}\right) \underline{\tilde{\boldsymbol{\theta}}}}_{g(\tilde{\boldsymbol{\theta}})} \geq \underbrace{\Re\left\{\left(\underline{\tilde{\boldsymbol{\theta}}}^\star_{(\ell)}\right)^\dagger \boldsymbol{\Xi}\left(\underline{\boldsymbol{a}}_{\text{fix}}\right) \underline{\tilde{\boldsymbol{\theta}}}\right\} + \text{const},}_{f\left(\tilde{\boldsymbol{\theta}}|\tilde{\boldsymbol{\theta}}^\star_{(\ell)}\right)} \tag{4.18}$$

where "const" represents terms that do not depend on $\underline{\tilde{\boldsymbol{\theta}}}$. Thus, the phase-shift optimization in each iteration of the AO process can be expressed as:

$$\mathcal{P}_5 : \tilde{\boldsymbol{\theta}}^\star_{(\ell+1)} = \underset{|\theta_m|=1,\, \theta_M=1}{\arg\max}\; \Re\left\{\left(\underline{\tilde{\boldsymbol{\theta}}}^\star_{(\ell)}\right)^\dagger \boldsymbol{\Xi}\left(\underline{\boldsymbol{a}}_{\text{fix}}\right) \underline{\tilde{\boldsymbol{\theta}}}\right\}. \tag{4.19}$$

The closed-form optimal solution to $\mathcal{P}_5$ is given by:

$$\angle\underline{\tilde{\theta}}^{\star}_{(\ell+1)} = \angle\left(\Xi\left(\underline{\boldsymbol{a}}_{\text{fix}}\right)\underline{\tilde{\theta}}^{\star}_{(\ell)}\right), \tag{4.20}$$

where $\angle(\cdot)$ denotes the entry-wise phase of the generic complex-valued vector.

This approach alternates between the closed-form updates in Eqs. (4.13) **[Step (A)]** and (4.20) **[Step (B)]**.

Algorithm 4.1 Joint fusion rule and RIS design via AO+MM (IS assumption)

1: **Initialize:** Construct initial guess $\tilde{\theta}^{\star}_{(0)}$, set $\ell = 0$
2: **repeat**
3:   **Step (A):** Keep $\tilde{\theta}^{\star}_{(\ell)}$ fixed and update $\underline{a}$ according to Eq. (4.13)
4:   **Step (B):** Keep $\underline{a}$ fixed and update $\tilde{\theta}^{\star}_{(\ell+1)}$ via Eq. (4.20)
5:   **Update:** Set $\ell \leftarrow \ell + 1$
6: **until** convergence

Thus, the IS-simplified deflection in Eq. (4.10) is ensured to grow monotonically and approach a local optimum by exploiting the structural characteristics of the AO procedure. For simplicity, the initial point $\tilde{\theta}^{\star}_{(0)}$ may be selected (other initialization strategies are possible) as randomly generated phase-shifts from a uniform distribution. The procedure presented is provably convergent in the value of the objective because the objective function grows monotonically with the iteration number. The overall procedure is summarized in Algorithm 4.1.

The computational complexity of the proposed IS-based joint design is $\mathcal{O}(N_{iter} \times (C_{fus} + C_{ris}))$, where N_{iter} represents the total number of iterations in the AO process. Here, C_{fus} and C_{ris} correspond to the costs of the WL vector fusion and RIS phase-shift matrix design, respectively. The specific costs for each step are detailed in Table 4.1.

Table 4.1 Summary of the computational complexity required needed for joint IS-based design for both fusion rule and RIS design steps involved in the AO.

Step in AO	Big-O complexity[a)]
Step (A): fusion rule design	$C_{fus} \rightarrow \mathcal{O}(M + MN + N)$
Step (B): RIS design	$C_{ris} \rightarrow \mathcal{O}((M+1)^2 + MN + N)$

a) M denotes the number of RIS elements, whereas N represents the number of antennas at the FC.

4.4 Performance Analysis

The design developed in the previous section is here validated in a DD scenario analogous to that considered by previous literature [Chen et al., 2006; Ciuonzo et al., 2012]. Specifically, the numerical setup considers a WSN made of $K = 10$ sensors. Regarding the sensing phase, local decisions regarding the phenomenon of interest (POI) are assumed to be conditionally independent and identically distributed. Specifically, it is assumed $(P_{D,k}, P_{F,k}) \triangleq (0.5, 0.05)$, $k \in \mathcal{K}$.

In the reporting phase, all sensors transmit their decisions with uniform energy, meaning $\alpha_k = 1$. The small-scale fading follows a Rayleigh distribution, while the path loss is characterized by the model $\mu, (d/d_0)^{-\nu}$, where $\mu = -30\,\text{dB}$. Here, $\mu = -30\,\text{dB}$ denotes the path loss at a reference distance of 1 m. The path loss exponent ν is set to 2 for the links between the WSN and the RIS, as well as the links from the RIS to the FC, whereas it is set to 4 for the links directly connecting the WSN to the FC. Sensor locations are uniformly and randomly spread in a $[0, 40] \times [0, 40]\,\text{m}^2$ square. The RIS is positioned at $[60, 20]$ m, while the FC is at $[65, 25]$ m. In the subsequent analysis, noise variance σ_w^2 is fixed at -80 dBm.

The WSN system performance is evaluated in terms of the global probabilities of false alarm $P_{F_0} \triangleq \Pr\{\Lambda > \gamma | H_0\}$ and detection $P_{D_0} \triangleq \Pr\{\Lambda > \gamma | H_1\}$.

For the sake of completeness, the analysis also includes the concept of the "observation bound," which signifies the performance of the optimal decision fusion rule under ideal channel conditions. This is expressed mathematically as follows:

$$P_{D_0}^{\text{obs-bound}} = \sum_{j=\nu}^{K} \binom{K}{j} (P_D)^j (1 - P_D)^{K-j}. \tag{4.21}$$

$$P_{F_0}^{\text{obs-bound}} = \sum_{j=\nu}^{K} \binom{K}{j} (P_F)^j (1 - P_F)^{K-j}.$$

Here, $\nu \in 0, \ldots, K$ serves as a discrete threshold. This bound provides a crucial reference point for evaluating (i) the degradation in detection performance due to interference from the distributed MIMO channel and (ii) the corresponding advantages gained from the support of the RIS.

In the numerical results reported, the joint IS-based fusion rule and RIS design is reported with "■" markers. By contrast, the IS-based rule not aided by an RIS [Ciuonzo et al., 2012, 2015] is represented with a "□" marker. For the sake of a complete analysis, the following performance analysis also includes

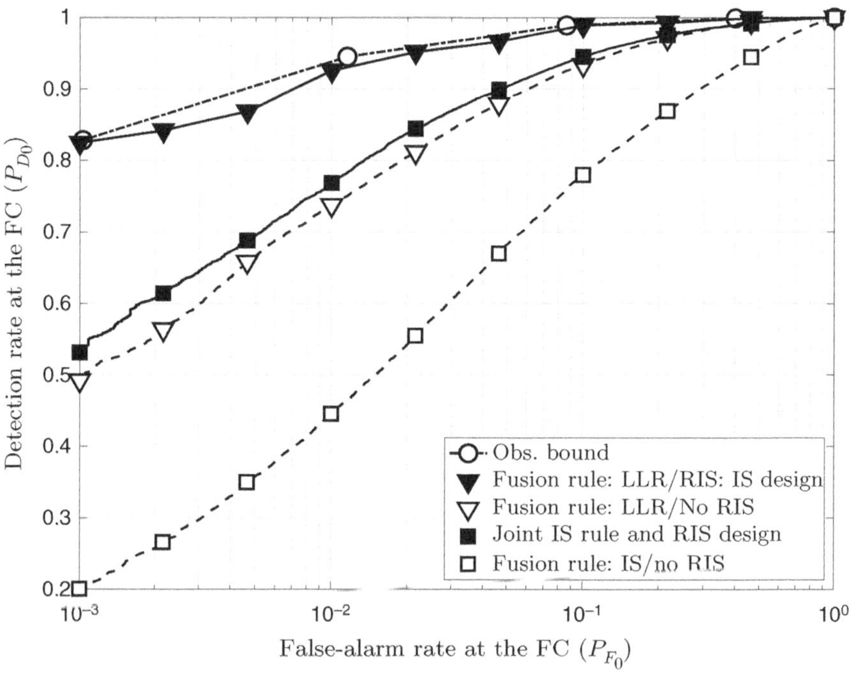

Figure 4.2 P_{D_0} versus P_{F_0} for the specified rule and RIS configurations. Configuration details: $N = 4$ antennas at the FC and RIS comprising $M = 20$ elements.

a configuration with an LLR-based fusion rule without an RIS and the same fusion rule with an RIS whose phase shifts are designed according to the IS-based criterion. The latter two counterparts a represented with "▽" and "▼" markers, respectively.

Figure 4.2 illustrates the receiver operating characteristic (ROC), depicting the relationship between P_{D_0} and P_{F_0} in a WSN featuring $N = 4$ antennas at the FC and an RIS with $M = 20$ elements. The primary aim of this analysis is to evaluate the performance enhancements achieved when an RIS supports the DD task. The presented set of curves clearly demonstrates that the IS-based joint design yields improved performance compared to an IS-based fusion rule that operates without the assistance of an RIS [Ciuonzo et al., 2012, 2015].

In this scenario, the joint design also surpasses the performance of the LLR without an RIS (denoted by "▽"). Furthermore, the IS-based RIS design

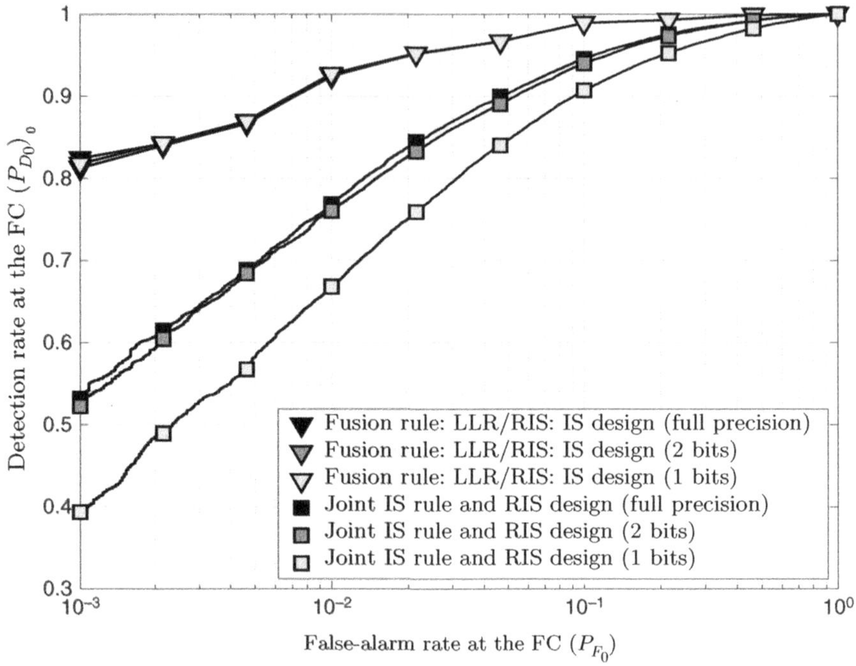

Figure 4.3 Evaluation of P_{D_0} versus P_{F_0} for the optimized rule and RIS configurations across different RIS codebook resolutions. Configuration details: $N = 4$ antennas at the FC and RIS comprising $M = 20$ elements.

significantly enhances the detection capabilities of the WSN, as demonstrated by the improved LLR when the proposed IS-based approach is used to configure the RIS phase-shifts (denoted by "▼").

Then, Figure 4.3 reports the ROC for the IS-based joint design by varying the number of quantization bits associated with the phase values of the RIS coefficients designed. The aim is to investigate the performance degradation due to finite resolution of the codebook associated to the RIS configuration. By looking at the results, it is apparent that 2 bits are sufficient to represent the designed phases of the RIS. This applies also to the case when the LLR is used as the relevant fusion rule, for which only one-bit is sufficient.

Figure 4.4 illustrates the relationship between P_{D_0} and the number of RIS elements M, with a fixed false-alarm rate $P_{F_0} = 0.01$. The primary aim of this analysis is to assess the detection improvements possible with a larger RIS. It is evident

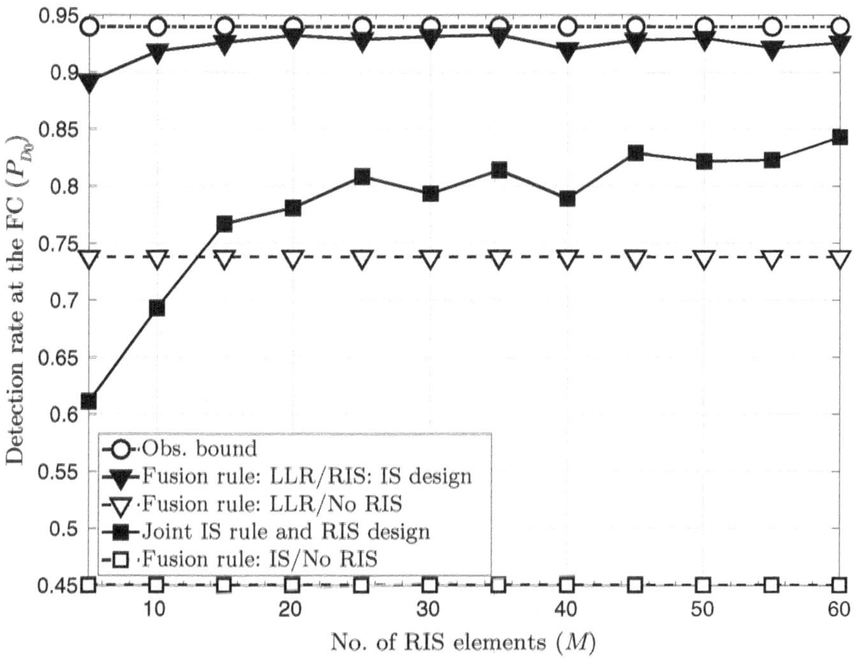

Figure 4.4 Analysis of P_{D_0} as a function of M for the evaluated rule and RIS configurations. The setup parameters include a false-alarm rate at the FC of $P_{F_0} = 0.01$ and $N = 4$ antennas at the FC receiver.

that the IS design enhances detection performance when the size M of the RIS increases. However, for the joint IS design, optimal performance levels off around $M \approx 40$ in this scenario. In comparison, the LLR achieves performance saturation at $M \approx 20$. This saturation for the LLR is mainly due to its reaching the maximum achievable performance as dictated by the observation bound.

Finally, Figure 4.5 reports the P_{D_0} versus the number of receive antennas N at the FC. The aim is to understand the interplay between the number of receive antennas and the RIS benefit for a relevant number of atoms. Performance highlights an obvious detection rate improvement with N. Still, it is apparent that the same improvement can be obtained with far less receive antennas when an optimized RIS is used, e.g. close to 80% detection rate is obtained with only $N = 4$ antennas. On the contrary, when no RIS is used, a larger number of antennas is required to obtain the same performance, e.g. $N = 64$ antennas are needed to obtain a detection rate ≥75% with an IS-based rule with no RIS assistance.

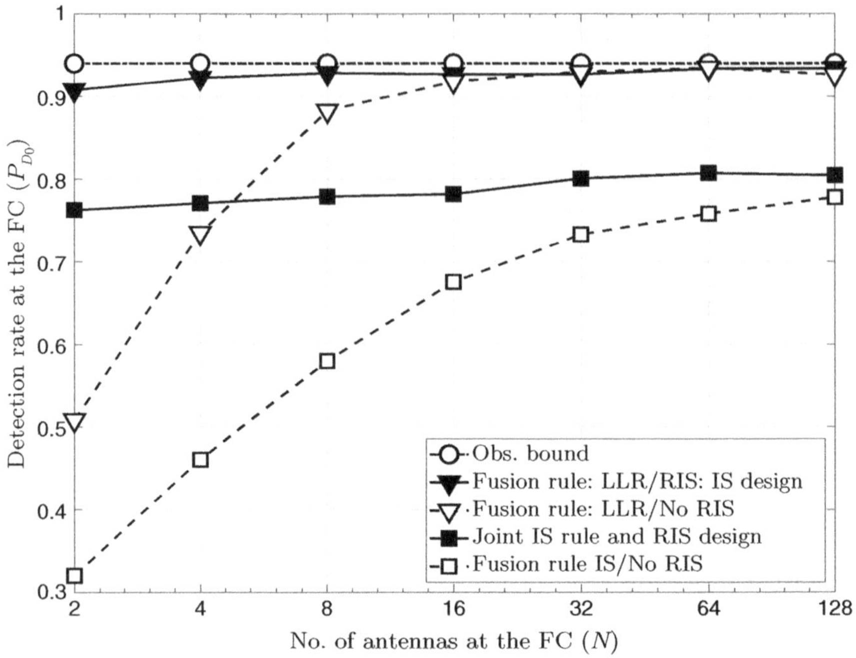

Figure 4.5 Evaluation of P_{D_0} as a function of N for the assessed rule and RIS configurations. The setup parameters include a false-alarm rate at the FC of $P_{F_0} = 0.01$ and an RIS comprising $M = 20$ elements.

4.5 Conclusions and Further Reading

This chapter centered on developing practical fusion rules and RIS designs to facilitate channel-aware decision fusion in a distributed MIMO framework through smart environments. Specifically, it presents a sensor-agnostic approach for a joint fusion rule and RIS design grounded in the IS assumption, aimed at overcoming the challenges posed by computational complexity and the absence of closed-form performance metrics for the LLR. The resulting non-convex optimization problem is approached through AO and MM techniques, yielding a simple ping-pong closed-form optimization method. While the proposed design may be suboptimal, numerical results highlight the significant advantages of employing an RIS to enhance the capabilities of WSNs in performing the DD task.

Finally, this section ends with suggestions for the interested reader. Literature on the use of smart surfaces for aiding decentralized inference is less developed [Fang et al., 2021; Ahmed et al., 2022] than that dealing with RIS used for communication purposes. The study in Fang et al. [2021] addresses over-the-air computation for generic nomographic functions, which are not designed for dependent and non-zero-mean decisions. The latter work tackles distributed estimation using a single-antenna FC while incorporating additional secrecy objectives [Ahmed et al., 2022]. Methodology wise, the use of deflection metric for optimization purposes in DF dates back to channel-aware literature in DF, see Ciuonzo et al. [2015]. Finally, from the optimization viewpoint, the principle of AO has been shown to be useful in several applications related to design of wireless transceivers [Sun et al., 2016].

Acknowledgments

The research of D. Ciuonzo and A. Zappone leading to these results has received funding from Project "GARDEN," CUP H53D23000480001, funded by EU in Next Generation EU plan, Mission 4 Component 1, through the Italian "Bando Prin 2022 – D.D. 104 del 02-02- 2022" by MUR. This chapter reflects only the authors' views and opinions and the Ministry cannot be considered responsible for them. The work of P. Salvo Rossi was partially supported by the Research Council of Norway under project ML4ITS within the IKTPLUSS framework. The work of M. Di Renzo was supported in part by the European Union through the Horizon Europe project COVER under grant agreement number 101086228, the Horizon Europe project UNITE under grant agreement number 101129618, the Horizon Europe project INSTINCT under grant agreement number 101139161, and the Horizon Europe project TWIN6G under grant agreement number 101182794, as well as by the Agence Nationale de la Recherche (ANR) through the France 2030 project ANR-PEPR Networks of the Future under grant agreement NF-YACARI 22-PEFT-0005, and by the CHIST-ERA project PASSIONATE under grant agreements CHIST-ERA-22-WAI-04 and ANR-23-CHR4-0003-01.

Bibliography

H. R. Ahmadi and A. Vosoughi. Channel aware sensor selection in distributed detection systems. In *IEEE 10th Workshop on Signal Processing Advances in Wireless Communications (SPAWC)*, 2009.

M. F. Ahmed, K. P. Rajput, N. K. D. Venkategowda, K. V. Mishra, and A. K. Jagannatham. Joint transmit and reflective beamformer design for secure estimation in IRS-aided WSNs. *IEEE Signal Processing Letters*, 29:692–696, 2022.

S. Appadwedula, V. V. Veeravalli, and D. L. Jones. Energy-efficient detection in sensor networks. *IEEE Journal on Selected Areas in Communications*, 23(4):693–702, 2005.

C. R. Berger, M. Guerriero, S. Zhou, and P. Willett. PAC vs. MAC for decentralized detection using noncoherent modulation. *IEEE Transactions on Signal Processing*, 57(9):3562–3575, 2009.

Z. Chair and P. K. Varshney. Optimal data fusion in multiple sensor detection systems. *IEEE Transactions on Aerospace and Electronic Systems*, AES-22(1):98–101, 1986.

A. Chawla, A. Patel, A. K. Jagannatham, and P. K. Varshney. Distributed detection in massive MIMO wireless sensor networks under perfect and imperfect CSI. *IEEE Transactions on Signal Processing*, 67(15):4055–4068, 2019.

A. Chawla, R. K. Singh, A. Patel, A. K. Jagannatham, and L. Hanzo. Distributed detection for centralized and decentralized millimeter wave massive MIMO sensor networks. *IEEE Transactions on Vehicular Technology*, 70(8):7665–7680, 2021.

Z. Chen and Y. Zhang. Providing spectrum information service using TV white space via distributed detection system. *IEEE Transactions on Vehicular Technology*, 68(8):7655–7667, 2019.

B. Chen, R. Jiang, T. Kasetkasem, and P. K. Varshney. Channel aware decision fusion in wireless sensor networks. *IEEE Transactions on Signal Processing*, 52(12):3454–3458, Dec 2004.

B. Chen, L. Tong, and P. K. Varshney. Channel-aware distributed detection in wireless sensor networks. *IEEE Signal Processing Magazine*, 23(4):16–26, 2006.

D. Ciuonzo, G. Romano, and P. Salvo. Channel-aware decision fusion in distributed MIMO wireless sensor networks: Decode-and-fuse vs. decode-then-fuse. *IEEE Transactions on Wireless Communications*, 11(8):2976–2985, Aug 2012.

D. Ciuonzo, P. S. Rossi, and S. Dey. Massive MIMO channel-aware decision fusion. *IEEE Transactions on Signal Processing*, 63(3):604–619, 2015.

D. Ciuonzo, G. Gelli, A. Pescapé, and F. Verde. Decision fusion rules in ambient backscatter wireless sensor networks. In *IEEE 30th Annual International Symposium on Personal, Indoor and Mobile Radio Communications (PIMRC)*, 2019.

M. Di Renzo, A. Zappone, M. Debbah, M.-S. Alouini, C. Yuen, J. De Rosny, and S. Tretyakov. Smart radio environments empowered by reconfigurable intelligent surfaces: How it works, state of research, and the road ahead. *IEEE Journal on Selected Areas in Communications*, 38(11):2450–2525, 2020.

W. Fang, Y. Jiang, Y. Shi, Y. Zhou, W. Chen, and K. B. Letaief. Over-the-air computation via reconfigurable intelligent surface. *IEEE Transactions on Communications*, 69(12):8612–8626, 2021.

C. Huang, A. Zappone, G. C. Alexandropoulos, M. Debbah, and C. Yuen. Reconfigurable intelligent surfaces for energy efficiency in wireless communication. *IEEE Transactions on Wireless Communications*, 18(8):4157–4170, 2019.

S. M. Kay. *Fundamentals of Statistical Signal Processing, Vol. II: Detection Theory. Signal Processing*. Prentice-Hall, Upper Saddle River, NJ, 1998.

A. Lei and R. Schober. Coherent max-log decision fusion in wireless sensor networks. *IEEE Transactions on Communications*, 58(5):1327–1332, 2010.

G. Mergen, V. Naware, and L. Tong. Asymptotic detection performance of type-based multiple access over multiaccess fading channels. *IEEE Transactions on Signal Processing*, 55(3):1081–1092, 2007.

N. Mudkey, D. Ciuonzo, A. Zappone, and P. S. Rossi. Wireless inference gets smarter: RIS-assisted channel-aware MIMO decision fusion. In *IEEE 12th Sensor Array and Multichannel Signal Processing Workshop (SAM)*, 2022.

B. Picinbono. On deflection as a performance criterion in detection. *IEEE Transactions on Aerospace and Electronic Systems*, 31(3):1072–1081, 1995.

Z. Quan, S. Cui, and A. H. Sayed. Optimal linear cooperation for spectrum sensing in cognitive radio networks. *IEEE Journal on Selected Topics in Signal Processing*, 2(1):28–40, Feb 2008.

A. R. Reibman and L. W. Nolte. Optimal detection and performance of distributed sensor systems. *IEEE Transactions on Aerospace and Electronic Systems*, AES-23(1):24–30, 1987.

Y. Sun, P. Babu, and D. P. Palomar. Majorization-minimization algorithms in signal processing, communications, and machine learning. *IEEE Transactions on Signal Processing*, 65(3):794–816, 2016.

G. Tabella, D. Ciuonzo, N. Paltrinieri, and P. S. Rossi. Spatio-temporal decision fusion for quickest fault detection within industrial plants: The oil and gas scenario. In *IEEE 24th International Conference on Information Fusion (FUSION)*, 2021.

A. Tarighati, J. Gross, and J. Jaldén. Decentralized hypothesis testing in energy harvesting wireless sensor networks. *IEEE Transactions on Signal Processing*, 65(18):4862–4873, 2017.

R. R. Tenney and N. R. Sandell. Detection with distributed sensors. *IEEE Transactions on Aerospace and Electronic Systems*, AES-17(4):501–510, 1981.

D. J. Warren and P. K. Willett. Optimal decentralized detection for conditionally independent sensors. In *IEEE American Control Conference (ACC)*, 1989.

S. Yiu and R. Schober. Nonorthogonal transmission and noncoherent fusion of censored decisions. *IEEE Transactions on Vehicular Technology*, 58(1):263–273, 2008.

A. Zappone, M. Di Renzo, and R. K. Fotock. Surface-based techniques for IoT networks: Opportunities and challenges. *IEEE Internet of Things Magazine*, 5(4):72–77, 2022.

X. Zhang, H. V. Poor, and M. Chiang. Optimal power allocation for distributed detection over MIMO channels in wireless sensor networks. *IEEE Transactions on Signal Processing*, 56(9):4124–4140, 2008.

5

Data Fusion in Millimeter Wave Massive MIMO Wireless Sensor Networks

Apoorva Chawla[1], Domenico Ciuonzo[2], Aditya K. Jagannatham[3], and Pierluigi Salvo Rossi[1]

[1] *Department of Electronic Systems, Norwegian University of Science and Technology, Trondheim, Norway*
[2] *Department of Electrical Engineering and Information Technologies (DIETI), University of Naples "Federico II", Napoli, Italy*
[3] *Electrical Engineering, Indian Institute of Technology Kanpur, Kanpur, India*

5.1 Introduction

Wireless sensor networks (WSNs) play a crucial role across various domains, such as disaster management, surveillance, military operations, healthcare, and more [Akyildiz et al., 2002]. Typically, sensors send their observations to a fusion center (FC) in these systems using a multiple access channel (MAC), where the FC utilizes a well-designed decision rule to determine the absence/presence of a specific phenomenon [Hall and Llinas, 1997]. However, the growing significance of these applications and the rapid proliferation of sensors have resulted in a significant bandwidth shortage in the sub-6 GHz spectrum, posing a considerable challenge for traditional cellular networks [Rappaport et al., 2013].

To combat this spectrum crisis, millimeter-wave (mmWave) communication has emerged as a promising solution, leveraging the unexplored spectrum from 30 to 300 GHz. MmWave communication is well-suited for applications requiring high-speed and substantial bandwidth in 5G (fifth-generation) and beyond 5G (B5G) WSNs, thereby alleviating congestion in the sub-6 GHz frequency bands. The mmWave spectrum is particularly appealing for deploying WSN due to its exceptionally short wavelength, enabling the compact arrangement of antennas within constrained physical dimensions. This enables the installation of a sizable antenna array at the FC [Rappaport et al., 2013], facilitating simultaneous connections and communication with numerous sensors, improving spectral efficiency, reducing sensor power consumption, and supporting highly directional beamforming to mitigate high-frequency propagation losses.

Wireless Sensor Networks in Smart Environments: Enabling Digitalization from Fundamentals to Advanced Solutions, First Edition. Edited by Domenico Ciuonzo and Pierluigi Salvo Rossi.

However, traditional fully digital transceiver architectures, which require a specific radio frequency (RF) chain for each antenna, are impractical for mmWave massive multiple-input multiple-output (MIMO) systems due to the substantial power consumption, complexity, and high cost. Consequently, hybrid combiners, which process signals in both analog and digital domains and necessitate fewer RF chains, are typically employed in these systems [El Ayach et al., 2012; Sohrabi and Yu, 2015].

In the existing works concerning massive MIMO technology, two architectures have gained popularity. The centralized (C-MIMO) configuration deploys an extensive antenna array at an FC, located at the cell center, offering lower deployment costs. Conversely, various FCs, each equipped with an extensive antenna array, are positioned at different geographical locations within a cell in the distributed (D-MIMO) topology. These FCs are interconnected via high-speed optical fibers for data processing and aggregation [Kerpez, 1996; Clark et al., 2001; Wang et al., 2008]. In D-MIMO systems, larger FC separation and shorter distances to the closest FC may likely result in enhanced performance and lower correlation.

In these architectures, transmitting local decisions to the FC can lead to performance degradation since the FC lacks complete information about the observed phenomenon. Moreover, complexity arises in optimizing local fusion rules to minimize detection error and evaluating optimal local thresholds for each sensor jointly with the global fusion rule. Hence, the complexity of finding optimal local detectors could potentially grow exponentially. Additionally, existing works have not utilized hybrid combiners at the FC to exploit the benefits of massive MIMO and mmWave technologies for data fusion, leaving their detection rules unknown.

To address these issues, this chapter investigates data fusion in both C- and D-MIMO-based mmWave massive MIMO WSNs. The generalized likelihood ratio test (GLRT) is employed to derive low-complexity decision rules based on hybrid combining for both configurations, considering unknown spatially correlated parameter. The chapter also characterizes closed-form expressions for the probabilities of false alarm (P_{FA}) and detection (P_D) under perfect channel state information (CSI) along with the development of optimal sensor gains for detection performance improvement. Considering the large-scale antenna regime, power scaling laws are established for both configurations, demonstrating reductions in sensor transmit power without compromising the system performance. Finally, the chapter discusses a novel sparse Bayesian learning (SBL) technique for channel estimation in a realistic scenario with CSI uncertainty. The decision rules are determined for the imperfect CSI scenario, along with simulation results.

5.2 System Model

We consider a WSN employing mmWave and massive MIMO technologies, where K single-antenna sensors are distributed randomly within a cell to monitor a signal of interest ($\Theta \in \mathbb{R}$), such that $\Theta \in \{0, \theta\}$. This setup can be framed as a binary hypothesis testing problem, where the null hypothesis $\mathcal{H}_0$ signifies $\Theta = 0$, indicating the signal's absence, while the alternative hypothesis $\mathcal{H}_1$ implies its presence. The amplified measurement of the kth sensor is

$$x_k = f_k(a_k\Theta + \eta_k), \quad k = 1, 2, \dots, K, \tag{5.1}$$

where a_k is the observation constant, $\eta_k \sim \mathcal{CN}(0, \sigma_\eta^2)$ is the additive white Gaussian noise (AWGN) and f_k is the complex gain. All K-amplified sensor measurements are transmitted to the FC over MAC.

5.2.1 C-MIMO System

The C-MIMO system involves K sensors communicating with the FC. The FC employs a huge M-element antenna array and $N_{RF} = K$ RF chains ($M \gg K$), as depicted in Figure 5.1a. Further, the FC utilizes a fully connected structure (FCS), where each RF chain is linked to all M antennas. Considering $\mathbf{a} = [a_1, a_2, \dots, a_K]^T$, the signal $\mathbf{z}_C \in \mathbb{C}^{M\times 1}$ can be expressed as

$$\mathbf{z}_C = \mathbf{GFa}\Theta + \mathbf{GF}\boldsymbol{\eta} + \mathbf{v}, \tag{5.2}$$

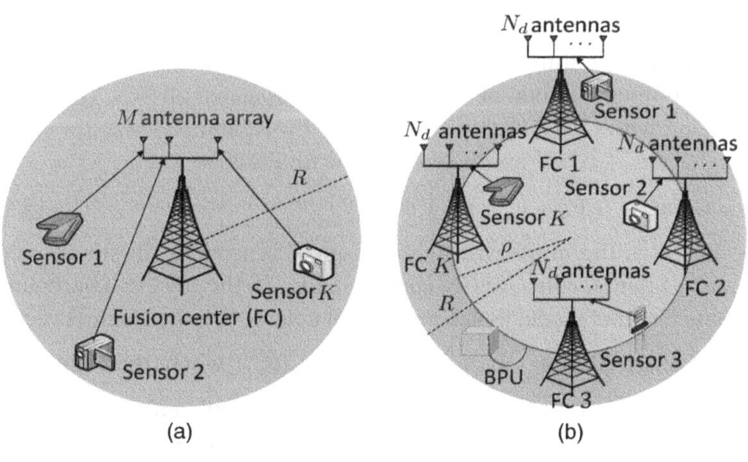

Figure 5.1 Different antenna topologies of the mmWave massive MIMO WSN: (a) centralized (C-MIMO) and (b) distributed (D-MIMO).

where $\mathbf{G} \in \mathbb{C}^{M \times K}$ is the mmWave channel from the sensors to the FC, $\boldsymbol{\eta} \in \mathbb{C}^{K \times 1}$ is the noise vector, $\mathbf{v} \sim \mathcal{CN}(\mathbf{0}, \sigma_v^2 \mathbf{I}_M)$ is the AWGN at the FC, and $\mathbf{F} = \text{diag}(f_1, f_2, \dots, f_K)$ is the complex gain matrix. Further, the received signal $\mathbf{z}_C$ at the FC follows the complex Gaussian distribution $\mathbf{z}_C \sim \mathcal{CN}(\mathbf{GFa}\Theta, \widetilde{\boldsymbol{\Sigma}})$, with covariance matrix as $\widetilde{\boldsymbol{\Sigma}} = \sigma_\eta^2 \mathbf{GFF}^H \mathbf{G}^H + \sigma_v^2 \mathbf{I}_M$. The channel $\mathbf{G}$ is modeled as $\mathbf{G} = \mathbf{HD}^{1/2}$, where $\mathbf{H}$ is the small-scale fading matrix and $\mathbf{D} = \text{diag}(\beta_1, \beta_2, \dots, \beta_K)$ is the large-scale fading matrix accounting for path loss and shadowing. The Saleh–Valenzuela channel model [Saleh and Valenzuela, 1987] is employed to describe the kth sensor small-scale fading vector, denoted by $\mathbf{h}_k$, as

$$\mathbf{h}_k = \sqrt{\frac{M}{L_k}} \sum_{i=1}^{L_k} \alpha_k^i \mathbf{a}_r(\theta_k^i), \tag{5.3}$$

where $\theta_k^i \in [0, 2\pi]$ is the angle of arrival (AoA) and $\alpha_k^i \sim \mathcal{CN}(0, 1)$ is the complex gain for the ith path. Further, the number of paths for the kth sensor L_k follows a discrete uniform distribution $L_k \sim \mathcal{DU}[1, L]$ and L denotes the maximum number of paths. Defining $v = \frac{2\pi}{\lambda} d$, λ as the carrier wavelength and d as the antenna spacing, the receive array response vector $\mathbf{a}_r(\theta_k^i) \in \mathbb{C}^{M \times 1}$ for an M-element uniform linear array (ULA) is given by $\mathbf{a}_r(\theta_k^i) = \frac{1}{\sqrt{M}} \left[1, e^{jv \sin(\theta_k^i)}, \dots, e^{jv(M-1)\sin(\theta_k^i)}\right]^T$.

5.2.2 D-MIMO System

In the D-MIMO system, J FCs are uniformly distributed on a circle of radius ρ, satisfying $r_{min} \ll \rho < R$, where R and r_{min} denote the cell radius and the minimum distance between the cell center and the FC, respectively. The FCs are interconnected through high-capacity optical fiber backhaul, as depicted in Figure 5.1b. Each FC is outfitted with an N_d-element antenna array and a single RF chain, contributing to a total of $N_T = JN_d$ antennas across all FCs. To ensure a fair comparison with C-MIMO, we set $N_T = M$, $J = K$, and $N_{RF} = K$ [Li et al., 2018]. The signal $\mathbf{z}_{D,j} \in \mathbb{C}^{N_d \times 1}$ associated with the jth FC is

$$\mathbf{z}_{D,j} = \mathbf{G}_j \mathbf{Fa}\Theta + \mathbf{G}_j \mathbf{F}\boldsymbol{\eta} + \mathbf{v}_j, \quad j = 1, 2, \dots, J, \tag{5.4}$$

where $\mathbf{G}_j \in \mathbb{C}^{N_d \times K}$ is the mmWave channel from K sensors to the jth FC and $\mathbf{v}_j \sim \mathcal{CN}(0, \sigma_v^2 \mathbf{I}_{N_d})$ denotes the AWGN corresponding to the jth FC. Leveraging the narrowband channel model [Saleh and Valenzuela, 1987], the channel vector $\mathbf{g}_{k,j}$ can be characterized as $\mathbf{g}_{k,j} = \sqrt{\frac{N_d \beta_{k,j}}{L_{k,j}}} \sum_{i=1}^{L_{k,j}} \alpha_{k,j}^i \mathbf{a}_r(\theta_{k,j}^i)$, where the definitions of $\beta_{k,j}$, $L_{k,j}$, $\alpha_{k,j}^i$, $\theta_{k,j}^i$, and $\mathbf{a}_r(\theta_{k,j}^i)$ for the jth FC and the kth sensor are similar to the centralized topology. Further, the received array response vector $\mathbf{a}_r(\theta_{k,j}^i)$ is given by $\mathbf{a}_r(\theta_{k,j}^i) = \frac{1}{\sqrt{N_d}} \left[1, e^{jv \sin(\theta_{k,j}^i)}, \dots, e^{jv(N_d-1)\sin(\theta_{k,j}^i)}\right]^T$ and the signal $\mathbf{z}_{D,j}$ in (5.4) follows the complex Gaussian distribution $\mathbf{z}_{D,j} \sim \mathcal{CN}(\mathbf{G}_j \mathbf{Fa}\Theta, \widetilde{\boldsymbol{\Sigma}}_j)$, with its covariance matrix as $\widetilde{\boldsymbol{\Sigma}}_j = \sigma_\eta^2 \mathbf{G}_j \mathbf{FF}^H \mathbf{G}_j^H + \sigma_v^2 \mathbf{I}_{N_d}$.

5.3 Problem Formulation

Considering the perfect CSI scenario, this section discusses the fusion rules based on the proposed architecture for the unknown parameter detection in C- and D-MIMO topologies.

5.3.1 C-MIMO: Fusion Rule for Perfect CSI

Leveraging the generalized likelihood ratio test (GLRT) framework in (5.2), the likelihood function $p(\mathbf{z}_C; \theta|\mathcal{H}_1)$ for hypothesis $\mathcal{H}_1$ can be given as

$$p(\mathbf{z}_C; \theta|\mathcal{H}_1) = \frac{1}{|\pi\tilde{\mathbf{\Sigma}}|} \exp\left(- (\mathbf{z}_C - \mathbf{GFa}\theta)^H \tilde{\mathbf{\Sigma}}^{-1} (\mathbf{z}_C - \mathbf{GFa}\theta) \right). \tag{5.5}$$

Further, the maximum-likelihood estimate (MLE) of the unknown parameter θ can be expressed as $\hat{\theta} = \frac{\Re(\mathbf{z}_C^H \tilde{\mathbf{\Sigma}}^{-1}\mathbf{GFa})}{\mathbf{a}^H\mathbf{F}^H\mathbf{G}^H\tilde{\mathbf{\Sigma}}^{-1}\mathbf{GFa}}$, determined by taking the derivative of the logarithm of the likelihood function in (5.5) with respect to θ and setting it to zero. The GLRT statistic $T_{C,UP}(\mathbf{z}_C)$ can be formulated as

$$T_{C,UP}(\mathbf{z}_C) = \ln \left[\frac{p(\mathbf{z}_C; \hat{\theta}|\mathcal{H}_1)}{p(\mathbf{z}_C|\mathcal{H}_0)} \right] \underset{\mathcal{H}_0}{\overset{\mathcal{H}_1}{\gtrless}} \gamma. \tag{5.6}$$

Employing the likelihood functions $p(\mathbf{z}_C; \hat{\theta}|\mathcal{H}_1)$ and $p(\mathbf{z}_C|\mathcal{H}_0)$ for hypotheses $\mathcal{H}_1$ and $\mathcal{H}_0$, respectively, the aforementioned test simplifies to

$$T_{C,UP}(\mathbf{z}_C) = \ln \left[\frac{\exp\left(-(\mathbf{z}_C - \mathbf{GFa}\hat{\theta})^H \widetilde{\mathbf{\Sigma}}^{-1} (\mathbf{z}_C - \mathbf{GFa}\hat{\theta}) \right)}{\exp\left(-\mathbf{z}_C^H \widetilde{\mathbf{\Sigma}}^{-1} \mathbf{z}_C \right)} \right]. \tag{5.7}$$

This test exhibits a high complexity of $\mathcal{O}(M^3 + M^2K)$, escalating with M due to the necessity of computing $\widetilde{\mathbf{\Sigma}}^{-1} \in \mathbb{C}^{M\times M}$. To address this challenge, a two-step architecture is exploited. The first step utilizes a hybrid combiner, comprising of an analog combiner $\mathbf{W}_{RF,C} \in \mathbb{C}^{M\times K}$ and a digital combiner $\mathbf{W}_{BB,C} \in \mathbb{C}^{K\times K}$, to process the signal $\mathbf{z}_C$. The measurements obtained after the hybrid combining are combined in the subsequent step to yield a global decision, as illustrated in Figure 5.2. The analog combiner $\mathbf{W}_{RF,C} = \left[\mathbf{a}_r(\theta_1^{i_1}), \dots, \mathbf{a}_r(\theta_K^{i_K})\right]$ is computed by stacking the receive array response vectors of K sensors linked with the maximum path gains. For the kth sensor, i_k denotes the path having the maximum gain $|\alpha_k^{i_k}|$ and $\theta_k^{i_k}$ is the corresponding AoA [Li et al., 2018]. Leveraging the asymptotic orthogonality characteristic of the mmWave massive MIMO channel [Zhou et al., 2017], it can be deduced that

$$\mathbf{a}_r^H\left(\theta_k^{i_s}\right)\mathbf{a}_r\left(\theta_l^{i_n}\right) = \begin{cases} 1, & s = n \text{ and } k = l, \\ 0, & s \neq n \text{ or } k \neq l, \end{cases} \tag{5.8}$$

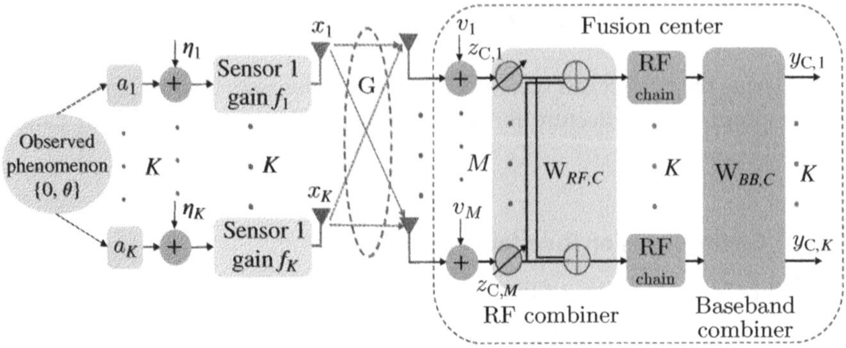

Figure 5.2 Block diagram showcasing data fusion in a mmWave massive C-MIMO based WSN.

when $L_k = o(M)$, $\forall k$ and $M \to \infty$. Consequently, the digital combiner for the corresponding baseband channel can be characterized as $\mathbf{W}_{BB,C} = \mathbf{W}_{RF,C}^H \mathbf{G}$, which becomes diagonal by exploiting the property stated in (5.8), such that $\left[\mathbf{W}_{BB,C}\right]_{k,k} = \sqrt{\frac{M\beta_k}{L_k}}\alpha_k^{i_k}$. After implementing the hybrid combining on the signal $\mathbf{z}_C$, the resulting signal $\mathbf{y}_C \in \mathbb{C}^{K\times 1}$ is expressed as

$$\mathbf{y}_C = \mathbf{W}_{BB,C}^H \mathbf{W}_{RF,C}^H (\mathbf{G}\mathbf{F}\mathbf{a}\Theta + \mathbf{G}\mathbf{F}\boldsymbol{\eta} + \mathbf{v}), \tag{5.9}$$

which on exploiting the property in (5.8) can be simplified as

$$\mathbf{y}_C = M\boldsymbol{\Psi}\mathbf{F}\mathbf{a}\Theta + M\boldsymbol{\Psi}\mathbf{F}\boldsymbol{\eta} + \tilde{\mathbf{v}}. \tag{5.10}$$

Here, $\boldsymbol{\Psi} = \text{diag}(\psi_1, \psi_2, \dots, \psi_K)$ is a diagonal matrix, where $\psi_k = \frac{\beta_k}{L_k}|\alpha_k^{i_k}|^2$, and the effective noise $\tilde{\mathbf{v}} = \mathbf{W}_{BB,C}^H \mathbf{W}_{RF,C}^H \mathbf{v}$ is distributed as $\tilde{\mathbf{v}} \sim \mathcal{CN}\left(\mathbf{0}, \mathbf{C}_{\tilde{\mathbf{v}}}\right)$, with $\mathbf{C}_{\tilde{\mathbf{v}}} = \sigma_v^2 M\boldsymbol{\Psi}$. Additionally, the distribution of $\mathbf{y}_C$ is given by $\mathbf{y}_C \sim \mathcal{CN}\left(M\boldsymbol{\Psi}\mathbf{F}\mathbf{a}\Theta, \boldsymbol{\Sigma}_C\right)$, where its diagonal covariance matrix is defined as $\boldsymbol{\Sigma}_C = \sigma_\eta^2 M^2 \boldsymbol{\Psi}\mathbf{F}\mathbf{F}^H\boldsymbol{\Psi}^H + \mathbf{C}_{\tilde{\mathbf{v}}}$ and $[\boldsymbol{\Sigma}_C]_{k,k} = \sigma_\eta^2 M^2 \psi_k^2 |f_k|^2 + \sigma_v^2 M\psi_k$. Employing these quantities, the GLRT for the hybrid combiner output $\mathbf{y}_C$ can be expressed as

$$T_{C,UP}(\mathbf{y}_C) = \Re(\mathbf{y}_C^H \boldsymbol{\Sigma}_C^{-1}\boldsymbol{\Psi}\mathbf{F}\mathbf{a}\hat{\theta}) = \left|\sum_{k=1}^{K} \Re\left(\frac{y_{C,k}^* f_k a_k}{M\psi_k |f_k|^2 \sigma_\eta^2 + \sigma_v^2}\right)\right| \underset{\mathcal{H}_0}{\overset{\mathcal{H}_1}{\gtrless}} \gamma', \tag{5.11}$$

where the MLE of θ, determined using (5.10), is given by $\hat{\theta} = \frac{\Re(\mathbf{y}_C^H \boldsymbol{\Sigma}_C^{-1}\boldsymbol{\Psi}\mathbf{F}\mathbf{a})}{M\mathbf{a}^H\mathbf{F}^H\boldsymbol{\Psi}^H\boldsymbol{\Sigma}_C^{-1}\boldsymbol{\Psi}\mathbf{F}\mathbf{a}}$. Note that the aforementioned test exhibits a low complexity of $\mathcal{O}(K)$, which is invariant of M. On the contrary, the test in (5.7) is characterized by a complexity of $\mathcal{O}(M^3 + M^2K)$, which grows with M. Notably, $T_{C,UP}(\mathbf{y}_C)$ is a weighted linear combination of complex Gaussian random variables $y_{C,k}$, resulting in its distribution being complex Gaussian. Further, the P_D and P_{FA} performance of $T_{C,UP}(\mathbf{y}_C)$ in (5.11) is described below.

Theorem 5.1 For mmWave massive C-MIMO WSN, the P_D and P_{FA} performance of the test $T_{C,UP}(\mathbf{y}_C)$ for unknown parameter detection is

$$P_D = Q\left(\frac{\gamma' - \mu_{T_{C,UP}|\mathcal{H}_1}}{\sigma_{T_{C,UP}|\mathcal{H}_1}}\right) + Q\left(\frac{\gamma' + \mu_{T_{C,UP}|\mathcal{H}_1}}{\sigma_{T_{C,UP}|\mathcal{H}_1}}\right),$$
$$P_{FA} = Q\left(\frac{\gamma' - \mu_{T_{C,UP}|\mathcal{H}_0}}{\sigma_{T_{C,UP}|\mathcal{H}_0}}\right) + Q\left(\frac{\gamma' + \mu_{T_{C,UP}|\mathcal{H}_0}}{\sigma_{T_{C,UP}|\mathcal{H}_0}}\right), \tag{5.12}$$

where the means $\mu_{T_{C,UP}|\mathcal{H}_1}$, $\mu_{T_{C,UP}|\mathcal{H}_0}$ and the standard deviations $\sigma_{T_{C,UP}|\mathcal{H}_1}$, $\sigma_{T_{C,UP}|\mathcal{H}_0}$ under alternate and null hypotheses, respectively, are given as

$$\mu_{T_{C,UP}|\mathcal{H}_0} = 0, \quad \mu_{T_{C,UP}|\mathcal{H}_1} = \sum_{k=1}^{K} \frac{M\psi_k|f_k|^2|a_k|^2\theta}{M\psi_k|f_k|^2\sigma_\eta^2 + \sigma_\upsilon^2}, \tag{5.13}$$

$$\sigma^2_{T_{C,UP}|\mathcal{H}_0} = \sigma^2_{T_{C,UP}|\mathcal{H}_1} = \sum_{k=1}^{K} \frac{M\psi_k|f_k|^2|a_k|^2}{2(M\psi_k|f_k|^2\sigma_\eta^2 + \sigma_\upsilon^2)}. \tag{5.14}$$

Proof: Refer to Chawla et al. [2022] for the proof.

5.3.2 D-MIMO: Fusion Rule for Perfect CSI

The location of the kth sensor and the jth FC in polar coordinates can be defined as (r_k, ν_k) and $(\rho_j, \phi_j) = \left(\rho, \frac{2\pi(j-1)}{J}\right)$, respectively. Thus, the distance of the kth sensor from the jth FC is $\delta_{k,j} = \sqrt{\rho_j^2 + r_k^2 - 2\rho_j r_k \cos(\phi_j - \nu_k)}$. The D-MIMO system performance can be improved by optimally assigning sensors to their respective FCs using the distance (D)-based FC selection approach [Li et al., 2018]. Let $\mathbf{Q} \in \mathbb{R}^{K\times K}$ denotes the binary correction matrix such that $\mathbf{Q}^H\mathbf{Q} = \mathbf{I}_K$ and $\boldsymbol{\Delta} \in \mathbb{R}^{K\times K}$ represent the distance matrix, where $[\boldsymbol{\Delta}]_{k,j} = \delta_{k,j}$. The (k, j)th entry in $\boldsymbol{\Delta}$ is 1, i.e. $[\mathbf{Q}]_{k,j} = 1$, if the kth sensor is allocated to the jth FC using the minimum distance criterion, otherwise $[\mathbf{Q}]_{k,j} = 0$.

We assume that FCs are sufficiently separated to ensure the independence of observation vectors $\mathbf{z}_{D,j}$, $\forall j$. Akin to the C-MIMO case, the received signals are processed using a two-step processing architecture, as depicted in Figure 5.3. In the initial step, the observation vector is processed at each FC using an RF combiner. The RF combiner at the jth FC is chosen as the array response vector with the maximum path gain $|\alpha^{i_{k_j}}_{k_j,j}|$, i.e. $\mathbf{w}_{RF,j} = \mathbf{a}_r(\theta^{i_{k_j}}_{k_j,j}) \in \mathbb{C}^{N_d\times 1}$. Here, i_{k_j} signifies the ith path for the k_j th sensor and $\theta^{i_{k_j}}_{k_j,j}$ denotes the associated AoA. Leveraging the asymptotic orthogonality property in (5.8), the RF combiner output $\tilde{z}_{D,j}$ for the jth FC is

$$\tilde{z}_{D,j} = \mathbf{w}^H_{RF,j}(\mathbf{G}_j\mathbf{F}\mathbf{a}\Theta + \mathbf{G}_j\mathbf{F}\boldsymbol{\eta} + \mathbf{v}_j). \tag{5.15}$$

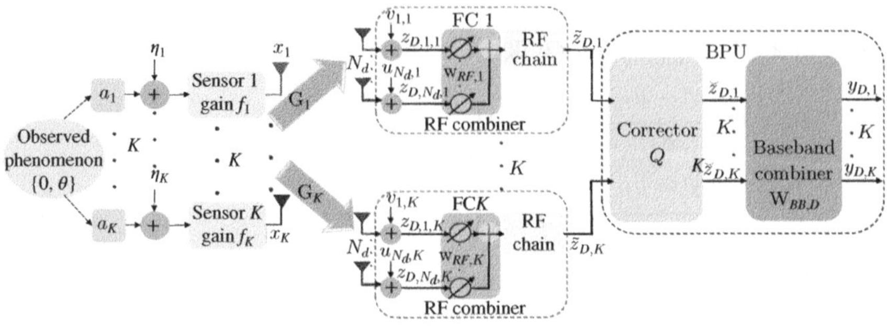

Figure 5.3 Block diagram depicting data fusion in a mmWave massive D-MIMO based WSN.

All J RF combiner outputs are stacked together at the baseband processing unit (BPU) to obtain $\tilde{\mathbf{z}}_D = [\tilde{z}_{D,1}, \tilde{z}_{D,2}, \dots, \tilde{z}_{D,K}]^T$, which is processed using $\mathbf{Q}$ to yield the rearranged soft sensor decisions as

$$\check{\mathbf{z}}_D = \mathbf{Q}\tilde{\mathbf{z}}_D = \mathbf{Q}\mathbf{W}_{RF,D}^H(\tilde{\mathbf{G}}\mathbf{F}\mathbf{a}\Theta + \tilde{\mathbf{G}}\mathbf{F}\boldsymbol{\eta} + \check{\mathbf{v}}), \tag{5.16}$$

where $\check{\mathbf{v}} = [\mathbf{v}_1^T, \mathbf{v}_2^T, \dots, \mathbf{v}_K^T]^T \in \mathbb{C}^{KN_d \times 1}$, $\tilde{\mathbf{G}} = [\mathbf{G}_1^T, \mathbf{G}_1^T, \dots, \mathbf{G}_K^T]^T \in \mathbb{C}^{KN_d \times K}$ and $\mathbf{W}_{RF,D} = \mathrm{diag}(\mathbf{w}_{RF,1}, \mathbf{w}_{RF,2}, \dots, \mathbf{w}_{RF,K}) \in \mathbb{C}^{KN_d \times K}$. These outputs are forwarded to a central BPU for digital combining. The digital combiner is chosen as $\mathbf{W}_{BB,D} = \mathbf{Q}\mathbf{W}_{RF,D}^H\tilde{\mathbf{G}} \in \mathbb{C}^{K \times K}$, which is diagonal with $[\mathbf{W}_{BB,D}]_{k,k} = \sqrt{\frac{N_d\beta_{k,j_k}}{L_{k,j_k}}}\alpha_{k,j_k}^{i_k}$. The digital combined output $\mathbf{y}_D \in \mathbb{C}^{K \times 1}$ can be given as

$$\mathbf{y}_D = \mathbf{W}_{BB,D}^H\mathbf{Q}\mathbf{W}_{RF,D}^H(\tilde{\mathbf{G}}\mathbf{F}\mathbf{a}\Theta + \tilde{\mathbf{G}}\mathbf{F}\boldsymbol{\eta} + \check{\mathbf{v}}), \tag{5.17}$$

which can be further simplified using (5.8) as

$$\mathbf{y}_D = N_d\boldsymbol{\Psi}_D\mathbf{F}\mathbf{a}\Theta + N_d\boldsymbol{\Psi}_D\mathbf{F}\boldsymbol{\eta} + \mathbf{v}', \tag{5.18}$$

where the noise $\mathbf{v}' = \mathbf{W}_{BB,D}^H\mathbf{Q}\mathbf{W}_{RF,D}^H\check{\mathbf{v}}$ is distributed as $\mathbf{v}' \sim \mathcal{CN}(\mathbf{0}, \mathbf{C}_{\mathbf{v}'})$, with $\mathbf{C}_{\mathbf{v}'} = \sigma_v^2 N_d\boldsymbol{\Psi}_D$, $\boldsymbol{\Psi}_D = \mathrm{diag}(\psi_{1,j_1}, \psi_{2,j_2}, \dots, \psi_{K,j_K})$ and $\psi_{k,j_k} = \frac{\beta_{k,j_k}}{L_{k,j_k}}|\alpha_{k,j_k}^{i_k}|^2$. The vector $\mathbf{y}_D$ follows Gaussian distribution, i.e. $\mathbf{y}_D \sim \mathcal{CN}(N_d\boldsymbol{\Psi}_D\mathbf{F}\mathbf{a}\Theta, \boldsymbol{\Sigma}_D)$, where $\boldsymbol{\Sigma}_D = \sigma_\eta^2 N_d^2\boldsymbol{\Psi}_D\mathbf{F}\mathbf{F}^H\boldsymbol{\Psi}_D^H + \mathbf{C}_{\mathbf{v}'}$ is diagonal with $[\boldsymbol{\Sigma}_D]_{k,k} = \sigma_\eta^2 N_d^2\psi_{k,j_k}^2|f_k|^2 + \sigma_v^2 N_d\psi_{k,j_k}$. The hybrid combiner outputs are finally combined to form the global decision. Hence, the GLRT statistic $T_{D,UP}(\mathbf{y}_D)$ for the D-MIMO architecture can be formulated as

$$\begin{aligned} T_{D,UP}(\mathbf{y}_D) &= \ln\left[\frac{p(\mathbf{y}_D; \hat{\theta}|\mathcal{H}_1)}{p(\mathbf{y}_D|\mathcal{H}_0)}\right] \underset{\mathcal{H}_0}{\overset{\mathcal{H}_1}{\gtrless}} \gamma' \\ &= \Re(\mathbf{y}_D^H\boldsymbol{\Sigma}_D^{-1}\boldsymbol{\Psi}_D\mathbf{F}\mathbf{a}\hat{\theta}) = \left|\sum_{k=1}^{K}\Re\left(\frac{y_{D,k}^* f_k a_k}{N_d\psi_{k,j_k}|f_k|^2\sigma_\eta^2 + \sigma_v^2}\right)\right|, \end{aligned} \tag{5.19}$$

where the MLE can be expressed as $\hat{\theta} = \frac{\Re(\mathbf{y}_D^H \boldsymbol{\Sigma}_D^{-1} \boldsymbol{\Psi}_D \mathbf{F}\mathbf{a})}{N_d \mathbf{a}^H \mathbf{F}^H \boldsymbol{\Psi}_D^H \boldsymbol{\Sigma}_D^{-1} \boldsymbol{\Psi}_D \mathbf{F}\mathbf{a}}$. Observe that the detector $T_{D,UP}(\mathbf{y}_D)$ exhibits a complexity of $\mathcal{O}(K)$, which does not depend on N_d. On the contrary, the detector $T_{D,UP}(\mathbf{z}_D)$, which doesn't employ the hybrid combining architecture, has a complexity of $\mathcal{O}(N_d^3 + N_d^2 K)$. Next, we analyze the analytical performance of the above detector.

Theorem 5.2 For mmWave massive D-MIMO WSN, the unknown parameter detection performance of the test $T_{D,UP}(\mathbf{y}_D)$ in (5.19), characterized by probabilities of detection P_D and false alarm P_{FA}, can be determined as

$$P_D = Q\left(\frac{\gamma' - \mu_{T_{D,UP}|\mathcal{H}_1}}{\sigma_{T_{D,UP}|\mathcal{H}_1}}\right) + Q\left(\frac{\gamma' + \mu_{T_{D,UP}|\mathcal{H}_1}}{\sigma_{T_{D,UP}|\mathcal{H}_1}}\right), \tag{5.20}$$

$$P_{FA} = Q\left(\frac{\gamma' - \mu_{T_{D,UP}|\mathcal{H}_0}}{\sigma_{T_{D,UP}|\mathcal{H}_0}}\right) + Q\left(\frac{\gamma' + \mu_{T_{D,UP}|\mathcal{H}_0}}{\sigma_{T_{D,UP}|\mathcal{H}_0}}\right), \tag{5.21}$$

where the means $\mu_{T_{D,UP}|\mathcal{H}_0}, \mu_{T_{D,UP}|\mathcal{H}_1}$ and the standard deviations $\sigma_{T_{D,UP}|\mathcal{H}_0}, \sigma_{T_{D,UP}|\mathcal{H}_1}$ for the null and alternative hypotheses, respectively, can be given as

$$\mu_{T_{D,UP}|\mathcal{H}_0} = 0, \quad \mu_{T_{D,UP}|\mathcal{H}_1} = \sum_{k=1}^{K} \frac{N_d \psi_{k,j_k} |f_k|^2 |a_k|^2 \theta}{N_d \psi_{k,j_k} |f_k|^2 \sigma_\eta^2 + \sigma_v^2}, \tag{5.22}$$

$$\sigma^2_{T_{D,UP}|\mathcal{H}_1} = \sigma^2_{T_{D,UP}|\mathcal{H}_0} = \sum_{k=1}^{K} \frac{N_d \psi_{k,j_k} |f_k|^2 |a_k|^2}{2(N_d \psi_{k,j_k} |f_k|^2 \sigma_\eta^2 + \sigma_v^2)}. \tag{5.23}$$

Proof: On similar lines as Theorem 5.1.

5.4 Sensor Gain Optimization

This section discusses the framework to determine optimal sensor gains for C- and D-MIMO architectures.

5.4.1 Optimized Sensor Gains for C-MIMO

One can enhance the detection performance of the C-MIMO system through the maximization of the deflection coefficient [Kay, 1993a]. Leveraging Theorem 5.1, the deflection coefficient $d^2_{C,UP}$ can be expressed as

$$d^2_{C,UP} = \frac{(\mu_{T_{C,UP}|\mathcal{H}_1} - \mu_{T_{C,UP}|\mathcal{H}_0})^2}{\sigma^2_{T_{C,UP}|\mathcal{H}_0}} = \sum_{k=1}^{K} \frac{2M\psi_k |f_k|^2 |a_k|^2 \theta^2}{M\psi_k |f_k|^2 \sigma_\eta^2 + \sigma_v^2}. \tag{5.24}$$

Defining $p_k = |f_k|^2$ and P as the total transmit power, the optimization problem using (5.24) can be formulated as

$$\min_{p_k} \sum_{k=1}^{K} -\frac{M\psi_k p_k |a_k|^2}{M\psi_k p_k \sigma_\eta^2 + \sigma_v^2}$$
$$\text{s.t.} \sum_{k=1}^{K} p_k = P,$$
$$p_k \geq 0, 1 \leq k \leq K. \tag{5.25}$$

Its Lagrangian is expressed as

$$\mathcal{L}(p_k, \lambda, \mu_k) = \sum_{k=1}^{K} -\frac{M\psi_k p_k |a_k|^2}{M\psi_k p_k \sigma_\eta^2 + \sigma_v^2} + \lambda\left(\sum_{k=1}^{K} p_k - P\right) - \sum_{k=1}^{K} \mu_k p_k. \tag{5.26}$$

On leveraging the Karush–Kuhn–Tucker (KKT) conditions [Boyd and Vandenberghe, 2004], the optimal sensor power p_k is given as

$$p_k = \left(\sqrt{\frac{|a_k|^2 \sigma_v^2}{\lambda M \psi_k \sigma_\eta^4}} - \frac{\sigma_v^2}{M\psi_k \sigma_\eta^2}\right)^+, \tag{5.27}$$

where $(b)^+ = b$, for $b \geq 0$ and 0 otherwise. Additionally, λ is confined within the range $[\lambda_l, \lambda_u]$ and is selected to satisfy the constraint $\sum_{k=1}^{K} p_k = P$, where $\lambda_l = \min_k \left\{ \frac{M\psi_k |a_k|^2 \sigma_v^2}{(M\psi_k P \sigma_\eta^2 + \sigma_v^2)^2} \right\}$ and $\lambda_u = \max_k \left\{ \frac{M\psi_k |a_k|^2}{\sigma_v^2} \right\}$. The value of λ can be determined by employing a simple bisection search over $[\lambda_l, \lambda_u]$.

5.4.2 Optimized Sensor Gains for D-MIMO

Following a similar approach to that of C-MIMO, one can obtain the optimal sensor gains for the D-MIMO setup as

$$p_k = \left(\sqrt{\frac{|a_k|^2 \sigma_v^2}{\lambda N_d \psi_{k,j_k} \sigma_\eta^4}} - \frac{\sigma_v^2}{N_d \psi_{k,j_k} \sigma_\eta^2}\right)^+, \tag{5.28}$$

where the determination of λ involves employing a traditional bisection search over $[\lambda_l, \lambda_u]$, with $\lambda_l = \min_k \left\{ \frac{N_d \psi_{k,j_k} |a_k|^2 \sigma_v^2}{(N_d \psi_{k,j_k} P \sigma_\eta^2 + \sigma_v^2)^2} \right\}$ and $\lambda_u = \max_k \left\{ \frac{N_d \psi_{k,j_k} |a_k|^2}{\sigma_v^2} \right\}$.

5.5 Power Scaling Laws

This section delves into the asymptotic performance analysis and unveils the power scaling laws for data fusion in C- and D-MIMO based WSNs. The analysis demonstrates substantial power conservation and improved sensor longevity.

Next, we examine the performance of $T_{C,UP}(\mathbf{y}_C)$, derived in (5.11), in the large antenna regime considering uniform and optimal transmit gains.

5.5.1 Uniform Transmit Gains

Under the uniform transmit gain scheme, we allocate equal transmit power to all sensors, i.e. $p_k = |f_k|^2 = p = P/K$ and chose power scaling as $p = \tilde{p}/M$, with $\tilde{p}$ as the average transmit power. Using these quantities, the asymptotic performance of the detector $T_{C,UP}(\mathbf{y}_C)$ is discussed next.

Lemma 5.1 Considering uniform transmit gains and $M \to \infty$, the asymptotic performance, characterized by probabilities of detection $P^a_{D,u}$ and false alarm $P^a_{FA,u}$, for the C-MIMO detector in (5.11) can be expressed as

$$\begin{aligned} P^a_{D,u} &= Q(\gamma' - \mu^a_{T_{C,UP}|\mathcal{H}_1,u}) + Q(\gamma' + \mu^a_{T_{C,UP}|\mathcal{H}_1,u}), \\ P^a_{FA,u} &= Q(\gamma' - \mu^a_{T_{C,UP}|\mathcal{H}_0,u}) + Q(\gamma' + \mu^a_{T_{C,UP}|\mathcal{H}_0,u}), \end{aligned} \tag{5.29}$$

where the normalized asymptotic means $\mu^a_{T_{C,UP}|\mathcal{H}_0,u}$ and $\mu^a_{T_{C,UP}|\mathcal{H}_1,u}$ for the null and alternate hypotheses, respectively, are given as

$$\mu^a_{T_{C,UP}|\mathcal{H}_0,u} = 0, \quad \mu^a_{T_{C,UP}|\mathcal{H}_1,u} = \left(\sum_{k=1}^{K} \frac{2\tilde{p}\psi_k |a_k|^2 \theta^2}{\tilde{p}\psi_k \sigma_\eta^2 + \sigma_v^2} \right)^{1/2}. \tag{5.30}$$

Proof: Refer to Chawla et al. [2022] for the proof.

5.5.2 Optimal Transmit Gains

Considering sensor transmit power as $p_k = |f_k|^2$, power scaling as $p_k = \tilde{p}_k/M$, where $\tilde{p}_k = \left(\sqrt{\frac{|a_k|^2\sigma_v^2}{\tilde{\lambda}\psi_k\sigma_\eta^4}} - \frac{\sigma_v^2}{\psi_k\sigma_\eta^2} \right)^+$, and modified KKT multiplier as $\tilde{\lambda} = \lambda/M$, the value of $\tilde{\lambda}$ can be obtained through a straightforward bisection search over $[\tilde{\lambda}_l, \tilde{\lambda}_u]$, where $\tilde{\lambda}_l = \min_k . \left\{ \frac{\psi_k |a_k|^2 \sigma_v^2}{(P\psi_k\sigma_\eta^2 + \sigma_v^2)^2} \right\}$ and $\tilde{\lambda}_u = \max_k . \left\{ \frac{\psi_k |a_k|^2}{\sigma_v^2} \right\}$. Using the above quantities, the asymptotic probabilities, $P^a_{D,o}$ and $P^a_{FA,o}$, for the optimal sensor transmit gains can be determined by substituting $\tilde{p}$ with $\tilde{p}_k$ in (5.30).

For uniform and optimal sensor transmit gains, the asymptotic performance of the D-MIMO detector $T_{D,UP}(\mathbf{y}_D)$ in (5.19), as $N_d \to \infty$, mirrors that of the C-MIMO. From the results, it can be deduced that scaling down the sensor transmit power by $1/M$ and $1/N_d$ for C-MIMO and D-MIMO systems, respectively, does not degrade the detection performance. This allows for significant power savings, crucial for extending sensor lifetime in practical WSN deployments within future wireless networks.

5.6 SBL-Based CSI Estimation

In this section, we discuss the SBL framework [Wipf and Rao, 2004] for CSI acquisition and decision rules for both antenna topologies. We consider an M size angular grid to characterize the unknown CSI. The AoAs are selected as $\boldsymbol{\Phi}_R = \{\phi_v : \phi_v \in [0, \pi], \forall 1 \leq v \leq M\}$ such that $\sin(\phi_v) = \frac{2}{M}(v-1) - 1, \forall v$ [Lee et al., 2016]. Employing these quantized angles, the receive array response dictionary matrix $\mathbf{A}_{R,C} \in \mathbb{C}^{M \times M}$ can be defined as $\mathbf{A}_{R,C} = [\mathbf{a}_{R,C}(\phi_1), \mathbf{a}_{R,C}(\phi_2), \ldots, \mathbf{a}_{R,C}(\phi_M)]$, where $\mathbf{A}_{R,C}\mathbf{A}_{R,C}^H = \mathbf{A}_{R,C}^H\mathbf{A}_{R,C} = \mathbf{I}_M$. Consequently, the channel matrix can be characterized as $\mathbf{G} = \mathbf{A}_{R,C}\mathbf{H}_b$, with $\mathbf{H}_b = [\mathbf{h}_{b,1}, \mathbf{h}_{b,2}, \ldots, \mathbf{h}_{b,K}]$ being the equivalent beamspace channel matrix. The kth sensor channel vector is $\mathbf{g}_k = \mathbf{A}_{R,C}\mathbf{h}_{b,k}$ and the vectorized version of $\mathbf{G}$, denoted by $\mathbf{g} \in \mathbb{C}^{MK \times 1}$, can be expressed as $\mathbf{g} = \text{vec}(\mathbf{G}) = \boldsymbol{\Xi}_C\mathbf{h}_b$, where $\boldsymbol{\Xi}_C = [\mathbf{I}_K \otimes \mathbf{A}_{R,C}] \in \mathbb{C}^{MK \times MK}$. Moreover, the beamspace channel vector $\mathbf{h}_b = \text{vec}(\mathbf{H}_b)$ is obtained by the columnwise stacking of the vectors in $\mathbf{H}_b$.

For the D-MIMO architecture, we utilize an N_d size angular grid and AoAs are chosen from $\boldsymbol{\Phi}_R = \{\phi_n : \phi_n \in [0, \pi], \forall 1 \leq n \leq N_d\}$. The angle selection procedure is akin to the C-MIMO scenario. Further, the vectorized channel vector $\mathbf{g}_j \in \mathbb{C}^{N_d K \times 1}$ is expressed as $\mathbf{g}_j = \text{vec}(\mathbf{G}_j) = \boldsymbol{\Xi}_D\mathbf{h}_{b,j}$, where $\mathbf{h}_{b,j} = \text{vec}(\mathbf{H}_{b,j})$ denotes the beamspace channel vector at the jth FC, $\boldsymbol{\Xi}_D = [\mathbf{I}_K \otimes \mathbf{A}_{R,D}] \in \mathbb{C}^{N_d K \times N_d K}$ and $\mathbf{A}_{R,D} = [\mathbf{a}_{R,D}(\phi_1), \ldots, \mathbf{a}_{R,D}(\phi_{N_d})] \in \mathbb{C}^{N_d \times N_d}$ represents the receive array response dictionary matrix.

Considering $M_f = M/K$ combining frames for channel estimation, the training RF combiner $\mathbf{F}_C^{(l)} \in \mathbb{C}^{M \times K}$ during the lth frame is chosen as an $M \times K$-submatrix of the normalized discrete Fourier transform (DFT) matrix $\mathbf{F} \in \mathbb{C}^{M \times M}$. Hence, the RF combiner output $\mathbf{Y}^{(l)} \in \mathbb{C}^{K \times K}$ at the FC can be given as $\mathbf{Y}^{(l)} = \sqrt{p_p}\left(\mathbf{F}_C^{(l)}\right)^H \mathbf{G}\mathbf{X}_p + \left(\mathbf{F}_C^{(l)}\right)^H \mathbf{N}$, where p_p is the training power, $\mathbf{N} = [\mathbf{n}_1, \ldots, \mathbf{n}_K]$ is the AWGN matrix with $\mathbf{n}_k \sim \mathcal{CN}\left(\mathbf{0}, \sigma_n^2\mathbf{I}_M\right)$ and the training matrix $\mathbf{X}_p$ is chosen as $\mathbf{I}_K$. The concatenated RF combiner output $\mathbf{Y} = \left[(\mathbf{Y}^{(1)})^T, \ldots, (\mathbf{Y}^{(M_f)})^T\right]^T$, corresponding to M_f frames, can be expressed as

$$\mathbf{Y} = \sqrt{p_p}\mathbf{F}^H\mathbf{G}\mathbf{X}_p + \overline{\mathbf{N}}, \tag{5.31}$$

where $\mathbf{F} = \left[\mathbf{F}_C^{(1)}, \ldots, \mathbf{F}_C^{(M_f)}\right]$ and $\overline{\mathbf{N}} = \left[\mathbf{N}^H\mathbf{F}_C^{(1)}, \ldots, \mathbf{N}^H\mathbf{F}_C^{(M_f)}\right]^H$. Further, the vectorized version of the RF combiner output $\mathbf{Y}$ can be given as

$$\mathbf{y} = \sqrt{p_p}\boldsymbol{\Phi}\boldsymbol{\Xi}_C\mathbf{h}_b + \tilde{\mathbf{n}} = \sqrt{p_p}\mathbf{S}\mathbf{h}_b + \tilde{\mathbf{n}}, \tag{5.32}$$

where the equivalent noise $\tilde{\mathbf{n}} = \text{vec}(\mathbf{F}^H\mathbf{N})$ is distributed as $\tilde{\mathbf{n}} \sim \mathcal{CN}(\mathbf{0}, \mathbf{C}_{\tilde{\mathbf{n}}})$, with $\mathbf{C}_{\tilde{\mathbf{n}}} = \sigma_n^2\mathbf{I}_{MK}$, $\boldsymbol{\Phi} = (\mathbf{X}_p^T \otimes \mathbf{F}^H) \in \mathbb{C}^{MK \times MK}$ and $\mathbf{S} \triangleq \boldsymbol{\Phi}\boldsymbol{\Xi}_C \in \mathbb{C}^{MK \times MK}$ denotes the equivalent sensing matrix satisfying $\mathbf{S}^H\mathbf{S} = \mathbf{I}_{MK}$. Utilizing the received signal $\mathbf{y}$, the step-by-step procedure to obtain the SBL-based beamspace channel estimate $\hat{\mathbf{h}}_b$ is outlined in Algorithm 5.1.

Algorithm 5.1 SBL-based mmWave massive MIMO channel estimation

Input: Sensing matrix $\mathbf{S}$, pilot power p_p, pilot output $\mathbf{y}$ and stopping parameter ϵ

Initialization: $\widehat{\mathbf{\Gamma}}^{(0)} = \mathbf{I}_{MK}$
Fix $\widehat{\mathbf{\Gamma}}^{(-1)} = \mathbf{0}_{MK \times MK}$ and $m = 0$
while $\left\| \widehat{\mathbf{\Gamma}}^{(m)} - \widehat{\mathbf{\Gamma}}^{(m-1)} \right\|_F > \epsilon$ **do**
  E-step: Calculate *a posteriori* mean and covariance
    $\boldsymbol{\mu}^{(m)} = \sqrt{p_p}\,\mathbf{\Sigma}^{(m)}\mathbf{S}^H \mathbf{C}_{\tilde{\mathbf{n}}}^{-1}\mathbf{y};\ \mathbf{\Sigma}^{(m)} = \left(p_p\mathbf{S}^H\mathbf{C}_{\tilde{\mathbf{n}}}^{-1}\mathbf{S} + (\widehat{\mathbf{\Gamma}}^{(m)})^{-1}\right)^{-1}$
  M-step: Perform hyperparameter estimation
    for $i = 1, 2, \ldots, MK$ **do**
      $\left[\widehat{\mathbf{\Gamma}}^{(m+1)}\right]_{i,i} = |\boldsymbol{\mu}^{(m)}(i)|^2 + \left[\mathbf{\Sigma}^{(m)}\right]_{i,i}$
    end for
  $m \leftarrow m + 1$
end while

Output: $\hat{\mathbf{h}}_b = \boldsymbol{\mu}^{(m)}$

5.6.1 C-MIMO: Fusion Rule for Imperfect CSI

On employing Algorithm 5.1, the SBL-based beamspace channel estimate, upon convergence, for the centralized topology is given by $\hat{\mathbf{h}}_b = \boldsymbol{\mu}^{(m)}$, with $\boldsymbol{\mu}^{(m)}$ being the *a posteriori* mean. Additionally, the *a posteriori* covariance matrix $\mathbf{\Sigma} = \mathbf{\Sigma}^{(m)} \in \mathbb{C}^{MK \times MK}$ is diagonal because $\widehat{\mathbf{\Gamma}}^{(m)}$, $\mathbf{S}^H\mathbf{S} = \mathbf{I}_{MK}$ and $\mathbf{C}_{\tilde{\mathbf{n}}} = \sigma_n^2\mathbf{I}_{MK}$ are diagonal. Hence, the SBL estimate of $\mathbf{G}$ can be expressed as $\widehat{\mathbf{G}} = \mathbf{A}_{R,C}\widehat{\mathbf{H}}_b$, where $\widehat{\mathbf{H}}_b = \text{vec}^{-1}(\hat{\mathbf{h}}_b)$. Further, the estimation error $\boldsymbol{\mathcal{E}} = [\mathbf{e}_1, \ldots, \mathbf{e}_K]$ can be defined as $\boldsymbol{\mathcal{E}} = \widehat{\mathbf{G}} - \mathbf{G} = \mathbf{A}_{R,C}\boldsymbol{\mathcal{E}}_b$, where $\boldsymbol{\mathcal{E}}_b = \widehat{\mathbf{H}}_b - \mathbf{H}_b$ is the beamspace estimation error. The estimation error of the kth sensor is distributed as $\mathbf{e}_k \sim \mathcal{CN}(\mathbf{0}, \mathbf{C}_{\mathbf{e}_k})$, with $\mathbf{C}_{\mathbf{e}_k} = \mathbf{A}_{R,C}\mathbf{\Sigma}_k\mathbf{A}_{R,C}^H$ being the error covariance matrix of the kth sensor and $\mathbf{\Sigma}_k = \mathbf{\Sigma}\left[(k-1)M + 1 : kM, (k-1)M + 1 : kM\right]$. Hence, the received signal under imperfect CSI conditions can be reformulated as

$$\mathbf{z}_C = \widehat{\mathbf{G}}\mathbf{F}\mathbf{a}\Theta - \boldsymbol{\mathcal{E}}\mathbf{F}\mathbf{a}\Theta + \widehat{\mathbf{G}}\mathbf{F}\boldsymbol{\eta} - \boldsymbol{\mathcal{E}}\mathbf{F}\boldsymbol{\eta} + \mathbf{v}. \tag{5.33}$$

Even for the imperfect CSI scenario, we utilize a similar two-step architecture to minimize the detector's complexity. The first step involves a hybrid combiner, defined as $\mathbf{W}_C = \tilde{\mathbf{W}}_{RF,C}\tilde{\mathbf{W}}_{BB,C} \in \mathbb{C}^{M \times K}$, at the FC. The RF combiner $\tilde{\mathbf{W}}_{RF,C} = [\mathbf{a}_{R,C}(\phi_{l_1}), \mathbf{a}_{R,C}(\phi_{l_2}), \ldots, \mathbf{a}_{R,C}(\phi_{l_K})]$ is determined by concatenating K receive array response vectors having the maximum estimated path gains in

$\widehat{\mathbf{H}}_b$. Further, the index i_k corresponds to the maximum estimated path gain in the kth column of $\widehat{\mathbf{H}}_b$ and the baseband combiner $\widetilde{\mathbf{W}}_{BB,C}$ is defined as $\widetilde{\mathbf{W}}_{BB,C} = \widetilde{\mathbf{W}}_{RF,C}^H \widehat{\mathbf{G}} \in \mathbb{C}^{K \times K}$.

Leveraging the asymptotic orthogonality characteristic of the mmWave massive MIMO channel and considering distinct AoAs for all K sensors, the product of matrices $\widetilde{\mathbf{W}}_{RF,C}^H \mathbf{A}_{R,C}$ yields a permutation matrix. In each row of this matrix, K entries are unity, while the remaining entries are approximately 0. Consequently, the matrix $\widetilde{\mathbf{W}}_{BB,C}$ becomes diagonal and its kth diagonal element is $[\widetilde{\mathbf{W}}_{BB,C}]_{k,k} = \hat{h}_{i_k,k} = [\widehat{\mathbf{H}}_b]_{i_k,k}$. Following the hybrid combining, the resulting output signal $\tilde{\mathbf{y}}_C$ can be formulated as

$$\tilde{\mathbf{y}}_C = \check{\mathbf{G}}\mathbf{F}\mathbf{a}\Theta - \mathbf{W}_C^H \mathcal{E}\mathbf{F}\mathbf{a}\Theta + \tilde{\mathbf{v}},$$

where $\check{\mathbf{G}} = \mathbf{W}_C^H \widehat{\mathbf{G}}$ is the effective channel and $\tilde{\mathbf{v}} = \mathbf{W}_C^H(\widehat{\mathbf{G}}\mathbf{F}\boldsymbol{\eta} - \mathcal{E}\mathbf{F}\boldsymbol{\eta} + \mathbf{v})$. Under both hypotheses, the signal $\tilde{\mathbf{y}}_C$ is distributed as

$$\mathcal{H}_0 : \tilde{\mathbf{y}}_C \sim \mathcal{CN}\left(\mathbf{0}, \mathbf{C}_{\tilde{\mathbf{v}}}\right), \quad \mathcal{H}_1 : \tilde{\mathbf{y}}_C \sim \mathcal{CN}\left(\check{\mathbf{G}}\mathbf{F}\mathbf{a}\theta, \mathbf{C}_{\tilde{\boldsymbol{\eta}}}\right), \tag{5.34}$$

where $\mathbf{C}_{\tilde{\mathbf{v}}} = \sigma_\eta^2 \check{\mathbf{G}}\mathbf{F}\mathbf{F}^H \check{\mathbf{G}}^H + \sigma_\eta^2 \mathbf{W}_C^H \mathbf{C}_1 \mathbf{W}_C + \sigma_v^2 \mathbf{W}_C^H \mathbf{W}_C$, $\mathbf{C}_{\tilde{\boldsymbol{\eta}}} = \mathbf{W}_C^H \mathbf{C}_2 \mathbf{W}_C + \mathbf{C}_{\tilde{\mathbf{v}}}$, $\mathbf{C}_1 = \mathbf{A}_{R,C}(\sum_{k=1}^K |f_k|^2 \boldsymbol{\Sigma}_k)\mathbf{A}_{R,C}^H$, and $\mathbf{C}_2 = \mathbf{A}_{R,C}(\sum_{k=1}^K |a_k|^2 |f_k|^2 \boldsymbol{\Sigma}_k)\mathbf{A}_{R,C}^H$. The covariance matrices are diagonal, with $\mathbf{C}_{\tilde{\mathbf{v}}} = \text{diag}(\sigma_{\tilde{v},1}^2, \sigma_{\tilde{v},2}^2, \dots, \sigma_{\tilde{v},K}^2)$ and $\mathbf{C}_{\tilde{\boldsymbol{\eta}}} = \text{diag}(\sigma_{\tilde{\eta},1}^2, \sigma_{\tilde{\eta},2}^2, \dots, \sigma_{\tilde{\eta},K}^2)$, indicating the independence of observations across different sensors. Employing the MLE [Kay, 1993b], the unknown parameter θ is estimated. Using these quantities, the GLRT statistic can be formulated as

$$T_{C,UIP}(\tilde{\mathbf{y}}_C) = \tilde{\mathbf{y}}_C^H(\mathbf{C}_{\tilde{\mathbf{v}}}^{-1} - \mathbf{C}_{\tilde{\boldsymbol{\eta}}}^{-1})\tilde{\mathbf{y}}_C + 2\Re(\tilde{\mathbf{y}}_C^H \mathbf{C}_{\tilde{\boldsymbol{\eta}}}^{-1} \check{\mathbf{G}}\mathbf{F}\mathbf{a}\hat{\theta}) - \hat{\theta}^2 \mathbf{a}^H \mathbf{F}^H \check{\mathbf{G}}^H \mathbf{C}_{\tilde{\boldsymbol{\eta}}}^{-1} \check{\mathbf{G}}\mathbf{F}\mathbf{a}.$$

On substituting $\hat{\theta} = \frac{\Re(\tilde{\mathbf{y}}_C^H \mathbf{C}_{\tilde{\boldsymbol{\eta}}}^{-1} \check{\mathbf{G}}\mathbf{F}\mathbf{a})}{\mathbf{a}^H \mathbf{F}^H \check{\mathbf{G}}^H \mathbf{C}_{\tilde{\boldsymbol{\eta}}}^{-1} \check{\mathbf{G}}\mathbf{F}\mathbf{a}}$ in the above expression, it reduces to

$$T_{C,UIP}(\tilde{\mathbf{y}}_C) = \tilde{\mathbf{y}}_C^H(\mathbf{C}_{\tilde{\mathbf{v}}}^{-1} - \mathbf{C}_{\tilde{\boldsymbol{\eta}}}^{-1})\tilde{\mathbf{y}}_C + \frac{[\Re(\tilde{\mathbf{y}}_C^H \mathbf{C}_{\tilde{\boldsymbol{\eta}}}^{-1} \check{\mathbf{G}}\mathbf{F}\mathbf{a})]^2}{\mathbf{a}^H \mathbf{F}^H \check{\mathbf{G}}^H \mathbf{C}_{\tilde{\boldsymbol{\eta}}}^{-1} \check{\mathbf{G}}\mathbf{F}\mathbf{a}}. \tag{5.35}$$

It is mathematically challenging to characterize the detection performance of the above test [Chawla et al., 2021a]. Hence, it can be approximated as

$$T_{C,UIP}(\tilde{\mathbf{y}}_C) \approx \tilde{\mathbf{y}}_C^H \mathbf{X}_C \tilde{\mathbf{y}}_C \underset{\mathcal{H}_0}{\overset{\mathcal{H}_1}{\gtrless}} \tilde{\gamma}, \tag{5.36}$$

where $\mathbf{X}_C = (\mathbf{C}_{\tilde{\mathbf{v}}}^{-1} - \mathbf{C}_{\tilde{\boldsymbol{\eta}}}^{-1}) + \frac{\mathbf{C}_{\tilde{\boldsymbol{\eta}}}^{-1} \check{\mathbf{G}}\mathbf{F}\mathbf{a}\mathbf{a}^H \mathbf{F}^H \check{\mathbf{G}}^H \mathbf{C}_{\tilde{\boldsymbol{\eta}}}^{-1}}{\mathbf{a}^H \mathbf{F}^H \check{\mathbf{G}}^H \mathbf{C}_{\tilde{\boldsymbol{\eta}}}^{-1} \check{\mathbf{G}}\mathbf{F}\mathbf{a}}$. Further, the detection performance of $T_{C,UIP}(\tilde{\mathbf{y}}_C)$, characterized by P_D and P_{FA}, can be expressed as

$$P_D \approx Q_{\chi^2_{l_{C,UD}}(\lambda_{C,UD})}(\tilde{\gamma}), \quad P_{FA} \approx Q_{\chi^2_{l_{C,UF}}(\lambda_{C,UF})}(\tilde{\gamma}). \tag{5.37}$$

Here, $l_{C,UD}$, $l_{C,UF}$ and $\lambda_{C,UD}$, $\lambda_{C,UF}$ denote the degrees of freedom and the non-centrality parameters of the chi-squared random variables $\chi^2_{l_{C,UD}}(\lambda_{C,UD})$ and

$\chi^2_{l_{C,UF}}(\lambda_{C,UF})$, respectively. Utilizing the first four cumulants of $T_{C,UIP}(\tilde{\mathbf{y}}_C)$, these quantities can be obtained using the steps given in Chawla et al. [2022].

5.6.2 D-MIMO: Fusion Rule for Imperfect CSI

The estimated beamspace mmWave channel at the jth FC in the distributed architecture, upon convergence, can be expressed as $\hat{\mathbf{h}}_{b,j} = \boldsymbol{\mu}_j^{(m)}$, where $\boldsymbol{\mu}_j^{(m)}$ denotes the *a posteriori* mean. Further, the SBL estimate corresponds to the jth FC can be given as $\hat{\mathbf{G}}_j = \mathbf{A}_{R,D}\hat{\mathbf{H}}_{b,j}$, where $\hat{\mathbf{H}}_{b,j} = \text{vec}^{-1}(\hat{\mathbf{h}}_{b,j})$. The associated estimation error can be defined as $\mathcal{E}_j = [\mathbf{e}_{1,j}, \dots, \mathbf{e}_{K,j}] = \mathbf{A}_{R,D}\mathcal{E}_{b,j}$, where $\mathcal{E}_{b,j} = \hat{\mathbf{H}}_{b,j} - \mathbf{H}_{b,j}$ denotes the beamspace estimation error. Moreover, the kth sensor estimation error is distributed as $\mathbf{e}_{k,j} \sim \mathcal{CN}(\mathbf{0}, \mathbf{C}_{\mathbf{e}_{k,j}})$, where $\mathbf{C}_{\mathbf{e}_{k,j}} = \mathbf{A}_{R,D}\boldsymbol{\Sigma}_{k,j}\mathbf{A}_{R,D}^H$ and $\boldsymbol{\Sigma}_{k,j} = \boldsymbol{\Sigma}_j\left[(k-1)M+1 : kM, (k-1)M+1 : kM\right] \in \mathbb{C}^{N_d \times N_d}$ represents the *a posteriori* covariance matrix for the kth sensor and the jth FC. Using the above quantities, the received signal $\mathbf{z}_{D,j}$ at the jth FC is

$$\mathbf{z}_{D,j} = (\hat{\mathbf{G}}_j - \mathcal{E}_j)\mathbf{F}\boldsymbol{\eta} + (\hat{\mathbf{G}}_j - \mathcal{E}_j)\mathbf{F}\mathbf{a}\Theta + \mathbf{v}_j. \tag{5.38}$$

Next, the received signal at thc jth FC is processed using an RF combiner, defined as $\tilde{\mathbf{w}}_{RF,j} = \mathbf{a}_{R,D}(\phi_{i_{k_j}})$, where the index i_{k_j} corresponds to the maximum estimated gain in the k_jth column of $\hat{\mathbf{H}}_{b,j}$. Subsequently, all K RF combiner outputs are consolidated at the BPU and then processed using the correction matrix $\mathbf{Q}$ to rearrange the observations based on the sensors. These rearranged observations are then combined using the baseband combiner $\tilde{\mathbf{W}}_{BB,D} = \mathbf{Q}\tilde{\mathbf{W}}_{RF,D}^H\check{\mathbf{G}} \in \mathbb{C}^{K\times K}$, where $\check{\mathbf{G}} = [\hat{\mathbf{G}}_1^T, \hat{\mathbf{G}}_2^T, \dots, \hat{\mathbf{G}}_K^T]^T$ denotes the stacked channel matrix and $\tilde{\mathbf{W}}_{RF,D} = \text{diag}\{\tilde{\mathbf{w}}_{RF_1}, \tilde{\mathbf{w}}_{RF_2}, \dots, \tilde{\mathbf{w}}_{RF_K}\} \in \mathbb{C}^{N_d K\times K}$. Further, the baseband combiner becomes diagonal on leveraging the asymptotic orthogonality property of mmWave massive MIMO channel such that $[\tilde{\mathbf{W}}_{BB,D}]_{k,k} = \hat{h}_{i_k,k,j_k} = [\hat{\mathbf{H}}_{b,j_k}]_{i_k,k}$. The hybrid combiner output $\tilde{\mathbf{y}}_D$ is characterized as

$$\tilde{\mathbf{y}}_D = \tilde{\mathbf{W}}_D^H\left[(\check{\mathbf{G}} - \tilde{\mathcal{E}})\mathbf{F}\boldsymbol{\eta} + (\check{\mathbf{G}} - \tilde{\mathcal{E}})\mathbf{F}\mathbf{a}\Theta + \tilde{\mathbf{v}}\right], \tag{5.39}$$

where the quantities $\tilde{\mathcal{E}} = \left[\mathcal{E}_1^T, \mathcal{E}_1^T, \dots, \mathcal{E}_K^T\right]^T$, $\tilde{\mathbf{W}}_D = \tilde{\mathbf{W}}_{RF,D}\mathbf{Q}^H\tilde{\mathbf{W}}_{BB,D}$, and $\tilde{\mathbf{v}} = \left[\mathbf{v}_1^T, \mathbf{v}_2^T, \dots, \mathbf{v}_K^T\right]^T$. Defining $\check{\mathbf{G}}_D = \tilde{\mathbf{W}}_D^H\check{\mathbf{G}} \in \mathbb{C}^{K\times K}$ as the equivalent channel matrix and $\check{\mathbf{v}} = \tilde{\mathbf{W}}_D^H[(\check{\mathbf{G}} - \tilde{\mathcal{E}})\mathbf{F}\boldsymbol{\eta} + \tilde{\mathbf{v}}]$ as the equivalent noise, the signal in (5.39) is distributed as

$$\mathcal{H}_0 : \tilde{\mathbf{y}}_D \sim \mathcal{CN}\left(\mathbf{0}, \mathbf{C}_{\check{\mathbf{v}}}\right), \quad \mathcal{H}_1 : \tilde{\mathbf{y}}_D \sim \mathcal{CN}(\check{\mathbf{G}}_D\mathbf{F}\mathbf{a}\theta, \mathbf{C}_{\check{\boldsymbol{\eta}}}). \tag{5.40}$$

The covariance matrices $\mathbf{C}_{\check{\mathbf{v}}} = \sigma_\eta^2\tilde{\mathbf{W}}_D^H\mathbf{C}_3\tilde{\mathbf{W}}_D + \sigma_\eta^2\check{\mathbf{G}}_D\mathbf{F}\mathbf{F}^H\check{\mathbf{G}}_D^H + \sigma_v^2\tilde{\mathbf{W}}_D^H\tilde{\mathbf{W}}_D$ and $\mathbf{C}_{\check{\boldsymbol{\eta}}} = \tilde{\mathbf{W}}_D^H\mathbf{C}_4\tilde{\mathbf{W}}_D^H + \mathbf{C}_{\check{\mathbf{v}}}$ are diagonal, where $\mathbf{C}_3 = \boldsymbol{\Xi}_D\mathbf{C}_{\boldsymbol{\Sigma}}\boldsymbol{\Xi}_D^H$, $\mathbf{C}_{\boldsymbol{\Sigma}} = \text{diag}(\sum_{k=1}^K(|f_k|^2\boldsymbol{\Sigma}_{k,1}), \dots, \sum_{k=1}^K(|f_k|^2\boldsymbol{\Sigma}_{k,K}))$, and $\mathbf{C}_4$ is determined by substituting f_k with $a_k f_k$ in $\mathbf{C}_3$. Consequently, K sensors have independent soft decisions. Defining

$\mathbf{C}_{\check{\eta}} = \text{diag}(\sigma^2_{\check{\eta},1}, \dots, \sigma^2_{\check{\eta},K})$ and $\mathbf{C}_{\check{\mathbf{v}}} = \text{diag}(\sigma^2_{\check{v},1}, \dots, \sigma^2_{\check{v},K})$, the unknown parameter estimate is derived utilizing the MLE [Kay, 1993b]. On substituting $\hat{\theta}$, the GLRT statistic for the D-MIMO architecture can be given as

$$T_{D,UIP}(\tilde{\mathbf{y}}_D) \approx \tilde{\mathbf{y}}_D^H \mathbf{X}_D \tilde{\mathbf{y}}_D \underset{\mathcal{H}_0}{\overset{\mathcal{H}_1}{\gtrless}} \tilde{\gamma}, \tag{5.41}$$

where $\mathbf{X}_D = \left(\mathbf{C}_{\check{\mathbf{v}}}^{-1} - \mathbf{C}_{\check{\eta}}^{-1}\right) + \frac{\mathbf{C}_{\check{\eta}}^{-1}\check{\mathbf{G}}_D\mathbf{F}\mathbf{a}\mathbf{a}^H\mathbf{F}^H\check{\mathbf{G}}_D^H\mathbf{C}_{\check{\eta}}^{-1}}{\mathbf{a}^H\mathbf{F}^H\check{\mathbf{G}}_D^H\mathbf{C}_{\check{\eta}}^{-1}\check{\mathbf{G}}_D\mathbf{F}\mathbf{a}}$. Further, the system performance can be characterized in terms of P_D and P_{FA}, as

$$P_D = Q_{\chi^2_{l_{D,UD}}(\lambda_{D,UD})}(\tilde{\gamma}), \quad P_{FA} = Q_{\chi^2_{l_{D,UF}}(\lambda_{D,UF})}(\tilde{\gamma}), \tag{5.42}$$

where $\chi^2_{l_{D,UD}}(\lambda_{D,UD})$ and $\chi^2_{l_{D,UF}}(\lambda_{D,UF})$ denote the noncentral chi-squared random variables with degrees of freedom as $l_{D,UD}$ and $l_{D,UF}$ and noncentrality parameters as $\lambda_{D,UD}$ and $\lambda_{D,UF}$, respectively.

5.7 Simulation Results

This section demonstrates the effectiveness of the proposed C-MIMO and D-MIMO detectors in various scenarios. For the centralized architecture, consider a ring-shaped cell containing $K = 12$ sensors dispersed randomly within the region $[r_{min}, R]$. Here, cell radius R and the minimum distance between the FC and the sensors r_{min} are selected as $R = 200$ m and $r_{min} = 1$ m [Chawla et al., 2021b]. For the distributed topology, the sensors are assumed to be randomly distributed within the range $[0, \rho - r_{min}] \cup [\rho + r_{min}, R]$, with $\rho = 0.6R$ being the radius of the circle where the FCs are positioned [Li et al., 2018].

Defining z_k as log-normal random variable with mean $\mu_z = 4$ dB and standard deviation $\sigma_z = 2$ dB, the large-scale fading coefficient for C-MIMO can be defined as $\beta_k = z_k(r_{min}/r_k)^\nu$, where the path-loss exponent ν is chosen as $\nu = 2$ [Ngo et al., 2013]. For D-MIMO, the large-scale fading coefficient for the kth sensor and the jth FC is defined as $\beta_{k,j} = z_{k,j}(r_{min}/\delta_{k,j})^\nu$, where $z_{k,j}$ is log-normal distributed having mean μ_z and standard deviation σ_z. The carrier frequency f_c is chosen as $f_c =$ 28 GHz, $d = \lambda/2$ and the noise variances at the FC σ_v^2 and the sensors σ_η^2 are chosen as 0.7 and 0.2, respectively.

In Figure 5.4a, the receiver operating characteristic (ROC) comparison is illustrated for the C-MIMO detector, given in (5.11), for the unknown parameter detection. Notably, the simulation plots closely align with the analytical curves, affirming the validity of the analytical expressions. Additionally, an improved detector performance is observed, achieved through optimal sensor gain allocation, compared to the detector employing uniform transmit gains. Figure 5.4b analyzes the asymptotic performance of the proposed detector in (5.11).

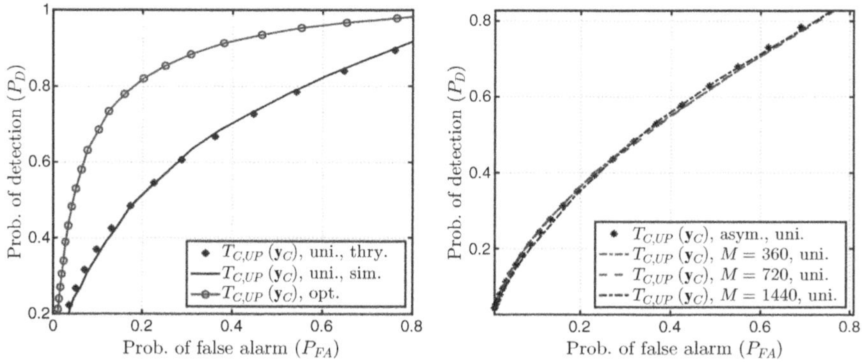

(a) (b)

Figure 5.4 P_D versus P_{FA} plots of $T_{C,UP}(\mathbf{y}_C)$ for comparing (a) theoretical and simulated performance of uniform and optimal sensor gain with $M = 250$ antennas; (b) asymptotic performance for uniform sensor transmit gains with $M \in \{360, 720, 1440\}$ and $P = 20$ dB.

The performance is contrasted against the associated asymptotic theoretical expressions determined in Lemma 5.1. The findings indicate that as the sensor transmit power is scaled as $p = \tilde{p}/M$, the simulated plots converge toward the asymptotic counterpart, promising substantial power savings in practical WSNs.

For the distributed architecture, Figure 5.5 investigates the detection performance of the proposed detector in (5.19). In Figure 5.5a, a close alignment is observed between the simulated plots and the analytical results. Additionally, optimal gain allocation leads to a substantial enhancement in detection performance

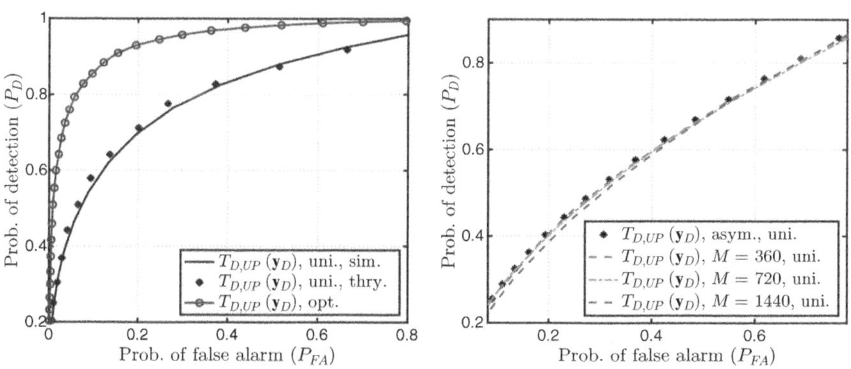

(a) (b)

Figure 5.5 P_D versus P_{FA} plots of $T_{D,UP}(\mathbf{y}_D)$ in (5.19) for comparing (a) theoretical and simulated performance of uniform and optimal sensor gain with $N_T = 480$; (b) asymptotic performance for uniform sensor transmit gains with $N_T \in \{360, 720, 1440\}$ and $P = 10$ dB.

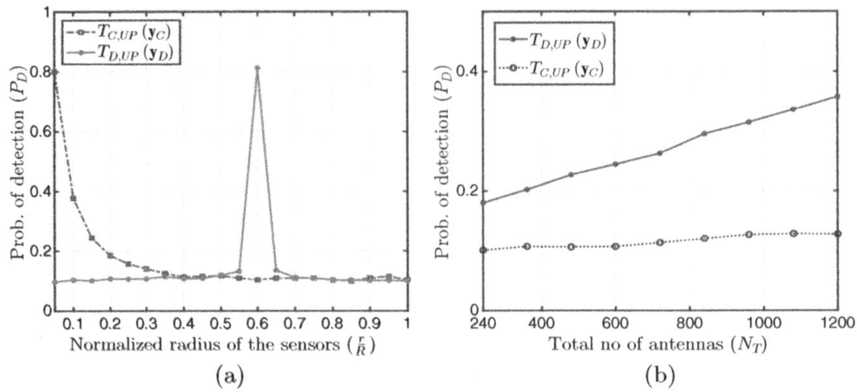

Figure 5.6 At $P_{FA} = 0.1$, detection performance is compared with (a) normalized sensor radius ρ/R for $P = 1$, $N_d = 40$, $K = 12$, $N_T = M = 480$; (b) total number of FC antennas (N_T), where sensors are distributed in the region $[0.5R, R]$.

compared to the uniform transmit gain allocation scenario. Figure 5.5b depicts the asymptotic performance of the D-MIMO detector with uniform transmit gains. Notably, the ROC plots converge toward the asymptotic counterpart as the sensor transmit gain is scaled as $p_k = \tilde{p}_k/N_d$, for $N_T \in \{360, 720, 1440\}$.

Figure 5.6a analyzes the impact of normalized radius r/R on detection performance, when sensors are uniformly distributed on a circle ($r_k = r, 1 \leq k \leq K$). A comparison is drawn between C-MIMO and D-MIMO detectors, derived in (5.11) and (5.19), respectively. Simulations, conducted with $M = N_T = 480$ and $P_{FA} = 0.1$, highlight the superiority of D-MIMO performance over C-MIMO in the region $[0.5R, R]$. Notably, D-MIMO peaks at $r/R = 0.6$ due to sensor proximity to their operating FCs located on the circle of radius $\rho = 0.6R$, leading to substantial enhancement in detection performance. In Figure 5.6b, the impact of increasing the number of antennas at the FC on the detection performance is investigated for both D-MIMO and C-MIMO detectors, considering sensor distribution in the region $[0.5R, R]$ at $P_{FA} = 0.1$. The results demonstrate that the D-MIMO detector exhibits improved performance as the total number of antennas N_T increases. This improvement is attributed to the allocation of more antennas to each FC (N_d increases as N_T increases). The performance is primarily influenced by the distribution range of the sensors. Notably, the improvement is less significant for C-MIMO, where sensors are distant from the FC located at the cell center, and even an increase in the number of antennas does not contribute substantially to the detection performance.

Considering signal to noise ratio (SNR) = 15 dB and $M = 240$, Figure 5.7a demonstrates the ROC plots for the C-MIMO tests, determined in (5.11) and (5.36), for perfect and imperfect CSI scenarios, respectively. Moreover, Figure 5.7b

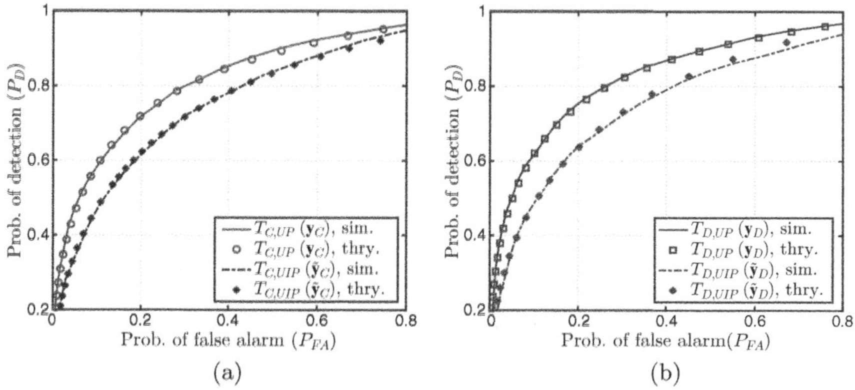

Figure 5.7 Theoretical and simulated performance is compared via ROC plots for (a) C-MIMO detectors in (5.11) and (5.36), with SNR = 15 dB and $M = 240$; (b) D-MIMO detectors in (5.19) and (5.41), for $N_T = 720$ and SNR = 15 dB.

analyzes the performance of the D-MIMO detectors in (5.19) and (5.41), for perfect and imperfect CSI scenarios, respectively, with SNR = 15 dB and $N_T = 720$. It can be concluded from both plots that the theoretical approximation of the likelihood ratio test (LRT) aligns closely with the simulated plot, thus substantiating the analytical results, even for the imperfect CSI scenario.

5.8 Conclusions

In this chapter, we investigated the hybrid combining-based low-complexity fusion rules for unknown parameter detection in the next-generation WSNs employing mmWave and massive MIMO technologies. The study explores scenarios with both perfect and imperfect CSI, encompassing both centralized and distributed antenna topologies. Simulation results demonstrated the superior performance of the D-MIMO detector over the C-MIMO detector. This advantage is primarily attributed to the proximity of the sensors to their respective FCs in the distributed architecture. The readers are encouraged to read related works [Chawla et al., 2018, 2021a,b–2022] to have a deeper understanding of the discussed topics.

Bibliography

I. F. Akyildiz, W. Su, Y. Sankarasubramaniam, and E. Cayirci. A survey on sensor networks. *IEEE Communications Magazine*, 40(8):102–114, 2002. doi: 10.1109/MCOM.2002.1024422.

S. Boyd and L. Vandenberghe. *Convex Optimization*. Cambridge University Press, New York, NY, USA, 2004. ISBN 0521833787.

A. Chawla, A. Patel, A. K. Jagannatham, and P. K. Varshney. Robust distributed detection in massive MIMO wireless sensor networks under CSI uncertainty. In *2018 IEEE 88th Vehicular Technology Conference (VTC-Fall)*, pages 1–5, Aug 2018. doi: 10.1109/VTCFall.2018.8690918.

A. Chawla, A. S. Sarode, A. K. Jagannatham, and L. Hanzo. Distributed parameter detection in massive MIMO wireless sensor networks relying on imperfect CSI. *IEEE Transactions on Wireless Communications*, 20(1):506–519, Jan 2021a. doi: 10.1109/TWC.2020.3025877.

A. Chawla, R. K. Singh, A. Patel, A. K. Jagannatham, and L. Hanzo. Distributed detection for centralized and decentralized millimeter wave massive MIMO sensor networks. *IEEE Transactions on Vehicular Technology*, 70(8):7665–7680, Aug 2021b. doi: 10.1109/TVT.2021.3089669.

A. Chawla, P. S. Kumar, S. Srivastava, and A. K. Jagannatham. Technical report: Centralized and distributed millimeter wave massive MIMO-based data fusion with perfect and Bayesian learning (BL)-based imperfect CSI. *IEEE Transactions on Communications*, 70(3):1777–1791, Mar 2022. doi: 10.1109/TCOMM.2022.3141411.

M. V. Clark, T. M. Willis, L. J. Greenstein, A. J. Rustako, V. Erceg, and R. S. Roman. Distributed versus centralized antenna arrays in broadband wireless networks. In *IEEE VTS 53rd Vehicular Technology Conference*, volume 1, pages 33–37, May 2001. doi: 10.1109/VETECS.2001.944798.

O. El Ayach, R. W. Heath, S. Abu-Surra, S. Rajagopal, and Z. Pi. The capacity optimality of beam steering in large millimeter wave MIMO systems. In *2012 IEEE 13th International Workshop on Signal Processing Advances in Wireless Communications (SPAWC)*, pages 100–104, Jun 2012. doi: 10.1109/SPAWC.2012.6292865.

D. L. Hall and J. Llinas. An introduction to multisensor data fusion. *Proceedings of the IEEE*, 85(1):6–23, 1997. doi: 10.1109/5.554205.

S. M. Kay. *Fundamentals of Statistical Signal Processing, volume 2: Detection Theory*. Prentice-Hall PTR, Upper Saddle River, NJ, USA, 1993a.

S. M. Kay. *Fundamentals of Statistical Signal Processing: Estimation Theory*. Prentice-Hall, Inc., Upper Saddle River, NJ, USA, 1993b. ISBN 0-13-345711-7.

K. J. Kerpez. A radio access system with distributed antennas. *IEEE Transactions on Vehicular Technology*, 45(2):265–275, May 1996. doi: 10.1109/25.492850.

J. Lee, G.-T. Gil, and Y. H. Lee. Channel estimation via orthogonal matching pursuit for hybrid MIMO systems in millimeter wave communications. *IEEE Transactions on Communications*, 64(6):2370–2386, Jun 2016. doi: 10.1109/TCOMM.2016.2557791.

J. Li, D.-W. Yue, and Y. Sun. Performance analysis of millimeter wave massive MIMO systems in centralized and distributed schemes. *IEEE Access*, 6:75482–75494, Nov 2018. doi: 10.1109/ACCESS.2018.2882003.

H. Q. Ngo, E. G. Larsson, and T. L. Marzetta. Energy and spectral efficiency of very large multiuser MIMO systems. *IEEE Transactions on Communications*, 61(4):1436–1449, Apr 2013. doi: 10.1109/TCOMM.2013.020413.110848.

T. S. Rappaport, S. Sun, R. Mayzus, H. Zhao, Y. Azar, K. Wang, G. N. Wong, J. K. Schulz, M. Samimi, and F. Gutierrez. Millimeter wave mobile communications for 5G cellular: It will work! *IEEE Access*, 1:335–349, May 2013. doi: 10.1109/ACCESS.2013.2260813.

A. A. M. Saleh and R. Valenzuela. A statistical model for indoor multipath propagation. *IEEE Journal on Selected Areas in Communications*, 5(2):128–137, Feb 1987. doi: 10.1109/JSAC.1987.1146527.

F. Sohrabi and W. Yu. Hybrid digital and analog beamforming design for large-scale MIMO systems. In *2015 IEEE International Conference on Acoustics, Speech and Signal Processing (ICASSP)*, pages 2929–2933, Aug 2015. doi: 10.1109/ICASSP.2015.7178507.

D. Wang, X. You, J. Wang, Y. Wang, and X. Hou. Spectral efficiency of distributed MIMO cellular systems in a composite fading channel. In *2008 IEEE International Conference on Communications*, pages 1259–1264, May 2008. doi: 10.1109/ICC.2008.245.

D. P. Wipf and B. D. Rao. Sparse Bayesian learning for basis selection. *IEEE Transactions on Signal Processing*, 52(8):2153–2164, Aug 2004. doi: 10.1109/TSP.2004.831016.

S. Zhou, W. Xu, H. Zhang, and X. You. Hybrid precoding for millimeter wave massive MIMO with analog combining. In *2017 9th International Conference on Wireless Communications and Signal Processing (WCSP)*, pages 1–5, Oct 2017. doi: 10.1109/WCSP.2017.8171060.

6

Software-Defined Radio (SDR)-Based Real-Time WLANs for Industrial Wireless Sensing and Control

Zelin Yun[1], Natong Lin[1], Shengli Zhou[2], and Song Han[1]

[1] *School of Computing, University of Connecticut, Storrs, CT, USA*
[2] *Department of Electrical and Computer Engineering, University of Connecticut, Storrs, CT, USA*

6.1 Introduction

Applying real-time wireless technologies in industrial control systems has been gaining popularity in recent years. Pervasively deployed sensors and actuators based on high-throughput wireless standards allow for increased network throughput, improved system mobility, and reduced maintenance costs. A key step of designing a wireless control system is to choose the most appropriate wireless protocol based on its desired control specifications. Some protocols based on low-data-rate physical layers (PHYs), like 802.15.4, focus on real-time packet delivery and reliable performance but are only suitable for low-speed control applications. Comparatively, IEEE 802.11 standard (WiFi) is designed for high-speed wireless local area networks (WLANs) [Tramarin et al., 2019].

Table 6.1 gives an overview of the evolution of IEEE 802.11 standards, which was first released in 1997 and designed for WLAN usage as part of the IEEE 802 family. After the first widely used version IEEE 802.11b (WiFi 1) in 1999, the 802.11 working group (WG) released the version of IEEE 802.11a (WiFi 2) and IEEE 802.11g (WiFi 3) supporting orthogonal frequency division multiplexing (OFDM). From IEEE 802.11n (WiFi 4), the single-user MIMO (SU-MIMO) is supported with multiple optional beamforming transmissions. In IEEE 802.11ac (WiFi 5), the multiuser MIMO (MU-MIMO) is added and the compressed channel feedback is the only method for MU-MIMO beamforming. IEEE 802.11ax (WiFi 6) supports orthogonal frequency-division multiple access (OFDMA), and the latest IEEE 802.11be standard (WiFi 7) has a maximum data rate of 46 Gbps.

Notably, the nondeterministic communication performance of standard 802.11 makes it incapable of mission- and safety-critical applications that require high

Wireless Sensor Networks in Smart Environments: Enabling Digitalization from Fundamentals to Advanced Solutions, First Edition. Edited by Domenico Ciuonzo and Pierluigi Salvo Rossi.

Table 6.1 An overview of 802.11 standard evolution.

Protocol		PHY name	Max.rate (Mbit/s)	Channel bandwidth (MHz)	Band (GHz)	Name
802.11	1997	DSSS	2	22	2.4	(Wi-Fi 1)
802.11b	1999	HR/DSSS	11	22	2.4	(Wi-Fi 2)
802.11a	1999	OFDM	54	20	5	—
802.11g	2003	ERP-OFDM	54	20	2.4	(Wi-Fi 3)
802.11n	2009	HT-MIMO	600	20/40	2.4/5	Wi-Fi 4
802.11ac	2013	VHT-MIMO	3466.8	20/40/80/160	5	Wi-Fi 5
802.11ax	2019	HE-OFDM	10,530	20/40@2.4GHz 20/40/80/160@5GHz	2.4/5	Wi-Fi 6
802.11be	2024	EHT-OFDM	46,120	20/40@2.4GHz 20/40/80/160@5GHz 80/160/320@6GHz	2.4/5/6	Wi-Fi 7

determinism and reliability. To address this issue, a systematic solution named RT-WiFi [Wei et al., 2013; Leng et al., 2014; Wei et al., 2018] was proposed to provide real-time data delivery for a range of wireless control systems. RT-WiFi is a time division multiple access (TDMA)-based data link layer (DLL) protocol built on IEEE 802.11 a/g PHY, providing deterministic timing guarantees for packet delivery with a configurable sampling rate of up to 6 kHz. RT-WiFi was implemented on AR9285, a commercial-off-the-shelf (COTS) 802.11 network interface card (NIC). This allows running existing applications on top of RT-WiFi with minimum modifications thus offering the advantage of much shortened development periods; however, it comes with the trade-off of limited flexibility in terms of radio technologies. For instance, the Atheros AR9285 is limited to compatibility with IEEE 802.11a/g, whereas many real-time wireless protocols are based on varied radio technologies and thus require different hardware platforms. It is thus a significant challenge to develop a uniform communication platform that integrates various real-time wireless technologies to maximize the existing hardware investments and software development efforts.

These limitations motivate us to develop an software-defined radio (SDR)-based configurable real-time wireless platform. This platform is programmable at both PHY and DLL layers to meet the diverse requirements of industrial control systems, including those with multiple operational modes. SRT-WiFi [Yun et al., 2022] is a SDR-based solution for RT-WiFi to serve this purpose. Its design and implementation leverage an advanced SDR platform (Xilinx Zynq-7000 and Analog Device AD9364), with radio functions programmed on an field programmable gate array (FPGA). SRT-WiFi can operate in hard real-time because its radio functions are executed by logic blocks in the FPGA running at oscillator-driven speeds, and thus support the essential functions needed for high-speed real-time communications and provide an open-source platform to accommodate the evolving IEEE 802.11 standards.

While SRT-WiFi provides real-time and reliable wireless communications for industrial control applications, its current version only supports IEEE 802.11a/g PHYs and SISO communications. Our ultimate goal is to develop SRT-WiFi into a full-blown SDR-based real-time wireless platform and support newer standards of WiFi, e.g. IEEE 802.11n/ac/ax, to enable both SU/MU MIMO and OFDMA. As the first step toward this goal, we extend SRT-WiFi on GNU Radio, a widely used open-source SDR platform [GNU Radio Foundation, 2007]. With GNU Radio and USRP, we can implement and evaluate the PHYs of newer 802.11 standards with a much shorter development period when compared to developing those PHYs directly on FPGA-based SDR platform. For simplicity of presentation, we call this GNU Radio-based implementation GR-WiFi to differentiate it from the SRT-WiFi system developed on FPGA-based SDR platform. Once GR-WiFi is fully developed and tested on GNU Radio, it will be ported on the FPGA-based SDR platform to make it full-blown and support hard real-time performance. In GR-WiFi, we have successfully implemented the PHYs of IEEE 802.11a/g/n/ac standards supporting the Legacy, high-throughput (HT), and very-high-throughput (VHT) PHY formats with SISO and 2×2 SU-MIMO and MU-MIMO. Both FPGA-based SRT-WiFi and GNU Radio-based GR-WiFi implementations, once mature, will be made public to the wireless communities to support a broad range of R&D activities. Table 6.2 summarizes the strengths and limitations of the three solutions reviewed in this chapter.

Table 6.2 Pros and cons of the three solutions.

	Pros	Cons
RT-WiFi [Wei et al., 2013]	• Timing guarantee on packet delivery • Flexible DLL configuration • Seamless integration with existing hardware	• Needs significant effort and hardware expertise to manage and upgrade COTS devices
SRT-WiFi [Yun et al., 2022]	• Full-stack configurability • Precise time synchronization and real-time communication	• Complexity of implementation and long developing period • Only support IEEE 802.11a/g
GR-WiFi	• Efficient queue management • Support multiple standards, including IEEE 802.11a/g/n/ac	• Not able to perform on real-time testbed yet

6.2 RT-WiFi Based on IEEE 802.11a/g

An RT-WiFi network consists of three primary components: RT-WiFi stations (STAs), which are devices equipped with 802.11-compatible hardware and the RT-WiFi protocol stack; RT-WiFi Access Points (APs), which function as intermediaries to support message exchange between the network manager and RT-WiFi STAs; and the network manager, a software module that configures the network, coordinates communication between APs and STAs, and adjusts the communication schedule when necessary. An RT-WiFi AP and its associated STAs are defined as a cluster. An RT-WiFi network with multiple APs is called a multi-cluster RT-WiFi network [Leng et al., 2019] (see Figure 6.1). In industrial practice, the placement of the RT-WiFi APs will be done through careful site survey, resulting in each AP and its associated STAs forming a star topology. In a multi-cluster RT-WiFi network, each AP is managed by an AP network manager responsible for its cluster. Also, a central network manager supervises all AP network managers and coordinates packet transmissions among different clusters. Since RT-WiFi is a TDMA-based communication protocol, the local clocks of all STAs and APs are synchronized [Wei et al., 2013].

6.2.1 RT-WiFi Protocol Design

The RT-WiFi protocol stack is the most essential building block of the RT-WiFi network. It enables real-time and high-speed data transmissions and customizable DLL configurations for diverse applications. The RT-WiFi protocol design takes into consideration the requirements of different types of control applications, enabling control designers to choose the communication behavior that fits their applications the best. At the same time, RT-WiFi design minimizes the modification of the original WiFi protocol so that it can be transparent to both the upper layer software stack and underlying hardware to provide the most compatibility and usability.

The architecture of RT-WiFi protocol is shown in Figure 6.2. At the very bottom, RT-WiFi utilizes IEEE 802.11 PHY, which is sufficiently fast for most wireless control systems. Control application users can easily implement the RT-WiFi DLL on COTS IEEE 802.11 hardware to support high-speed and real-time data transmissions. Above the IEEE 802.11 PHY layer is a TDMA-based DLL, which is the core of the RT-WiFi protocol. Combined with the centralized channel and time management schemes imposed by the RT-WiFi network manager, this DLL ensures collision-free and deterministic communications. Additionally, it offers a flexible abstraction for the upper layers, allowing seamless support for standard UDP/TCP-based applications.

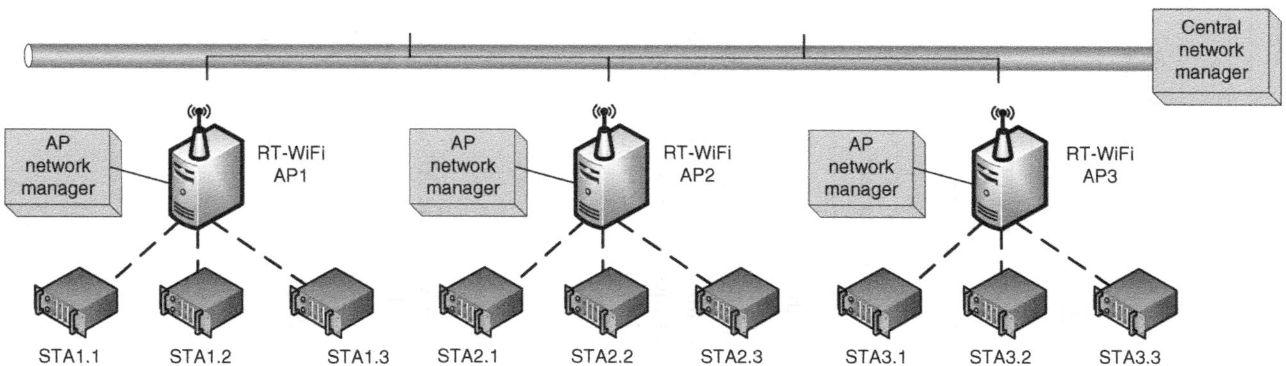

Figure 6.1 Overview of an RT-WiFi network with three clusters.

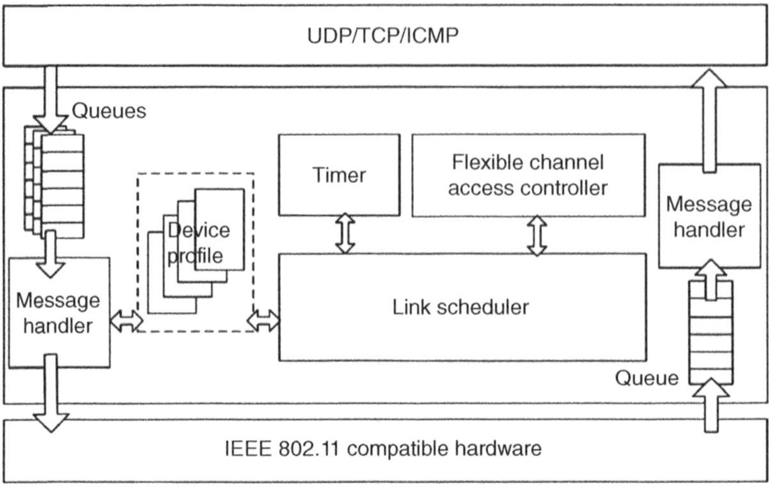

Figure 6.2 System architecture of the RT-WiFi protocol.

The RT-WiFi DLL comprises three main components: a timer that ensures global synchronization across all RT-WiFi nodes and initiates timing events; a link scheduler that manages media access and executes scheduled events at designated time points; and a flexible channel access controller that dynamically configures hardware parameters to execute timing events based on the target application's behavior.

6.2.2 Performance Evaluation

We set up a small-scale testbed with one AP and three STAs to evaluate the performance of the RT-WiFi DLL design (see Figure 6.3). Each device utilized the Atheros AR9285 NIC operating on the 802.11g protocol, though they were

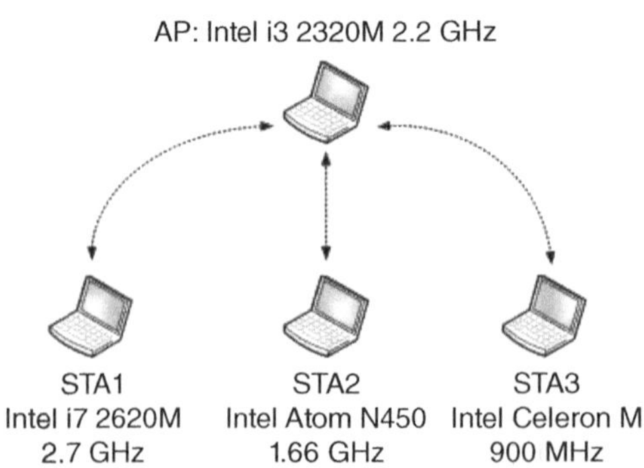

Figure 6.3 Testbed setup for RT-WiFi performance evaluation.

Table 6.3 Comparison of delay between RT-WiFi and regular WiFi networks.

	Max delay (μs)		Mean delay (μs)		Standard deviation (μs)	
Link	RT-WiFi	Wi-Fi	RT-WiFi	Wi-Fi	RT-WiFi	Wi-Fi
STA1 → AP	3865	100,078	176	401	25.86	1491.69
STA2 → AP	4193	81,499	171	348	27.62	1000.60
STA3 → AP	3861	75,298	174	429	25.16	1221.72
AP → STA1	1197	78,089	184	788	16.86	2861.42
AP → STA2	1342	78,923	189	790	15.19	2806.56
AP → STA3	2186	77,860	189	799	19.03	2855.89

powered by CPUs with varying computing capabilities. In the experiments, six pairs of UDP sockets were established between the STAs and the AP, with data published every 4 ms for each socket with a fixed payload of 460 bytes. The comparison results between WiFi and RT-WiFi are summarized in Table 6.3. It shows that the average latency variation in a WiFi network is up to 90 times greater than in an RT-WiFi network, with the maximum delay exceeding 30 times that of RT-WiFi. In contrast, RT-WiFi supports a sampling rate of up to 6 kHz, and less than 0.01% of packets have latency greater than 1 ms, meeting the requirements of most industrial control systems [Wei et al., 2013].

6.3 SRT-WiFi Based on IEEE 802.11a/g

The RT-WiFi protocol design offers several advantages, including deterministic timing guarantee on packet delivery, flexible data link layer configuration, and seamless integration with existing hardware. However, it has some limitations, such as limited flexibility in terms of radio technologies and no support for frequently updated wireless protocols. To address these limitations, SRT-WiFi [Yun et al., 2022] introduces an SDR-based configurable real-time solution. In contrast to RT-WiFi, which relies on COTS hardware, the SDR platform offers programmability at both the PHY and DLL levels. This flexibility allows it to accommodate the needs of various industrial control systems with multiple operational modes.

SRT-WiFi is built upon Openwifi [Jiao et al., 2020], a SoftMAC IEEE 802.11 design compatible with the Linux MAC80211 subsystem. As shown in Figure 6.4, the Openwifi system has two major components: the processing system (PS) and the programmable logic (PL). The PS handles the majority of the MAC layer and all higher layers. The PL is an FPGA-based embedded system responsible for the

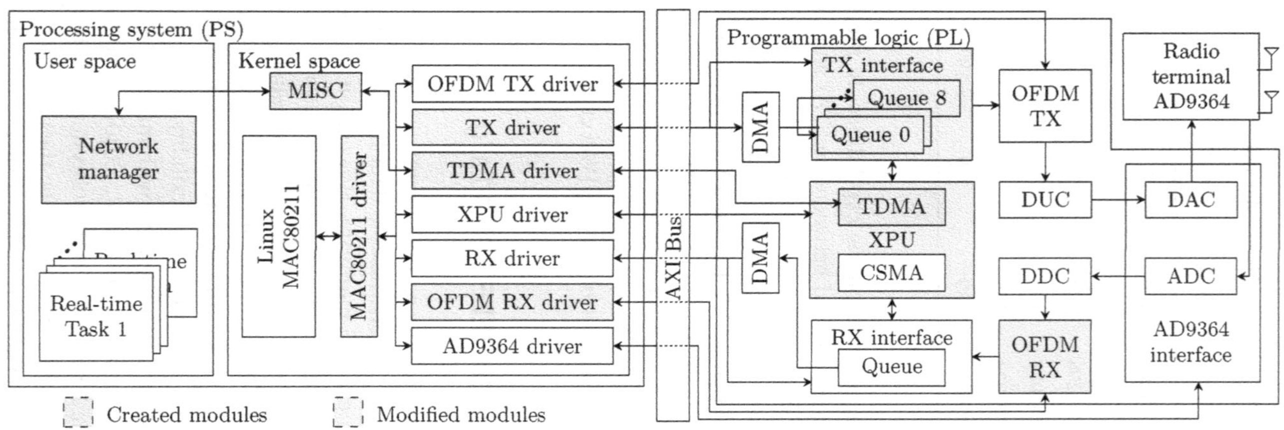

Figure 6.4 Overview of the SRT-WiFi system architecture. It highlights the created and modified modules in SRT-WiFi based on the Openwifi architecture.

real-time portion of the MAC and PHY layers. Both PS and PL are implemented on the Zynq-7000 SoC, which includes an FPGA for the PL and an ARM processor for the PS. Data exchange between PL and PS occurs through the Advanced eXtensible Interface (AXI) bus, supporting direct memory access as well as register reading and writing. Additionally, the PL is connected to an AD9364 radio terminal from Analog Devices for signal transceiving.

Figure 6.4 shows three main modules of PL on the right side: the TX interface (TXI), the XPU (application-specific processing unit), and the RX interface (RXI). The TXI module manages packet transmission, while the RXI module handles packet reception. The XPU module controls channel access by using IEEE 802.11 distributed coordination function (DCF) [IEEE 802.11 Working Group, 2021]. Leveraging concurrent processing ability in FPGA, the radio terminal can operate its transmitter (TX) and receiver (RX) modules at the same time. Besides, PL modules equipped with registers allow configurations of operation modes and parameters.

In PS, a Linux OS is operating on an ARM processor. As the platform adopts SoftMAC architecture, most MAC functionalities are integrated into the Linux kernel (MAC80211 subsystem [Mur, 2011]), excluding the real-time MAC and PHY that are being implemented in PL. Between the Linux MAC80211 subsystem and the wireless adapter (PL), the MAC80211 driver is in place to facilitate communication. Sub-drivers (depicted on the left side of Figure 6.4) ensure data communications between the MAC80211 driver and PL. MAC80211 driver interacts with PL by calling APIs provided by sub-drivers. Additionally, TX and RX drivers manage the transmission and reception of data packets between the PS and PL, respectively, using direct memory access (DMA).

Building on top of Openwifi, SRT-WiFi aims to achieve several key objectives: enabling precise network-wide time synchronization and facilitating multi-cluster real-time communications with effective rate adaptation at run time. The design details of the modified PL and PS components of SRT-WiFi are presented below.

6.3.1 Programmable Logic (PL) in SRT-WiFi

The PL component of SRT-WiFi is designed to (i) achieve real-time transmissions with high-precision time synchronization, (ii) enhance queue management efficiency, and (iii) measure precise link reception SNR as a reference for rate adaptation. We now describe how to achieve these functions in SRT-WiFi PL.

6.3.1.1 TDMA Block Design in SRT-WiFi PL

To improve real-time performance, a TDMA block is designed in XPU to supplement the carrier sense multiple access (CSMA) block. SRT-WiFi can seamlessly switch between TDMA and CSMA modes during runtime.

The TDMA mode in SRT-WiFi is designed to transmit and receive frames at designated times, coordinating communication between APs and STAs to prevent collisions. With this objective in mind, all transmissions follow a schedule that includes the transmitting times of the links with a specified time duration called superframe. A superframe consists of consecutive time slots, with each slot specifying the transmission state (TX, RX, or Idle) and the corresponding sender or receiver. At run time, the superframe is continuously generated to schedule the transmissions. Each time slot of the superframe has an atomic slot as the basic time unit. The length of time slots varies along with the rate to support rate adaptation, as a lower rate requires more time, namely more atomic slots, to transmit the same packet. In SRT-WiFi, superframe lengths, time slots, and atomic slots are fully customized. In most cases, application's requirements decide superframe length and selected data rate in the PHY link configure time slot and atomic slot lengths.

Figure 6.5 illustrates a superframe example in an SRT-WiFi network. The superframe has 127 atomic slots, with Slot0 and Slot1 allocated to AP1 and AP2 for sending beacons. Shared slots shown in Slot2 and Slot3 are available for any links that are used for the association process. The remaining atomic slots are either used for specific link communications or left idle. In the example, each link has the same maximum transmission unit (MTU) but operates at different rates, requiring varied time slot lengths and different numbers of atomic slots to transmit.

In the TDMA block, a register page is implemented for the TDMA driver can store and maintain schedule information. A scheduler timer located in the TDMA block will trigger the transmissions based on schedule. Each time slot attached a link assignment at the beginning will be retrieved by the TDMA block. The TXI module, which has the queue ready for each link, will send a frame to the next module as soon as it detects a loaded queue. The OFDM TX module then processes the frame by modulating the bit stream into the digital signal stream and handing it over to the digital-to-analog converter (DAC) interface for final signal emission through the radio terminal's antenna.

6.3.1.2 TDMA Time Synchronization Design

SRT-WiFi has an accurate time synchronization feature among the devices in the network. The SRT-WiFi network contains multiple clusters with an AP and several STAs, which may share the same channel. STAs and AP within a cluster running

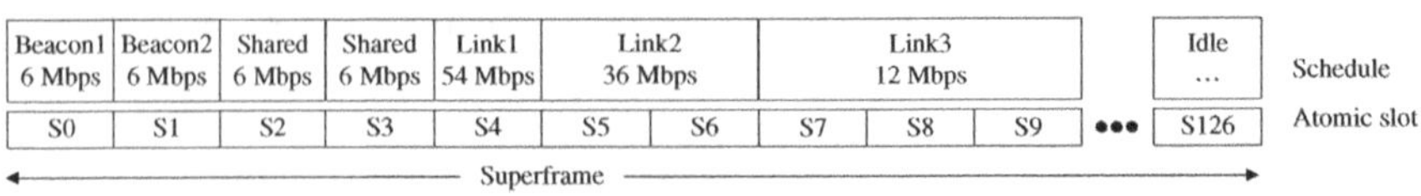

Figure 6.5 The timing diagram of an example superframe in multi-cluster SRT-WiFi with 127 time slots.

at the same channel need to be synchronized to avoid potential collisions. To solve this, a novel synchronization method is implemented at the PHY layer on the SDR device. In the scenario that APs run on the same channel, one AP is selected as the main AP (MAP), and the rest are the subordinate APs (SAPs). The setting assumes that all SAPs can receive signals from the MAP, which serves as the provider of the reference clock. The SAPs synchronize with the MAP, while all STAs synchronize with their respective APs. For instance, as depicted in Figure 6.1, AP2 serves as the MAP while AP1 and AP3 are SAPs synchronizing to AP2. This synchronization mechanism prevents devices from relying on the timer in a non-real-time operating system, leveraging instead the timer in hard real-time PL. To achieve this, a scheduler timer with nanosecond precision is added into the TDMA block to trigger transmissions.

In the PL layer, OFDM RX module performs PHY demodulation, and demodulated status and symbols are passed to RXI and XPU. In the TDMA block, a synchronization function is added to synchronize time with a specific AP by utilizing the demodulated result. The synchronization procedure has the following steps.

The synchronization function waits for the MAC packet header from the OFDM RX module and checks the packet's content. If it's a beacon packet, the function continually waits for the service set ID (SSID) in the following packet payload. Upon reading the SSID, the synchronization function compares it against the target SSID provided by the TDMA driver through the registers. If two SSIDs match, the buffered time is updated to the schedule timer; if not, the synchronization function waits for the next packet, and the schedule timer continues to run as usual without any update. Notably, this synchronization method is also compatible with other protocol versions like IEEE 802.11n/ac/ax.

Using this synchronization mechanism, our experimental results demonstrate that the synchronization time drift of the SRT-WiFi devices can be maintained within 0.2 μs, which outperforms the COTS hardware. Figure 6.6 presents the results of two experiments from Yun et al. 2022. In the first experiment, two SAPs, AP2 and AP3, synchronize with a MAP AP1, and their synchronization time error is measured, as shown in Figure 6.6a. The maximum error observed is 0.2 μs. In the second experiment, AP2, acting as an SAP, synchronizes with the MAP AP1, while an STA synchronizes with AP2. The maximal synchronization error of the STA in multi-cluster SRT-WiFi networks is measured, which is within 1 μs, as depicted in Figure 6.6b. This improvement in time synchronization accuracy can help reduce guard time and support shorter time slot lengths, thereby improving the sampling rates.

6.3.1.3 Queue Management

In SRT-WiFi, packets from the PS are pushed into queues before transmission. Unlike COTS hardware-based RT-WiFi, where the queue number is fixed and

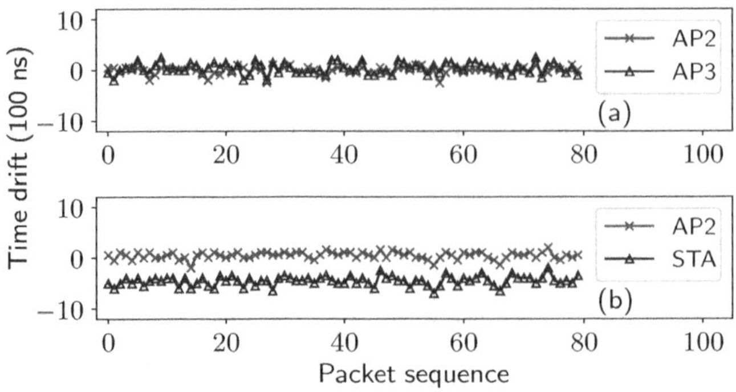

Figure 6.6 Synchronization performance of APs and STAs in a multi-cluster SRT-WiFi network. (a) Time synchronization error between SAPs and MAP and (b) time synchronization error between STA and MAP via an intermediate SAP.

cannot be changed, SRT-WiFi provides greater flexibility in queue configuration. For instance, AR9285 used in RT-WiFi [Wei et al., 2013] has only eight queues. In SRT-WiFi real-time transmissions, each link has its own queue stack to guarantee the desired timing and throughput performance. However, if the number of STAs exceeds the number of available queues in the AP, packets from different links need to share a queue, potentially causing timing violations.

Figure 6.7a shows a device with 10 links but has 8 available queues, causing STA9 and STA10 to share a queue with others. When multiple packets from different links enter the same queue, they may wait until the packet at the queue head being sent, even if their assigned time slots in the superframe occur earlier. AP will encounter this time violation issue when managing real-time transmissions for multiple STAs.

In SRT-WiFi, the TDMA block initiates packet transmission. Queues can be assigned to different links, and the packets from different links are put into the corresponding queues, as shown in Figure 6.7b. The TDMA block's schedule determines which queue to trigger for each slot. This flexibility is a feature of SDR-based systems, as the queues can be configured in software rather than hard programmed, as in COTS hardware. Given sufficient FPGA resources, the supported queues can be extended to accommodate any number.

To address the time violation issue when the available AP queues are less than supported STAs, a dynamic buffer is designed in SRT-WiFi as shown in Figure 6.7c. A dynamic buffer contains a series of slots, each of which stores one packet at most. When a packet arrives, the TXI pushes it into an unused buffer slot. Next, based on the schedule, the TDMA module checks the link information located at the beginning of each time slot and scans the whole buffer to check if a packet for that link is present. If it finds one, it triggers the transmission from the corresponding

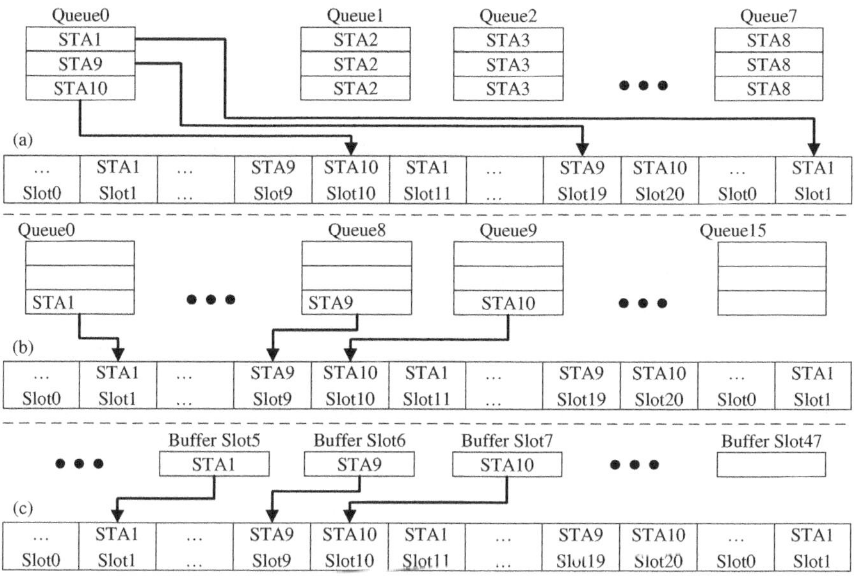

Figure 6.7 Queue management issues in real-time networks with shared queues. (a) Queue contention when 10 stations share only 8 queues, (b) assigned queues per link using an SDR-based design, and (c) dynamic buffer design using single-packet slots.

buffer slot. Since each buffer slot holds only one packet, more buffer slots can be implemented with the same FPGA resources.

Table 6.4 compares the performance of assigned and dynamic queue management using 16 links and a variable number of queues. AP periodically generates a packet that requires only one atomic slot for transmission for each link. In assigned queue management, the packet is pushed into this pre-assigned and handled according to the schedule. While in dynamic queue management, the packet is pushed into an available queue when it arrives. All transmissions follow a randomly generated schedule where the throughput of each link is guaranteed and

Table 6.4 Assigned and dynamic queue management maximum and average packet delay (slot number) with 16 links.

Number of queues	8	10	12	14	16
Assigned maximum delay (slot)	2816	2358	1707	1125	82
Dynamic maximum delay (slot)	591	162	106	104	103
Assigned average delay (slot)	336	236	159	87	16
Dynamic average delay (slot)	271	42	16	16	16

the length of superframe is fixed. During the scheduled transmission, packet delay is recorded, and no packets are dropped due to delay. The results show that even a small disparity between the number of queues and links in the assigned queue management can lead to a significantly high maximum and average delay. On the other hand, the dynamic queue management method can manage more links with the same number of queues and maintain a lower max. and avg. packet delays. However, it does not completely eliminate delays since all queues are shared.

6.3.1.4 Link Quality Measurement

The rate adaptation function in SRT-WiFi requires precise SNR measurement. There are two practical methods developed to meet the SNR requirement. Both methods leverage the long training field (LTF) in the Legacy physical layer protocol data unit (PPDU) preamble of the 802.11 PHY signal. The first method involves calculating the autocorrelation [Lee and Messerschmitt, 2012] of the LTF. The LTF contains a half symbol followed by two repeated symbols, which correspond to 160 samples at a 20 MHz sampling rate. So, the LTF shows the same pattern in every 64 samples [IEEE 802.11 Working Group, 2021]. We utilize 128 consecutive samples of the LTF to calculate the autocorrelation, denoted as ρ, and the SNR value (in dB) can then be determined as follows:

$$\mathrm{SNR} = 10\log_{10}\left(\frac{\rho}{1-\rho}\right), \tag{6.1}$$

where we assume that $\rho < 1$. Two 64 repeated samples are used, but not the first 32 samples because of transient effects that can occur at the start of a transmission in the sender's hardware.

In the second method, the LTF and a piece of background noise before the data symbol are buffered after the packet arrival. The power of the background noise can be measured before packet arrival, and the power of LTF signal includes noise power plus the signal power. Then, the SNR (dB) is computed in this way:

$$\mathrm{SNR} = 10\log_{10}\left(\frac{P_{\mathrm{LTF}} - P_{\mathrm{noise}}}{P_{\mathrm{noise}}}\right), \tag{6.2}$$

where P_{LTF} is the signal power of LTF and P_{noise} is the power of the background noise before the packet. We assume that P_{LTF} is larger than P_{noise}.

Both SNR measurement methods are integrated into the OFDM RX module of SRT-WiFi. An SNR value is calculated each time a packet is received. If the received packet contains a source address, the computed SNR value is stored along with the source address. The MAC80211 driver forwards the SNR information to the TDMA driver, who manages the scheduling in the system and utilizes the SNR data for scheduling decisions. The device manager on each device interacts with the TDMA driver to access the SNR information and forwards the SNR information to the central network manager. The central network manager uses this SNR data to determine the appropriate data rate for each link based on the quality of the wireless connections and creates the network schedule

6.3.2 Processing System (PS) in SRT-WiFi

The two main components of the PS design in SRT-WiFi are the drivers and the network manager. The drivers act as the interface between the PL and Linux (see Figure 6.4), serving two primary functions: (i) configuring parameters in the PL modules to support various working modes and functions and (ii) managing the packet exchange between the PL and the OS. Each PL module has a corresponding driver connected to the kernel because each PL module contains a register page used to set or read status. For example, TXI uses a register to determine if a packet requires an ACK and another register to report the packet delivery status. On the kernel side, sub-drivers handle configuration tasks within the PL component and interact with the OS by encapsulating read and write functions into APIs for the MAC80211 driver. The TDMA block in the XPU also has a register page containing three parts. The first part is for schedule allocation, including superframe and atomic slot length. The second part is used for PL synchronization by acquiring AP'SSID for stations to synchronize with. The third part is a mode switch for toggling between defaulted CSMA and customized TDMA modes. Since the TDMA mode's functions are incompatible with the MAC80211 subsystem, configuring the TDMA block through MAC80211 is challenging. Therefore, the TDMA driver is implemented as a miscellaneous character driver (MISC), providing basic read/write functions for user space. In user space, the network manager configures the TDMA block by calling the APIs of the TDMA driver so that it can adjust the schedule, set parameters, and switch modes as needed.

In SRT-WiFi, there are three types of network managers forming a hierarchical structure: the central network manager (CNM) managing all network resources, cluster managers (CM) operating on the APs, and device managers (DM) operating on the STAs. During the process of joining an SRT-WiFi network, the CNM starts first, awaiting TCP connections from CMs to distribute schedules to the links. Following this, CMs initiate the cluster networks, with slave APs synchronizing with the master AP on the same channel, and then await STA connections. To simplify the synchronization and the joining process, beacon and shared slots remain fixed throughout system operation, and this information is broadcasted among all devices. Once an STA powers on, it scans the channels, synchronizes with the designated AP, and joins the network. Once the network is joined, the DM on the STA establishes a TCP connection with the CM on the AP to receive and update the schedule. Until the schedule is received, the STA uses shared slots to complete the joining process. Unlike assigned slots, shared slots are contention based, where each sender first undergoes a random backoff, similar to CSMA mode, and then senses the channel. During operation, DMs and CMs on each device monitor link qualities and interference. This channel information is collected by the CNM, which then determines and updates schedules and data rates for the DMs and CMs to adapt to current channel conditions, ensuring stable transmissions.

A unique aspect of SRT-WiFi network management is its capability to dynamically adjust slot lengths within the schedule to support rate adaptation during real-time transmission. While the MTU for an individual link remains fixed, the data rate may vary depending on interference levels. A lower data rate requires more time to transmit a packet of the same length, which can exceed the time slot boundary and lead to collisions. SRT-WiFi addresses this issue with dynamic slot lengths. In the schedule, an atomic slot (AS) is defined as the shortest slot length that can support transmitting an MTU-sized packet at the highest rate. When transmitting at a lower rate, a packet can use multiple consecutive atomic slots without preemption. Therefore, by selecting different rates during runtime, the packet transmission can occupy varying atomic slots.

6.3.3 Performance Evaluation

A testbed of multi-cluster SRT-WiFi is implemented to perform a comprehensive evaluation. This network configuration setting includes two APs, AP1 and AP2 for Cluster1 and Cluster2, operating on a single channel, with each AP having two STAs connected, STA1 and STA2 in Cluster1, STA3 and STA4 in Cluster2. Figure 6.8 provides a testbed overview with a total of four links. CNM and APs are connected to a router in the testbed, forming a backbone network, and stations connect to their APs accordingly. Also, a USRP device is employed to create interference via an antenna positioned near AP2.

Due to the page limit, this chapter presents a single experiment to showcase the rate adaptation function in SRT-WiFi. In this experiment, interference is introduced to the testbed. Each device measures its reception SNR and reports it to the CNM. The CNM then constructs the schedule and selects appropriate data rate.

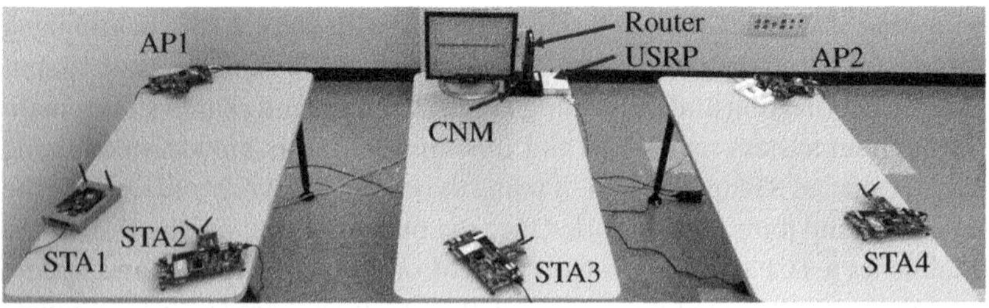

Figure 6.8 An overview of the multi-cluster SRT-WiFi testbeds. Source: Yun et al. [2022]/IEEE.

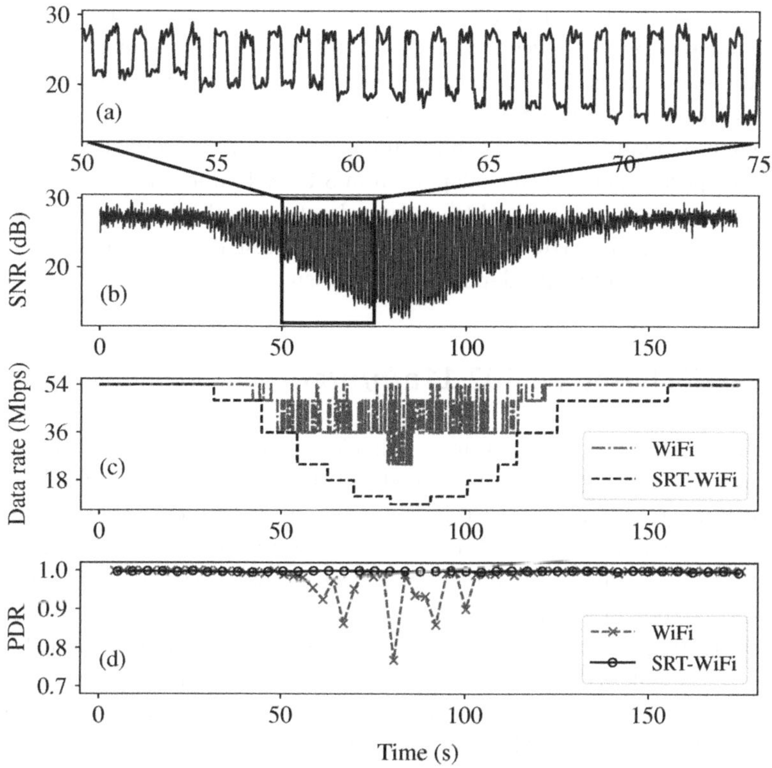

Figure 6.9 Data rate and throughput comparison between SRT-WiFi and regular WiFi in the presence of interference. (a) Zoomed-in SNR measurements, (b) overall measured SNR, (c) data rates for SRT-WiFi and standard WiFi, and (d) PDR comparison.

In the experiment, we add the interference on the AP side and let the station both send UDP packets to the AP and measure the packet delivery ratio (PDR) and SNR. The level of interference is not fixed but varied every 0.5 seconds, meaning that in the first half of each second, the interference rises to a set level while in the next half of that second, the interference shuts down so that the interference varies fast.

Figure 6.9b shows the measured SNR of the channel and Figure 6.9a provides a closer look at a portion of the measured SNR to illustrate the interference variations. The SNR values decrease from 27 to 12 dB and rise gradually. Figure 6.9c presents the data rates for both SRT-WiFi and standard WiFi, the latter of which uses the Minstrel algorithm [Xia et al., 2013] for rate adaptation based on transmission history. The corresponding PDR is shown in Figure 6.9d. The experiment result reveals that standard WiFi cannot maintain stable transmissions when the SNR value falls below 20 dB. In contrast, SRT-WiFi employing a rate adaptation method can provide stable transmissions under different SNR conditions.

The CNM buffers the measured SNR values over a time window and adjusts the data rate based on the lowest SNR value in the buffer; once a lower SNR is detected, the data rate is immediately reduced and does not increase until all buffered SNR values exceed the threshold for a higher rate. This method wastes some resources when the channel condition is good, but it ensures stable transmissions. The performance of rate adaptation in SRT-WiFi under interference is shown in Figure 6.9c, characterized by a stepped pattern without rapid changes. Figure 6.9d illustrates the PDR of SRT-WiFi during the test, demonstrating stable performance thanks to the rate adaptation mechanism.

6.4 GR-WiFi Based on 802.11a/g/n/ac

SRT-WiFi currently provides real-time, reliable wireless communication for industrial control applications but is limited to IEEE 802.11a/g PHYs and SISO communications. Our long-term goal is to enhance SRT-WiFi to support newer WiFi standards such as IEEE 802.11n/ac/ax. As the first step toward this goal, we extend SRT-WiFi on GNU Radio, a popular open-source SDR platform, and introduced GR-WiFi, a GNU Radio-based open-source platform for IEEE 802.11 research. Note that there is already a GNU Radio implementation for IEEE 802.11/a/g/p available as referenced in Bloessl et al. [2013]. In this work, we implement PHYs of 802.11a/g/n/ac standards on GR-WiFi, which can support the Legacy, high-throughput (HT) and very-high-throughput (VHT) PHY formats with SISO and up to 2×2 SU-MIMO and MU-MIMO. Figure 6.7 summarizes the three supported PHY formats (Legacy, HT and VHT) in GR-WiFi and detail reference can be found at IEEE 802.11 standards [IEEE 802.11 Working Group, 2021]. In Section 6.4, we present the design of the packet transmission and reception functions in GR-WiFi (Sections 6.4.1 and 6.4.2), followed by the implementation and evaluation details, including key blocks and performance analysis (Sections 6.4.3.1 and 6.4.3.2).

6.4.1 Packet Transmission Design

The packet transmission function in GR-WiFi follows these steps to generate the packets of different supported formats in 802.11a/g/n/ac standards.

In Legacy format, beginning with the preparation of training field, transmitter converts the given orthogonal frequency division multiplexing (OFDM) training symbol from frequency to time domain waveform by applying inverse fast Fourier transform (IFFT), scales the waveform amplitude with a tone scaling factor, and inserts the guard interval (GI). In the Legacy signal field, 24-bit Legacy signal bits, which contain information on packet length and modulation and coding scheme (MCS), go through binary convolutional coding (BCC) and interleave process. The interleaved bits are then modulated with binary phase shift keying (BPSK) and converted to a scaled time domain waveform. The data payload that is distributed into data symbols will go through the same steps as Legacy signal symbols but with an additional bit of scrambling before the BCC.

The HT format's packet transmission is more complex due to the MIMO and beamforming support. First, the signal is generated for multiple spatial streams (SSs) for MIMO transmissions. Each SS has the same packet format as shown in Figure 6.10. Before each SS is modulated and emitted into the air through antennas, a cyclic shift is applied to each SS to prevent constructive interference and unintentional beamforming, which happen when the same signal is transmitted through different transmit chains. After applying cyclic shift, the modulation and scaling steps are the same as the Legacy packet. The Legacy signal field in HT packet has the rate set to lowest, and the length is computed to cover the duration of the following HT portion transmission. The HT signal field is the same as Legacy signal field but occupies two symbols. The HT portion starts from the HT training field. Besides the cyclic shift, there is also the spatial mapping for each subcarrier to apply the required phase for beamforming, called the Q matrix in the standard. HT-STF and each of the HT-LTFs have one symbol. The number of the HT-LTFs is determined by the SS number to provide sufficient channel information for the receiver to estimate the channel(s). In HT data symbols, after coding, stream parsing is performed to separate the coded bits into multiple spatial streams. Interleave is also applied for each spatial stream but with different phase rotations as specified in the standard. Cyclic shift, waveform scaling, and GI insertion are necessary steps before the packet is ready to be transmitted through the air. The signal generation for the VHT format is similar to HT's, but it supports 256 QAM, up to eight spatial streams, and a 160 MHz bandwidth.

6.4.2 Packet Reception Design

To design the packet reception in GR-WiFi, we begin by addressing the packet reception trigger. As shown in Figure 6.10 from sample 10 to 170, the STF has

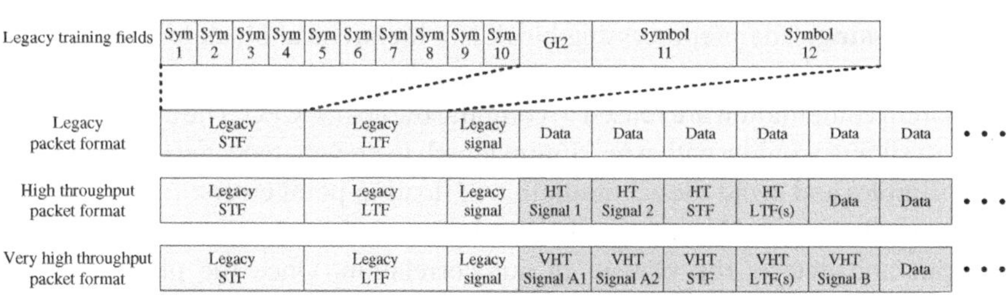

Figure 6.10 IEEE 802.11a/g/n/ac physical layer packet formats.

10 repeated symbols last for 0.8 μs with 16 samples. We utilize an autocorrelation method that leverages the 10 times repetition of the STF symbol. This approach is robust against multipath propagation effects carrier frequency offset (CFO) distortion during the reception. The autocorrelation output forms a plateau. Once the output passes the threshold, we measure the length of the continuous plateau to detect the STF and initiate the reception process.

Once packet reception is triggered, the next steps involve packet synchronization and fine-tune the timing. To fine-tune the timing, we find the maximum autocorrelation within a specified time window and locate two shoulder indices at 80% of the maximum value on both the left and right sides. As illustrated in Figure 6.10, the long training field (LTF) repeats 2.5 times. The correlation reaches its peak at the start of the LTF Guard Interval 2 (GI2) and drops at the end of the LTF GI2, making the center of this correlation correspond to the middle of LTF GI2.

With the correct timing, we estimate channel and CFO. The CFO of the following samples will be compensated. For each data symbol, it will be converted to frequency domain and then be equalized with channel. The recovered data symbols are demodulated using quadrature amplitude modulation (QAM) constellations and then decoded. Currently, the proposed receiver only supports the binary convolutional coding (BCC) with a soft-input Viterbi decoder.

6.4.3 Implementation and Evaluation

We now present the implementation details of GR-WiFi on the GNU Radio. As shown in Figure 6.11, we implement both SISO and 2×2 MIMO receivers. The 2×2 MIMO can also handle SISO packet reception. However, we design a separate SISO block because each port in this block processes samples in parallel. Using an MIMO receiver to decode SISO input will keep the second port running without processing any samples. For the transmission, the transmitter only generates packet samples, so the waste is negligible.

6.4.3.1 Key Blocks in GR-WiFi Implementation

The GR-WiFi implementation on GNU Radio includes the following key blocks:

Preprocessing: The preprocessing block is a hierarchical block to compute the autocorrelation of the input samples. The values obtained during the autocorrelation computation are reused to compute the coarse CFO. The average blocks use a sliding window with a maximum length to prevent repeated computations of samples and avoid the accumulation of floating-point errors.

Trigger: The trigger block takes the continuous autocorrelation samples as the input and detects the plateau of autocorrelation. Once the plateau length

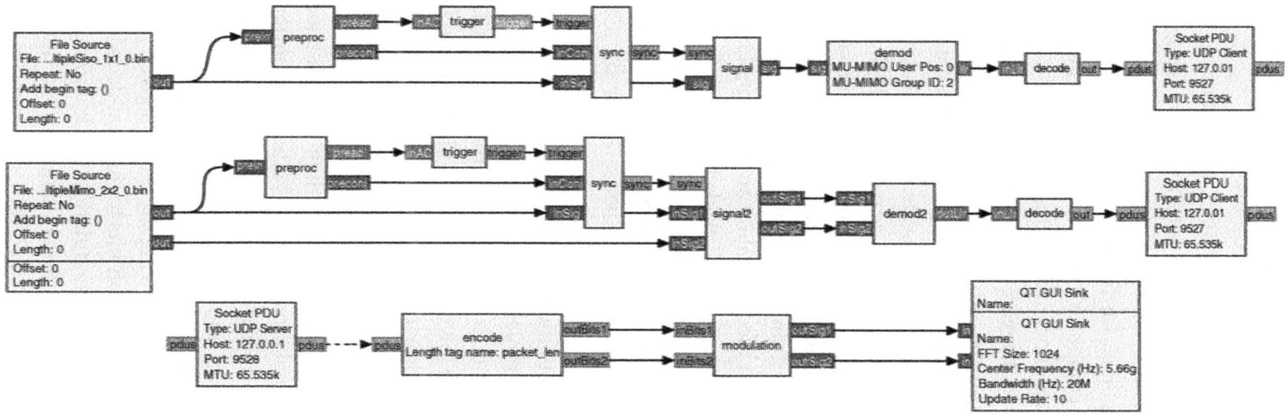

Figure 6.11 GNU Radio implementations of IEEE 802.11a/g/n/ac SISO and 2 × 2 MIMO receivers and transmitter.

meets a threshold, it generates output flag to trigger the synchronization block. The trigger block also has a state machine to avoid multiple triggers for one packet to avoid wasting computation resources.

Synchronization: The synchronization block is triggered to use autocorrelation of LTF and identify the start of LTF, with the index being passed to the next block. At the same time, the synchronization block compensates the coarse CFO from the preprocessing block for the LTF samples, re-estimates the CFO with two LTF symbols, and computes the accurate CFO value. This accurate CFO value is passed tp the next block using a tag.

Signal: The signal block is triggered by the synchronization block to get the timing of the packet and the estimated CFO. It first compensates the CFO for LTF and Legacy Signal to estimate the Legacy channel and demodulate and decode the Legacy Signal. If the Legacy Signal is correctly decoded, this packet is identified as at least a Legacy packet with the maximum possible timing length corresponding to the Legacy length. The signal block computes the symbol and sample number according to the MCS and packet length. The following input samples within the sample number will be compensated with CFO and passed to the next block. That means the signal block chops the input sample stream and only keeps useful packet samples for further processing. For MIMO receivers, the Signal2 block is used, with the primary difference being that Signal2 also chops samples from the second sample stream to match the length of the first stream and compensates for the CFO of the second stream.

Demodulation: The demodulation block converts the OFDM symbols to QAM constellations and disassembles them into soft bits. A state machine is used to determine packet format. It first updates the Legacy channel and checks whether legacy MCS is the lowest. The lowest MCS leads to further demodulation and decodes on the following two symbols to check the HT Signal and VHT Signal A, which will decide the following demodulation. If the Non-Legacy checking fails, the packet is demodulated as Legacy. If it is either HT or VHT, the channel is re-estimated, and packet is then demodulated accordingly. For the OFDM part, the channel is compensated after fast Fourier transform (FFT), and pilots are used to correct the residual CFO. The QAM constellations are disassembled into soft bits and deinterleaved. Some steps are simplified to speed up the processing, such as deinterleaving using a predefined lookup table for each symbol but not following the method given in the standards. The demodulation block outputs the soft bits to the next decoding block.

The difference between the demodulation block and demodulation-2 block lies in their design purposes and functionalities. The demodulation block is intended for SISO and MU-MIMO receivers. It can perform channel sounding and receive MU-MIMO packets with group number and position in the group to estimate the corresponding channel and demodulate accordingly. On the other hand, the demodulation-2 block is designed for the AP side in MU-MIMO, which can handle two stream inputs simultaneously. To simplify processing, channel sounding function is removed. However, we are considering adding full functionality to all blocks in future developments.

Decoding: The decoding block processes a bit stream input and outputs a message stream, which is the required input type for the socket PDU block. It takes the input soft bits with a specified trellis length, decodes the packet, and checks the cyclic redundancy check 32 (CRC32). The Viterbi decoder performs a forward update for the whole packet and then traces back from the very end, which is the 6-bit-zero tail. This approach is used because GNU Radio accumulates some samples and then provides them to the blocks so that the proposed receiver does not aim for real-time performance in the communication stack. If the packet is correct, decoding block passes the packet to the Python MAC layer through a UDP message for further customized processing. In case of a null data packet (NDP) used for channel sounding that has no bits to decode, the decoder simply packages the channel information into a UDP message and sends it to the MAC layer.

6.4.3.2 Performance Evaluation

We now present the performance evaluation of GR-WiFi through simulations and real-world testbed experiments.

We generate simulation signal samples and have the receiver demodulate. The receiver successfully receives the packet under low SNR conditions, indicating its enhanced performance in handling lower-quality signals. A higher SNR value means a better wireless link that can support higher MCS, leading to higher data rates. Figure 6.12 presents the packet delivery ratio (PDR) of GR-WiFi for three different format packets under different SNR conditions with different MCS. The figure reveals that a VHT format may require a higher SNR than the other two formats to achieve the same PDR. Figure 6.13 presents the PDR of GR-WiFi VHT and HT for SU-MIMO transmissions. With the additive white Gaussian noise (AWGN) channel simulation, the performance is similar to SISO. However, the performance drops in real-world multipath propagation conditions.

Figure 6.14 shows our MU-MIMO testbed, which aims to demonstrate the simultaneous transmissions between AP and two stations. Equipment deployed for the testbed includes one USRP B210 with 2×2 TRX antenna array and two USRP B200 with 1×1 TRX antennas. One complete transmission involves the following steps: the AP first broadcasts NDP packets to have all stations' attention. Stations receive and capture the two LTFs in the NDP packets. Each station sends its two received LTFs back to AP. AP gathers all the LTFs information from stations and calculates the steering matrix accordingly. Based on the steering matrix,

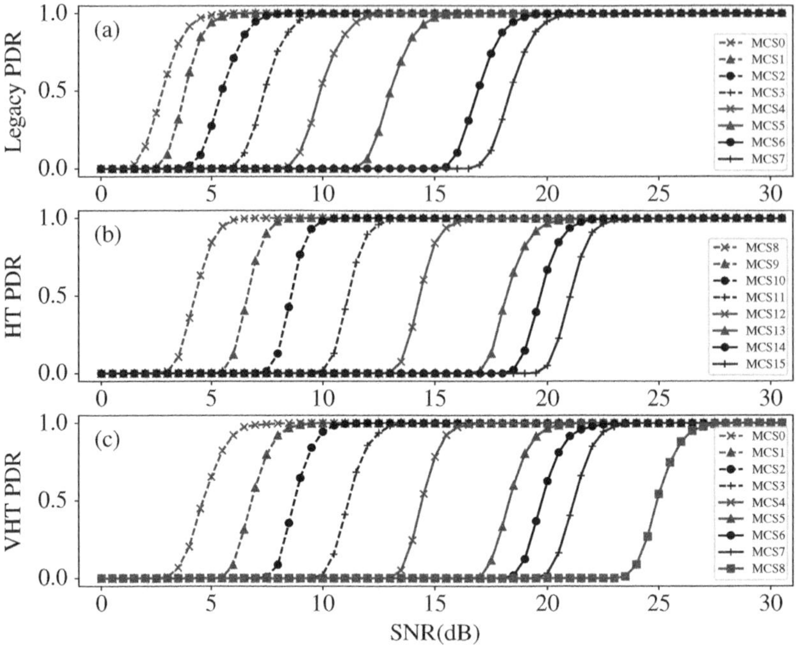

Figure 6.12 GR-WiFi packet delivery ratio against signal-to-noise ratio. (a) Legacy format packet, (b) HT format packet, and (c) VHT packet.

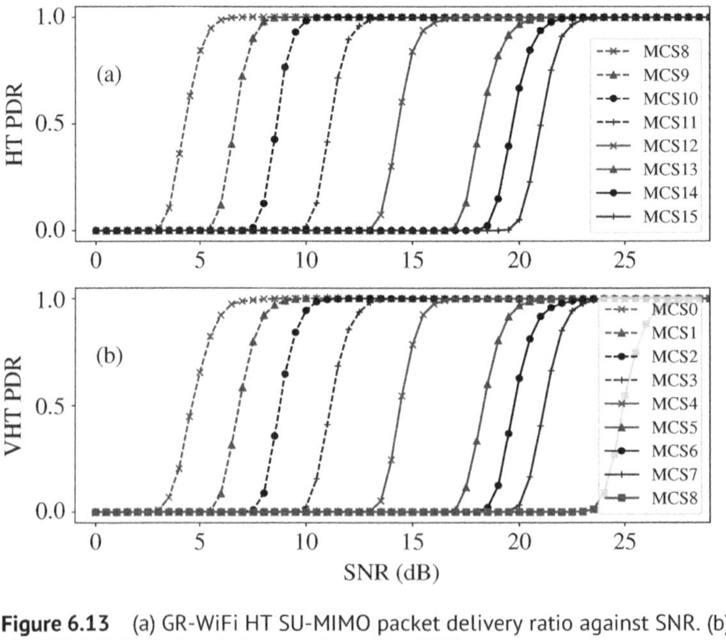

Figure 6.13 (a) GR-WiFi HT SU-MIMO packet delivery ratio against SNR. (b) GR-WiFi VHT SU-MIMO packet delivery ratio against SNR.

Figure 6.14 GR-WiFi MU-MIMO testbed with 1 AP (laptop) and two STAs (desktops) where real channel response is shown at the AP side.

AP can then transmit packets biased in a particular direction based on where the station is. In our testbed, we demonstrate the successful reception in both stations.

The current implementation of GR-WiFi has certain limitations, primarily due to its inability to fully connect to a real-world WiFi network. This is because the receiver cannot return the acknowledgment (ACK) to the sender within the designated ACK timeout period. Nonetheless, it can operate as a passive receiver to intercept packets transmitted over the air. Despite its limitations, GR-WiFi is a crucial step in exploring new protocols, and its successful deployment on GNU Radio offers a valuable reference for future implementations on SRT-WiFi.

6.5 Conclusion and Future Work

In this chapter, we review three different WiFi implementations for supporting high-speed and real-time industrial control applications. The RT-WiFi solution is based on COTS hardware. SRT-WiFi is implemented on an advanced SDR platform where the radio functions are programmed on FPGA to support hard real-time performance. GR-WiFi is implemented on GNU Radio-based SDR platform, supporting a much shorter development period. Extensive experiments have been conducted on both SRT-WiFi and GR-WiFi for functional validation and performance evaluation.

As future work, we will port the implementations of 802.11n/ac PHYs from GR-WiFi to SRT-WiFi to make it full-blown and support hard real-time performance. We will add newer standards to SRT-WiFi, such as IEEE 802.11ax.

Both FPGA-based SRT-WiFi and GNU Radio-based GR-WiFi implementations, once mature, will be made public to the wireless communities to support a broad range of research and development (R&D) activities.

Bibliography

B. Bloessl, M. Segata, C. Sommer, and F. Dressler. Decoding IEEE 802.11a/g/p OFDM in software using GNU radio. In *Proceedings of the 19th Annual International Conference on Mobile Computing & Networking*, MobiCom'13, pages 159–162, New York, NY, USA, 2013.

GNU Radio Foundation. GNU-Radio, 2007. URL https://www.gnuradio.org/.

IEEE 802.11 Working Group. IEEE Standard for Information Technology-Telecommunications and Information Exchange between Systems-Local and Metropolitan Area Networks–Specific Requirements Part 11: Wireless LAN Medium Access Control and Physical Layer Specifications. *IEEE Std 802.11-2020 (Revision of IEEE Std 802.11-2016)*, pages 1–7524, 2021.

X. Jiao, W. Liu, M. Mehari, M. Aslam, and I. Moerman. openwifi: A free and open-source IEEE 802.11 SDR implementation on SoC. In *IEEE 91st Vehicular Technology Conference*, pages 1–2, 2020.

E. A. Lee and D. G. Messerschmitt. *Digital Communication*. Springer Science & Business Media, 2012.

Q. Leng, Y.-H. Wei, S. Han, A. K. Mok, W. Zhang, and M. Tomizuka. Improving control performance by minimizing jitter in RT-WiFi networks. In *2014 IEEE Real-Time Systems Symposium*, pages 63–73. IEEE, 2014.

Q. Leng, W.-J. Chen, P.-C. Huang, Y.-H. Wei, A. K. Mok, and S. Han. Network management of multicluster RT-WiFi networks. *ACM Transactions on Sensor Networks (TOSN)*, 15(1):1–26, 2019.

D. C. Mur. Linux Wi-Fi open source drivers-mac80211, ath9k/ath5k, 2011. URL https://www.yumpu.com/en/document/view/39298312/linux-wi-fi-open-source-drivers-daniel-camps-mur.

F. Tramarin, A. K. Mok, and S. Han. Real-time and reliable industrial control over wireless LANs: Algorithms, protocols, and future directions. *Proceedings of the IEEE*, 107(6):1027–1052, 2019.

Y.-H. Wei, Q. Leng, S. Han, A. K. Mok, W. Zhang, and M. Tomizuka. RT-WiFi: Real-time high-speed communication protocol for wireless cyber-physical control applications. In *2013 IEEE 34th Real-Time Systems Symposium*, pages 140–149. IEEE, 2013.

Y.-H. Wei, Q. Leng, W.-J. Chen, A. K. Mok, and S. Han. Schedule adaptation for ensuring reliability in RT-WiFi-based networked embedded systems. *ACM Transactions on Embedded Computing Systems (TECS)*, 17(5):1–23, 2018.

D. Xia, J. Hart, and Q. Fu. Evaluation of the Minstrel rate adaptation algorithm in IEEE 802.11 g WLANs. In *2013 IEEE International Conference on Communications (ICC)*, pages 2223–2228. IEEE, 2013.

Z. Yun, P. Wu, S. Zhou, A. K. Mok, M. Nixon, and S. Han. RT-WiFi on software-defined radio: design and implementation. In *2022 IEEE 28th Real-Time and Embedded Technology and Applications Symposium (RTAS)*, pages 254–266, 2022.

Part III

Cyber-Security in Wireless Sensor Networks

7

Security and Privacy in Distributed Kalman Filtering

Naveen K. D. Venkategowda[1], Ashkan Moradi[2], and Stefan Werner[2]

[1] *Department of Science and Technology, Linköping University, Norrköping, Sweden*
[2] *Department of Electronic Systems, NTNU, Trondheim, Norway*

7.1 Introduction

Multi-agent networks such as Internet of Things (IoT) and networked cyber-physical systems (CPS) are vital in Industry 4.0 and smart infrastructure and used in applications such as structural health monitoring, climate monitoring, smart cities, and digital twins [Kim and Kumar, 2012; Da Xu et al., 2014; Yu and Xue, 2016; Xu et al., 2018]. These networks comprise agents such as sensors, actuators, and controllers distributed over a wide area to monitor and control physical phenomena by collecting data from agents for learning and decision-making. To that end, multi-agent systems rely on distributed signal processing and learning algorithms where the network agents exchange information with their neighbors. This enables agents to cooperatively perform tasks such as event detection, tracking, and parameter estimation. One such framework is Kalman filters, which have found wide applications in positioning [Feng et al., 2020], smart grids [Yang et al., 2013], localization [Rezaei and Sengupta, 2007], target tracking [Liggins et al., 1997], and dead reckoning [Brossard et al., 2020].

However, the limited computational and energy resources available to agents, coupled with the decentralized nature of networks, make distributed Kalman filters susceptible to cybersecurity threats and malicious attacks from adversaries [Humayed et al., 2017; Wolf and Serpanos, 2018]. Attacks on multi-agent networks can be divided into active and passive. Specifically, passive attacks involve an adversary eavesdropping on communications between agents to gather information [Kapetanovic et al., 2015], whereas active attacks, such as denial-of-service (DoS) and integrity attacks, aim to disrupt the normal functioning of the network.

Wireless Sensor Networks in Smart Environments: Enabling Digitalization from Fundamentals to Advanced Solutions, First Edition. Edited by Domenico Ciuonzo and Pierluigi Salvo Rossi.

During a DoS attack, communication between agents is hindered due to jamming [Chang, 2002], while integrity attacks involve adversaries or malicious agents injecting false information into the network [Vempaty et al., 2013]. In data-falsification attacks, an adversary might modify the data stream of an agent or gain control of multiple agents to degrade the overall system performance. In these situations, the primary challenges for distributed algorithms include verifying the trustworthiness of local information and ensuring robust inference to achieve system objectives despite the presence of adversaries.

Distributed algorithms rely on agents sharing their local information with adjacent (i.e. neighboring) agents, which can potentially compromise privacy. While exchanging information fosters collaboration among network agents, it also raises concerns about privacy due to the risk of exposing private data to adversaries. For example, in the context of smart grids, multiple generators need to reach a consensus on costs without disclosing their individual data [Yang et al., 2013]. The multi-agent rendezvous problem involves agents agreeing on a meeting location without revealing their initial positions [Lin et al., 2007].

In this chapter, we explore distributed Kalman filtering (DKF) algorithms and address the following questions: (i) How can we ensure robustness against data-falsification attacks without significantly degrading performance? (ii) How can we enhance privacy for agents and improve the overall privacy-accuracy tradeoff for all agents without imposing a high computational load? The rest of the chapter is organized as follows. In Section 7.2, we provide an overview of a generic distributed Kalman filter. We present the data-falsification attack and Byzantine-robust Kalman filter model in Section 7.3. Section 7.4 focuses on privacy and privacy-preserving Kalman filters.

Mathematical Notations: Scalars, vectors, and matrices are represented by lowercase, bold lowercase, and bold uppercase letters, whereas $\mathbf{I}_m$, $\mathbf{0}_m$, and $\mathbf{1}_m$ denote an $m \times m$ identity matrix, an $m \times m$ zero matrix, and a column vector of ones with m elements. The transpose, inverse, and expectation operators are represented by $(\cdot)^T$, $(\cdot)^{-1}$, and $\mathbb{E}\{\cdot\}$. The trace operator is $\text{tr}(\cdot)$, and $\text{Blockdiag}(\{\mathbf{A}_i\}_{i=1}^N)$ denotes a block diagonal matrix with $\mathbf{A}_i$s on the main diagonal. The Kronecker product and Hadamard product of two matrices are represented by $\otimes$ and $\odot$, respectively, whereas $\text{diag}(\mathbf{a})$ denotes a diagonal matrix with elements of vector $\mathbf{a}$ on the diagonal. The notation $\mathbf{x}(k) \sim \mathcal{N}(\mathbf{0}, \mathbf{\Sigma})$ indicates that $\mathbf{x}(k)$ is a white Gaussian sequence with covariance $\mathbf{\Sigma}$, and $\dagger$ represents the Moore–Penrose pseudoinverse. The greater than and less than symbols in the scalar inequalities are represented by $>$ and $<$, respectively, while $\mathbf{A} \succcurlyeq 0$ indicates a positive semidefinite matrix and $\mathbf{A} \succcurlyeq \mathbf{B}$ implies $\mathbf{A} - \mathbf{B}$ is positive semidefinite. The (i,j)th element of matrix $\mathbf{A}$ is $[\mathbf{A}]_{ij}$, and $\mathcal{A} \subseteq \mathcal{B}$ indicates set $\mathcal{A}$ is a subset of $\mathcal{B}$.

Element-wise inequality is denoted by $\mathbf{A} \leq \mathbf{B}$. The maximum and minimum eigenvalues of square matrix $\mathbf{A}$ are $\lambda_{max}(\mathbf{A})$ and $\lambda_{min}(\mathbf{A})$. Operator vec($\mathbf{A}$) converts matrix $\mathbf{A}$ to a column vector and $\text{vec}^{-1}(\cdot)$ is its inverse. Half vectorization of a symmetric matrix $\mathbf{M} \in \mathbb{R}^{m\times m}$ is represented by $\text{vec}_h(\mathbf{M}) \in \mathbb{R}^{m(m+1)/2}$, where $\text{vec}_h(\mathbf{M}) = [M_{1,1}, \dots, M_{1,m}, M_{2,2}, \dots, M_{2,m}, \dots, M_{m,m}]^T$ with M_{ij} as the (i,j)th element of $\mathbf{M}$, and $\text{vec}_h^{-1}(\cdot)$ is its inverse.

7.2 Distributed Kalman Filter

We examine a network of N that observes a dynamic process, which is modeled by a linear time-varying state equation

$$\mathbf{x}_{n+1} = \mathbf{A}\mathbf{x}_n + \mathbf{w}_n, \quad n = 1, 2 \dots, \tag{7.1}$$

where, for time instant n, $\mathbf{x}_n \in \mathbb{R}^m$ is the state vector, $\mathbf{A} \in \mathbb{R}^{m\times m}$ is the state-transition matrix, and $\mathbf{w}_n$ is the process noise. At time n, agent i observes $\mathbf{y}_{i,n} \in \mathbb{R}^q$ given by

$$\mathbf{y}_{i,n} = \mathbf{H}_i\mathbf{x}_n + \mathbf{v}_{i,n}, \tag{7.2}$$

where $\mathbf{H}_i \in \mathbb{R}^{q\times m}$ and $\mathbf{v}_{i,n} \in \mathbb{R}^q$ are the observation matrix and observation noise, respectively. The state and observation noises, $\mathbf{w}_n$ and $\mathbf{v}_{i,n}$, are independent Gaussian processes with zero mean and covariances $\mathbf{Q} \in \mathbb{R}^{m\times m}$ and $\mathbf{R}_i \in \mathbb{R}^{q\times q}$, respectively.

The agents form an interconnected network modeled as an undirected graph $\mathcal{G}(\mathcal{N}, \mathcal{E})$. Here, the set $\mathcal{N}$ includes all agents, totaling $|\mathcal{N}| = N$, while the edge set $\mathcal{E}$ represents agents that can communicate with each other. The open neighbor set $\mathcal{N}_i$ of agent i, with cardinality $|\mathcal{N}_i|$, includes all its adjacent neighbors. The network adjacency matrix, $\mathbf{E}$, has $e_{ij} = 1$ whenever nodes i and j are adjacent, and $e_{ij} = 0$ otherwise. Finally, the degree matrix $\mathbf{D} \triangleq \text{diag}(\{|\mathcal{N}_i|\}_{i=1}^N)$ contains the node degrees on its main diagonal.

In this setup, agents aim to collaboratively estimate the system state through information exchange with their adjacent agents without coordination from a central entity. To that end, let $\mathbf{y}_n = [\mathbf{y}_{1,n}^T, \mathbf{y}_{2,n}^T, \dots, \mathbf{y}_{N,n}^T]^T$ denote the collection of observations from all the agents at time n, the augmented observation matrix be defined as $\mathbf{H} = [\mathbf{H}_1^T, \mathbf{H}_2^T, \dots, \mathbf{H}_N^T]^T$ and the network-wide observation noise $\mathbf{v}_n = [\mathbf{v}_{1,n}^T, \mathbf{v}_{2,n}^T, \dots, \mathbf{v}_{N,n}^T]^T$ with covariance matrix defined as $\mathbf{R} = Block\text{diag}\left(\{\boldsymbol{R}_i\}_{i=1}^N\right)$. Let $\hat{\mathbf{x}}_{n|n-1}$ and $\hat{\mathbf{x}}_{n|n}$ denote the a priori and a posteriori estimates of $\mathbf{x}_n$, respectively, of the state vector defined in (7.1). Then the Kalman filter in the information form [Olfati-Saber, 2005; Talebi and Werner, 2019], can be written as

$$\begin{aligned}\hat{\mathbf{x}}_{n|n-1} &= \mathbf{A}\hat{\mathbf{x}}_{n-1|n-1}\\ \mathbf{P}_{n|n-1} &= \mathbf{A}\mathbf{P}_{n-1|n-1}\mathbf{A}^{\mathrm{T}} + \mathbf{Q}\\ \mathbf{P}^{-1}_{n|n} &= \mathbf{P}^{-1}_{n|n-1} + \mathbf{H}^T\mathbf{R}^{-1}\mathbf{H}\\ \hat{\mathbf{x}}_{n|n} &= \hat{\mathbf{x}}_{n|n-1} + \mathbf{P}_{n|n}\mathbf{H}^T\mathbf{R}^{-1}\left(\mathbf{y}_n - \mathbf{H}\hat{\mathbf{x}}_{n|n-1}\right),\end{aligned} \tag{7.3}$$

where $\mathbf{P}_{n|n-1}$ and $\mathbf{P}_{n-1|n-1}$ are the a priori and a posteriori error covariances, respectively. Let us assume that the above computations are performed at every node. At agent i, the local copy of a priori and a posteriori state estimates are denoted $\hat{\mathbf{x}}_{i,n|n-1}$ and $\hat{\mathbf{x}}_{i,n|n}$, respectively, whereas the a priori and a posteriori error covariances are represented by $\mathbf{P}_{i,n|n-1}$ and $\mathbf{P}_{i,n-1|n-1}$, respectively. The local copy of state estimates and covariance matrices at agent i can be written as

$$\begin{aligned}\hat{\mathbf{x}}_{i,n|n-1} &= \mathbf{A}\hat{\mathbf{x}}_{i,n-1|n-1}\\ \mathbf{P}_{i,n|n-1} &= \mathbf{A}\mathbf{P}_{i,n-1|n-1}\mathbf{A}^T + \mathbf{Q}.\end{aligned} \tag{7.4}$$

Substituting the augmented observation matrix $\mathbf{H}$ and observation noise covariance matrix $\mathbf{R}$ in (7.3), the local error covariance can be rewritten as

$$\mathbf{P}^{-1}_{i,n|n} = \mathbf{P}^{-1}_{n|n} = \mathbf{P}^{-1}_{i,n|n-1} + \sum_{j=1}^{N}\mathbf{H}_j^T\mathbf{R}_j^{-1}\mathbf{H}_j = \frac{1}{N}\sum_{j=1}^{N}\mathbf{\Gamma}_{j,n}, \tag{7.5}$$

where the local intermediate covariance information $\mathbf{\Gamma}_{i,n}$ at agent i and time n is defined as

$$\mathbf{\Gamma}_{i,n} = \mathbf{P}^{-1}_{i,n|n-1} + N\mathbf{H}_i^T\mathbf{R}_i^{-1}\mathbf{H}_i. \tag{7.6}$$

Similarly, at agent i, the local copy $\hat{\mathbf{x}}_{i,n|n}$ of state estimate given in (7.3) can be rewritten as

$$\hat{\mathbf{x}}_{i,n|n} = \hat{\mathbf{x}}_{i,n|n-1} + \sum_{i=1}^{N}\mathbf{P}_{i,n|n}\mathbf{H}_i^T\mathbf{R}_i^{-1}\left(\mathbf{y}_{i,n} - \mathbf{H}_i\hat{\mathbf{x}}_{i,n|n-1}\right) = \frac{1}{N}\sum_{i=1}^{N}\boldsymbol{r}_{i,n}, \tag{7.7}$$

where

$$\boldsymbol{r}_{i,n} = \hat{\mathbf{x}}_{i,n|n-1} + N\mathbf{P}_{i,n|n}\mathbf{H}_i^T\mathbf{R}_i^{-1}\left(\mathbf{y}_{i,n} - \mathbf{H}_i\hat{\mathbf{x}}_{i,n|n-1}\right). \tag{7.8}$$

From (7.5) and (7.7), it can be seen that the Kalman filter in (7.3) reduces to computing the network-wide average of quantities $\mathbf{\Gamma}_{i,n}$ and $\boldsymbol{r}_{i,n}$ at each agent. These averages can be computed in a distributed manner via peer-to-peer communication by employing some average consensus algorithm.

Consider an iterative average consensus algorithm given by

$$\mathbf{\Theta}_{i,n}(k) = \mathbf{\Theta}_{i,n}(k-1) + \sum_{j\in\mathcal{N}_i} w_{i,j}\left(\mathbf{\Theta}_{j,n}(k-1) - \mathbf{\Theta}_{i,n}(k-1)\right), \quad k = 1, 2, \ldots, \tag{7.9}$$

Algorithm 7.1 Distributed Kalman Filtering algorithm

For each agent $i \in \mathcal{N}$

Initialize: $\hat{\mathbf{x}}_{i,0|0}$ and $\mathbf{P}_{i,0|0}$

$\hat{\mathbf{x}}_{i,n|n-1} = \mathbf{A}\hat{\mathbf{x}}_{i,n-1|n-1}$

$\mathbf{P}_{i,n|n-1} = \mathbf{A}\mathbf{P}_{i,n-1|n-1}\mathbf{A}^T + \mathbf{Q}$

$\mathbf{\Gamma}_{i,n} = \mathbf{P}^{-1}_{i,n|n-1} + N\mathbf{H}_i^T\mathbf{R}_i^{-1}\mathbf{H}_i$

$\mathbf{P}^{-1}_{i,n|n} \leftarrow \boxed{\text{Avg. Consensus}} \leftarrow \{\mathbf{\Gamma}_{j,n},\ \forall j \in \mathcal{N}_i \cup i\}$

$\boldsymbol{r}_{i,n} = \hat{\mathbf{x}}_{i,n|n-1} + N\mathbf{P}_{i,n|n}\mathbf{H}_i^T\mathbf{R}_i^{-1}\left(\mathbf{y}_{i,n} - \mathbf{H}_i\hat{\mathbf{x}}_{i,n|n-1}\right)$

$\hat{\mathbf{x}}_{i,n|n} \leftarrow \boxed{\text{Avg. Consensus}} \leftarrow \{\boldsymbol{r}_{j,n},\ \forall j \in \mathcal{N}_i \cup i\}$

with the initial conditions $\mathbf{\Theta}_{j,n}(0), j = 1, \dots, N$. If the weights $w_{i,j}$ are chosen such that the matrix $\mathbf{W}$ with elements $[\mathbf{W}]_{i,j} = w_{i,j}$ is doubly stochastic, then at every agent we have

$$\lim_{k\to\infty} \mathbf{\Theta}_{i,n}(k) = \frac{1}{N}\sum_{j=1}^{N} \mathbf{\Theta}_{j,n}(0), \quad i = 1, 2, \dots, N.$$

At every time instant n, we can set $\mathbf{\Theta}_{i,n}(0) = \mathbf{\Gamma}_{i,n}$ to obtain (7.5) or set $\mathbf{\Theta}_{i,n}(0) = \boldsymbol{r}_{i,n}$ to obtain (7.7) and then employing the average consensus algorithm, the agents can estimate the state in a distributed manner. In the remainder of this chapter, we illustrate the average consensus algorithm using the below diagram:

$$\mathbf{\Theta}_{i,n}(k) \leftarrow \boxed{\text{Avg. Consensus}} \leftarrow \{\mathbf{\Theta}_{j,n}(0),\ \forall j \in \mathcal{N}_i \cup i\}, \tag{7.10}$$

where, for agent i and instant n, $\mathbf{\Theta}_{j,n}(0)$, $j \in \mathcal{N}_i \cup i$ serves as the input to the average consensus algorithm, while $\mathbf{\Theta}_{i,n}(k)$ is the resulting output given by (7.9). We can now summarize the steps involved in the distributed Kalman filtering as Algorithm 7.1.

It can be observed in the above discussion that agents share local information with their immediate neighbors. Therefore, the efficacy of the algorithm depends on agents following Algorithm 7.1 faithfully. In a scenario where the agents are compromised, and the information exchanged by the compromised nodes is unreliable, the performance of the filtering algorithm is adversely affected. Further, in many applications, agents possess local data that contain sensitive information requiring privacy protection. Adversaries can deduce the local state or correlated information of other agents by analyzing the data received from neighboring agents. Therefore, in this chapter, we examine distributed Kalman filtering algorithms considering two types of adversaries within the network:

- A *Byzantine agent* that injects false information into the estimation process to impair the overall network performance.

- An *honest-but-curious agent* that contributes faithfully to the estimation task but is interested in inferring private information of other agents from messages received from its neighbors.

7.3 Security in Distributed Kalman Filter

Consider a subset of Byzantine agents, denoted as $\mathcal{B} \subset \mathcal{N}$. Unlike regular agents, these Byzantine agents share falsified information with neighbors, aiming to degrade network-wide performance [Vempaty et al., 2013], as shown in Figure 7.1. During a linear data falsification attack, these malicious agents perturb the shared messages by introducing noise. Let $\bar{\mathbf{x}}_{j,n}$ denote the information transmitted by agent j at each time instant n, which can be expressed as

$$\bar{\mathbf{x}}_{j,n} = \begin{cases} \hat{\mathbf{x}}_{j,n} + \boldsymbol{\delta}_{j,n} & j \in \mathcal{B} \\ \hat{\mathbf{x}}_{j,n} & j \notin \mathcal{B}, \end{cases} \tag{7.11}$$

where $\boldsymbol{\delta}_{j,n} \in \mathbb{R}^m$ is noise sequence added by the Byzantine agent, assumed here Gaussian with zero mean and covariance $\boldsymbol{\Sigma}_i \in \mathbb{R}^{m \times m}$ [Bai et al., 2017; Chen et al., 2018a], which is also the optimal sequence that maximizes stealthiness, i.e. evade attack detection. In addition, we assume that the Byzantine agents co-operate to design the attack sequence that maximizes the deterioration in the network performance. To that end, in contrast to the attack sequences being independent, i.e. $\mathbb{E}\{\boldsymbol{\delta}_{i,n}\boldsymbol{\delta}_{j,n}^T\} = \mathbf{0}$ for all $i \neq j$, a coordinated attack is given by correlated attack

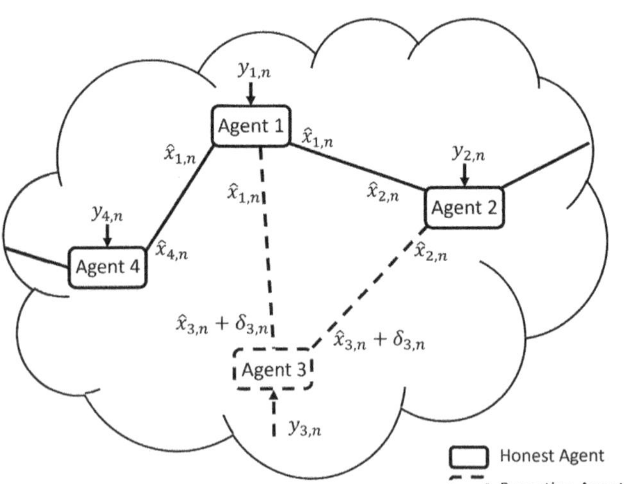

Figure 7.1 Information exchanged in a network containing Byzantine agents.

sequences with covariance matrix $\boldsymbol{\Sigma} = \mathbb{E}\{\boldsymbol{\delta}_n \boldsymbol{\delta}_n^T\}$, where the network-wide perturbation vector is defined by $\boldsymbol{\delta}_n = [\boldsymbol{\delta}_{1,n}^T, \dots, \boldsymbol{\delta}_{N,n}^T]^T$ with $\boldsymbol{\delta}_{j,n} = \mathbf{0}$ if $j \notin \mathcal{B}$.

The effect of Byzantine agents can be minimized using statistical approaches proposed in Kailkhura et al. [2016], He et al. [2022], and Chen et al. [2018b–2019]. These techniques primarily focus on adjusting the weights of measurements received from neighbors. Measurements suspected to be from Byzantine agents are given lower weights, thereby minimizing their influence on the state estimates. To provide robustness against adversaries, homomorphic encryption-based schemes [Fauser and Zhang, 2020, 2021; Ni et al., 2021; Zhou et al., 2022], randomization-based methods [Lin et al., 2019], and redundancy-based approaches [Mitra and Sundaram, 2019; Rajput et al., 2019; Krishnamurthy and Khorrami, 2021; Mitra et al., 2021] have also been proposed. In contrast, this chapter focuses on distributed Kalman filtering algorithms using regularization and distributed optimization to achieve robustness against Byzantine attacks.

7.3.1 Byzantine Robust Distributed Kalman Filter

Here, the distributed Kalman filter is formulated as a maximum-likelihood estimation problem that bridges traditional Kalman filtering principles Olfati [2009] with optimization techniques [Ryu and Back, 2019; Moradi et al., 2023]. In particular, as shown in Moradi et al. [2023], the *a posteriori* state estimates are obtained by solving the following problem:

$$\begin{aligned}
&\min_{\{\mathbf{x}_{i,n}\}_{i=1}^{N}} \sum_{i=1}^{N} f_i(\mathbf{x}_{i,n}) \\
&\text{s.t. } \mathbf{x}_{i,n} = \mathbf{x}_{j,n}, \quad \forall j \in \mathcal{N}_i, i \in \mathcal{N},
\end{aligned} \tag{7.12}$$

where $f_i(\mathbf{x}_{i,n})$ are local objectives given by

$$\begin{aligned}
f_i(\mathbf{x}_{i,n}) = \frac{1}{2} \Big(&(\mathbf{y}_{i,n} - \mathbf{H}_i \mathbf{x}_{i,n})^T \mathbf{R}_i^{-1} (\mathbf{y}_{i,n} - \mathbf{H}_i \mathbf{x}_{i,n}) \\
&+ \frac{1}{N} (\mathbf{x}_{i,n} - \hat{\mathbf{x}}_{i,n|n-1})^T \mathbf{P}_{i,n|n-1}^{-1} (\mathbf{x}_{i,n} - \hat{\mathbf{x}}_{i,n|n-1}) \Big),
\end{aligned} \tag{7.13}$$

and the constraints in (7.12) enforce consensus among network agents. The DKF problem finds optimal solutions for $\mathbf{x}_{i,n}^*$, $i \in \mathcal{N}$, in (7.12). This leads to a posteriori state estimates $\hat{\mathbf{x}}_n = [\hat{\mathbf{x}}_{1,n}^T, \dots, \hat{\mathbf{x}}_{N,n}^T]^T$ where $\hat{\mathbf{x}}_{i,n} = \mathbf{x}_{i,n}^*$.

Inspired by Ben-Ameur et al. [2016] and Peng et al. [2021], the constraints in (7.12) are approximated using a TV-norm penalty, enhancing resilience against data falsification attacks. Hence, (7.12) is reformulated as:

$$\overline{\mathbf{x}}_n^* = \min_{\{\mathbf{x}_{i,n}\}_{i=1}^{N}} \sum_{i=1}^{N} \left(f(\mathbf{x}_{i,n}) + \frac{\lambda_{\text{tv}}}{2} \sum_{j \in \mathcal{N}_i} \|\mathbf{x}_{i,n} - \mathbf{x}_{j,n}\|_1 \right), \tag{7.14}$$

where vector $\overline{\mathbf{x}}_n^* = [(\mathbf{x}_{1,n}^*)^T, \dots, (\mathbf{x}_{N,n}^*)^T]^T$ comprises all local optimal parameters, and penalty parameter λ_{tv} ensures that estimates $\mathbf{x}_{i,n}$ and $\mathbf{x}_{j,n}$ are kept close together. Increasing λ_{tv} results in greater similarity between these estimates. Nonetheless, the TV-norm penalty allows certain pairs of estimates to remain distinct, which is essential whenever Byzantines are present.

The optimization problem in (7.14) can be solved using a subgradient method [Peng et al., 2021; Moradi et al., 2023], resulting in the following local state estimate update at agent i:

$$\mathbf{x}_{i,n}^{l+1} = \mathbf{x}_{i,n}^{l} - \alpha_n \left(\nabla_{\mathbf{x}_{i,n}} f(\mathbf{x}_{i,n}^{l}) + \lambda_{\text{tv}} \sum_{j \in \mathcal{N}_i} \text{sign}(\mathbf{x}_{i,n}^{l} - \mathbf{x}_{j,n}^{l}) \right). \tag{7.15}$$

Here, $\alpha_n > 0$ is the step size, and $\mathbf{x}_{i,n}^{l}$ represents the local-state estimate during iteration l of the subgradient method. The element-wise sign function, denoted as sign$(\cdot)$, assigns sign$(x) = 1$ for $x > 0$, and sign$(x) = -1$ for $x < 0$. If $x = 0$, sign(x) takes an arbitrary value within the range $[-1, 1]$. Assuming the presence of a group of Byzantine agents, $\mathcal{B} \subset \mathcal{N}$ and substituting the gradient $\nabla_{\mathbf{x}_{i,n}} f(\mathbf{x}_{i,n}^{l})$, the state update is obtained as:

$$\begin{aligned} \mathbf{x}_{i,n}^{l+1} = \mathbf{x}_{i,n}^{l} - \alpha_n \Bigg(& \mathbf{\Omega}_{i,n} \mathbf{x}_{i,n}^{l} - \boldsymbol{\theta}_{i,n} + \lambda_{\text{tv}} \sum_{j \in \mathcal{R}_i} \text{sign}(\mathbf{x}_{i,n}^{l} - \mathbf{x}_{j,n}^{l}) \\ & + \lambda_{\text{tv}} \sum_{j \in \mathcal{B}_i} \text{sign}(\mathbf{x}_{i,n}^{l} - \tilde{\mathbf{x}}_{j,n}^{l}) \Bigg), \end{aligned} \tag{7.16}$$

where, for agent i, $\mathcal{R}_i$ is the set of honest neighbors and $\mathcal{B}_i$ is the set of Byzantine neighbors; thus, the corresponding Byzantine states in (7.16) are given by $\tilde{\mathbf{x}}_{j,n}^{l} = \mathbf{x}_{j,n}^{l} + \delta_j^l$. Finally, we have

$$\begin{aligned} \mathbf{\Omega}_{i,n} &= \mathbf{H}_i^T \mathbf{R}_i^{-1} \mathbf{H}_i + \frac{1}{N} \mathbf{P}_{i,n|n-1}^{-1} \\ \boldsymbol{\theta}_{i,n} &= \mathbf{H}_i^T \mathbf{R}_i^{-1} \mathbf{y}_{i,n} + \frac{1}{N} \mathbf{\Omega}_{i,n|n-1} \hat{\mathbf{x}}_{i,n|n-1}, \end{aligned} \tag{7.17}$$

with $\mathbf{\Omega}_{i,n|n-1} = \mathbf{P}_{i,n|n-1}^{-1}$. Notice that elements of sign$(\mathbf{x}_{i,n}^{l} - \tilde{\mathbf{x}}_{j,n}^{l})$ are constrained to the interval $[-1, 1]$. Consequently, the final term in (7.16) mitigates the effect of corrupted data from Byzantine agents, enhancing the robustness of the state update.

Similarly, updating the error covariance necessitates formulating an optimization problem to achieve consensus on the information matrices $N\mathbf{\Omega}_{i,n}$ across agents. The optimization problem for updating the error covariance is expressed as follows:

$$\begin{aligned} &\min_{\{\zeta_i\}_{i=1}^N} \sum_{i=1}^{N} \|\zeta_i - \mathrm{vec}_h(N\mathbf{\Omega}_{i,n})\|_2^2 \\ &\text{s.t.} \qquad \zeta_i = \zeta_j, \ \ \forall j \in \mathcal{N}_i, i \in \mathcal{N}. \end{aligned} \tag{7.18}$$

The optimal solution to (7.18), $\zeta^* = [(\zeta_1^*)^T, \dots, (\zeta_N^*)^T]^T$, provides the average of $\mathrm{vec}_h(N\mathbf{\Omega}_{i,n})$ across the network. Consequently, the error covariance can be updated using $\mathbf{P}_{i,n} = (\mathrm{vec}_h^{-1}(\zeta_i^*))^{-1}$. Drawing from the TV-norm-penalized optimization problem in (7.14), we adjust (7.18) as follows:

$$\zeta^* = \min_{\{\zeta_i\}_{i=1}^N} \sum_{i=1}^{N} \left(\|\zeta_i - \mathrm{vec}_h(N\mathbf{\Omega}_{i,n})\|_2^2 + \frac{\lambda_{\mathrm{tv}}}{2} \sum_{j\in\mathcal{N}_i} \|\zeta_i - \zeta_j\|_1 \right). \tag{7.19}$$

Using a subgradient method similar to the one in (7.15), we obtain the following update rule:

$$\zeta_i^{l+1} = \zeta_i^l - \gamma_n \left(\zeta_i^l - \mathrm{vec}_h(N\mathbf{\Omega}_{i,n}) + \lambda_{\mathrm{tv}} \sum_{j\in\mathcal{N}_i} \mathrm{sign}(\zeta_i^l - \zeta_j^l) \right), \tag{7.20}$$

where $\gamma_n > 0$ is the step size. After sufficient iterations, denoted by l^*, the suboptimal solutions in (7.16) and (7.20) converge to $(\mathbf{x}_{i,n}^{l^*}, \zeta_i^{l^*})$. This results in updated a posteriori state estimate $\hat{\mathbf{x}}_{i,n} = \mathbf{x}_{i,n}^{l^*}$ and error covariance $\mathbf{P}_{i,n} = (\mathrm{vec}_h^{-1}(\zeta_i^{l^*}))^{-1}$.

Algorithm 7.2 outlines the steps of the Byzantine-robust distributed Kalman filter (BR-DKF) under the assumption that Byzantine agents only manipulate state estimates. Specifically, these agents alter state estimate $\mathbf{x}_{i,n}^l$ at iteration l to $\mathbf{x}_{i,n}^l + \boldsymbol{\delta}_i^l$, where $\boldsymbol{\delta}_i^l \sim \mathcal{N}(\mathbf{0}, \mathbf{\Sigma}_i)$. Here, $\mathbf{\Sigma}_i$ denotes the covariance of the noise-injection sequence at agent $i \in \mathcal{B}$.

7.3.2 Performance Analysis

We will now show that the formulation in (7.14) provides a feasible solution for sufficiently large penalty parameter λ_{tv}. Moreover, the suboptimal solution in (7.16) is within a bounded radius to the optimal solution of (7.14), even with Byzantine agents present. To facilitate future calculations, we define the node-edge incidence matrix $\mathcal{A} = [a_{ij}] \in \mathbb{R}^{N\times|\mathcal{E}|}$, where $a_{ei} = 1$ and $a_{je} = -1$ for each edge $e = (i,j) \in \mathcal{E}$ with $i < j$; otherwise, elements of $\mathcal{A}$ are zero. As shall be seen, we can establish that the solution in (7.14) is optimal and matches the centralized Kalman filter (CKF) solution $\hat{\mathbf{x}}_n^*$ from Ryu and Back [2019]. Further, a lower bound for the penalty parameter λ_{tv} can be derived, ensuring convergence of the solution in (7.14) to the centralized solution in Ryu and Back [2019].

Algorithm 7.2 Byzantine-Robust Distributed Kalman Filter

For each agent $i \in \mathcal{N}$
Initialize: $\hat{\mathbf{x}}_{i,0}$ and $\mathbf{P}_{i,0}$
for all $n > 0$ **do**
$\hat{\mathbf{x}}_{i,n|n-1} = \mathbf{A}\hat{\mathbf{x}}_{i,n-1}$
$\mathbf{P}_{i,n|n-1} = \mathbf{A}\mathbf{P}_{i,n-1}\mathbf{A}^T + \mathbf{Q}$
$\boldsymbol{\Omega}_{i,n|n-1} = \mathbf{P}^{-1}_{i,n|n-1}$
$\boldsymbol{\Omega}_{i,n} = \mathbf{H}_i^T\mathbf{R}_i^{-1}\mathbf{H}_i + \frac{1}{N}\boldsymbol{\Omega}_{i,n|n-1}$
$\boldsymbol{\theta}_{i,n} = \mathbf{H}_i^T\mathbf{R}_i^{-1}\mathbf{y}_{i,n} + \frac{1}{N}\boldsymbol{\Omega}_{i,n|n-1}\hat{\mathbf{x}}_{i,n|n-1}$
Set $\mathbf{x}^1_{i,n} = \mathbf{0}$ and $\zeta_i^1 = \mathbf{0}$
for $l = 1$ to l^* **do**
Share $\mathbf{x}^l_{i,n} + \delta^l_i$ with neighbors if $i \in \mathcal{B}$
$\mathbf{x}^{l+1}_{i,n} = \mathbf{x}^l_{i,n} - \alpha_n\left(\boldsymbol{\Omega}_{i,n}\mathbf{x}^l_{i,n} - \boldsymbol{\theta}_{i,n} + \lambda_{\text{tv}}\sum_{j\in\mathcal{N}_i}\text{sign}(\mathbf{x}^l_{i,n} - \tilde{\mathbf{x}}^l_{j,n})\right)$
$\zeta_i^{l+1} = \zeta_i^l - \gamma_n\left(\zeta_i^l - \text{vec}_h(N\boldsymbol{\Omega}_{i,n}) + \lambda_{\text{tv}}\sum_{j\in\mathcal{N}_i}\text{sign}(\zeta_i^l - \zeta_j^l)\right)$
end for
$\hat{\mathbf{x}}_{i,n} = \mathbf{x}^{l^*}_{i,n}$
$\mathbf{P}_{i,n} = (\text{vec}_h^{-1}(\zeta_i^{l^*}))^{-1}$
end for

In particular, given a connected network topology, if $\lambda_{\text{tv}} \geq \lambda_0$ where

$$\lambda_0 = \frac{\sqrt{N}}{\sigma_{min}(\mathcal{A})}\max_{\forall n}\max_{i\in\mathcal{N}}\|\boldsymbol{\Omega}_{i,n}\mathbf{x}^*_{i,n} - \boldsymbol{\theta}_{i,n}\|_\infty, \tag{7.21}$$

with $\sigma_{min}(\mathcal{A})$ denoting the smallest nonzero singular value of $\mathcal{A}$, $\boldsymbol{\Omega}_{i,n}$, and $\boldsymbol{\theta}_{i,n}$ as defined in (7.17); then, for the optimal solution $\overline{\mathbf{x}}^*_n$ in (7.14) and the optimal solution of the CKF problem $\hat{\mathbf{x}}^*_n$ in Ryu and Back [2019], we have $\overline{\mathbf{x}}^*_n = \mathbf{1}_N \otimes \hat{\mathbf{x}}^*_n$.

With the assumptions stated above and $\lambda_{\text{tv}} \geq \lambda_0$, the solution in (7.16), for each agent $i \in \mathcal{N}$, including the presence of Byzantine agents, remains close to the optimal solution $\overline{\mathbf{x}}^*_n = \mathbf{1}_N \otimes \mathbf{x}^*_{i,n}$ in (7.14) with a radius of

$$\lim_{l\to\infty}\mathbb{E}_l\{\|\mathbf{x}^{l+1}_{i,n} - \mathbf{x}^*_{i,n}\|^2\} \leq \frac{\lambda^2_{\text{tv}}\alpha_n(4\alpha_n + \frac{1}{\varepsilon})(4|\mathcal{R}_i|^2 + |\mathcal{B}_i|^2)m}{1 - \|\left(1 + 2\alpha_n^2\|\boldsymbol{\Omega}_{i,n}\|^2 + 2\varepsilon\alpha_n\right)\mathbf{I} - 2\alpha_n\boldsymbol{\Omega}_{i,n}\|}, \tag{7.22}$$

where $0 \leq \varepsilon \leq \lambda_{min}(\boldsymbol{\Omega}_{i,n})$, and $\mathcal{R}_i$ and $\mathcal{B}_i$ are, as in (7.16), the partitions of neighbor set $\mathcal{N}_i$ containing honest and Byzantine agents, respectively. Step size α_n must satisfy

$$\alpha_n \leq \min_{i\in\mathcal{N}}\left\{\frac{\lambda_{min}(\boldsymbol{\Omega}_{i,n}) - \varepsilon}{\|\boldsymbol{\Omega}_{i,n}\|^2}\right\}. \tag{7.23}$$

The error gap in (7.22) demonstrates that the BR-DKF effectively mitigates the influence of attack amplitude, primarily through the sign(·)-terms, although the number of Byzantines still has an influence.

To assess the performance of the BR-DKF algorithm, we consider an undirected random connected graph with $N = 25$ agents, including $B = 5$ Byzantines selected among those agents having the most neighbors. We consider the following linear dynamic system:

$$\mathbf{x}_{n+1} = \begin{bmatrix} 0.4 & 0.9 & 0 & 0 \\ -0.9 & 0.4 & 0 & 0 \\ 0 & 0 & 0.5 & 0.8 \\ 0 & 0 & -0.8 & 0.5 \end{bmatrix} \mathbf{x}_n + \mathbf{w}_n, \quad \text{and} \quad \mathbf{y}_{i,n} = \begin{bmatrix} 1 & 0 & 0 & 0 \\ 1 & 1 & 0 & 0 \\ 0 & 0 & 1 & 1 \\ 0 & 0 & 1 & 0 \end{bmatrix} \mathbf{x}_n + \mathbf{v}_{i,n},$$

with $\mathbf{Q} = 0.1\mathbf{I}$, and $\mathbf{R}_i = \text{diag}(0.1, 0.2, 0.3, 0.1)$. We compare the BR-DKF algorithm against the CKF, the distributed Kalman filter (DKF) as per [Ryu and Back, 2019], DKF under Byzantine attack (B-DKF), and the BR-DKF under Byzantine attack. The state and error covariance of the BR-DKF were obtained by $l^* = 25$ iterations of the subgradient method, and the results were averaged over 500 realizations to benchmark performance.

Figure 7.2 displays the tracking performance for the various state elements. The shaded colors represent the estimated values of agents, and the solid curves indicate their averages. We see that the BR-DKF decreases the uncertainty significantly and tracks the true state more accurately than the B-DKF algorithm.

In the following, we consider the average MSE across agents, $\text{MSE} = \frac{1}{N}\sum_{i=1}^{N} \|\mathbf{x}_n - \hat{\mathbf{x}}_{i,n}\|^2$, as the performance measure. For the case when Byzantines are absent, parameters α_n, γ_n, and λ_{tv} are adjusted to achieve an MSE as close as possible to the DKF algorithm. However, even in the absence of a Byzantine attack, the performance of BR-DKF is inferior to that of DKF due to the presence

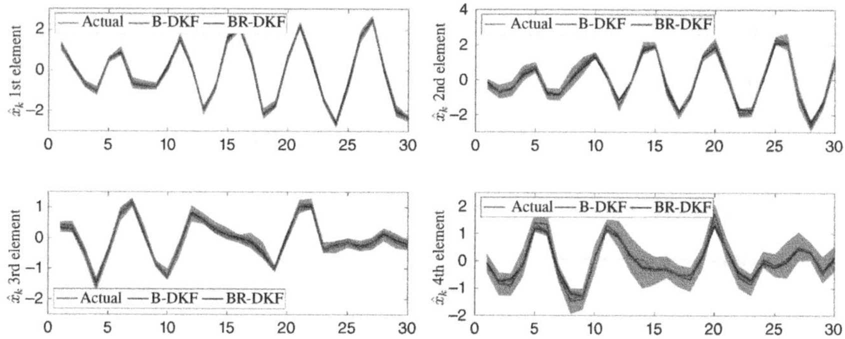

Figure 7.2 Tracking performance for various state elements.

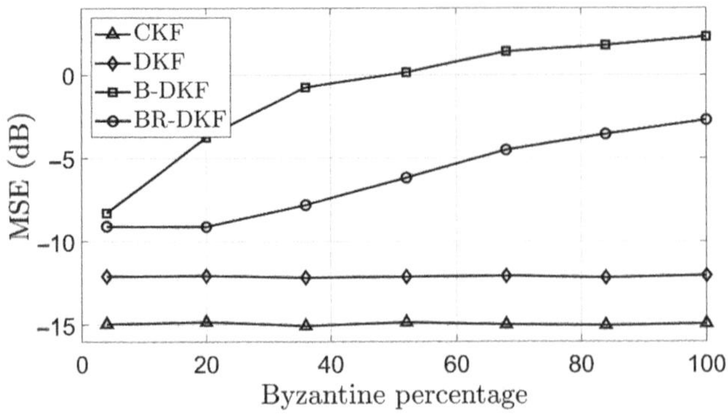

Figure 7.3 Steady-state MSE versus the proportion of Byzantines.

of sign(·)-terms in the update. Here, Byzantine agents execute a coordinated attack, where $\mathbf{\Sigma}_i$ represents the covariance of the perturbation sequence injected by Byzantine agent $i \in \mathcal{B}$.

Figure 7.3 shows MSE against the proportion of Byzantines for various algorithms. It reveals a direct correlation between the increase in Byzantine agents and the MSE. However, the sensitivity of BR-DKF to this percentage is notably lower than that of B-DKF. Figure 7.4, on the other hand, demonstrates the MSE against the attack power (i.e. the trace of the perturbation covariance). We see that when the attack covariance has a low trace, the sign(·)-terms in the BR-DKF constrain the updated values, leading to a higher MSE compared to the DKF. However, as Byzantine agents introduce more noise, the BR-DKF maintains stable performance, whereas the MSE of the B-DKF increases significantly.

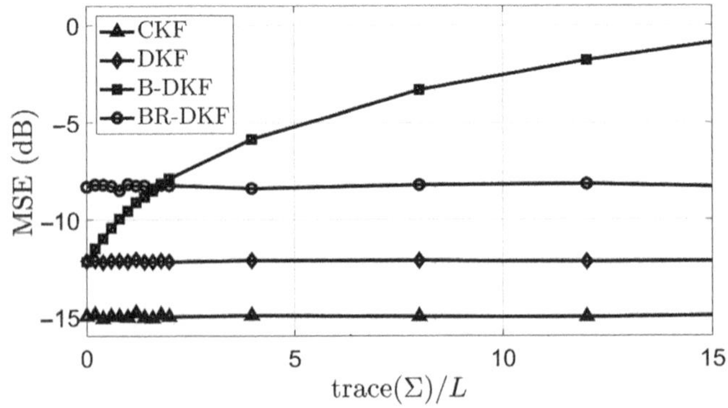

Figure 7.4 Steady-state MSE versus Byzantine attack power.

7.4 Privacy in Distributed Kalman Filters

From the generic distributed Kalman filtering algorithm in Algorithm 7.1, we can observe that agents exchange information with neighbors to estimate the state vector collectively using average consensus. However, the messages exchanged, i.e. local state $\boldsymbol{r}_{i,n}$, contain information about local measurements which includes private or sensitive information about the agent. The messages shared with neighbors improve the state estimation accuracy but, at the same time, provide information to potential adversaries to infer or extract private/sensitive information about an agent. This results in contradictory objectives of maximizing the state estimation accuracy while minimizing the private information leakage. In other words, we encounter an inherent trade-off in accuracy and privacy when designing distributed Kalman filtering algorithms. To understand the trade-off, we need a metric that quantifies privacy leakage. In practice, the privacy leakage is measured through metrics such as differential privacy, mutual information, KL divergence, and uncertainty at the adversary.

7.4.1 Privacy Measures

A commonly used privacy metric is the (ϵ, δ)-differential privacy. A randomized algorithm $\mathcal{M}$ is considered (ϵ, δ)-differentially private if for any two neighboring datasets $\mathcal{S}$ and $\mathcal{S}'$ that differ by a single data sample, and for any subset of outputs $\mathcal{O} \subseteq \text{range}(\mathcal{M})$, the following condition holds:

$$\Pr[\mathcal{M}(\mathcal{S}) \in \mathcal{O}] \leq e^{\epsilon} \Pr[\mathcal{M}(\mathcal{S}') \in \mathcal{O}] + \delta. \tag{7.24}$$

This implies that the ratio of the probability distributions of $\mathcal{M}(\mathcal{S})$ and $\mathcal{M}(\mathcal{S}')$ is bounded by e^{ϵ}. Here, ϵ and δ are privacy parameters that determine the level of privacy guaranteed by the algorithm. Smaller values of ϵ or δ indicate stronger privacy protection.

In differential privacy, the data shared are curated such that inferring about a single entity is prevented and can guarantee that an adversary cannot differentiate an individual datum by removing or adding database entries. Differentially private learning algorithms utilize perturbation mechanisms, i.e. adding controlled noise to the shared data, to maintain privacy at the expense of accuracy. For example, to prevent other agents or external eavesdroppers from inferring individual data, local messages are perturbed with uncorrelated random sequences [Nozari et al., 2017; He et al., 2020; Le Ny, 2020]. Solutions for differentially private Kalman filtering, which aim to minimize MSE under differential privacy constraints, are discussed in Wang et al. [2018], Degue and Le Ny [2017], and Le Ny and Pappas [2014].

DP-based privacy methods [Dwork, 2008] are widely applied in distributed learning and estimation algorithms when the specifics of the attack model and adversary's information set are unknown. Generally considered a conservative metric, these methods provide a worst-case privacy measure. Consequently, they necessitate the use of perturbation noise with higher variance, which leads to a notable decrease in performance.

In distributed Kalman filtering applications, an adversary may have access to various types of information, such as state distributions, noise statistics, observation dynamics, and network details. We can model the prior knowledge and the specific type of reconstruction attack an adversary might employ to extract sensitive data. In these situations, traditional probabilistic indistinguishability metrics like differential privacy or mutual information-based methods are inadequate to capture the extent of private information known to adversaries.

In DKF applications, another privacy measure is the MSE of the adversaries' estimates of the private information. The primary goal of DKF is to prevent adversaries from inferring private data, and with a specified attack model and information set, the MSE metric is more appropriate. This metric evaluates the level of uncertainty faced by adversaries in estimating the private information of agents based on the data they can access [Braca et al., 2016; Mo and Murray, 2017; Wagner and Eckhoff, 2018].

7.4.2 Privacy-Preserving Distributed Kalman Filter

Consider the DKF algorithm described in Section 7.2. We assume, without loss of generality, that the local states $\boldsymbol{r}_{i,n}$, which contain information about observations, are private. Therefore, the objective is to safeguard the private information $\boldsymbol{r}_{i,n}$ from being inferred by an adversary. Privacy leakage can be reduced through two mechanisms [Moradi et al., 2022]. First, the amount of information exchanged or shared with other agents can be reduced to restrict the information available to the adversary. Second, we can reduce reconstruction accuracy or knowledge by adding noise to the information exchanged.

To implement the first mechanism, we decompose the local state into a public substate, denoted by $\boldsymbol{\alpha}_{i,n} \in \mathbb{R}^m$, that is shared with neighbors, and a private substate, denoted by $\boldsymbol{\beta}_{i,n} \in \mathbb{R}^m$, that evolves locally and is not visible to neighbors, as illustrated in Figure 7.5. To avoid privacy leakage of state $\boldsymbol{r}_{i,n}$ during information passing within a neighborhood, the initial values of the substates, $\boldsymbol{\alpha}_{i,n}(0)$ and $\boldsymbol{\beta}_{i,n}(0)$, are randomly drawn from the set of real numbers while requiring to satisfy the relation $\boldsymbol{\alpha}_{i,n}(0) + \boldsymbol{\beta}_{i,n}(0) = 2\boldsymbol{r}_{i,n}$ [Wang, 2019]. Hence, although $\boldsymbol{\beta}_{i,n}$ is hidden from neighbors, it directly influences the evolution of $\boldsymbol{\alpha}_{i,n}$.

Next, to increase the uncertainty of the adversary, public states are perturbed by noise before being shared with neighbors; see, e.g. Mo and Murray [2017]. In particular, agent i modifies its shared public state by adding a correlated noise

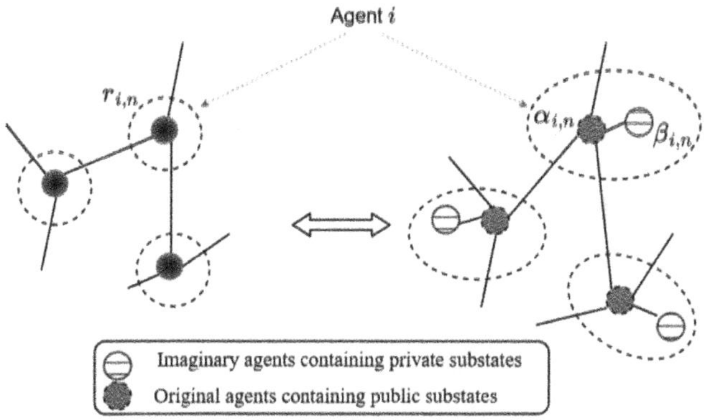

Figure 7.5 Local state $r_{i,n}$ decomposed into a public substate $\boldsymbol{\alpha}_{i,n}$ and a private substate $\boldsymbol{\beta}_{i,n}$.

sequence $\boldsymbol{\omega}_i(k)$, i.e. $\tilde{\boldsymbol{\alpha}}_{i,n}(k) = \boldsymbol{\alpha}_{i,n}(k) + \boldsymbol{\omega}_i(k)$, defined as follows:

$$\boldsymbol{\omega}_i(k) = \begin{cases} \boldsymbol{v}_i(0) & k = 0 \\ \phi^k \boldsymbol{v}_i(k) - \phi^{k-1}\boldsymbol{v}_i(k-1) & \text{o.w.} \end{cases} \tag{7.25}$$

with constant $\phi \in (0, 1)$ being shared among agents and $\boldsymbol{v}_i(k) \in \mathbb{R}^m \sim \mathcal{N}(\mathbf{0}, \sigma^2\mathbf{I}_m)$ being spatially and temporally independent (for each k and i).

Consequently, during consensus iteration k, agent i updates its public and private substates as:

$$\begin{aligned} \boldsymbol{\alpha}_{i,n}(k+1) &= \boldsymbol{\alpha}_{i,n}(k) + \varepsilon \mathbf{U}_i(k)\left(\boldsymbol{\beta}_{i,n}(k) - \boldsymbol{\alpha}_{i,n}(k)\right) \\ &\quad + \varepsilon \sum_{j \in \mathcal{N}_i} w_{ij}(k)\left(\tilde{\boldsymbol{\alpha}}_{j,n}(k) - \boldsymbol{\alpha}_{i,n}(k)\right) \\ \boldsymbol{\beta}_{i,n}(k+1) &= \boldsymbol{\beta}_{i,n}(k) + \varepsilon \mathbf{U}_i(k)\left(\boldsymbol{\alpha}_{i,n}(k) - \boldsymbol{\beta}_{i,n}(k)\right), \end{aligned} \tag{7.26}$$

where ε is a step size, chosen in range $(0, \frac{1}{\Delta+1}]$, $\Delta \triangleq \max_i N_i$, $w_{ij}(k) = w_{ji}(k)$ is the interaction weight with agent j, and coupling matrix $\mathbf{U}_i(k) \triangleq \text{diag}(\mathbf{u}_i(k))$ contains the coupling weight vector, $\mathbf{u}_i(k) \in \mathbb{R}^m$, on its diagonal matrix. To ensure privacy, agents independently select the interaction weights and all elements of the coupling weight vector at iteration $k = 0$, i.e. $w_{ij}(0) = w_{ji}(0)$ and $\mathbf{u}_i(0)$, from the set of real numbers. For subsequent consensus iterations, i.e. for $k > 0$, these values are restricted to fall in the range $[\eta, 1)$, with $0 < \eta < 1$. This condition ensures that each agent appropriately weights the information received from its neighbors. From (7.26), it is evident that the public substate is the only parameter that directly incorporates information from neighbors during its update, while the private substate update relies solely on information specific to the node.

In the following analysis, the coupling and interaction weights are randomly selected at $k = 0$ and kept fixed for $k > 0$, adhering to the mechanism described in Wang [2019]. For simplicity, all interaction weights are compiled into matrix $\mathbf{W} \triangleq [w_{ij}]$ for $k \geq 1$.

The privacy-preserving average consensus mechanism in (7.26) asymptotically converges to the true average state, which is expressed as

$$\lim_{k\to\infty} \boldsymbol{\alpha}_{i,n}(k) = \lim_{k\to\infty} \boldsymbol{\beta}_{i,n}(k) = \frac{1}{N}\sum_{i-1}^{N} \boldsymbol{r}_{i,n}.$$

In practice, the updates in (7.26) are performed for a fixed number of iterations, say K, after which local state, $\hat{\mathbf{x}}_{i,n|n}$, is set to:

$$\hat{\mathbf{x}}_{i,n|n} = \boldsymbol{\alpha}_{i,n}(K) \ \ \forall i \in \mathcal{N}.$$

Algorithm 7.3 summarizes the steps of the PP-DKF algorithm. Given that the number of consensus iterations used with PP-DKF is finite in practical scenarios, it is important to address concerns regarding its impact on performance, convergence behavior, and resulting privacy. These aspects are discussed next.

Algorithm 7.3 PP-DKF algorithm

For each agent $i \in \mathcal{N}$

Initialize: $\hat{\mathbf{x}}_{i,0|0}$ and $\mathbf{P}_{i,0|0}$

$\hat{\mathbf{x}}_{i,n|n-1} = \mathbf{A}\hat{\mathbf{x}}_{i,n-1|n-1}$

$\mathbf{P}_{i,n|n-1} = \mathbf{A}\mathbf{P}_{i,n-1|n-1}\mathbf{A}^T + \mathbf{Q}$

$\boldsymbol{\Gamma}_{i,n} = \mathbf{P}^{-1}_{i,n|n-1} + N\mathbf{H}_i^T\mathbf{R}_i^{-1}\mathbf{H}_i$

$\mathbf{P}^{-1}_{i,n|n} \leftarrow \boxed{\text{Avg. Consensus}} \leftarrow \{\boldsymbol{\Gamma}_{j,n},\ \forall j \in \mathcal{N}_i \cup i\}$

$\boldsymbol{r}_{i,n} = \hat{\mathbf{x}}_{i,n|n-1} + N\mathbf{P}_{i,n|n}\mathbf{H}_i^T\mathbf{R}_i^{-1}\left(\mathbf{y}_{i,n} - \mathbf{H}_i\hat{\mathbf{x}}_{i,n|n-1}\right)$

Privacy-Preserving Mechanism:

Choose $\boldsymbol{\alpha}_{i,n}(0)$, and set $\boldsymbol{\beta}_{i,n}(0) = 2\boldsymbol{r}_{i,n} - \boldsymbol{\alpha}_{i,n}(0)$

Choose weights $w_{ij}(k)$, $\mathbf{u}_i(k)$

Share weights $w_{ij}(k)$ within neighborhood

Generate $\{\boldsymbol{\omega}_i(k), k = 0, 1, \dots, K\}$ based on (7.25)

Share $\tilde{\boldsymbol{\alpha}}_{i,n}(0) = \boldsymbol{\alpha}_{i,n}(0) + \boldsymbol{\omega}_i(0)$

for $k = 1$ **to** K **do**

  Receive $\tilde{\boldsymbol{\alpha}}_{j,n}(k-1)$, $\forall j \in \mathcal{N}_i$

  Update $\boldsymbol{\alpha}_{i,n}(k)$ and $\boldsymbol{\beta}_{i,n}(k)$, as given in (7.26)

  Share $\tilde{\boldsymbol{\alpha}}_{i,n}(k) = \boldsymbol{\alpha}_{i,n}(k) + \boldsymbol{\omega}_i(k)$

end for

$\hat{\mathbf{x}}_{i,n|n} = \boldsymbol{\alpha}_{i,n}(K)$

7.4.3 Privacy Guarantees

This section explores the privacy protections provided by the PP-DKF in the presence of an HBC agent. For this analysis, we designate agent N as the HBC agent, which uses local and neighboring information to infer details of other agents. According to Algorithm 7.3, the data available to the HBC agent after k consensus iterations is as follows:

$$\mathcal{I}_{\text{HBC}}(k) = \{\boldsymbol{\alpha}_{N,n}(l), \boldsymbol{\beta}_{N,n}(l), \boldsymbol{\omega}_N(l), \mathbf{u}_N(l), w_{Nj}(l), \tilde{\boldsymbol{\alpha}}_{j,n}(l) : \forall j \in \mathcal{N}_N\}_{l=0}^{k}$$

We can observe that the level of privacy for an agent depends on the adversary's prior knowledge about the interaction and coupling weights in use. To evaluate the worst-case attack scenario, we assume the HBC agent has full access to weight matrix $\mathbf{W}$ and an estimate $\hat{\mathbf{U}}$ of the coupling weight matrix $\mathbf{U}$. Consequently, the adversary's information set can be expressed as $\tilde{\mathcal{I}}_{\text{HBC}}(k) = \mathcal{I}_{\text{HBC}}(k) \cup \{\mathbf{W}(l), \hat{\mathbf{U}}(l)\}_{l=0}^{k}$.

Here, we consider the local estimate $\mathbf{r}j, n$ as private since it captures the local a posteriori estimate, which holds more node-specific details compared to the global a posteriori state estimate $\hat{\mathbf{x}}j, n|n$. With the information set $\tilde{\mathcal{I}}_{\text{HBC}}(k)$, the HBC agent aims to estimate the initial substates of all network agents and, with those at hand, infer the local state information using the known relationship $\boldsymbol{r}_n = 0.5(\boldsymbol{\alpha}_n(0) + \boldsymbol{\beta}_n(0))$.

Following the approach in Wagner and Eckhoff [2018] and Mo and Murray [2017], we consider that the adversary employs an estimator to determine the private states of agents at time n, specifically, $\boldsymbol{r}_{j,n}$. Accordingly, we use the MSE of the adversary's estimate as a measure of privacy. Let $\hat{\boldsymbol{r}}_{j,n}(k)$ denote the estimated private information of agent j at time n and after k consensus iterations. The corresponding privacy loss, denoted by $\mathcal{E}_{j,n}(k)$, is defined as

$$\mathcal{E}_{j,n}(k) \triangleq \mathbb{E}\{\|\boldsymbol{r}_{j,n} - \hat{\boldsymbol{r}}_{j,n}(k)\|^2\}. \tag{7.27}$$

To compute the privacy value $\mathcal{E}_{j,n}(k)$, we model the observation vector at the adversary, incorporating the shared information from neighbors and the HBC agent's data available at iteration k, as follows:

$$\mathbf{y}_n(k) = \mathbf{C}\mathbf{z}_n(k) + \mathbf{C}_\alpha \boldsymbol{\omega}(k), \tag{7.28}$$

where $\mathbf{z}_n(k) = [\boldsymbol{\alpha}_n^T(k), \boldsymbol{\beta}_n^T(k)]^T$ with $\boldsymbol{\alpha}_n(k) = [\boldsymbol{\alpha}_{1,n}^T(k), \dots, \boldsymbol{\alpha}_{N,n}^T(k)]^T$ and $\boldsymbol{\beta}_n(k) = [\boldsymbol{\beta}_{1,n}^T(k), \dots, \boldsymbol{\beta}_{N,n}^T(k)]^T$. To capture the relevant set of information, we define $\mathbf{C} \triangleq [\mathbf{C}_\alpha, \mathbf{C}_\beta]$ with $\mathbf{C}_\beta = \left[\mathbf{0}, \mathbf{e}_N\right]^T \otimes \mathbf{I}_m$ capturing the private substate of the HBC agent and $\mathbf{C}_\alpha = \left[\mathbf{e}_{j_1}, \dots, \mathbf{e}_{j_{N_N}}, \mathbf{e}_N\right]^T \otimes \mathbf{I}_m$ capturing the public substates of neighbors and the HBC agent. Vector $\mathbf{e}_j \in \mathbb{R}^N$ is the kth standard basis, and $\mathcal{N}_N = \{j_1, \dots, j_{N_N}\}$ represents the HBC agent's adjacency set with cardinality N_N.

By substituting the network-wide update equations in (7.26) into (7.28), the observation model at iteration kth is given by:

$$\mathbf{y}_n(k) = \mathbf{C}\mathbf{G}^k\mathbf{z}_n(0) + \mathbf{C}_\alpha\left(\sum_{t=0}^{k-1} C_{k-1-t}\mathcal{B}\boldsymbol{\omega}(t) + \boldsymbol{\omega}(k)\right), \tag{7.29}$$

where $C_k = \begin{bmatrix}\mathbf{I} & \mathbf{0}\end{bmatrix}\mathbf{G}^k\begin{bmatrix}\mathbf{I} & \mathbf{0}\end{bmatrix}^T$, $\boldsymbol{\omega}(k) = [\boldsymbol{\omega}_1^T(k), \dots, \boldsymbol{\omega}_N^T(k)]^T$, and $\mathcal{B} = \varepsilon(\mathbf{W} \otimes \mathbf{I}_m)$. Given that $\boldsymbol{v}(k) = [\boldsymbol{v}_1^T(k), \dots, \boldsymbol{v}_N^T(k)]^T$ is a zero-mean i.i.d. sequence, stacking all observations up till iteration k into a vector yields:

$$\begin{bmatrix}\frac{\sum_{t=0}^{0}\mathbf{y}_n(t)}{\phi^0}\\ \frac{\sum_{t=0}^{1}\mathbf{y}_n(t)}{\phi^1}\\ \vdots\\ \frac{\sum_{t=0}^{k}\mathbf{y}_n(t)}{\phi^k}\end{bmatrix} = \underbrace{\begin{bmatrix}\frac{\mathbf{C}}{\phi^0}\\ \frac{\mathbf{C}(\mathbf{I}+\mathbf{G})}{\phi^1}\\ \vdots\\ \frac{\mathbf{C}(\mathbf{I}+\sum_{t=1}^{k}\mathbf{G}^t)}{\phi^k}\end{bmatrix}}_{\mathbf{H}(k)}\mathbf{z}_n(0) + \underbrace{\begin{bmatrix}\hat{\mathbf{F}}_0 & \mathbf{0} & \cdots & \mathbf{0}\\ \hat{\mathbf{F}}_1 & \hat{\mathbf{F}}_0 & \cdots & \mathbf{0}\\ \vdots & \vdots & \ddots & \vdots\\ \hat{\mathbf{F}}_{k-1} & \hat{\mathbf{F}}_{k-2} & \cdots & \hat{\mathbf{F}}_0\end{bmatrix}}_{\mathbf{F}(k)}\begin{bmatrix}\boldsymbol{v}(0)\\ \boldsymbol{v}(1)\\ \vdots\\ \boldsymbol{v}(k)\end{bmatrix}, \tag{7.30}$$

where $\hat{\mathbf{F}}_0 = \mathbf{C}_\alpha$ and $\hat{\mathbf{F}}_k = \frac{\varepsilon}{\phi^{k+1}}\mathbf{C}_\alpha C_k(\mathbf{W} \otimes \mathbf{I}_m)$ for $k \geq 1$.

With the above at hand, we can now quantify the worst-case privacy loss. In particular, assuming an HBC agent gains access to the above information set $\tilde{\mathcal{I}}_{\mathrm{HBC}}(k)$, comprising estimate $\hat{\mathbf{U}} = \mathbf{U} + \boldsymbol{\Delta}_{\mathbf{U}}$, with $\boldsymbol{\Delta}_{\mathbf{U}}$ denoting the uncertainty, $\{\mathbf{W}(l)\}_{l=0}^{k}$, and messages exchanged by its neighbors, the resulting error covariance for the estimate of $[\boldsymbol{\alpha}_n^T(0), \boldsymbol{\beta}_n^T(0)]^T$ is given by

$$\tilde{\mathbf{P}}_n(k) = \overline{\mathbf{P}}_{\mathrm{n}}(k) + \mathbb{E}_{\mathbf{U}}\left\{\varepsilon^2\mathbf{H}^\dagger(k)\boldsymbol{\Delta}_{\mathbf{H}}(k)\tilde{\boldsymbol{\Pi}}_n\boldsymbol{\Delta}_{\mathbf{H}}^T(k)(\mathbf{H}^\dagger(k))^T\right\}, \tag{7.31}$$

Here, $\tilde{\boldsymbol{\Pi}}_n = \mathbf{1}_{2N}\mathbf{1}_{2N}^T \otimes \mathbb{E}\{\mathbf{x}_n\mathbf{x}_n^T\}$, and

$$\begin{aligned}\overline{\mathbf{P}}_{\mathrm{n}}(k) = \mathbb{E}_{\mathbf{U}}\{&\varepsilon^2\mathbf{H}^\dagger(k)\boldsymbol{\Delta}_{\mathbf{H}}(k)\tilde{\boldsymbol{\Sigma}}_n\boldsymbol{\Delta}_{\mathbf{H}}^T(k)(\mathbf{H}^\dagger(k))^T\\ &+ \sigma^2(\mathbf{I} - \varepsilon\mathbf{H}^\dagger(k)\boldsymbol{\Delta}_{\mathbf{H}}(k))\mathbf{H}^\dagger(k)\mathbf{F}(k)\mathbf{F}^T(k)(\mathbf{H}^\dagger(k))^T\\ &(\mathbf{I} - \varepsilon\mathbf{H}^\dagger(k)\boldsymbol{\Delta}_{\mathbf{H}}(k))^T\},\end{aligned} \tag{7.32}$$

with $\tilde{\boldsymbol{\Sigma}}_n$ being the error covariance, while $\mathbf{H}(k)$ and $\mathbf{F}(k)$ are defined in (7.30), and

$$\boldsymbol{\Delta}_{\mathbf{H}}(k) = \begin{bmatrix}\mathbf{0}\\ \phi^{-1}\mathbf{C}\boldsymbol{\Delta}_{\mathbf{G}_1}\\ \vdots\\ \phi^{-k}\mathbf{C}\sum_{t=1}^{k}\boldsymbol{\Delta}_{\mathbf{G}_t}\end{bmatrix},$$

with $\boldsymbol{\Delta}_{\mathbf{G}_k} = \sum_{t=1}^{k}\frac{k!\varepsilon^{t-1}}{(k-t)!t!}\mathbf{G}^{k-t}\boldsymbol{\Delta}_{\mathbf{G}_1}^t$, $\boldsymbol{\Delta}_{\mathbf{G}_1} = -\mathcal{L}^T\boldsymbol{\Delta}_{\mathbf{U}}\mathcal{L}$, and $\mathcal{L} = [-\mathbf{I}, \mathbf{I}]$.

It can be shown that the first term in (7.31) converges to the fixed matrix $\overline{\mathbf{P}}_{\mathrm{LB}}(k) = \lim_{n\to\infty}\overline{\mathbf{P}}_{\mathrm{n}}(k)$ as $\lim_{n\to\infty}\tilde{\boldsymbol{\Sigma}}_n = \tilde{\boldsymbol{\Sigma}}$, while the second term diverges with

$\lim_{n\to\infty} \mathrm{tr}\left(\mathbb{E}\{\mathbf{x}_n\mathbf{x}_n^T\}\right) = \infty$ [Moradi et al., 2022]. Thus, the lower bound of the privacy leakage at agent j following k consensus iterations can be expressed as:

$$\overline{\mathcal{E}}_j(k) = \mathrm{tr}\left((\mathbf{e}_j^T \otimes \mathbf{I}_m)\mathbf{P}(k)(\mathbf{e}_j \otimes \mathbf{I}_m)\right) \tag{7.33}$$

where $\mathbf{P}(k) = \frac{1}{4}\left[\mathbf{I}\ \mathbf{I}\right]\overline{\mathbf{P}}_{\mathrm{LB}}(k)\left[\mathbf{I}\ \mathbf{I}\right]^T$. In the extreme case where the HBC agent has perfect knowledge of $\mathbf{U}$, $\tilde{\mathbf{P}}_n(k)$ in (7.31) becomes independent of n and reduces to:

$$\tilde{\mathbf{P}}(k) = \sigma^2\left(\mathbf{H}^T(k)\left(\mathbf{F}(k)\mathbf{F}^T(k)\right)^{-1}\mathbf{H}(k)\right)^{-1}. \tag{7.34}$$

7.4.4 Simulation Results

In this section, we demonstrate the performance of the PP-DKF algorithm through numerical simulations and theoretical bounds, as presented in (7.34). We consider an arbitrary, undirected, and connected network comprising $N = 25$ agents. The PP-DKF is applied to track the velocity and position of a moving target in a two-dimensional space according to the following state equation:

$$\mathbf{x}_n = \begin{bmatrix} \mathbf{I}_2 & \Delta T\mathbf{I}_2 \\ \mathbf{0}_2 & \mathbf{I}_2 \end{bmatrix} \mathbf{x}_{n-1} + \begin{bmatrix} \frac{1}{2}(\Delta T)^2\mathbf{I}_2 \\ \Delta T\mathbf{I}_2 \end{bmatrix} \mathbf{w}_n$$

where state vector $\mathbf{x}_n = [X_n, Y_n, \dot{X}_n, \dot{Y}_n]^T$ comprises positions $\{X_n, Y_n\}$ and velocities $\{\dot{X}_n, \dot{Y}_n\}$ in the horizontal and vertical directions, respectively, $\mathbf{w}_n = [\ddot{X}_n, \ddot{Y}_n]^T$ represents the unknown acceleration in both directions, and $\Delta T = 0.04$ is the sampling interval. The acceleration is modeled as a Gaussian process with zero mean and covariance $\mathbb{E}\{\mathbf{w}_n\mathbf{w}_n^T\} = 1.44\,\mathbf{I}_2$. The observation parameters for each agents $i \in \mathcal{N}$ are given by:

$$\mathbf{H}_i = \left[\mathbf{I}_2\ \mathbf{0}_2\right] \quad \text{and} \quad \mathbf{R}_i = \begin{bmatrix} 0.0416 & 0.008 \\ 0.008 & 0.04 \end{bmatrix}.$$

For comparison, we also consider a DKF that utilizes only the noise-injection-based average consensus technique proposed in Mo and Murray [2017], where the injected noise follows (7.25). We will henceforth refer to this method as the noise-injection-based privacy-preserving DKF (NIP-DKF). The consensus parameter ε is set at $1/4$, and the noise parameter ϕ is assigned a value of 0.9. The interaction weight matrix $\mathbf{W}$ is specified as $0.75\,\mathbf{E}$, where $\mathbf{E}$ denotes the adjacency matrix. The components of the coupling weight vector $\mathbf{u}_i$ are independently sampled from the distribution $\mathcal{U}(\eta, 1)$ with $\eta = 0.4$. The number of consensus iterations employed is fixed at $K = 30$ for the duration of the experiment.

Figure 7.6 shows the perturbation variance σ^2 versus MSE for the different Kalman filtering algorithms. As anticipated, higher perturbation noise variance leads to a degradation in MSE performance compared to the conventional DKF in Algorithm 7.1. However, the PP-DKF demonstrates a more gradual

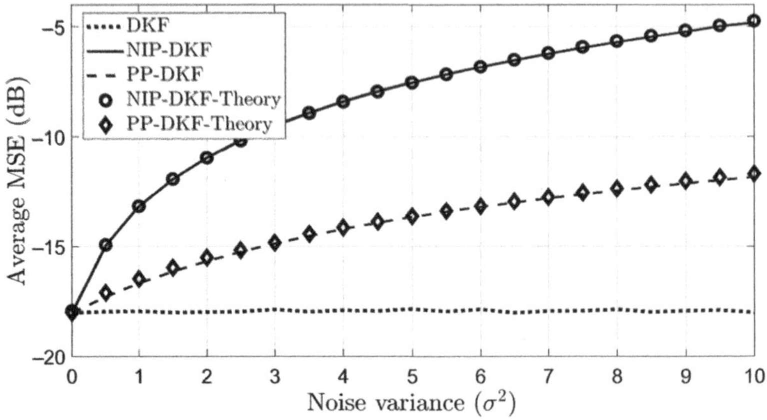

Figure 7.6 Average MSE of the DKF as a function of noise variance σ^2, comparing theoretical and simulation results.

increase in MSE than NIP-DKF, showcasing its better resilience to injected noise. Furthermore, the results predicted by theory for both algorithms closely match simulations.

To gain deeper insights into the privacy preservation of the PP-DKF in the presence of an HBC agent, we consider a smaller network with five agents, as shown in Figure 7.7, where the fifth agent is considered to be the HBC agent, indicated by the dark gray. The HBC agent does not have access to the coupling weights of the other agents, whereas a legitimate network agent is aware of the parameter η. For the estimation purpose, the HBC agent employs an average value $\overline{\mathbf{U}}$, based on the distribution of the coupling weights and accounting for the uncertainty $\boldsymbol{\Delta}_{\mathrm{U}} = \mathbf{U} - \overline{\mathbf{U}}$.

Figure 7.8 illustrates the lower bound on the privacy in (7.33) against the variance of the noise injected by Byzantines. The results indicate that using the NIP-DKF compromises the privacy of agent four due to its limited number of neighbors, which includes only the HBC agent. Consequently, the HBC agent can accurately determine the initial state of agent four with minimal error. The PP-DKF method, on the other hand, considerably improves the privacy of all agents, even when only a minimal amount of noise is added.

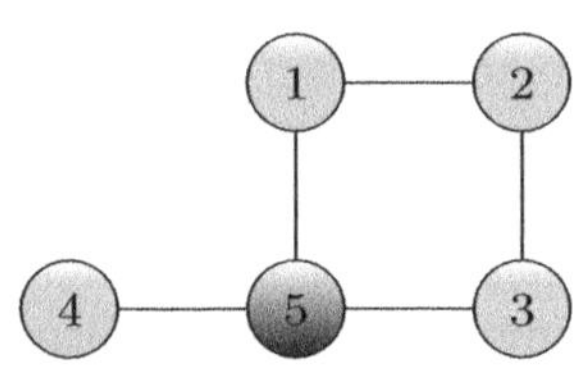

Figure 7.7 Diagram showing the employed network topology having $N = 5$ interconnected agents.

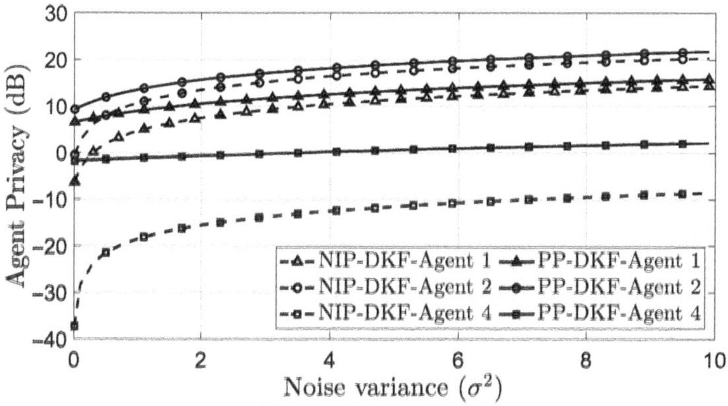

Figure 7.8 Privacy of agents versus noise variance (σ^2). Results are shown only for agent 1 due to symmetric topology, which gives agents 1 and 3 identical privacy levels.

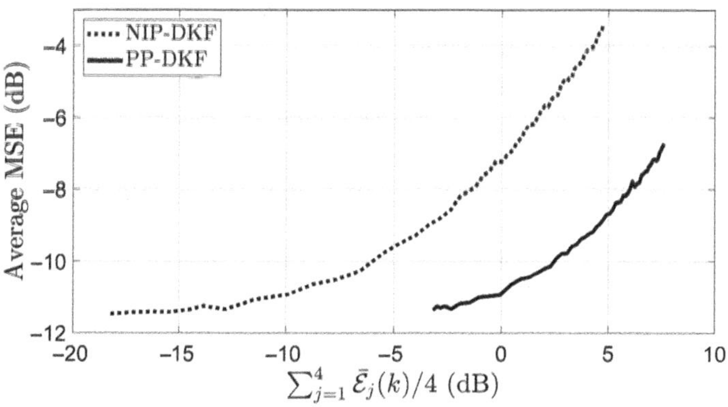

Figure 7.9 Privacy-accuracy tradeoff.

Figure 7.9 illustrates the tradeoff between filtering accuracy and average privacy $\sum_{j=1}^{4} \overline{\mathcal{E}}_j(k)/4$. It specifically shows the privacy-MSE tradeoff for different values of injected noise variance σ^2. An increase in privacy guarantees leads to a decrease in filtering accuracy, as indicated by a higher MSE. Furthermore, it can be observed that for a given privacy guarantee, PP-DKF achieves the lowest MSE; this is because it only injects noise into the public substate, in contrast to NIP-DKF, which perturbs the entire state.

Bibliography

C.-Z. Bai, V. Gupta, and F. Pasqualetti. On Kalman filtering with compromised sensors: Attack stealthiness and performance bounds. *IEEE Transactions on Automatic Control*, 62(12):6641–6648, Dec 2017.

W. Ben-Ameur, P. Bianchi, and J. Jakubowicz. Robust distributed consensus using total variation. *IEEE Transactions on Automatic Control*, 61(6):1550–1564, June 2016.

P. Braca, R. Lazzeretti, S. Marano, and V. Matta. Learning with privacy in consensus +obfuscation. *IEEE Signal Processing Letters*, 23(9):1174–1178, Sept 2016.

M. Brossard, A. Barrau, and S. Bonnabel. Ai-IMU dead-reckoning. *IEEE Transactions on Intelligent Vehicles*, 5(4):585–595, 2020. doi: 10.1109/TIV.2020.2980758.

R. K. C. Chang. Defending against flooding-based distributed denial-of-service attacks: A tutorial. *IEEE Communications Magazine*, 40(10):42–51, Oct 2002.

Y. Chen, S. Kar, and J. M. F. Moura. Optimal attack strategies subject to detection constraints against cyber-physical systems. *IEEE Transactions on Control of Network Systems*, 5(3):1157–1168, Sept 2018a.

Y. Chen, S. Kar, and J. M. F. Moura. Resilient distributed estimation: Sensor attacks. *IEEE Transactions on Automatic Control*, 64(9):3772–3779, Sept 2018b.

Y. Chen, S. Kar, and J. M. F. Moura. Resilient distributed parameter estimation with heterogeneous data. *IEEE Transactions on Signal Processing*, 67(19):4918–4933, Oct 2019.

L. Da Xu, W. He, and S. Li. Internet of Things in industries: A survey. *IEEE Transactions on Industrial Informatics*, 10(4):2233–2243, 2014. doi: 10.1109/TII.2014.2300753.

K. H. Degue and J. Le Ny. On differentially private Kalman filtering. In *Proceedings of the 5th IEEE Global Conference on Signal and Information Processing*, pages 487–491, 2017.

C. Dwork. Differential privacy: A survey of results. In *Proceedings Springer International Conference on Theory and Applications of Models of Computation*, pages 1–19, 2008.

M. Fauser and P. Zhang. Resilience of cyber-physical systems to covert attacks by exploiting an improved encryption scheme. In *Proceedings of the 59th IEEE Conference on Decision and Control*, pages 5489–5494, 2020.

M. Fauser and P. Zhang. Resilient homomorphic encryption scheme for cyber-physical systems. In *Proceedings of the 60th IEEE Conference Decision and Control (CDC)*, pages 5634–5639, 2021.

D. Feng, C. Wang, C. He, Y. Zhuang, and X.-G. Xia. Kalman-filter-based integration of IMU and UWB for high-accuracy indoor positioning and navigation. *IEEE Internet of Things Journal*, 7(4):3133–3146, 2020. doi: 10.1109/JIOT.2020.2965115.

J. He, L. Cai, and X. Guan. Differential private noise adding mechanism and its application on consensus algorithm. *IEEE Transactions on Signal Processing*, 68:4069–4082, Jul 2020.

X. He, X. Ren, H. Sandberg, and K. H Johansson. How to secure distributed filters under sensor attacks. *IEEE Transactions on Automatic Control*, 67(6):2843–2856, 2022.

A. Humayed, J. Lin, F. Li, and B. Luo. Cyber-physical systems security—a survey. *IEEE Internet of Things Journal*, 4(6):1802–1831, 2017. doi: 10.1109/JIOT.2017.2703172.

B. Kailkhura, S. Brahma, and P. K. Varshney. Data falsification attacks on consensus-based detection systems. *IEEE Transactions on Signal and Information Processing over Networks*, 3(1):145–158, Mar 2016.

D. Kapetanovic, G. Zheng, and F. Rusek. Physical layer security for massive MIMO: An overview on passive eavesdropping and active attacks. *IEEE Communications Magazine*, 53(6):21–27, June 2015.

K.-D. Kim and P. R. Kumar. Cyber–physical systems: A perspective at the centennial. *Proceedings of the IEEE*, 100(Special Centennial Issue):1287–1308, 2012. doi: 10.1109/JPROC.2012.2189792.

P. Krishnamurthy and F. Khorrami. Resilient redundancy-based control of cyber–physical systems through adaptive randomized switching. *Systems & Control Letters*, 158:105066, Dec 2021.

J. Le Ny. Differentially private Kalman filtering. In J. Le Ny, editor, *Differential Privacy for Dynamic Data*, pages 55–75. Springer International Publishing, 2020.

J. Le Ny and G. J. Pappas. Differentially private filtering. *IEEE Transactions on Automatic Control*, 59(2):341–354, Feb 2014.

M.E. Liggins, C.-Y. Chong, I. Kadar, M.G. Alford, V. Vannicola, and S. Thomopoulos. Distributed fusion architectures and algorithms for target tracking. *Proceedings of the IEEE*, 85(1):95–107, 1997. doi: 10.1109/JPROC.1997.554211.

J. Lin, A. S. Morse, and B. D. O. Anderson. The multi-agent rendezvous problem. Part 1: The synchronous case. *SIAM Journal on Control and Optimization*, 46(6):2096–2119, 2007. doi: 10.1137/040620552.

H. Lin, Z. T. Kalbarczyk, and R. K. Iyer. RAINCOAT: Randomization of network communication in power grid cyber infrastructure to mislead attackers. *IEEE Transactions on Smart Grid*, 10(5):4893–4906, Sept 2019.

A. Mitra and S. Sundaram. Byzantine-resilient distributed observers for LTI systems. *Elsevier Automatica*, 108:108487, Oct 2019.

A. Mitra, F. Ghawash, S. Sundaram, and W. Abbas. On the impacts of redundancy, diversity, and trust in resilient distributed state estimation. *IEEE Transactions on Control of Network Systems*, 8(2):713–724, June 2021.

Y. Mo and R. M Murray. Privacy preserving average consensus. *IEEE Transactions on Automatic Control*, 62(2):753–765, Feb 2017.

A. Moradi, N. K. D. Venkategowda, S. P. Talebi, and S. Werner. Privacy-preserving distributed Kalman filtering. *IEEE Transactions on Signal Processing*, 70:3074–3089, 2022. doi: 10.1109/TSP.2022.3182590.

A. Moradi, N. K. D. Venkategowda, and S. Werner. Total variation-based distributed Kalman filtering for resiliency against byzantines. *IEEE Sensors Journal*, 23(4):4228–4238, 2023. doi: 10.1109/JSEN.2022.3233700.

Y. Ni, J. Wu, L. Li, and L. Shi. Multi-party dynamic state estimation that preserves data and model privacy. *IEEE Transactions on Information Forensics and Security*, 16:2288–2299, Jan 2021.

E. Nozari, P. Tallapragada, and J. Cortés. Differentially private average consensus: obstructions, trade-offs, and optimal algorithm design. *Automatica*, 81:221–231, Jul 2017.

R. Olfati. Kalman-consensus filter: Optimality, stability, and performance. In *Proceedings of the 48th IEEE Conference on Decision and Control*, pages 7036–7042, 2009.

R. Olfati-Saber. Distributed Kalman filter with embedded consensus filters. In *Proceedings of the 44th IEEE Conference on Decision and Control*, pages 8179–8184, 2005.

J. Peng, W. Li, and Q. Ling. Byzantine-robust decentralized stochastic optimization over static and time-varying networks. *Elsevier Signal Process.*, 183:108020, June 2021.

S. Rajput, H. Wang, Z. Charles, and D. Papailiopoulos. DETOX: A redundancy-based framework for faster and more robust gradient aggregation. In *Proceedings of NIPS*, volume 32, page 10320–10330, 2019.

S. Rezaei and R. Sengupta. Kalman filter-based integration of DGPS and vehicle sensors for localization. *IEEE Transactions on Control Systems Technology*, 15(6):1080–1088, 2007. doi: 10.1109/TCST.2006.886439.

K. Ryu and J. Back. Distributed Kalman-filtering: Distributed optimization viewpoint. In *Proceedings of the 58th IEEE Conference Decision and Control*, pages 2640–2645, 2019.

S. P. Talebi and S. Werner. Distributed Kalman filtering and control through embedded average consensus information fusion. *IEEE Transactions on Automatic Control*, 64(10):4396–4403, Oct 2019.

A. Vempaty, L. Tong, and P. K. Varshney. Distributed inference with Byzantine data: State-of-the-art review on data falsification attacks. *IEEE Signal Processing Magazine*, 30(5):65–75, Sept. 2013.

I. Wagner and D. Eckhoff. Technical privacy metrics: A systematic survey. *ACM Computing Surveys*, 51(3):1–38, Jun 2018.

Y. Wang. Privacy-preserving average consensus via state decomposition. *IEEE Transactions on Automatic Control*, 64(11):4711–4716, Nov 2019.

J. Wang, R. Zhu, and S. Liu. A differentially private unscented Kalman filter for streaming data in IoT. *IEEE Access*, 6:6487–6495, Mar 2018.

M. Wolf and D. Serpanos. Safety and security in cyber-physical systems and internet-of-things systems. *Proceedings of the IEEE*, 106(1):9–20, 2018. doi: 10.1109/JPROC.2017.2781198.

H. Xu, W. Yu, D. Griffith, and N. Golmie. A survey on industrial Internet of Things: A cyber-physical systems perspective. *IEEE Access*, 6:78238–78259, 2018. doi: 10.1109/ACCESS.2018.2884906.

S. Yang, S. Tan, and J.-X. Xu. Consensus based approach for economic dispatch problem in a smart grid. *IEEE Transactions on Power Systems*, 28(4):4416–4426, 2013. doi: 10.1109/TPWRS.2013.2271640.

X. Yu and Y. Xue. Smart grids: A cyber–physical systems perspective. *Proceedings of the IEEE*, 104(5):1058–1070, 2016. doi: 10.1109/JPROC.2015.2503119.

J. Zhou, W. Ding, and W. Yang. A secure encoding mechanism against deception attacks on multi-sensor remote state estimation. *IEEE Transactions on Information Forensics and Security*, 17:1959–1969, 2022. doi: 10.1109/TIFS.2022.3175617.

8

Event-Triggered and Privacy-Preserving Anomaly Detection for Smart Environments*

Yasin Yilmaz[1], *Mehmet Necip Kurt*[2], *and Xiaodong Wang*[2]

[1]*Department of Electrical Engineering, University of South Florida, Tampa, FL, USA*
[2]*Department of Electrical Engineering, Columbia University, New York, NY, USA*

8.1 Introduction

Anomaly detection deals with finding abnormal data patterns [Chandola et al., 2009]. Its applications are seen in a wide range of areas, such as cybersecurity [Xiang et al., 2011; Cui et al., 2019], hardware security [Elnaggar et al., 2019], medical health care [Zhang et al., 2018], surveillance videos [Ravanbakhsh, 2019], and aviation [Matthews, 2019]. Anomaly detection is a crucial task since an anomaly in the observed data may be a precursor of an unwanted and often actionable event such as failure and malicious activity in the system. In many real-time systems, timely and accurate detection of unexpected data patterns is critical and will allow proper countermeasures to be taken to prevent potential damage. Although anomaly detection has long been studied [Chandola et al., 2009], today's complex networks exhibit new challenges, including:

1) **Low-latency requirements**: Timely detection of anomalies in an intricate network is essential since local anomalies can quickly spread in the network causing large-scale problems, due to complex interactions between nodes and interdependency between networks. For instance, Internet of Things (IoT) enables real-time interaction and optimization between critical infrastructures, such as smart power grid and intelligent transportation system (ITS), forming a smart city network.
2) **Data size**: High dimensionality (e.g. a large number of devices in IoT or surveillance video streaming) renders learning baseline and online monitoring difficult. Hence, computationally efficient approximate methods are needed.

* This work is funded by US National Science Foundation under the Grant #2040572.

Wireless Sensor Networks in Smart Environments: Enabling Digitalization from Fundamentals to Advanced Solutions, First Edition. Edited by Domenico Ciuonzo and Pierluigi Salvo Rossi.

3) **System dynamics**: Due to the cyber-physical nature of today's complex networks (e.g. IoT, smart grid, and ITS) and sophisticated anomalies such as attacks from skilled adversaries, the dynamics (i.e. change in time) of baselines and anomalies should be taken into account for effective detection and mitigation of threats.
4) **Unknown distributions**: Due to disparate devices or data types (e.g. numerical and categorical) in the network, probability distributions can be quite complicated such that model-based anomaly detection approaches are not suitable. For instance, in an IoT network with different device types (e.g. smart home appliances, wearable devices, and smart vehicles), it is very resource-demanding (computation, energy, and time) to accurately estimate the joint probability distribution of measurements collected in the network. Furthermore, the myriad of vulnerabilities makes it intractable to estimate the anomaly distributions.
5) **Distributed nature**: Complex networks in general consist of several components which need to carry out some tasks locally due to either resource constraints (e.g. energy and communication bandwidth) or privacy concerns. This is especially relevant in networks for critical infrastructure, such as smart grid, as well as networks with human users, such as social networks and IoT networks.
6) **Privacy**: In networks with sensitive information, such as personal power consumption in smart grid or data consumption in IoT network, network-wide anomaly detection should be performed while preserving users' privacy.

These challenges and the solution methods presented here are generally applicable to a variety of complex systems, such as surveillance camera network and smart home network. The main motivation of this chapter is to study real-time data-driven anomaly detection for complex systems under the challenges explained above. Specifically, we study resource and privacy constraints common to decentralized/distributed systems, such as energy and communication bandwidth constraints. We propose *event-triggered sampling* methods to effectively summarize observed local data, and at the same time achieve accurate and timely detection. We also propose a privacy-preserving mechanism for online (i.e. real-time) anomaly detection in terms of *distributed differential privacy*.

8.2 Background and Literature Review

Consider a network consisting of N nodes where each node $n \in \{1, 2, \dots, N\}$ has an observation $\mathbf{x}_{t,n} \in \mathbb{R}^{m_n}$ at each discrete time $t \in \mathbb{Z}$, where $m_n \gg 1$ denotes the data dimensionality. At an unknown time τ, an unexpected event (anomaly) happens in the network, e.g. a cyber-attack, and the network deviates from its nominal operation. A change in the statistical properties of the data-generating process is expected due to the anomaly. For the network-wide data $\mathbf{x}_t \triangleq [\mathbf{x}_{t,1}^T, \mathbf{x}_{t,2}^T, \dots, \mathbf{x}_{t,N}^T]^T$, with ^{T} denoting the transpose, we can write

$$\mathbf{x}_t \sim f_0^{\mathbf{x}}, \text{ if } t < \tau, \text{ and } \mathbf{x}_t \sim f_1^{\mathbf{x}} \neq f_0^{\mathbf{x}}, \text{ if } t \geq \tau,$$

where $f_0^{\mathbf{x}}$ denotes the probability density function (pdf) of $\mathbf{x}_t$ under nominal conditions and $f_1^{\mathbf{x}}$ denotes the pdf of $\mathbf{x}_t$ after the anomaly occurs.

The objective is to detect network-wide anomalies in a timely fashion using the observed data sequence, which corresponds to a sequential change detection (SCD) problem [Basseville and Nikiforov, 1993]. When new observation arrives, a binary decision is made: declare a change (anomaly) or continue receiving more data in the next time interval. The aim is to detect the changes as quickly as possible after they occur while satisfying a constraint on the false alarm probability. In the literature, there are two main approaches for modeling the change-point. It is modeled as either a deterministic unknown quantity (minimax formulation [Lorden, 1971; Pollak, 1985]) or a random variable with a known geometric distribution (Bayesian formulation Shiryaev [1978]). For instance, Lorden's well-known minimax problem aims to minimize the worst-case average detection delay (ADD) subject to an upper bound on the false alarm probability (FAP) Lorden [1971]. In cases where the probabilistic data models $f_0^{\mathbf{x}}$ and $f_1^{\mathbf{x}}$ are known and the network-wide data $\{\mathbf{x}_t\}$ is accessible to a decision maker, the cumulative sum (CUSUM) algorithm is the optimal solution to such minimax problem. Furthermore, if the data models are known except for some unknown parameters, the generalized CUSUM algorithm, making use of the estimates of unknown parameters, has asymptotic optimality properties [Basseville and Nikiforov, 1993, Sec. 5.3]. However, for large-scale heterogeneous networks, such as IoT networks, usually the nominal pdf $f_0^{\mathbf{x}}$ might be difficult to model or intractable to estimate [Laxhammar and Falkman, 2014]. It is especially challenging to model the anomalous pdf $f_1^{\mathbf{x}}$ because of the myriad of possible anomaly forms depending on the type and cause of the anomalies [Kurt et al., 2019a]. Hence, in this chapter, we assume both $f_0^{\mathbf{x}}$ and $f_1^{\mathbf{x}}$ are unknown, and focus on the nonparametric methods that are free of statistical data model assumptions, robust to the data model mismatch, and widely applicable.

Early nonparametric SCD methods have various limitations such as normality assumption [Zou et al., 2009; Xie et al., 2013, 2015; Enikeeva and Harchaoui, 2013], window-based operation [Kifer et al., 2004; Desobry et al., 2005; Harchaoui et al., 2009; Liu et al., 2013; Zou et al., 2014; Li et al., 2015b], and specific change structures [Aue et al., 2009; Zou et al., 2009; Enikeeva and Harchaoui, 2013; Xie and Siegmund, 2013; Banerjee et al., 2015; Banerjee and Hero, 2016]. Several recent generic nonparametric methods suffer from high computational complexity [Chen, 2018; Zambon et al., 2018a,2018b]. Moreover, distributed SCD has also been discussed, albeit in a parametric setting with the independence assumption [Veeravalli et al., 1993; Hussain, 1994; Tartakovsky and Veeravalli, 2004, 2008; Mei, 2008, 2010; Fellouris and Moustakides, 2011; Xie and Siegmund, 2013; Jirak, 2015]. Hence, the existing SCD methods cannot address resource constraints and unknown probability distributions in complex distributed systems, such as large-scale heterogeneous IoT networks.

Neural network-based methods have become very popular in various machine learning tasks including anomaly detection [Sabokrou et al., 2018; Chatillon and Ballester, 2019; Ravanbakhsh, 2019]. However, they are not suitable for online/continual learning as their deep neural network classifiers typically require a long time to update with a new batch of samples [Liu et al., 2018; Ionescu et al., 2019]. Moreover, differential privacy has been attracting significant interest in machine learning and signal processing [Dwork et al., 2006a; Shi et al., 2011; Jain et al., 2012; Fan and Xiong, 2013; Le Ny and Pappas, 2013; Sarwate and Chaudhuri, 2013; Song et al., 2013; Dwork and Roth, 2014; Leontiadis et al., 2014; Ghassemi et al., 2016; Agarwal et al., 2018; Cummings et al., 2018; Degue and Le Ny, 2018; Li et al., 2018; Zhang and Zhu, 2018; Keshk et al., 2019; Canonne et al., 2019]. In particular, there are a few works focusing on change detection under privacy constraints. While Keshk et al. [2019] claimed practical privacy benefits without rigorous privacy analysis, Cummings et al. [2018] provided provable privacy guarantees, but for known pre- and post-change data models.

8.3 Event-Triggered Anomaly Detection

In large-scale distributed networks, sometimes it may not be feasible to transmit raw observations collected at the network edge (i.e. nodes) to the network center. One reason is the *resource constraints* (e.g. energy and communication bandwidth constraints), which is typical in distributed systems where nodes are battery-powered devices and the transmission medium is wireless, such as IoT networks. Another reason is the so-called *big data challenge*: in a complex network, vast amount of heterogeneous data needs to be processed in real time. Instead of gathering and processing all raw data in the network center, an effective way is to summarize the local data at the nodes and let the center deal with the gist of collected data. Yet another reason could be the *privacy of users* in the network. To preserve the privacy of the raw data, some salient features relevant to the considered task, such as some statistics useful for anomaly detection, can be sent to the center.

In all these cases, the method that is used to summarize data needs to be accurate while satisfying the potential resource (e.g. computation, energy, and bandwidth) constraints. This can be achieved by detecting characteristic "events" in the observations which typically summarize the observed data well with regard to the considered task, and have binary representations denoting presence/absence. In addition to preserving data privacy, event-based processing also provides secure transmission against jamming and eavesdropping. Specifically, the presence of an event can be reported by transmitting a single pulse using ultra-wide band (UWB) communications [Sahinoglu et al., 2008], whose wide spectrum of frequency bands provides robustness to jamming attacks. Moreover, since data is encoded in events, an eavesdropper needs to know the event type to understand the transmitted bits.

Considering model-based methods for several resource-constrained networks, transmitting local sufficient statistics (e.g. log-likelihood ratio) using a family of simple but effective techniques called *event-triggered sampling* [Yilmaz et al., 2015] have been studied. Specifically, *level-triggered sampling* (LTS) and *level-crossing sampling* (LCS), for which the underlying event is the crossing of some levels by the signal to be transmitted, as shown in Figure 8.1, have been extensively studied. LTS and LCS only differ in their treatment of the overshoot values ($\epsilon_{i,n}$ in Figure 8.1),[1] and they coincide for continuous-time signals (no overshoot case).

LTS/LCS was used to report local decision statistic $y_{t,n}$ from node n over time $t = 1, 2, \ldots$ to the network center (a.k.a. fusion center) [Yılmaz et al., 2012, 2013, Feb. 2016; Yılmaz and Wang, 2014a,2014b; Li et al., 2015a]. In such event-based schemes, sampling and transmission occur at random *event times* (denoted with $\tau_{i,n}$ in Figure 8.1), as opposed to the traditional uniform-in-time sampling which periodically transmits at fixed times. Each level crossing at time $\tau_{i,n}$ is reported to the network center by node n as a binary code $e_{i,n} \in \{-1, 1\}$, where $-1/1$ denotes

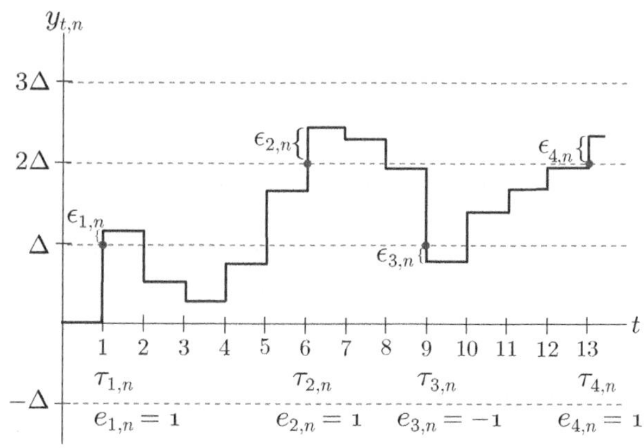

Figure 8.1 Level-crossing sampling.

1 LTS dynamically sets the sampling levels according to the overshoot values Yılmaz et al. [2012, 2015], as opposed to LCS.

negative/positive Δ change in the level of $y_{t,n}$. Note that infrequent transmission of such binary information satisfies resource constraints; and the network center is able to track the local decision statistic $y_{t,n}$ with an error bounded by Δ.

In previous work on LTS/LCS, a single event type (level crossings by signal magnitude) was considered at distributed nodes, and a single model-based data processing strategy at the network center with known probability distributions. However, in general, different event definitions are needed for nodes in different complex systems, such as appearance of an object (e.g. vehicle in walkway) or suspicious action (e.g. running) in the surveillance camera network. Similarly, different data processing techniques are needed at the network center for different tasks, such as detecting anomalies in the event sequences of surveillance camera networks. Hence, we investigate novel event types for nodes, and novel model-based and model-free data processing techniques for the network center.

8.3.1 Event Definitions at Nodes

Events are defined for the entity to be transmitted, i.e. raw data or a processed form such as detection statistic. In general, domain-specific events can be defined directly for the raw data to be detected at the network edge where it is observed. Such edge processing of raw data for detecting important events will also prevent error propagation and information loss which occurs when the network center processes the quantized raw data received from the network edge. LTS/LCS can be classified under the *magnitude-based* approach which defines event of interest based on the magnitude of the signal to be transmitted. A number of other approaches can be considered to develop a general framework for *event-based anomaly detection*.

i) **Frequency-based events**: Changes in the frequency content of the observed data may define an important event in some applications, such as denial-of-service (DoS) attack, in which data rate may gradually increase possibly from a number of users – distributed DoS (DDoS).
ii) **Object-based events**: In a network of surveillance cameras where continuous transmission of observed image/video to the network center is not feasible (e.g. unmanned aerial vehicle (UAV) networks), object-based processing can greatly facilitate network-wide anomaly detection. With the fast advancing hardware (computational power) and software (computational efficiency) technologies, it is now possible to do real-time object detection at a network node Redmon and Farhadi [2016]. Network-wide anomaly detection can be efficiently achieved with low-rate event stream (i.e. presence of objects) at the center.
iii) **Novelty-based events**: It is not obvious how to define a characteristic event if a node itself receives high-dimensional data at each time from its possibly

heterogeneous sensors, such as in a vehicular ad hoc network (VANET). In this case, model-free novelty/anomaly detection algorithms can be used to detect novel sets of measurements, which are sent to the network center that performs network-wide anomaly detection. Although not as low rate as binary event reports, this novelty-based reporting approach can ensure the transmission of only the informative data instances and help satisfy resource constraints.

8.3.2 Parametric Processing at Network Center

As an event sequence $e_{t,n}$ sequentially arrives from each node n to the network center, the network center may try to fit a probability distribution to the nominal (and possibly anomalous) training event sequences, and look for anomalies in received test sequences by sequentially comparing the likelihood under the nominal model (or the log-likelihood ratio) with a threshold, which is set to satisfy a false alarm probability constraint. In our preliminary work [Yılmaz et al., 2013], the times when $e_{t,n} \neq 0$ are sensed by the center, and positive ($e_{t,n} = 1$) or negative ($e_{t,n} = -1$) Δ change (Figure 8.1) is modeled using Bernoulli distribution. It was shown that, by computing the likelihoods of the received event sequences it is possible to achieve strong asymptotic optimality in terms of average detection delay while satisfying strict error probability constraints [Yilmaz et al., 2012, 2015]. Optimum communication schemes were designed when $e_{t,n}$ is reported through different noisy channel models [Yılmaz et al., 2013].

i) **Temporal anomalies**: The above parametric approach will fail for the challenging cases where the reported events look nominal, but anomalies exist in other aspects, such as timing of events. Renewal processes can be used to model the waiting times between event arrivals and also the propagation times of certain events through the nodes. In Figure 8.2, the event arrivals (shown by vertical lines) may correspond to the occurrence of the same type of events at a node or throughout the network. In general, for independent waiting times with distribution $w_i \sim f_0(\theta_0)$, we have a renewal process, and we can test for possible changes in the parameter or the distribution itself by considering an alternative distribution $f_1(\theta_1)$. Specifically, we can apply a CUSUM test [Basseville and Nikiforov, 1993] using the log-likelihood ratio

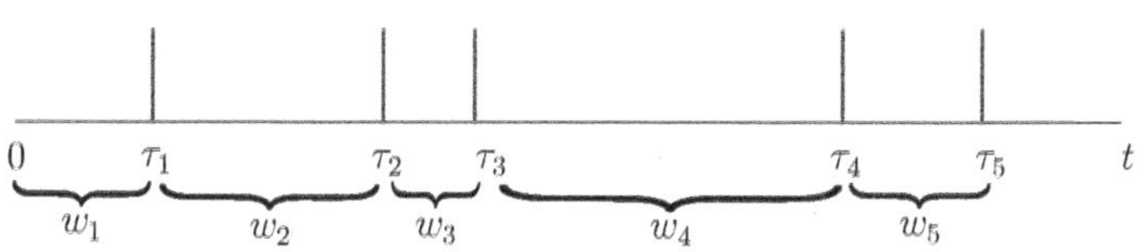

Figure 8.2 Event times and waiting times.

$\log[f_1(w_i|\theta_1)/f_0(w_i|\theta_0)]$. Domain knowledge and training sets can be used to estimate $f_0(\theta_0)$ and $f_1(\theta_1)$. For instance, when the waiting times are independent and exponentially distributed, i.e. $w_i \sim f_0(\theta_0) = \exp(\lambda_0)$, we have Poisson process for the number of events in a period of time, i.e. $N(t, t+h) \sim \text{Poisson}(\lambda_0 h)$. In such a case, depending on the domain knowledge, an increase or a decrease in the arrival rate can be tested, i.e. $f_1(\theta_1) = \exp(\lambda_1)$, $\lambda_1 \neq \lambda_0$. Moreover, multiple event sequences $\mathbf{e}_t = [e_{t,1} \dots e_{t,N}]$ can be monitored jointly. Assuming independence over time the number of points in a region in the N-dimensional Euclidean space defines a (nonstationary in general) Poisson point process, where the probability distribution of $\mathbf{e}_t$ determines the possibly non-stationary rate $\lambda(\mathbf{e}_t)$.

ii) **Markov process**: Denoting the network state at time t with $\mathbf{e}_t$, we can learn the state transition probabilities $p_{i,j} = P(\mathbf{e}_{t+1} = \text{state } j | \mathbf{e}_t = \text{state } i)$. Then using the learned Markov process we can sequentially detect anomalies via a CUSUM-like algorithm using the log-likelihood ratio between $p_{i,j}$ and a constant critical value p_α, which can be taken as the $1-\alpha$ percentile of all state transition probabilities. This corresponds to testing the nominal distribution against the uniform distribution that has $1/p_\alpha$ accessible states.

8.3.3 Nonparametric Processing at Network Center

For network-wide monitoring of events, we need to consider the joint probability of event states $\mathbf{e}_t = [e_{t,1} \dots e_{t,N}]$. When N is large, it becomes intractable to estimate the high-dimensional joint distribution or even to fit a parametric model as the number of training samples needed for accurate estimation grows quickly with the number of dimensions (i.e. nodes). Similarly, monitoring the event sequences over time without the Markov assumption to capture the multi-step temporal correlations also requires the estimation of the joint probability distribution of the event sequences. Hence, for such high-dimensional problems, we resort to model-free, i.e. nonparametric, methods.

The ODIT algorithm described in Mozaffari et al. [2022] may be used for monitoring network-wide events $\mathbf{e}_t$ at each time since no specific data type is assumed in ODIT. ODIT-like distance-based approaches can be studied for monitoring event sequences. A natural question is: What is a suitable similarity/dissimilarity metric for event sequences? In the case of multi-event sequences where $e_{t,n} \in \{0, 1, \dots, K\}$, where $K \gg 1$, some results from the time series analysis literature, e.g. Serrà and Arcos [2014], Batista et al. [2011], Kaya and Gündüz-Öğüdücü [2015], and Lhermitte et al. [2011], can be borrowed as in this case event sequence looks more like a typical time series. However, in a binary event sequence where $e_{t,n} \in \{0, 1\}$, this question becomes important. Recently in Moser and Natschlager [2014] and Moser [2017], a rigorous answer to this question has been provided, where it is shown that Weyl's discrepancy norm $\|e_{t,1} - e_{t,2}\|_D = \sup_{t_1,t_2 \in \mathbb{N}: t_1 \leq t_2} \left| \sum_{t=t_1}^{t_2} e_{t,1} - e_{t,2} \right|$ is a suitable metric for comparing binary event sequences $e_{t,1}$ and $e_{t,2}$.

In Figure 8.3, the top figure shows a sample path of $e_{t,1} \in \{0, 1\}$, and the other two figures plot the discrepancy norm $\|e_{t,1} - e_{t,2}\|_D$, Euclidean distance, and Hamming distance between $e_{t,1}$ and $e_{t,2}$. All distances are normalized by the sequence length 100 and averaged over 1000 trials. In the middle figure, $e_{t,2} = e_{t-\tau,1}$ (i.e. delayed) and $e_{t,1} = 0$ for $t < 0$, whereas in the bottom figure $e_{t,2} \neq e_{t,1}$ (i.e. flipped) with probability p. These cases respectively corresponds to practical communication channels with delay and bit error. While the discrepancy norm smoothly reacts to increasing delay, the Euclidean distance and especially Hamming distance overreacts to small values of delay, which are highly probable in practical communication channels. After a sharp increase for the smallest delay value of 1, Euclidean and Hamming distances are not responsive to changing

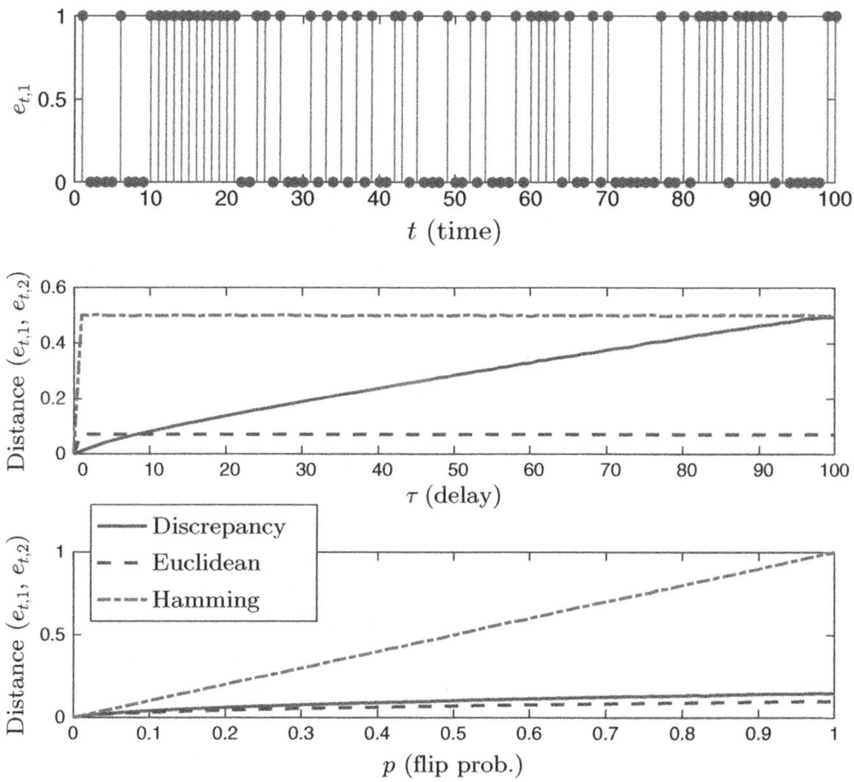

Figure 8.3 Discrepancy reacts smoothly to common communication nonidealities.

delay values. In the bottom figure which corresponds to a binary symmetric channel, it is also seen in the bottom figure that Hamming distance overreacts to possible bit errors while discrepancy norm and Euclidean distance react smoothly to this common type of communication error.

8.4 Privacy-Preserving Anomaly Detection

In distributed networks where each node holds privacy-sensitive data, data-driven statistical inference process has the risk of violating the privacy of data providers [Sarwate and Chaudhuri, 2013; Dwork and Roth, 2014]. For example, in distribution smart grids, user electricity consumption patterns can be used to build sensitive user profiles for malicious purposes. Hence, usually a balance between data utilization and data privacy needs to be sought. Differential privacy (DP) [Dwork and Roth, 2014] is a probabilistic framework based on the notion of indistinguishability. Specifically, it cannot be inferred whether any specific node/user/device contributed to the data by observing an output of a differentially private algorithm, which provides almost the same level of privacy to each data provider. In other words, the output likelihood of a differentially private algorithm is not significantly changed by the change/removal of the data of any single data provider. In this framework, privacy is mainly attained by randomizing the released statistics from a database, where the worst-case privacy risk can be quantified and adjusted with the level of randomization [Sarwate and Chaudhuri, 2013]. We will study real-time distributed anomaly detection over IoT networks under DP constraints.

In distributed settings where the data is privacy sensitive, nodes may only be willing to disclose some minimal information aligned with the anomaly detection task in an encrypted form because of privacy concerns, e.g. local differential privacy [Kasiviswanathan et al., 2011]. Considering such a setting, let every node n, based on its observation $\mathbf{x}_{t,n}$, share a univariate signal $z_{t,n}$ with the network center at each time t. The network center then receives $\mathbf{z}_t \triangleq [z_{t,1}, z_{t,2}, \dots, z_{t,N}]$ and decides on the anomaly based on the sequence of $\{\mathbf{z}_t\}$. Under this general setup, our goal is to introduce a novel method that provides (i) model-free real-time processing of the observed local data stream and disclosure of minimal task-oriented information at each node, (ii) differentially private aggregation of the node messages at the center, and (iii) quick and reliable network-wide anomaly detection.

We propose a solution scheme consisting of local data processing, private stream aggregation, and real-time anomaly detection (see Figure 8.4). First, sensitive data is analyzed and processed locally at each node, and a minimal task-oriented univariate statistic is extracted for the anomaly detection task, where the local data processing is free of data model assumptions. The output of the local data

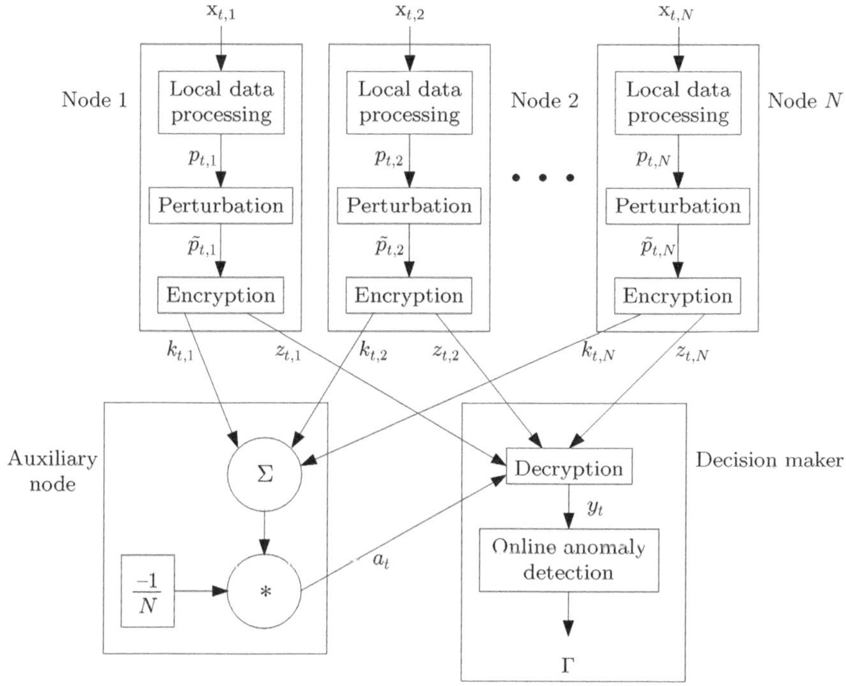

Figure 8.4 The proposed solution scheme for DP-enabled network-wide anomaly detection.

processing of node n is the p-value $p_{t,n}$ of observations at time t, which is the percentage of the training instances whose task-oriented univariate statistics are higher than that of the observations at time t (i.e. one minus cdf). Second, for DP, instead of releasing the extracted information directly, it is first perturbed by additive noise, e.g. $v_{t,n} \sim \mathcal{N}(0, \sigma^2)$, resulting in the perturbed statistic

$$\tilde{p}_{t,n} = p_{t,n} + v_{t,n}.$$

Next, a form of cryptographic communication is used between the nodes and the network center to ensure that the network center can only decrypt an aggregate statistic (p-value) over the entire network but not the individual node information. Specifically, each node n produces a private positive number $k_{t,n}$ at time t, which cannot be tracked by outsiders (e.g. a time-varying random variable); adds it to $\tilde{p}_{t,n}$, and obtains the signal

$$z_{t,n} = \tilde{p}_{t,n} + k_{t,n}$$

transmitted to the network center. Moreover, an auxiliary node helps for the cryptographic communication, as described in Leontiadis et al. [2014]. It receives the keys from the nodes and transmits the negative average of the keys

$$a_t = -\frac{1}{N}\sum_{n=1}^{N} k_{t,n}$$

to the network center.

After receiving $\{z_{t,n}, n = 1, 2, \dots, N\}$ from the nodes, and a_t from the auxiliary node, the network operator takes the average of the node messages, then sums the average with a_t, and obtains y_t:

$$\begin{aligned} y_t &= a_t + \frac{1}{N}\sum_{n=1}^{N} z_{t,n} = a_t + \frac{1}{N}\sum_{n=1}^{N}(\tilde{p}_{t,n} + k_{t,n}) \\ &= \underbrace{a_t + \frac{1}{N}\sum_{n=1}^{N} k_{t,n}}_{0} + \frac{1}{N}\sum_{n=1}^{N}\tilde{p}_{t,n} = \frac{1}{N}\sum_{n=1}^{N}\tilde{p}_{t,n}. \end{aligned}$$

For distributed DP, cryptographic communication is used to ensure that only an aggregate statistic over the network but nothing else about the individual nodes can be learned by the network operator. Interested readers are referred to Leontiadis et al. [2014] for details on cryptographic communication protocol.

Although a different cryptographic communication protocol can be designed with secret key sharing without needing the auxiliary node [Leontiadis et al., 2014], it requires coordination and peer-to-peer communication between nodes. Moreover, such a protocol is not robust to node failures and dynamic networks because it needs a redesign when any node joins or leaves.

8.4.1 Online Network Anomaly Detection

We start by analyzing the distribution of the aggregated statistic y_t at the network operator in both nominal and anomaly cases. Then, the proposed online network anomaly detection method is explained.

Distribution of y_t: The information aggregated at the network operator at time t can be rewritten as

$$\begin{aligned} y_t &= \frac{1}{N}\sum_{i=1}^{N}\tilde{p}_{t,n} = \frac{1}{N}\sum_{i=1}^{N}(p_{t,n} + v_{t,n}) \\ &= \underbrace{\frac{1}{N}\sum_{i=1}^{N} p_{t,n}}_{\overline{p}_t} + \underbrace{\frac{1}{N}\sum_{i=1}^{N} v_{t,n}}_{\overline{v}_t} \\ &= \overline{p}_t + \overline{v}_t, \end{aligned} \tag{8.1}$$

where $\overline{v}_t \sim \mathcal{N}(0, \sigma^2/N)$.

Recalling that $p_{t,n} \sim \mathcal{U}[0,1], \forall n \in \{1,2,\dots,N\}$ for $t < \tau$ and assuming $p_{t,n}$ is i.i.d. over time and space,[2] the central limit theorem yields, asymptotically

$$\overline{p}_t \sim \mathcal{N}\left(0.5, \frac{1}{12N}\right), \quad t < \tau. \tag{8.2}$$

Since $\overline{p}_t$ and $\overline{v}_t$ are independent, we can write

$$y_t \sim \mathcal{N}\left(0.5, \frac{\sigma^2 + 1/12}{N}\right), \quad t < \tau. \tag{8.3}$$

Moreover, we approximately have

$$y_t \sim \mathcal{N}(0.5, \sigma^2/N), \; t < \tau, \tag{8.4}$$

if $\sigma^2 \gg 1/12$.

If there is a network anomaly (i.e. for $t \geq \tau$), anomalous nodes will observe more frequent outliers, thus smaller p-values. This results in a decrease in the mean of y_t. Therefore, it can be argued that the mean of y_t is $0.5 - \gamma_t$ for $t \geq \tau$, where $\gamma_t \geq 0$ denotes the unknown and possibly time-varying mean decrease. Furthermore, when there is an anomaly, $p_{t,n} \sim \mathcal{U}[0,1]$ is not valid. The pdf of $p_{t,n}$ is unknown for $t \geq \tau$. However, $p_{t,n}$ is always between 0 and 1 and for a random variable taking values in this range, its variance is upper bounded[3] by 1/4. Hence, for $t \geq \tau$, $p_{t,n}$ has a mean $\mu_{t,n} \in [0, 0.5]$ and a variance $\sigma_{t,n}^2 \in [0, 1/4], \forall n \in \{1,2,\dots,N\}$.

There are different versions of the central limit theorem, which prove convergence to the normal distribution for nonidentical or dependent distributions under certain conditions. In large-scale networks (i.e. large N), for $t \geq \tau$, $p_{t,n}$ can be considered as nearly independent but nonidentically distributed across the nodes, where the Lindeberg central limit theorem [Billingsley, 1986, p. 369] can be utilized. Given

$$s_{t,N}^2 \triangleq \sum_{n=1}^{N} \sigma_{t,n}^2,$$

if for every $\varepsilon > 0$, the condition

$$\lim_{N\to\infty} \frac{1}{s_{t,N}^2} \sum_{n=1}^{N} \mathrm{E}[(p_{t,n} - \mu_{t,n})^2 \mathbf{1}\{|p_{t,n} - \mu_{t,n}| > \varepsilon\, s_{t,N}\}] = 0 \tag{8.5}$$

---

2 An i.i.d. $p_{t,n}$ stream can be achieved through local data processing. For instance, in Principal Component Analysis, if the linear approximation fits the data well, the residual term mostly corresponds to i.i.d. noise. Moreover, in large-scale networks, local data can be assumed independent from the majority of other data observed in the network.

3 The variance of a $x \in [0,1]$ is written as $\sigma_x^2 \triangleq \mathbb{E}[x^2] - (\mathbb{E}[x])^2 \leq \mathbb{E}[x] - (\mathbb{E}[x])^2$, where the inequality is because $x^2 \leq x$ for $x \in [0,1]$. Denoting $m \triangleq \mathbb{E}[x]$, we have $\sigma_x^2 \leq f(m) \triangleq m - m^2$ where $m \in [0,1]$. Finally, $\sigma_x^2 \leq 1/4$ because the maximum value of the function $f(m)$ is 1/4 at $m = 1/2$.

is satisfied, then

$$\frac{1}{s_{t,N}}\sum_{n=1}^{N}(p_{t,n}-\mu_{t,n})$$

converges to the standard normal distribution $\mathcal{N}(0,1)$. Under the condition above, we can asymptotically write

$$\overline{p}_t = \frac{1}{N}\sum_{n=1}^{N}p_{t,n} \sim \mathcal{N}\left(\frac{\sum_{n=1}^{N}\mu_{t,n}}{N}, \frac{s_{t,N}^2}{N^2}\right).$$

Since

$$0 \leq \sum_{n=1}^{N}\mu_{t,n} \leq N/2,$$

and

$$0 \leq s_{t,N}^2 \leq N/4,$$

the mean of $\overline{p}_t$ is between 0 and 0.5 and the variance of $\overline{p}_t$ is between 0 and $\frac{1}{4N}$. Hence, if $\sigma^2 \gg 1/4$, it approximately holds that, see Eq. (8.1),

$$y_t = \overline{p}_t + \overline{v}_t \sim \mathcal{N}(0.5-\gamma_t, \sigma^2/N),\ t \geq \tau. \tag{8.6}$$

Note that the condition in Eq. (8.5) is satisfied in our case. The indicator in Eq. (8.5) tends to 0 as $N \to \infty$ since the term $|p_{t,n}-\mu_{t,n}|$ is bounded because of $p_{t,n}, \mu_{t,n} \in [0,1]$. On the other hand, $s_{t,N} \to \infty$ as $N \to \infty$.

Finally, we can write

$$y_t \sim \begin{cases} \mathcal{N}(0.5, \sigma^2/N), & \text{if } t < \tau, \\ \mathcal{N}(0.5-\gamma_t, \sigma^2/N), & \text{if } t \geq \tau. \end{cases} \tag{8.7}$$

if $\sigma^2 \gg 1/4$ (see Eqs. (8.4) and (8.6)). In Eq. (8.7), we convert the original high-dimensional problem to a simpler problem where the goal is to sequentially detect a mean decrease in a univariate Gaussian data stream. This allows us to use a parametric setting, as explained next, instead of the original nonparametric setting.

Online anomaly detection: Let f_0^y and f_1^y denote the nominal and anomalous pdfs of y_t, respectively, and $\theta \triangleq \sigma/\sqrt{N}$. Then, $f_0^y \sim \mathcal{N}(0.5, \theta^2)$ and $f_1^y \sim \mathcal{N}(0.5-\gamma_t, \theta^2)$, where f_1^y has an unknown and possibly time-varying parameter γ_t, see Eq. (8.7). We next propose the following generalized CUSUM algorithm for online anomaly detection at the network operator:

$$\Gamma = \inf \left\{ m \in \mathbb{N} : \underbrace{\max_{1 \le j \le m} \sum_{t=j}^{m} \underbrace{\log \frac{\sup_{\gamma_t \ge \eta} f_1^y(y_t \mid \gamma_t)}{f_0^y(y_t)}}_{\beta_t}}_{g_m} \ge h \right\}, \tag{8.8}$$

where η denotes the minimum change of interest, indicating the detector sensitivity, and h denotes the test threshold. Further, β_t and g_t denote the generalized log-likelihood ratio (GLLR) and the decision statistic at time t, respectively, where the decision statistic can be written in the following recursive form, see Basseville and Nikiforov [1993, Sec. 2.2]:

$$g_t = (g_{t-1} + \beta_t)^+. \tag{8.9}$$

Furthermore, the GLLR β_t can be computed as follows:

$$\begin{aligned} \beta_t &= \frac{1}{2\theta^2} \sup_{\gamma_t \ge \eta} (1 - 2y_t)\gamma_t - \gamma_t^2 \\ &= \begin{cases} \frac{1}{2\theta^2}(0.5 - y_t)^2, & \text{if } y_t \le 0.5 - \eta, \\ \frac{1}{2\theta^2}(1 - 2y_t)\eta - \frac{\eta^2}{2\theta^2}, & \text{if } y_t > 0.5 - \eta. \end{cases} \end{aligned} \tag{8.10}$$

Note that η is only a detector parameter and not part of an anomaly model. The only difference between the conventional CUSUM algorithm [Basseville and Nikiforov, 1993, Sec. 2.2] and the proposed algorithm is the online estimation of γ_t.

Note that the proposed generalized CUSUM algorithm is not the same as the well-known GLR test [Basseville and Nikiforov, 1993, Sec. 5.3]. We cannot write the GLR test in a recursive form since it is assumed that the unknown pdf parameters are fixed over time. Whereas, the proposed algorithm can be written in a recursive form since the unknown pdf parameter γ_t is assumed to be time-varying because of unknown anomaly and it is estimated at each time t separately.

8.4.2 Experimental Results

We performed experiments over the N-BaIoT dataset [Meidan et al., 2018], which is collected from a real IoT network consisting of nine devices. The data $\mathbf{x}_{t,n} \in \mathbb{R}^{115}$ at each node represents network traffic statistics at that node. Each node employs the PCA-based local data processing method described in Kurt

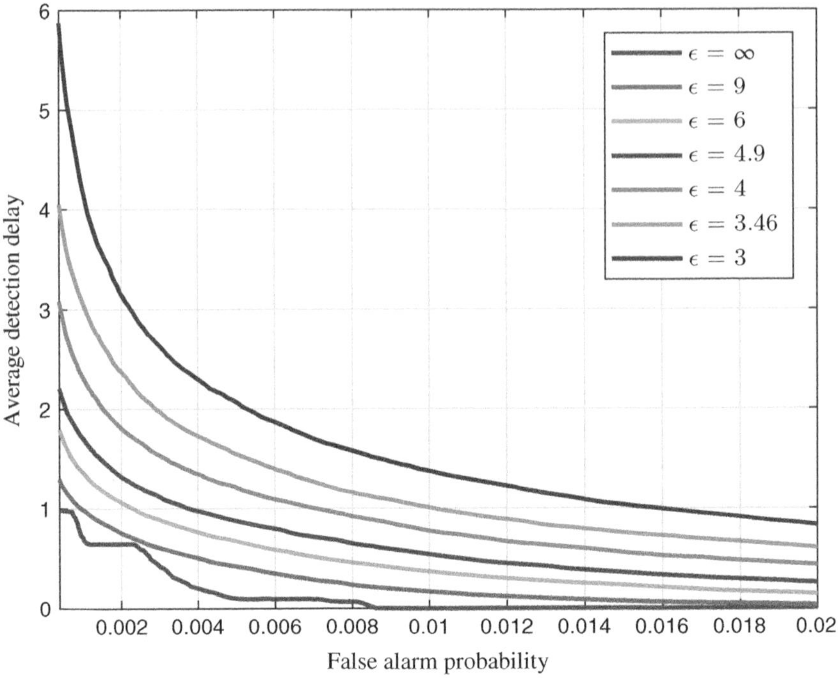

Figure 8.5 ADD versus FAP of the proposed detector in case of a spam attack for various DP levels.

et al. [2019b] and estimates the p-value $p_{t,n}$ of the residual of $\mathbf{x}_{t,n}$ at each time t. Each private number $k_{t,n} \sim \mathcal{N}(0, \sigma^2)$. Figure 8.5 presents the ADD versus false alarm probability (FAP) curves of the proposed detector for various levels of DP. To obtain various DP levels, we vary the variance of the local Gaussian perturbation noise σ^2. Figure 8.5 clearly shows the privacy-security tradeoff in the network-wide anomaly detection problem, i.e. stronger privacy guarantees (lower ϵ) worsen the anomaly detection performance.

8.4.3 DP Techniques

Following the Gaussian perturbation results for DP Dwork and Roth [2014], how much perturbation would be needed to prove DP for the proposed architecture can be derived [Kurt et al., 2022]. Such a proof for the proposed distributed scheme is facilitated by the parallel composition rule of the DP [McSherry, 2009], which states that if differentially private mechanisms are employed over disjoint subsets of a database, then the overall mechanism also achieves DP.

Other perturbation techniques such as adding Laplacian noise [Dwork et al., 2006b] can be also considered.

8.4.4 Anomaly Detection Performance

An asymptotic approximation and a lower bound for the FAP of the considered solution scheme can be derived leveraging the asymptotic distribution of aggregate p-value [Kurt et al., 2022]. Similarly, an asymptotic approximation and an upper bound for the ADD of the considered solution scheme can be derived. Since a CUSUM-type algorithm is employed in the network center, for the FAP and ADD analysis of the considered solution scheme, Wald's approximations and Siegmund's approximations as well as the theoretical upper and lower bounds can be presented for the average run length (ARL) of the general CUSUM-type algorithms in Basseville and Nikiforov [1993, Sec. 5.3]. Next, by using the derived FAP and ADD approximations, the analytical privacy-security tradeoff in the network-wide anomaly detection problem, which is controlled via the variance of perturbation noise, can be shown as in Kurt et al. [2022].

8.4.5 Differentially Private Event-Triggered Anomaly Detection

The event-triggered sampling, described in Section 8.3, has practical advantages in terms of preserving the privacy of local sensitive data as the node messages are coded into specific event definitions. However, this cannot be considered as a perfect privacy solution since releasing these messages may still lead to side information leakages and reconstruction attacks [Dwork and Roth, 2014]. Therefore, we need additional techniques such as DP to ensure privacy. Since in the event-triggered mechanism, a bit stream is transmitted from nodes, transmitted bits can be randomized via flipping them with certain probability in order to achieve DP. One can analyze the DP guarantees of the corresponding network-wide anomaly detection scheme, as well as its effects on the anomaly detection performance in terms of the FAP and ADD. The aim of this analysis is to establish the analytical tradeoff between the DP level, controlled via the bit flipping ratio, and the real-time anomaly detection performance. As shown in the bottom figure of Figure 8.3, higher flip probabilities will cause larger distance between nominal event sequences and in turn higher false alarm rates while at the same time providing better DP guarantees. Probabilistically flipping the bits to be transmitted from nodes corresponds to the stochastic encryption framework, which is used as a physical layer security tool [Aysal and Barner, 2008; Soosahabi and Naraghi-Pour, 2012; Soosahabi et al., 2014]. It is known that stochastic encryption causes significant information loss to an eavesdropper while minimally reducing the network center's inference performance. However, the privacy, in particular DP, aspect of stochastic encryption has not been studied yet. The stochastic encryption results [Aysal and Barner, 2008; Soosahabi and Naraghi-Pour, 2012; Soosahabi et al., 2014] can be leveraged, together with theoretical tools of sequential change detection – in particular asymptotic ADD and FAP analysis – to study the DP guarantees and analytical anomaly detection-DP tradeoff.

Bibliography

N. Agarwal, A. T. Suresh, F. X. X. Yu, S. Kumar, and B. McMahan. cpSGD: Communication-efficient and differentially-private distributed SGD. In *Advances in Neural Information Processing Systems*, pages 7564–7575, 2018.

A. Aue, S. Hörmann, L. Horváth, and M. Reimherr. Break detection in the covariance structure of multivariate time series models. *The Annals of Statistics*, 37(6B):4046–4087, 2009.

T. C. Aysal and K. E. Barner. Sensor data cryptography in wireless sensor networks. *IEEE Transactions on Information Forensics and Security*, 3(2):273–289, 2008.

T. Banerjee and A. O. Hero. Quickest hub discovery in correlation graphs. In *Signals, Systems and Computers, 2016 50th Asilomar Conference on*, pages 1248–1255. IEEE, 2016.

T. Banerjee, H. Firouzi, and A. O. Hero. Non-parametric quickest change detection for large scale random matrices. In *Information Theory (ISIT), 2015 IEEE International Symposium on*, pages 146–150. IEEE, 2015.

M. Basseville and I. V. Nikiforov. *Detection of Abrupt Changes: Theory and Application*. Prentice-Hall, Inc., Upper Saddle River, NJ, USA, 1993. ISBN 0-13-126780-9.

G. E. A. P. A. Batista, X. Wang, and E. J. Keogh. A complexity-invariant distance measure for time series. In *Proceedings of the 2011 SIAM International Conference on Data Mining*, pages 699–710. SIAM, 2011.

P. Billingsley. *Probability and Measure*. Wiley series in Probability and Mathematical Statistics. Wiley, 1986. ISBN 9780471804789.

C. L. Canonne, G. Kamath, A. McMillan, A. Smith, and J. Ullman. The structure of optimal private tests for simple hypotheses. In *Proceedings of the 51st Annual ACM SIGACT Symposium on Theory of Computing*, pages 310–321, 2019.

V. Chandola, A. Banerjee, and V. Kumar. Anomaly detection: A survey. *ACM computing surveys (CSUR)*, 41(3):15, 2009.

P. Chatillon and C. Ballester. History-based anomaly detector: An adversarial approach to anomaly detection. *arXiv preprint arXiv:1912.11843*, 2019.

H. Chen. Sequential change-point detection based on nearest neighbors. *arXiv preprint arXiv:1604.03611*, 2018.

M. Cui, J. Wang, and M. Yue. Machine learning-based anomaly detection for load forecasting under cyberattacks. *IEEE Transactions on Smart Grid*, 10(5):5724–5734, 2019.

R. Cummings, S. Krehbiel, Y. Mei, R. Tuo, and W. Zhang. Differentially private change-point detection. In *Advances in Neural Information Processing Systems*, pages 10825–10834, 2018.

K. H. Degue and J. Le Ny. On differentially private Gaussian hypothesis testing. In *2018 56th Annual Allerton Conference on Communication, Control, and Computing (Allerton)*, pages 842–847. IEEE, 2018.

F. Desobry, M. Davy, and C. Doncarli. An online kernel change detection algorithm. *IEEE Transactions on Signal Processing*, 53(8):2961–2974, 2005.

C. Dwork and A. Roth. The algorithmic foundations of differential privacy. *Foundations and Trends®. Theoretical Computer Science*, 9(3–4):211–407, 2014.

C. Dwork, K. Kenthapadi, F. McSherry, I. Mironov, and M. Naor. Our data, ourselves: Privacy via distributed noise generation. In *Annual International Conference on the Theory and Applications of Cryptographic Techniques*, pages 486–503. Springer, 2006a.

C. Dwork, F. McSherry, K.i Nissim, and A. Smith. Calibrating noise to sensitivity in private data analysis. In *Theory of Cryptography Conference*, pages 265–284. Springer, 2006b.

R. Elnaggar, K. Chakrabarty, and M. B. Tahoori. Hardware Trojan detection using changepoint-based anomaly detection techniques. *IEEE Transactions on Very Large Scale Integration (VLSI) Systems*, 27(12):2706–2719, 2019.

F. Enikeeva and Z. Harchaoui. High-dimensional change-point detection with sparse alternatives. *arXiv preprint arXiv:1312.1900*, 2013.

L. Fan and L. Xiong. Differentially private anomaly detection with a case study on epidemic outbreak detection. In *2013 IEEE 13th International Conference on Data Mining Workshops*, pages 833–840. IEEE, 2013.

G. Fellouris and G. V. Moustakides. Decentralized sequential hypothesis testing using asynchronous communication. *IEEE Transactions on Information Theory*, 57(1):534–548, 2011.

M. Ghassemi, A. D. Sarwate, and R. N. Wright. Differentially private online active learning with applications to anomaly detection. In *Proceedings of the 2016 ACM Workshop on Artificial Intelligence and Security*, pages 117–128, 2016.

Z. Harchaoui, E. Moulines, and F. R. Bach. Kernel change-point analysis. In *Advances in Neural Information Processing Systems*, pages 609–616, 2009.

A. M. Hussain. Multisensor distributed sequential detection. *IEEE Transactions on Aerospace and Electronic Systems*, 30(3):698–708, 1994.

R. T. Ionescu, F. S. Khan, M.-I. Georgescu, and L. Shao. Object-centric auto-encoders and dummy anomalies for abnormal event detection in video. In *Proceedings of the*

IEEE Conference on Computer Vision and Pattern Recognition, pages 7842–7851, 2019.

P. Jain, P. Kothari, and A. Thakurta. Differentially private online learning. In *Conference on Learning Theory*, pages 24–1, 2012.

M. Jirak. Uniform change point tests in high dimension. *The Annals of Statistics*, 43(6):2451–2483, 2015.

S. P. Kasiviswanathan, H. K. Lee, K. Nissim, S. Raskhodnikova, and A. Smith. What can we learn privately? *SIAM Journal on Computing*, 40(3):793–826, 2011. doi: 10.1137/090756090.

H. Kaya and Ş. Gündüz-Öğüdücü. A distance based time series classification framework. *Information Systems*, 51:27–42, 2015.

M. Keshk, E. Sitnikova, N. Moustafa, J. Hu, and I. Khalil. An integrated framework for privacy-preserving based anomaly detection for cyber-physical systems. *IEEE Transactions on Sustainable Computing*, 6(1):66–79, 2019.

D. Kifer, S. Ben-David, and J. Gehrke. Detecting change in data streams. In *Proceedings of the 13th International Conference on Very Large Data Bases-Volume 30*, pages 180–191. VLDB Endowment, 2004.

M. N. Kurt, O. Ogundijo, C. Li, and X. Wang. Online cyber-attack detection in smart grid: A reinforcement learning approach. *IEEE Transactions on Smart Grid*, 10(5):5174–5185, Sept 2019a. ISSN 1949-3061.

M. N. Kurt, Y. Yilmaz, and X. Wang. Sequential model-free anomaly detection for big data streams. In *2019 57th Annual Allerton Conference on Communication, Control, and Computing (Allerton)*, pages 421–425, Sept 2019b.

M. N. Kurt, Y. Yılmaz, X. Wang, and P. J Mosterman. Online privacy-preserving data-driven network anomaly detection. *IEEE Journal on Selected Areas in Communications*, 40(3):982–998, 2022.

R. Laxhammar and G. Falkman. Online learning and sequential anomaly detection in trajectories. *IEEE Transactions on Pattern Analysis and Machine Intelligence*, 36(6):1158–1173, June 2014. ISSN 1939-3539. doi: 10.1109/TPAMI.2013.172.

J. Le Ny and G. J. Pappas. Differentially private filtering. *IEEE Transactions on Automatic Control*, 59(2):341–354, 2013.

I. Leontiadis, K. Elkhiyaoui, and R. Molva. Private and dynamic time-series data aggregation with trust relaxation. In *International Conference on Cryptology and Network Security*, pages 305–320. Springer, 2014.

S. Lhermitte, J. Verbesselt, W. W. Verstraeten, and P. Coppin. A comparison of time series similarity measures for classification and change detection of ecosystem dynamics. *Remote Sensing of Environment*, 115(12): 3129–3152, 2011.

S. Li, Y. Yilmaz, and X. Wang. Quickest detection of false data injection attack in wide-area smart grids. *IEEE Transactions on Smart Grid*, 6(6): 2725–2735, Nov 2015a. ISSN 1949-3053. doi: 10.1109/TSG.2014.2374577.

S. Li, Y. Xie, H. Dai, and L. Song. M-statistic for kernel change-point detection. In *Advances in Neural Information Processing Systems*, pages 3366–3374, 2015b.

T. Li, J. Li, Z. Liu, P. Li, and C. Jia. Differentially private Naive Bayes learning over multiple data sources. *Information Sciences*, 444:89–104, 2018.

S. Liu, M. Yamada, N. Collier, and M. Sugiyama. Change-point detection in time-series data by relative density-ratio estimation. *Neural Networks*, 43:72–83, 2013.

W. Liu, W. Luo, D. Lian, and S. Gao. Future frame prediction for anomaly detection–a new baseline. In *Proceedings of the IEEE Conference on Computer Vision and Pattern Recognition*, pages 6536–6545, 2018.

G. Lorden. Procedures for reacting to a change in distribution. *The Annals of Mathematical Statistics*, 42(6):1897–1908, 1971. doi: 10.1214/aoms/1177693055.

B. Matthews. Automatic anomaly detection with machine learning. 2019.

F. D. McSherry. Privacy integrated queries: An extensible platform for privacy-preserving data analysis. In *Proceedings of the 2009 ACM SIGMOD International Conference on Management of data*, pages 19–30, 2009.

Y. Mei. Asymptotic optimality theory for decentralized sequential hypothesis testing in sensor networks. *IEEE Transactions on Information Theory*, 54(5):2072–2089, 2008.

Y. Mei. Efficient scalable schemes for monitoring a large number of data streams. *Biometrika*, 97(2):419–433, 2010.

Y. Meidan, M. Bohadana, Y. Mathov, Y. Mirsky, A. Shabtai, D. Breitenbacher, and Y. Elovici. N-BaIoT: Network-based detection of IoT botnet attacks using deep autoencoders. *IEEE Pervasive Computing*, 17(3):12–22, Jul 2018. ISSN 1558-2590. doi: 10.1109/MPRV.2018.03367731.

B. A. Moser. Similarity recovery from threshold-based sampling under general conditions. *IEEE Transactions on Signal Processing*, 65(17):4645–4654, Sept 2017. ISSN 1053-587X. doi: 10.1109/TSP.2017.2712121.

B. A. Moser and T. Natschlager. On stability of distance measures for event sequences induced by level-crossing sampling. *IEEE Transactions on Signal Processing*, 62(8):1987–1999, Apr 2014. ISSN 1053-587X. doi: 10.1109/TSP.2014.2305642.

M. Mozaffari, K. Doshi, and Ya. Yilmaz. Online multivariate anomaly detection and localization for high-dimensional settings. *Sensors*, 22(21):8264, 2022.

M. Pollak. Optimal detection of a change in distribution. *The Annals of Statistics*, 13(1):206–227, Mar 1985.

M. Ravanbakhsh. Generative models for novelty detection: Applications in abnormal event and situational change detection from data series. *arXiv preprint arXiv:1904.04741*, 2019.

J. Redmon and A. Farhadi. Yolo9000: Better, faster, stronger. *arXiv preprint arXiv:1612.08242*, 2016.

M. Sabokrou, M. Khalooei, M. Fathy, and E. Adeli. Adversarially learned one-class classifier for novelty detection. In *Proceedings of the IEEE Conference on Computer Vision and Pattern Recognition*, pages 3379–3388, 2018.

Z. Sahinoglu, S. Gezici, and I. Guvenc. *Ultra-Wideband Positioning Systems: Theoretical Limits, Ranging Algorithms, and Protocols*. Cambridge University Press, New York, NY, USA, 2008.

A. D. Sarwate and K. Chaudhuri. Signal processing and machine learning with differential privacy: Algorithms and challenges for continuous data. *IEEE Signal Processing Magazine*, 30(5):86–94, Sept 2013. doi: 10.1109/MSP.2013.2259911.

J. Serrà and J. L. Arcos. An empirical evaluation of similarity measures for time series classification. *Knowledge-Based Systems*, 67(Supplement C):305–314, 2014. ISSN 0950-7051.

E. Shi, T. H. H. Chan, E. Rieffel, R. Chow, and D. Song. Privacy-preserving aggregation of time-series data. In *Proceedings of NDSS*, volume 2, pages 1–17. Citeseer, 2011.

A. N. Shiryaev. *Optimal Stopping Rules*. Springer-Verlag, New York, NY, USA, 1978.

S. Song, K. Chaudhuri, and A. D. Sarwate. Stochastic gradient descent with differentially private updates. In *2013 IEEE Global Conference on Signal and Information Processing*, pages 245–248. IEEE, 2013.

R. Soosahabi and M. Naraghi-Pour. Scalable PHY-layer security for distributed detection in wireless sensor networks. *IEEE Transactions on Information Forensics and Security*, 7(4):1118–1126, 2012.

R. Soosahabi, M. Naraghi-Pour, D. Perkins, and M. A. Bayoumi. Optimal probabilistic encryption for secure detection in wireless sensor networks. *IEEE Transactions on Information Forensics and Security*, 9(3):375–385, 2014.

A. G. Tartakovsky and V. V. Veeravalli. Change-point detection in multichannel and distributed systems. *Applied Sequential Methodologies: Real-World Examples with Data Analysis*, 173:339–370, 2004.

A. G. Tartakovsky and V. V. Veeravalli. Asymptotically optimal quickest change detection in distributed sensor systems. *Sequential Analysis*, 27(4):441–475, 2008.

V. V. Veeravalli, T. BaSar, and H. V. Poor. Decentralized sequential detection with a fusion center performing the sequential test. *IEEE Transactions on Information Theory*, 39(2):433–442, 1993.

Y. Xiang, K. Li, and W. Zhou. Low-rate DDoS attacks detection and traceback by using new information metrics. *IEEE Transactions on Information Forensics and Security*, 6(2):426–437, 2011.

Y. Xie and D. Siegmund. Sequential multi-sensor change-point detection. *Ann. Statist.*, 41(2):670–692, Apr 2013. doi: 10.1214/13-AOS1094.

Y. Xie, J. Huang, and R. Willett. Change-point detection for high-dimensional time series with missing data. *IEEE Journal of Selected Topics in Signal Processing*, 7(1):12–27, 2013.

Y. Xie, M. Wang, and A. Thompson. Sketching for sequential change-point detection. In *2015 IEEE Global Conference on Signal and Information Processing (GlobalSIP)*, pages 78–82. IEEE, 2015.

Y. Yılmaz and X. Wang. Sequential distributed detection in energy-constrained wireless sensor networks. *IEEE Transactions on Signal Processing*, 62(12):3180–3193, June 2014a. ISSN 1053-587X. doi: 10.1109/TSP.2014.2320458.

Y. Yilmaz and X. Wang. Sequential decentralized parameter estimation under randomly observed fisher information. *IEEE Transactions on Information Theory*, 60(2):1281–1300, Feb 2014b. ISSN 0018-9448. doi: 10.1109/TIT.2013.2292062.

Y. Yılmaz, G. V. Moustakides, and X. Wang. Cooperative sequential spectrum sensing based on level-triggered sampling. *IEEE Transactions on Signal Processing*, 60(9):4509–4524, Sept 2012. ISSN 1053-587X. doi: 10.1109/TSP.2012.2202657.

Y. Yılmaz, G. V. Moustakides, and X. Wang. Channel-aware decentralized detection via level-triggered sampling. *IEEE Transactions on Signal Processing*, 61(2):300–315, Jan 2013. ISSN 1053-587X. doi: 10.1109/TSP.2012.2222401.

Y. Yilmaz, G. V. Moustakides, X. Wang, and A. O. Hero. Event-based statistical signal processing. In Marek Miskowicz, editor, *Event-Based Control and Signal Processing*, Chapter 20, pages 457–485. CRC Press, 2015.

Y. Yilmaz, G. V. Moustakides, and X. Wang. Sequential and decentralized estimation of linear-regression parameters in wireless sensor networks. *IEEE Transactions on Aerospace and Electronic Systems*, 52(1):288–306, Feb. 2016. ISSN 0018-9251. doi: 10.1109/TAES.2015.140665.

D. Zambon, C. Alippi, and L. Livi. Concept drift and anomaly detection in graph streams. *IEEE Transactions on Neural Networks and Learning Systems*, pages 1–14, 2018a. ISSN 2162-237X. doi: 10.1109/TNNLS.2018.2804443.

D. Zambon, L. Livi, and C. Alippi. Anomaly and change detection in graph streams through constant-curvature manifold embeddings. *arXiv preprint arXiv:1805.01360*, 2018b.

T. Zhang and Q. Zhu. Distributed privacy-preserving collaborative intrusion detection systems for VANETs. *IEEE Transactions on Signal and Information Processing over Networks*, 4(1):148–161, 2018.

H. Zhang, J. Liu, and N. Kato. Threshold tuning-based wearable sensor fault detection for reliable medical monitoring using Bayesian network model. *IEEE Systems Journal*, 12(2):1886–1896, 2018.

C. Zou, P. Qiu, and D. Hawkins. Nonparametric control chart for monitoring profiles using change point formulation and adaptive smoothing. *Statistica Sinica*, 19(3):1337–1357, 2009.

S. Zou, Y. Liang, H. V. Poor, and X. Shi. Nonparametric detection of anomalous data via Kernel mean embedding. *arXiv preprint arXiv:1405.2294*, 2014.

9

Decision-Making in Energy-Efficient Ordered Transmission-Based Networks Under Byzantine Attacks

Chen Quan[1] *and Pramod K. Varshney*[2]

[1]*Faculty of Electrical Engineering, Mathematics, and Computer Science, Delft University of Technology, Delft, Netherlands*
[2]*Department of Electrical Engineering and Computer Science, Syracuse University, Syracuse, NY, USA*

9.1 Introduction

Over the years, wireless communication systems have evolved into the most widely used framework for communication devices and networks. Many applications have benefited from wireless sensor networks (WSNs) such as military surveillance, autonomous driving systems and smart-homes. One important factor to consider when designing WSNs is the limited power supply. Reducing energy consumption due to radio communication is key to sustainability and longevity of WSNs since radio communication is the major component of WSNs that consumes large amounts of energy.

Radio communications' energy consumption is mainly determined by the amount of data transmitted throughout the entire network. Therefore, by reducing the amount of data transmitted and optimizing communication processes, energy-efficiency can be achieved. For example, it can be done through data compression, power-saving modes, and efficient routing algorithms. In the literature, some promising frameworks have been proposed for improving the energy efficiency of the WSNs such as censoring, ordered transmission and compressive sensing (e.g. [Baraniuk, 2007; Appadwedula et al., 2008; Blum and Sadler, 2008; Candès and Wakin, 2008]). Energy efficiency is achieved by either reducing the number of transmissions in the networks (e.g. censoring and ordered transmission) or compressing the data sent by the sensors (e.g. quantization and compressive sensing).

This chapter is concerned with the security aspects of one kind of promising energy-efficient framework mentioned earlier, namely ordered transmission. In ordered transmission-based (OT-based) schemes, energy efficiency is achieved

Wireless Sensor Networks in Smart Environments: Enabling Digitalization from Fundamentals to Advanced Solutions, First Edition. Edited by Domenico Ciuonzo and Pierluigi Salvo Rossi.

by omitting transmission of less informative data. The security threat we are particularly interested in is Byzantine attacks, which are one of the most significant security threats faced by WSNs. In Byzantine attacks, a fraction of the sensors within the network may be compromised and completely controlled by adversaries, while the fusion center (FC) is unaware of the behavioral identity of the malicious sensors. This can lead to the intentional transmission of corrupted data that degrades system performance. A number of studies have been conducted in the literature to design various defense mechanisms for mitigating the effect of Byzantine attacks on the overall system performance. However, energy-efficient OT-based frameworks make it more challenging to address security issues since only a portion of sensors' data is transmitted to the FC during each decision-making interval. Since FC does not receive data from all the sensors all the time, it cannot fully learn the behavior of all the sensors.

In this chapter, we conduct an investigation into the impact of Byzantine attacks on several energy-efficient OT-based schemes used in WSNs. The organization of this chapter is as follows. In Section 9.2, some existing threats from Byzantine attacks on conventional wireless sensor networks are briefly discussed, along with some possible defense mechanisms. Section 9.3 introduces the conventional OT-based (COT-based) scheme and investigates the impact of Byzantine attacks on this system. In Section 9.4, a communication-efficient ordered transmission-based (CEOT-based) scheme is presented, and an analysis of the effect of Byzantine attacks on the overall system is conducted. Section 9.5 compares the resilience of CEOT-based with COT-based systems. Finally, a summary of the chapter and some challenges related to the security of energy efficient OT-based schemes employed in WSNs are presented in Section 9.6.

9.2 Byzantine Attack Model

Byzantine attack is a type of attack that occurs at the physical layer. This type of attack can be traced back to the issue of Byzantine generals, first introduced by Pease et al. [1982], where traitors attempted to mislead other loyal generals by providing false information. In WSNs, this term specifically refers to the malicious behaviors that occur within WSNs when certain sensors are compromised and transmit false data within the network. There are different types of Byzantine attacks such as data modification attack, data omission attack, and delayed attack. Malicious nodes can selectively delay data (delayed attack), drop data (data omission attack), or alter data packets directly (data modification attack) to manipulate the network.

In the existing literature, there have been studies of distributed WSNs under Byzantine attacks (e.g. [Kailkhura et al., 2018; Vempaty et al., 2018; Quan et al., 2022b]). The interactions between Byzantines and the WSNs can be viewed as games between attackers and the detection systems. Byzantines aim to undermine the integrity of data transmitted, thereby lowering the reliability of wireless sensor networks. Correspondingly, the FC can enhance the reliability of the network by identifying the Byzantines and making suitable use of information coming from Byzantines for mitigation purposes. Hence, strategic Byzantines attackers strive to maximize their attack gains while attempting to avoid detection by the defense system.

The level of effort required for an effective attack varies depending on the data fusion system architecture. In centralized fusion, the system can better evaluate the behavior of all the sensors so that the attacks can be mitigated, especially when the majority of nodes are honest and the FC is trustworthy. However, in decentralized fusion, each sensor can only communicate with its neighbors to gather additional information regarding the phenomenon of interest before making a decision. This decentralized approach makes the system more susceptible to attacks since false data can be stealthily incorporated into the decisions of neighboring nodes and diffused throughout the network.

9.2.1 Typical Attack Model in WSNs

There are several factors that can be used to classify Byzantine attacks in WSNs. One such factor is the availability of additional information besides the sensing results at the Byzantine nodes. If no extra information is available, the attacks are referred to as *independent attacks*, meaning that the Byzantine nodes can only rely on their own sensing capabilities. On the other hand, if the attacks involve the acquisition of extra information by the Byzantine nodes, such as the current sensing results of other malicious nodes, fusion rules, and defense strategies, they are referred to as *dependent attacks*. The exchange of information in dependent attacks allows malicious nodes to increase their accuracy in sensing and the success rate of their attacks, making their collusion more effective. One approach to defend against these types of attacks is to use statistical methods to detect and identify malicious nodes that are demonstrating anomalous behavior [Quan et al., 2022c]. Another factor is the manner in which the attacks are executed. If the attacks are launched with a certain probability, they are referred to as *probabilistic attacks*. Defense algorithms for these types of attacks usually identify attackers by analyzing the consistency of their attack behavior over time, such as reputation-based schemes (e.g. [Quan et al., 2022c]) and cluster-based schemes (e.g. [Chen and Wang, 2019]). Conversely, if the attacks are launched based on specific conditions, such as when their posterior probability of being a malicious node exceeds a certain threshold, they are referred to as *non-probabilistic attacks*. These attacks are much harder to model compared to probabilistic attacks and can be very difficult to defend against, as the Byzantine nodes are intentionally trying to appear normal while causing disruptions.

9.2.2 Existing Defense Schemes

There are a number of studies exploring various defense mechanisms in the literature aimed at mitigating the impact of Byzantine attacks on the overall system performance. They either directly identify and isolate Byzantine attackers or design the system parameters to mitigate the effects of attacks on the system. There are many promising methods for dealing with Byzantine attacks in networks, such as game-theoretic techniques [Liu et al., 2020], reputation-based methods [Nadendla et al., 2014; Quan et al., 2022c], and machine learning techniques [Yang et al., 2020]. Several consensus-based algorithms have been used in decentralized fusion to improve their robustness under attack. Efforts have been made to exclude nodes with significant deviations from consensus (e.g. [Liu et al., 2012]) and to design network parameters to mitigate the impact of data falsification attacks (e.g. [Kailkhura et al., 2017; Mustafa et al., 2021]).

The previously discussed works have made strides in improving the resilience of systems against Byzantine attacks, however, they still have limitations in detecting distributed attacks when a large number of sensor nodes are attacked. Some works, such as Hashlamoun et al. [2017, 2018] and Zhang et al. [2018], have successfully reduced the impact of Byzantine attacks on wireless sensor networks (WSNs) even when the majority of sensors are malicious. There are also some works that have used the idea of quickest change detection to detect the presence of anomalous measurements due to the malicious sensors in the networks. A model of quickest change detection problems was proposed to detect the presence of Byzantines. The malicious behavior of Byzantines is characterized by distributions before and after the change time (e.g. [Fellouris et al., 2017; Huang et al., 2021]). Aside from works that deal with performance analysis and robust design of networks with fixed sample sizes, there are also studies that deal with performance analysis and robust design of networks with unknown sample sizes, such as sequential hypothesis testing (e.g. [Wu et al., 2018; Li et al., 2021]).

In the literature, there appears to be a large body of literature on performance analysis and robust design of networks that utilize data from all the sensors to make final decisions under Byzantine attacks, while the literature on performance analysis and the robust design of energy-efficient networks with an unknown number of sensors still needs more effort on methods, such as censoring-based schemes, ordered transmission-based schemes, and sleep scheduling algorithms. The reduced number of sensors needed to reach a final decision appears to meet the increasing demand for low-energy consumption and long-lasting WSNs for a variety of applications. Compressed sensing is also an energy-efficient framework in which data sent by the sensors is compressed using quantization and compressed sensing. These energy-efficient frameworks are still under investigation in terms of their robustness and their robust design in the context of error-prone environments and under attacks. In the rest of this chapter, we will focus on one promising framework which is ordered transmission-based framework.

9.3 COT-Based System

The COT-based scheme was initially introduced by Blum and Sadler [2008]. In this scheme, the FC receives the log-likelihood ratios (LLRs) instead of raw data from only a subset of sensors. This approach has been utilized in a wide range of detection problems to enhance energy efficiency while preserving the detection performance of networks. For instance, Rawas et al. [2011] applied the OT-based framework to the detection of noncoherent signals. Hesham et al. [2012] applied the OT framework to sequential detection problems. Chen et al. [2020a, 2021] applied the OT framework to quickest change detection problem. Gupta et al. [2020] applied the OT framework to energy-harvesting systems. In addition, researchers have explored the use of the OT-based framework in various applications beyond detection tasks for transmission-saving purposes. For instance, Chen et al. [2020b] proposed a scheme that applies the idea of ordered transmission to the gradient descent approach. This new scheme guarantees the same order of convergence rate but with fewer transmissions. Yang et al. [2019] applied the concept of ordered transmission to the discretized estimation problem and demonstrated its ability to significantly decrease latency without deteriorating estimation accuracy. The above literature demonstrates the versatility and efficiency of the OT-based framework in various applications.

In this section, the COT-based system proposed by Blum and Sadler [2008] is discussed. An evaluation of such a system under Byzantine attacks, as well as a discussion of possible defense strategies, will be conducted.

9.3.1 System Model of COT-Based System

To elucidate the basic concepts of COT schemes, a binary hypothesis testing problem is considered. The observations at the N sensors in the system indicate either the presence ($\mathcal{H}_1$) or absence ($\mathcal{H}_0$) of a phenomenon of interest (PoI). In COT-based frameworks, only a subset of sensors needs to transmit their LLRs to the FC. Once the FC gathers sufficient observations for a final decision of desired quality, it broadcasts a stop signal to halt further sensor transmissions for the current decision interval. Sensors that have not yet transmitted reset their timers for the next decision interval upon receiving the stop signal.

For sensor $i \in \{1, 2, \dots, N\}$, its observation y_i is modeled as

$$y_i = \begin{cases} n_i, & \text{under } \mathcal{H}_0, \\ s + n_i, & \text{under } \mathcal{H}_1, \end{cases} \tag{9.1}$$

where n_i is the Gaussian noise with zero mean and variance σ^2 and s is the signal strength at each sensor. n_i and s are assumed to be independent, and all the observations are assumed to be independent and identically distributed (i.i.d) conditioned on the hypotheses. Each sensor $i \in \{1, 2, \dots, N\}$ computes its LLR, which is given by $L_i = \log\left(\frac{f_{Y_i}(y_i|\mathcal{H}_1)}{f_{Y_i}(y_i|\mathcal{H}_0)}\right)$, and sends it to the FC. Here, $f_{Y_i}(y_i|\mathcal{H}_h)$ is the probability density function (PDF) of y_i given hypothesis $\mathcal{H}_h$, for $h = 0, 1$. The sensors send their LLRs to the FC according to the magnitude of their respective LLRs.[1] Based on this setup, the optimal decision rule is expressed as [Blum and Sadler, 2008]

$$\begin{cases} \sum_{i=1}^{k} L_{[i]} > \xi + (N-k)|L_{[k]}|, & \text{decide } \mathcal{H}_1, \\ \sum_{i=1}^{k} L_{[i]} < \xi - (N-k)|L_{[k]}|, & \text{decide } \mathcal{H}_0, \end{cases} \tag{9.2}$$

where $\xi = \log\left(\frac{\pi_0}{\pi_1}\right)$ is the threshold used by the FC. $L_{[i]}$ is the ith largest LLR and $\pi_h = P(\mathcal{H}_h)$ is the prior probabilities of hypothesis $\mathcal{H}_h$ for $h \in \{0, 1\}$.

9.3.1.1 Attack Model

One possible security threat for an OT-based framework in an error-prone environment is the order altering Byzantine attack (OA-Byzantine attack) [Quan et al., 2022a]. If attackers launch OA-Byzantine attacks, they can manipulate both the order and the data in the binary hypothesis testing problem. We assume that every sensor in the network has a probability α of being OA-Byzantine (B) and a probability $1-\alpha$ of being honest (H). Additionally, OA-Byzantine sensors are assumed to possess perfect knowledge of the underlying true hypothesis.[2] If sensor $i \in \{1, 2, \ldots, N\}$ is OA-Byzantine, the falsified observation is expressed as

$$\tilde{y}_i = \begin{cases} s + n_i - D, & \text{under } \mathcal{H}_1, \\ n_i + D, & \text{under } \mathcal{H}_0, \end{cases} \tag{9.3}$$

where $D \in \mathbb{R}^+$ is the attack strength. The above attack strategy involves generating falsified observations from an altered distribution, achieved by shifting the mean of the original distribution, and it is commonly utilized and discussed in the literature [Zhang et al., 2014; Kailkhura et al., 2017]. If sensor i is honest, its observation y_i as given in Eq. (9.1).

9.3.2 Performance Analysis

The effect of OA-Byzantine sensors on the FC's performance can be evaluated from two perspectives: the system's detection performance and the average percentage of transmission savings.

---

1 Specifically, if $|L[1]| > |L[2]| > \cdots > |L_{[N]}|$, then the sensor with the largest absolute LLR value, $L_{[1]}$, transmits first, followed by $L_{[2]}$, and so forth.

2 Although difficult to achieve in practice, considering this scenario is still useful as it illustrates the worst-case impact of OA-Byzantines.

9.3.2.1 Detection Performance

The following lemma stated in Quan et al. [2022a] always holds for COT-based systems:

Lemma 9.1 Under the optimum Bayesian decision rule, the detection performance remains the same whether or not the system uses the COT-based scheme in the presence of OA-Byzantine sensors.

Lemma 9.1 establishes that the COT-based system can achieve equivalent detection performance compared to the unordered system, irrespective of the presence of OA-Byzantine sensors. Hence, to assess the detection performance of the COT-based system under OA-Byzantine attacks, we analyze the detection performance of the corresponding unordered distributed system. According to Quan et al. [2022a], the detection probability P_d^{FC} and false alarm probability P_f^{FC} at the FC of an OT-based system are given as

$$P_d^{FC} = \sum_{t=1}^{2^N} (1-\alpha)^{N-b_t} \alpha^{b_t} Q\left(\frac{\xi - (\mu_1)_{A_t}}{\sqrt{N\beta}}\right) \tag{9.4}$$

and

$$P_f^{FC} = \sum_{t=1}^{2^N} (1-\alpha)^{N-b_t} \alpha^{b_t} Q\left(\frac{\xi - (\mu_0)_{A_t}}{\sqrt{N\beta}}\right), \tag{9.5}$$

respectively, where $Q(.)$ is the tail distribution function of the standard normal distribution. Here, $\beta = \frac{s^2}{\sigma^2}$ and A_t is the tth subset of the set $\{1, \dots, N\}$. The cardinality of set A_t is denoted by b_t, i.e. $b_t = |A_t|$, and $(\mu_h)_{A_t} = \mu_h |A_t| + \eta_h (N - |A_t|)$ for $h = 0, 1$, where $\mu_1 = -\mu_0 = \frac{s^2}{2\sigma^2}$, and $\eta_0 = -\eta_1 = \frac{s^2 - 2Ds}{2\sigma^2}$. The optimal attack strength D^* that attackers can adopt is given by [Quan et al., 2022a]

$$D^* = \frac{s}{2\alpha}, \tag{9.6}$$

which indicates the attack strategy to blind the FC, i.e. achieving a probability of error of 1/2 so that the FC does not derive any information from the observations.

9.3.2.2 Average Number of Transmissions Saved Under OA-Byzantine Attacks

In the COT-based system, the average number of transmissions (ANT) saved $\overline{N_t}$ under OA-Byzantine attacks can be calculated as

$$\overline{N_t} = \sum_{k=1}^{N} \pi_1 \Pr(\ell \geq k | \mathcal{H}_1) + \pi_0 \Pr(\ell \geq k | \mathcal{H}_0), \tag{9.7}$$

where ℓ is the minimum number of transmissions required to make a final decision without any loss of detection performance. Hence, to calculate $\overline{N_t}$, we need to first calculate $\Pr(\ell \geq k|\mathcal{H}_h)$ for $h = 0, 1$. According to Quan et al. [2022a], we have

$$\Pr(\ell \geq k|\mathcal{H}_h) = E_{\mathbf{L}_{k-1}}\left[F_{|L_{k-1}|}(L_{k-1}|\mathcal{H}_h)^{N-k+1}\mathbf{1}_{\{\mathcal{R}\}}\frac{N!}{(N-k+1)!}\right], \tag{9.8}$$

where $F_{|L|}(l_i|\mathcal{H}_h) = \alpha\left(Q\left(\frac{-l_i-\eta_h}{\sqrt{\beta}}\right) - Q\left(\frac{l_i-\eta_h}{\sqrt{\beta}}\right)\right) + (1-\alpha)\left(Q\left(\frac{-l_i-\mu_h}{\sqrt{\beta}}\right) - Q\left(\frac{l_i-\mu_h}{\sqrt{\beta}}\right)\right)$ is the cumulative distribution function (CDF) of $|L_i|$ for $h = 0, 1$, and $\mathbf{1}_{\{\mathcal{R}\}}$ is an indicator function that is equal to 1 if $\mathbf{L}_{k-1} = \{L_1, L_2, \dots, L_{k-1}\}$ lies in the region $\mathbf{1}_{\{\mathcal{R}\}}$ and equal to 0 otherwise. Here, $\mathcal{R}$ is a hyperplane with $k-1$ dimensions formed by the intersection of three hyperplanes, $\mathcal{R} = \mathcal{S} \cap \mathcal{L} \cap \mathcal{D}$, where $\mathcal{S} = \left\{\mathbf{L}_{k-1} : \sum_{i=1}^{k-1} L_i \leq \xi + (N-k+1)|L_{k-1}|\right\}$, $\mathcal{L} = \{\mathbf{L}_{k-1} : \sum_{i=1}^{k-1} L_i \geq \xi - (N-k+1)|L_{k-1}|\}$, and $\mathcal{D} = \left\{\mathbf{L}_{k-1} : |L_1| > |L_2| > \dots > |L_{k-1}|\right\}$. The set $\mathcal{S}$ consists of $\mathbf{L}_{k-1}$ for which the FC is unable to determine hypothesis $\mathcal{H}_1$. Similarly, the set $\mathcal{L}$ consists of $\mathbf{L}_{k-1}$ for which the FC is unable to determine hypothesis $\mathcal{H}_0$. The set $\mathcal{D}$ consists of $\mathbf{L}_{k-1}$ such that $L_1, L_2, \dots, L_{k-1}$ are ordered by magnitude.

For a given k, (9.8) can be numerically evaluated using the Monte Carlo approach. However, when the number of sensors N increases, the Monte Carlo approach requires a significant increase in the number of samples to accurately evaluate (9.8). It can become quite time-consuming to obtain the ANT saved under attack with a substantial number of sensors deployed in the network. While, for sufficiently large values of N, the upper bound (UB) (given in (9.9)) and lower bound (LB) (given in (9.10)) for the ANT saved can be obtained [Quan et al., 2022a].

$$\begin{aligned}\overline{N_s}^U = \sum_{k=1}^{N-1}\sum_{h=0}^{1} \pi_h \Bigg[&\Pr\left(|L_{[k]}| \leq \frac{q_U - \xi}{N-k}|\mathcal{H}_h\right) + \Pr\left(|L_{[k]}| \leq \frac{\xi - q_L}{N-k}|\mathcal{H}_h\right) \\ &- \Pr\left(|L_{[k]}| \leq \min\left(\frac{q_U - \xi}{N-k}, \frac{\xi - q_L}{N-k}\right)|\mathcal{H}_h\right)\Bigg]\end{aligned} \tag{9.9}$$

$$\overline{N_s}^L = \sum_{k=1}^{N-1}\sum_{h=0}^{1} \pi_h \left[\Pr\left(|L_{[k]}| < \frac{q_L - \xi}{(N-k)}|\mathcal{H}_h\right) + \Pr\left(|L_{[k]}| < \frac{\xi - q_U}{(N-k)}|\mathcal{H}_h\right)\right] \tag{9.10}$$

q_L and q_U in (9.9) and (9.10) are given as $q_L = -\left[\sum (a_i - k/N)^2 N\zeta_h^2\right]^{\frac{1}{2}} + k\delta_h$ and $q_U = \left[\sum (a_i - k/N)^2 N\zeta_h^2\right]^{\frac{1}{2}} + k\delta_h$, respectively. Here, the value of a_i for $i = 1, 2, \dots, N$ changes with the value of k, that is $a_i = 1$ if $i \leq k$ and $a_i = 0$ if $i > k$. δ_h and ζ_h^2 are given as $\delta_h = \alpha\eta_h + (1-\alpha)\mu_h$ and $\zeta_h^2 = \beta + \alpha\eta_h^2 + (1-\alpha)\mu_h^2 - \delta_h^2$, respectively. Since the pdf of $f_{|L_{[k]}|}(l_{[k]}|\mathcal{H}_h)$ is

$$f_{|L_{[k]}|}(l_{[k]}|\mathcal{H}_h) = \begin{cases} f_{L_{[k]}}(l_{[k]}|\mathcal{H}_h) - f_{L_{[k]}}(-l_{[k]}|\mathcal{H}_h), & \text{if } l_{[k]} \geq 0, \\ 0, & \text{if } l_{[k]} < 0, \end{cases} \tag{9.11}$$

where $f_{L_{[k]}}(l_{[k]}|\mathcal{H}_h) = Nf_L(l_{[k]}|\mathcal{H}_h)\binom{N-1}{k-1}F_L(l_{[k]}|\mathcal{H}_h)^{(N-k)}(1-F_L(l_{[k]}|\mathcal{H}_h))^{(k-1)}$ and $f_L(l_i|\mathcal{H}_h) = \alpha\mathcal{N}(\eta_h,\beta) + (1-\alpha)\mathcal{N}(\mu_h,\beta)$, we are able to compute $\Pr\left(|L_{[k]}|<W|\mathcal{H}_h\right)$ for $W \in \left\{\frac{q_U-\xi}{N-k}, \frac{\xi-q_L}{N-k}, \min\left(\frac{q_U-\xi}{N-k}, \frac{\xi-q_L}{N-k}\right), \frac{q_L-\xi}{N-k}, \frac{\xi-q_U}{N-k}\right\}$ in (9.10) and (9.9), which is given as $\Pr\left(|L_{[k]}| < E|\mathcal{H}_h\right) = \int_{-E}^{E} f_{|L_{[k]}|}(l_{[k]}|\mathcal{H}_h)\mathrm{d}l_{[k]}$.

9.4 CEOT-Based System

In this section, we discuss another OT-based framework, called CEOT-based scheme. It was first proposed by Sriranga et al. [2018] where sensors transmit binary decisions instead of LLRs to the FC. The performance of the CEOT-based system under two types of Byzantine sensors was evaluated: decision-falsifying (DF-Byzantine) sensors, which perform pure decision-flipping, and OA-Byzantine sensors, which not only flip decisions but also change the transmission order.

To demonstrate the basic concepts of CEOT-based schemes, we again consider a binary hypothesis testing problem. Based on the local observations $\{y_i\}_{i=1}^N$, each sensor $i \in \{1, \dots, N\}$ makes a binary decision $v_i \in \{0, 1\}$ regarding the true hypothesis using the LLR test $L_i \underset{v_i=0}{\overset{v_i=1}{\gtrless}} \log\left(\frac{\pi_0}{\pi_1}\right)$. Notably, sensor transmissions remain ordered according to the magnitude of their LLRs. Specifically, if the magnitudes of the LLRs are sorted as $|L_{[1]}| > |L_{[2]}| > \cdots > |L_{[N]}|$, the sensors send their local binary decisions to the FC in the order of $v_{[1]}, v_{[2]}, \dots, v_{[N]}$. Note that $v_{[k]}$ is the kth transmitted local decision which comes from the sensor with the kth largest LLR magnitude.

Thus, the optimal decision rule is given by [Sriranga et al., 2018]

$$\begin{cases} \sum_{i=1}^{k} v_{[i]} \geq T, & \text{decide } \mathcal{H}_1, \\ \sum_{i=1}^{k} v_{[i]} < T-(N-k), & \text{decide } \mathcal{H}_0, \end{cases} \tag{9.12}$$

which follows the T out of N counting rule.

9.4.1 Attack Model

Two possible types of security threats are considered in this framework: DF-Byzantine attacks and OA-Byzantine attacks. In the former, the Byzantine sensors perform pure decision flipping, while in the latter, the sensors not only alter decisions but also alter the sequence of the transmitted decisions. Each sensor is assumed to have a probability α of being a Byzantine sensor, and the Byzantines are assumed to possess perfect knowledge of the underlying true hypothesis.

- **DF-Byzantine attacks** For a DF-Byzantine sensor $i \in \{1, 2, \dots, N\}$, we have

$$\begin{cases} v_i = 1 - x_i, & \text{with probability } \alpha p, \\ v_i = x_i, & \text{with probability } 1 - \alpha p, \end{cases} \tag{9.13}$$

  where x_i represents the actual local decision made by sensor i, v_i represents the local decision transmitted to the FC by sensor i and p is the probability that the sensor i flips its local decisions. If sensor i is honest, x_i is the same as v_i.
- **OA-Byzantine attack** Attackers falsify data in the same way as stated in Section 9.3.1.

9.4.2 CEOT-Based System with DF-Byzantines

The performance of the CEOT-based system with DF-Byzantine sensors is analyzed in terms of the detection performance and the number of transmissions saved in the network. Next, we will first discuss the detection performance under DF-Byzantine attacks, and subsequently evaluate and analyze the number of transmissions saved.

9.4.2.1 Detection Performance

The following lemma stated in Quan et al. [2023] always holds for the CEOT-based scheme under DF-Byzantine attacks:

Lemma 9.2 When the FC follows the Bayesian decision rule, the detection performance of systems with and without the use of the CEOT-based scheme is the same in the presence of data falsification attacks.

The lemma shows that the CEOT-based system can achieve equivalent detection performance under DF-Byzantine attacks as an unordered system. Thus, the detection performance of the CEOT-based system under DF-Byzantine attacks can be evaluated by analyzing the detection performance of the corresponding unordered system. The detection probability $P^{FC}_{d,CEOT}$ and the false alarm probability $P^{FC}_{f,CEOT}$ of the FC are respectively given below as (see [Quan et al., 2023])

$$P^{FC}_{d,CEOT} = \sum_{i=T+1}^{N} \binom{N}{i} Q\left(\frac{\xi - \mu_1}{\sqrt{\beta}}\right)^i \left[1 - Q\left(\frac{\xi - \mu_1}{\sqrt{\beta}}\right)\right]^{N-i} \tag{9.14}$$

and

$$P^{FC}_{f,CEOT} = \sum_{i=T+1}^{N} \binom{N}{i} Q\left(\frac{\xi - \mu_0}{\sqrt{\beta}}\right)^i \left[1 - Q\left(\frac{\xi - \mu_0}{\sqrt{\beta}}\right)\right]^{N-i}. \tag{9.15}$$

The optimal threshold T^* that can be utilized by the FC is given by

$$T^* = \left[\log\left(\frac{\pi_0}{\pi_1}\right) + N \log\left(\frac{1 - \tilde{\pi}_{1,0}}{1 - \tilde{\pi}_{1,1}}\right)\right] / \log\left(\frac{\tilde{\pi}_{1,1}(1 - \tilde{\pi}_{1,0})}{\tilde{\pi}_{1,0}(1 - \tilde{\pi}_{1,1})}\right). \tag{9.16}$$

9.4.2.2 Average Number of Transmissions Saved Under DF-Byzantine Attacks

Let $\ell_L = \arg\min_{1\leq k\leq N}\left\{\sum_{i=1}^{k} v_{[i]} \geq T\right\}$ and $\ell_U = \arg\min_{1\leq k\leq N}\left\{\sum_{i=1}^{k} v_{[i]} < T-(N-k)\right\}$ define the minimum number of transmissions required to decide $\mathcal{H}_1$ and $\mathcal{H}_0$, respectively, under DF-Byzantine attacks. Then, the ANT saved when the FC decides $\mathcal{H}_1$ is expressed as [Quan et al., 2023]

$$\overline{N}_{s,1}(\beta) = E(N-\ell_L) = \sum_{k=1}^{N}(N-k)P(\ell_L = k) \tag{9.17a}$$

$$\geq \sum_{k=1}^{\lceil T\rceil+\beta}(N-k)P(\ell_L = k) \tag{9.17b}$$

$$\geq (N-\lceil T\rceil-\beta)P(\ell_L \leq \lceil T\rceil+\beta), \tag{9.17c}$$

where $\lceil T\rceil$ denotes the ceiling function, which rounds up T to the closest integer that is greater than or equal to T. In going from (9.17a) to (9.17b), some positive terms are dropped. To tighten the LB, an appropriate β should be selected since the difference between the actual ANT saved and its LB depends on the omitted terms. According to the definition of ℓ_L, $P(\ell_L \leq \lceil T\rceil+\beta)$ in (9.17c) can be expressed as $P(\ell_L \leq \lceil T\rceil+\beta) = \Pr\left(\sum_{k=1}^{\lceil T\rceil+\beta} v_{[k]} \geq T\right)$. Furthermore, (9.17c) can be lower bounded by $\overline{N}_{s,1}(\beta)^L = (N-\lceil T\rceil-\beta)P(\sum_{k=1}^{\lceil T\rceil+\beta} v_{[k]} \geq T|\sum_{i=1}^{\lceil T\rceil+\beta} v_{[i]} \geq T, \mathcal{H}_1)\pi_1$ due to the fact that any extra condition added will maintain or reduce the probability. When a sufficiently large signal is considered, $P\left(\sum_{k=1}^{\lceil T\rceil+\beta} v_{[k]} \geq T|\sum_{i=1}^{\lceil T\rceil+\beta} v_{[i]} \geq T, \mathcal{H}_1\right) = \sum_{i=0}^{\beta}\binom{\lceil T\rceil+\beta}{i}(\alpha p)^i(1-\alpha p)^{\lceil T\rceil+\beta-i}$ as given in Quan et al. [2023]. Similarly, for an adequately large signal, the ANT saved when the FC decides $\mathcal{H}_0$ is expressed as

$$\overline{N}_{s,2}(\beta) \geq (\lceil T\rceil-\beta)P\left(\sum_{k=1}^{N-\lceil T\rceil+\beta} v_{[k]} < \kappa|\sum_{i=1}^{N-\lceil T\rceil+\beta} v_{[i]} < T, \mathcal{H}_0\right)\pi_0 = \overline{N}_{s,2}(\beta)^L, \tag{9.18}$$

where $\lfloor T\rfloor$ denotes the floor function, which rounds down T to the nearest integer that is less than or equal to it. We can easily obtain $P\left(\sum_{k=1}^{N-\lceil T\rceil+\beta} v_{[k]} < \kappa|\sum_{i=1}^{N-\lceil T\rceil+\beta} v_{[i]} < T, \mathcal{H}_0\right) = \sum_{i=0}^{\lfloor T\rfloor-\lceil T\rceil+\beta}\binom{N-\lceil T\rceil+\beta}{i}(\alpha p)^i(1-\alpha p)^{N-\lceil T\rceil+\beta-i}$, where $\kappa = T-(\lceil T\rceil-\beta)$. Hence, a tight LB can be found by solving the following optimization problem:

$$\max_{\beta} \quad \overline{N}_{s,1}(\beta)^L + \overline{N}_{s,2}(\beta)^L \tag{9.19a}$$

$$\text{s.t.} \quad 0 \leq \beta \leq \min(N-\lceil T\rceil, \lceil T\rceil) \tag{9.19b}$$

$$\beta \in \mathbb{Z}, \tag{9.19c}$$

where $\mathbb{Z}$ denotes the set of integers. The constraint in (9.19b) arises because the upper index of the summations in the expressions for $\overline{N}_{s,1}(\beta)^L$ and $\overline{N}_{s,2}(\beta)^L$ must be less than or equal to N. Even though the optimization problem in (9.19) is a non-convex optimization problem, the optimal solution can be obtained by analyzing the property of its objective function and the corresponding discussion can be found in Quan et al. [2023].

9.4.3 CEOT-Based System with OA-Byzantines

9.4.3.1 Detection Performance

Similar to the case of DF-Byzantine system, the CEOT-based system achieves the same detection performance as the corresponding unordered system under OA-Byzantine attacks. To evaluate the detection performance of the CEOT-based system under OA-Byzantine attacks, we can analyze the detection performance of the corresponding unordered distributed system. According to Quan et al. [2022a], the detection probability $P_{d,CEOT}^{FC}$ and the false alarm probability $P_{f,CEOT}^{FC}$ of a CEOT-based system are, respectively, given by $P_{d,CEOT}^{FC} = \sum_{i=T}^{N} \binom{N}{i} \pi_{1,1}^{i} \pi_{0,1}^{N-i}$ and $P_{f,CEOT}^{FC} = \sum_{i=T}^{N} \binom{N}{i} \pi_{1,0}^{i} \pi_{0,0}^{N-i}$, where $\pi_{1,h} = P(v_i = 1|\mathcal{H}_h) = \alpha Q\left(\frac{\xi-\eta_h}{\sqrt{\beta}}\right) + (1-\alpha)Q\left(\frac{\xi-\mu_h}{\sqrt{\beta}}\right)$.

9.4.3.2 Average Number of Transmissions Saved Under OA-Byzantine Attacks

Let $\overline{N}_{s,CEOT}$ denote the ANT saved in the CEOT-based scheme given as

$$\overline{N}_{s,CEOT} = E(N - \ell) = \sum_{k=1}^{N} (N-k) \Pr(\ell = k) = \sum_{k=1}^{N-1} \Pr(\ell \le k). \tag{9.20}$$

Here, ℓ is again the minimum number of transmissions required to reach a final decision with desired accuracy. Since computing $\Pr(\ell \le k)$ is intractable, we instead derive the UB and LB of $\overline{N}_{s,CEOT}$. Let $\Lambda = \sum_{i=1}^{N} v_i$ denote the global statistic of the distributed unordered system. Note that the final decisions of the system, whether ordered or unordered, remain the same under OA-Byzantine attacks. Consequently, $\Lambda < T$ indicates that there exists an ℓ such that $\sum_{i=1}^{\ell} v_{[i]} < T - (N-\ell)$, while $\Lambda \ge T$ indicates that there exists an ℓ such that $\sum_{i=1}^{\ell} v_{[i]} \ge T$. To determine the LB and UB of $\overline{N}_{s,CEOT}$, we consider the worst-case and best-case scenarios, respectively. They are shown as the following [Quan et al., 2022a]:

- **The worst case**: To determine the maximum ℓ required to reach a final decision based on a set of local decisions $\{v_i\}_{i=1}^{N}$. In this case, there are two possibilities that need to be considered: (i) $\Lambda < T$; (ii)$\Lambda \ge T$. Given $\Lambda < T$, the worst-case scenario would be if the local decisions are ordered in descending magnitude, i.e.

$$|r_{[1]}| \ge |r_{[2]}| \cdots \ge |r_{[N]}|, \tag{9.21}$$

where $r_{[k]} \in \{0,1\}$ is the kth largest local decision.[3] Similarly, given $\Lambda \geq T$, the worst-case scenario would occur if

$$|r_{(1)}| \leq |r_{(2)}| \cdots \leq |r_{(N)}|, \tag{9.22}$$

where $z_{(k)} \in \{0,1\}$ is the kth smallest local decision. Note that $r_{[k]}$ and $z_{(k)}$ are not the same as $v_{[k]}$. The values $v_{[1]}, v_{[2]}, \dots, v_{[N]}$ are ordered by the magnitude of their corresponding LLRs, while $r_{[1]}, r_{[2]}, \dots, r_{[N]}$ (or $r_{(1)}, \dots, r_{(N)}$) are ordered by the magnitude of local decisions.

- **The best case**: To determine the minimum ℓ required to reach a final decision based on a set of local decisions $\{v_i\}_{i=1}^N$. We still need to consider two possibilities: (i) $\Lambda < T$; (ii)$\Lambda \geq T$. The best-case scenario when $\Lambda < T$ would occur if the magnitude of local decisions are ordered as (9.22). The best case given $\Lambda \geq T$ would occur if the magnitude of local decisions are ordered as (9.21).

Hence, the UB $\overline{N}_{s,CEOT}^U$ and the LB $\overline{N}_{s,CEOT}^L$ are given by [Quan et al., 2022a]

$$\begin{aligned}\overline{N}_{s,CEOT}^U = & \sum_{h=0}^{1}\sum_{k=1}^{N-1}\pi_h \left[P\left(\ell_0 \leq k|\Lambda \geq T, \mathcal{H}_h\right) P(\Lambda \geq T|\mathcal{H}_h)\right. \\ & \left. + P\left(\ell_1 \leq k|\Lambda < T, \mathcal{H}_h\right) P(\Lambda < T|\mathcal{H}_h)\right],\end{aligned} \tag{9.23}$$

$$\begin{aligned}\overline{N}_{s,CEOT}^L = & \sum_{h=0}^{1}\sum_{k=1}^{N-1}\pi_h \left[P\left(\ell_1 \leq k|\Lambda \geq T, \mathcal{H}_h\right) P(\Lambda \geq T|\mathcal{H}_h)\right. \\ & \left. + P\left(\ell_0 \leq k|\Lambda < T, \mathcal{H}_h\right) P(\Lambda < T|\mathcal{H}_h)\right],\end{aligned} \tag{9.24}$$

where $P(\Lambda \geq T|\mathcal{H}_h) = \sum_{i=T}^{N}\binom{N}{i}\pi_{1,h}^i\pi_{0,h}^{N-i}$, $P(\Lambda < T|\mathcal{H}_h) = 1 - P(\Lambda \geq T|\mathcal{H}_h)$, and

$$P\left(\ell_0 \leq k|\Lambda \geq T, \mathcal{H}_h\right) = \sum_{i=0}^{N-T}\binom{N}{i}\pi_{0,h}^i\pi_{1,h}^{N-i}, \tag{9.25}$$

$$P\left(\ell_1 \leq k|\Lambda \geq T, \mathcal{H}_h\right) = \sum_{i=0}^{\min(N-T,k-T)}\binom{N}{i}\pi_{0,h}^i\pi_{1,h}^{N-i}, \tag{9.26}$$

when $k \geq T$, and

$$P\left(\ell_1 \leq k|\Lambda < T, \mathcal{H}_h\right) = \sum_{i=0}^{T-1}\binom{N}{i}\pi_{1,h}^i\pi_{0,h}^{N-i}, \tag{9.27}$$

$$P\left(\ell_0 \leq k|\Lambda < T, \mathcal{H}_h\right) = \sum_{i=0}^{\min(T-1,k-(N-T+1))}\binom{N}{i}\pi_{1,h}^i\pi_{0,h}^{N-i}, \tag{9.28}$$

3 $\Lambda < T$ implies that the total number of "1"s is less than T.

when $k > N - T$. Here, ℓ_0 and ℓ_1 represent the minimum number of transmissions required to reach a final decision for local decisions ordered in descending and ascending order, respectively.

9.5 Comparison of COT-Based and CEOT-Based Systems Under Attack

In this section, we compare the performance of COT-based and CEOT-based systems under OA-Byzantine and DF-Byzantine attacks. Some simulation results regarding the performance of the OT-based systems under attacks are presented below.

9.5.1 Effect of OA-Byzantine Attacks on the COT-Based and CEOT-Based Systems

Figure 9.1 demonstrates that, for the same attack parameters, the CEOT-based system demonstrates greater resilience to OA-Byzantine attacks than the COT-based system with respect to error probabilities. It shows that the quantization procedure in OT-based frameworks improves the robustness of OT-based frameworks. Figure 9.2 displays the nature of the average percentage of transmission savings with respect to the attack parameter D. Initially, the average percentage of transmission savings decreases as D increases. However, with further increases in D, the FC begins to make more erroneous decisions, resulting in a decrease in the number of transmissions needed to reach a final decision and an increase in savings. We can observe the existence of a minimum average percentage of transmission savings, which corresponds to the optimal attack strength D^* obtained from (9.6). The LB obtained from (9.10) and the UB obtained from (9.9) are also shown in Figure 9.2, and they track the changes in the actual saved ANT. The LB performs better in tracking changes compared to the UB, allowing us to infer the optimal attack strategy for the attacker, i.e. the value of D that the attacker will use to cause the most damage to the system. The UB, on the other hand, offers insights into the maximum number of transmissions saved on average in the network and alerts us to the presence of outliers.[4] These trends give us an insight into how to improve the system's robustness. Figure 9.3 shows that OA-Byzantine sensors can have less impact on the final decision-making in the CEOT-based system. As D increases sufficiently, Byzantine sensors are most likely to provide the first several local decisions received by the FC. This scenario represents the worst-case performance of the system. As D continues to increase, the influence of Byzantine sensors on the number of transmissions saved in the network does not increase further, due to

4 For example, if the ANT saved exceeds the maximum value of the UB, it indicates potential outliers where data significantly deviates from the actual, suggesting attackers are employing an excessively large value of D.

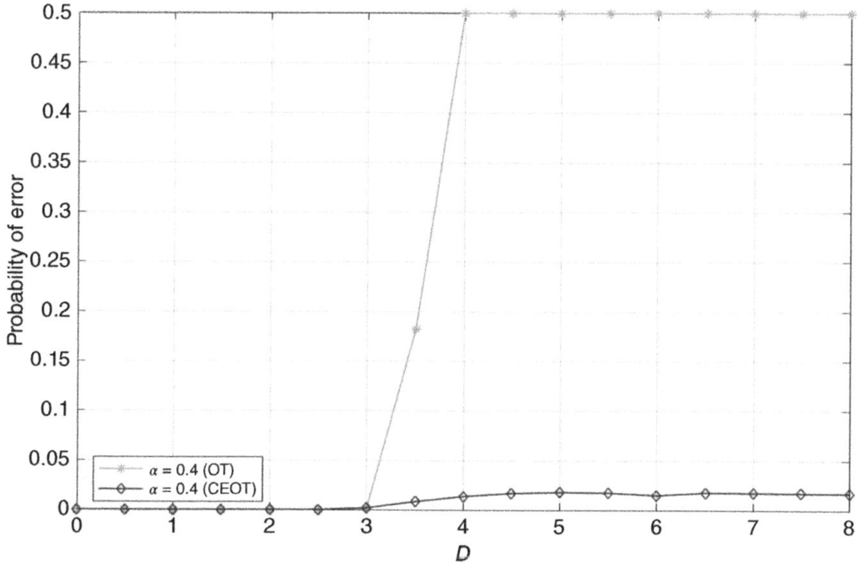

Figure 9.1 P_e as a function of D in the COT-based system and the CEOT-based system when $N = 300$, $\sigma^2 = 1$, $s = 3$ and $\pi_1 = \pi_0 = 0.5$. Source: Adapted from Quan et al. [2022a]. Please see the online version for the colored version of the figure.

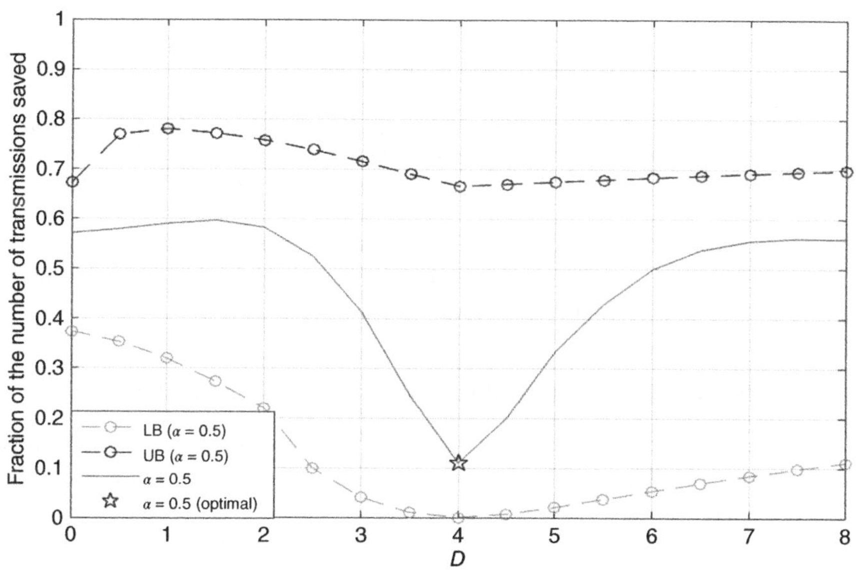

Figure 9.2 UB and LB for $\overline{N_s}/N$ as a function of D when $s = 4$, $\alpha = 0.5$, $N = 300$, $\sigma^2 = 1$ and $\pi_1 = \pi_0 = 0.5$ in the COT-based system. Source: Adapted from Quan et al. [2022a]. Please see the online version for the colored version of the figure.

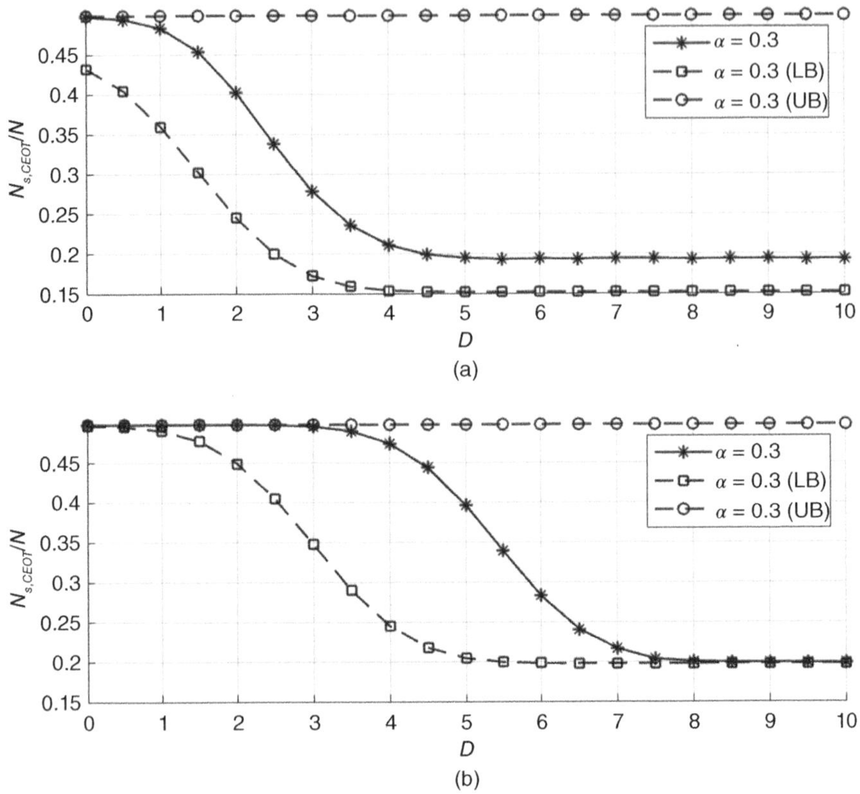

Figure 9.3 Benchmarking upper and lower bounds for $\overline{N}_{s,CEOT}/N$ as a function of D given $N = 300$ in the CEOT-based system (a) when $s = 3$; (b) when $s = 6$. Source: Adapted from Quan et al. [2022a]. Please see the online version for the colored version of the figure.

the quantization of LLRs, thereby mitigating their effects on the system. By comparing Figure 9.3a,b, we can also observe that the LB obtained in (9.24) becomes tighter as the signal strength s increases.

9.5.2 Effect of DF-Byzantine Attacks on the CEOT-Based System

Figure 9.4 shows that the optimal threshold T^* obtained from (9.16) results in the lowest possible error probability for the system.[5] Additionally, it can be

5 The threshold closest to the optimal threshold T^* is 49.5 in Figure 9.4 when $\pi_1 = 0.5$ and $\pi_1 = 0.3$.

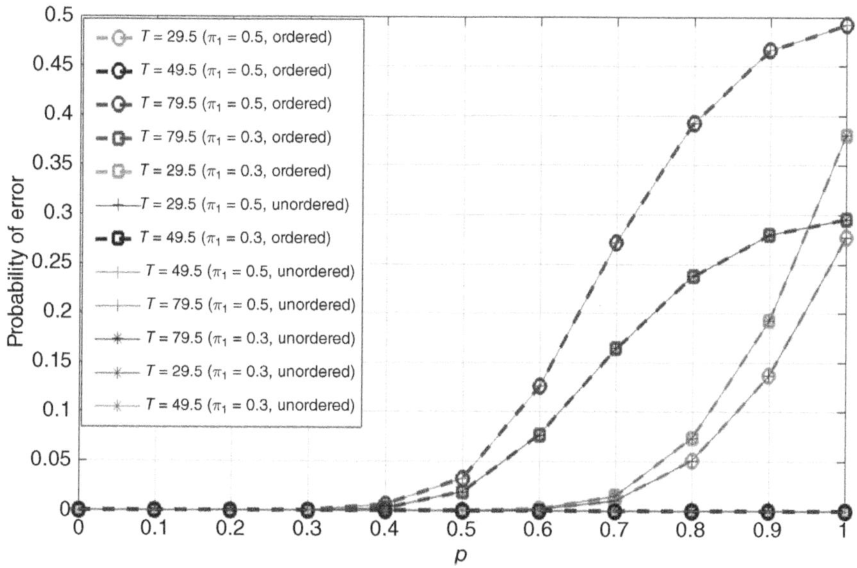

Figure 9.4 P_e as a function of p with different values of T for the CEOT-based system for $\pi_1 = 0.3$, $\pi_1 = 0.5$ and $N = 100$. Source: Adapted from Quan et al. [2023]. Please see the online version for the colored version of the figure.

observed that both the CEOT-based and unordered systems exhibit identical error probabilities for the given parameter values. This finding aligns with Lemma 9.2. In Figure 9.5, we can observe that a relatively tight UB is obtained compared with the UB obtained in the past work [Sriranga et al., 2018] for the average fraction of the number of transmissions required in the CEOT-based system. The effect of different values of prior probability and T on the performance of the CEOT-based system is shown in Figure 9.6. From Figure 9.6, it is evident that as T approaches T^* (approximately $N/2$), the system tends to require the maximum number of transmissions in the network when both hypotheses have prior probabilities of 0.5. However, variations in the prior probabilities can influence the optimal T that minimizes transmissions. A smaller T increases the transmissions needed for deciding $\mathcal{H}_0$ but reduces those for deciding $\mathcal{H}_1$. With a lower $\pi_1 < 0.5$, the likelihood of the FC deciding $\mathcal{H}_0$ increases. Consequently, a system using $T = 29.5$ requires more transmissions compared to one using $T = 49.5$, given $\pi_1 = 0.3$. Thus, the ANT needed is related to the prior probabilities of hypotheses in the system. According to Figures 9.4 and 9.6, it's possible to save transmissions while ensuring the quality of decisions by designing an appropriate threshold at the FC.

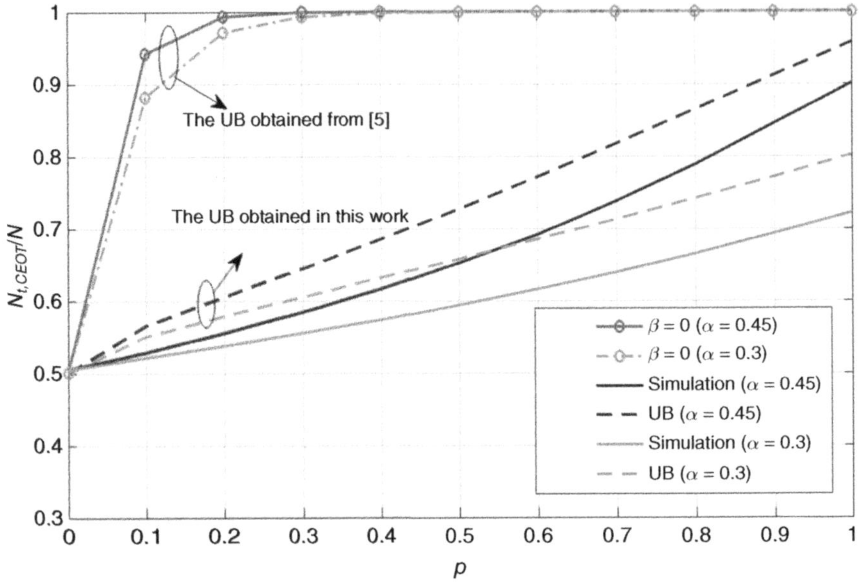

Figure 9.5 UBs for the fraction of the number of transmissions required $N_{t,CEOT}/N$ as a function of p with different values of α when $\pi_1 = 0.5$ and $N = 100$. Source: Quan et al. [2023]/IEEE. Please see the online version for the colored version of the figure.

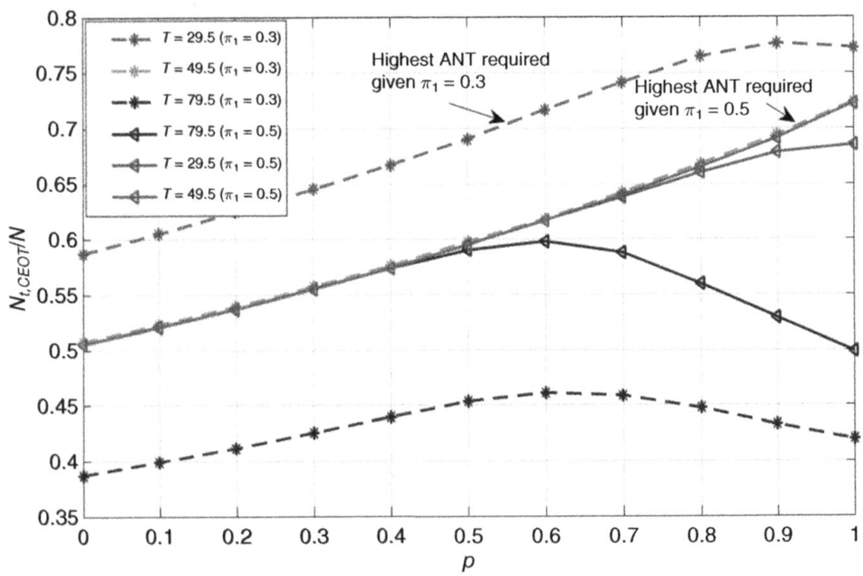

Figure 9.6 $N_{t,CEOT}/N$ as a function of p with different values of T when $\alpha = 0.3$, $\pi_1 = 0.3, 0.5$ and $N = 100$. Source: Adapted from Quan et al. [2023]. Please see the online version for the colored version of the figure.

9.5.3 Discussion

In this section, we presented simulation results illustrating the impact of OA-Byzantine attacks on both the COT-based and CEOT-based systems, as well as the effect of DF-Byzantine attacks specifically on the performance of the CEOT-based system. These performance comparisons provide valuable insights into the design of robust OT-based systems. For example, CEOT-based systems exhibit greater resilience to both OA-Byzantine and DF-Byzantine attacks compared to COT-based systems. By designing appropriate thresholds, CEOT-based systems can minimize the impact of both types of attacks on system performance. When implementing an OT-based system to optimize network transmissions, careful consideration of these factors is essential.

9.6 Conclusion

This chapter discussed several energy-efficient OT-based schemes for distributed detection in the presence of Byzantine attacks. It has been assumed in the past that OT-based framework operates in an attack-free environment. The designs for optimal decision fusion rules at the FC were discussed, and they were found to be highly effective under this assumption. However, some of the assumptions made in those works may be violated in the presence of attacks. In this chapter, the effect of Byzantine attacks on the performance of the COT-based system and the CEOT-based system were investigated in Sections 9.3 and 9.4. Moreover, a comparison of the resilience of CEOT-based and COT-based systems was made in Section 9.5, shedding light on how to employ OT-based frameworks in an attack-prone environment. Some possible countermeasures to mitigate the impact of Byzantines on OT-based systems were also discussed.

There are many open and challenging problems remain and require further research. For instance, analyzing the performance of the OT-based schemes with non-i.i.d. assumption under attacks is a very difficult problem. Designing robust OT-based schemes in distributed detection is also challenging since a fraction of multiple sensor data information is lost. Additionally, balancing energy efficiency and security in distributed systems is an open question. There are also some other promising energy-efficient schemes proposed in the literature (e.g. OT-based energy harvesting scheme [Gupta et al., 2020], censoring-based scheme [Gu et al., 2020]) by not receiving data from all the sensors all the time, which makes it difficult for the traditional intrusion detection systems to track the identity of each sensor in the networks. Therefore, further work on robust and energy-efficient schemes will be valuable.

Bibliography

S. Appadwedula, V. V. Veeravalli, and D. L. Jones. Decentralized detection with censoring sensors. *IEEE Transactions on Signal Processing*, 56(4):1362–1373, 2008.

R. G. Baraniuk. Compressive sensing [lecture notes]. *IEEE Signal Processing Magazine*, 24(4):118–121, 2007.

R. S. Blum and B. M. Sadler. Energy efficient signal detection in sensor networks using ordered transmissions. *IEEE Transactions on Signal Processing*, 56(7):3229–3235, 2008.

E. J. Candès and M. B. Wakin. An introduction to compressive sampling. *IEEE Signal Processing Magazine*, 25(2):21–30, 2008.

W.-N. Chen and I.-H. Wang. Anonymous heterogeneous distributed detection: Optimal decision rules, error exponents, and the price of anonymity. *IEEE Transactions on Information Theory*, 65(11):7390–7406, 2019.

Y. Chen, R. S. Blum, and B. M. Sadler. Optimal quickest change detection in sensor networks using ordered transmissions. In *2020 IEEE 21st International Workshop on Signal Processing Advances in Wireless Communications (SPAWC)*, pages 1–5. IEEE, 2020a.

Y. Chen, B. M. Sadler, and R. S. Blum. Ordered gradient approach for communication-efficient distributed learning. In *2020 IEEE 21st International Workshop on Signal Processing Advances in Wireless Communications (SPAWC)*, pages 1–5. IEEE, 2020b.

Y. Chen, R. S. Blum, and B. M. Sadler. Ordering for communication-efficient quickest change detection in a decomposable graphical model. *IEEE Transactions on Signal Processing*, 69:4710–4723, 2021.

G. Fellouris, E. Bayraktar, and L. Lai. Efficient Byzantine sequential change detection. *IEEE Transactions on Information Theory*, 64(5):3346–3360, 2017.

Y. Gu, Y. Jiao, X. Xu, and Q. Yu. Request–response and censoring-based energy-efficient decentralized change-point detection with IoT applications. *IEEE Internet of Things Journal*, 8(8):6771–6788, 2020.

S. S. Gupta, S. K. Pallapothu, and N. B. Mehta. Ordered transmissions for energy-efficient detection in energy harvesting wireless sensor networks. *IEEE Transactions on Communications*, 68(4):2525–2537, 2020.

W. Hashlamoun, S. Brahma, and P. K. Varshney. Mitigation of Byzantine attacks on distributed detection systems using audit bits. *IEEE Transactions on Signal and Information Processing Over Networks*, 4(1):18–32, 2017.

W. Hashlamoun, S. Brahma, and P. K. Varshney. Audit bit based distributed Bayesian detection in the presence of Byzantines. *IEEE Transactions on Signal and Information Processing Over Networks*, 4(4):643–655, 2018.

L. Hesham, A. Sultan, M. Nafie, and F. Digham. Distributed spectrum sensing with sequential ordered transmissions to a cognitive fusion center. *IEEE Transactions on Signal Processing*, 60(5):2524–2538, 2012.

Y.-C. Huang, Y.-J. Huang, and S.-C. Lin. Asymptotic optimality in Byzantine distributed quickest change detection. *IEEE Transactions on Information Theory*, 67(9):5942–5962, 2021.

B. Kailkhura, S. Brahma, and P. K. Varshney. Data falsification attacks on consensus-based detection systems. *IEEE Transactions on Signal and Information Processing Over Networks*, 3(1):145–158, 2017. doi: 10.1109/TSIPN.2016.2607119.

B. Kailkhura, A. Vempaty, and P. K. Varshney. Collaborative spectrum sensing in the presence of Byzantine attacks. In P. Djuric and C. Richard (Eds.), *Cooperative and Graph Signal Processing*, pages 505–522. Elsevier, 2018.

Z. Li, Y. Mo, and F. Hao. Distributed sequential hypothesis testing with Byzantine sensors. *IEEE Transactions on Signal Processing*, 69:3044–3058, 2021.

S. Liu, H. Zhu, S. Li, X. Li, C. Chen, and X. Guan. An adaptive deviation-tolerant secure scheme for distributed cooperative spectrum sensing. In *2012 IEEE Global Communications Conference (GLOBECOM)*, pages 603–608. IEEE, 2012.

X. Liu, T. J. Lim, and J. Huang. Optimal Byzantine attacker identification based on game theory in network coding enabled wireless ad hoc networks. *IEEE Transactions on Information Forensics and Security*, 15:2570–2583, 2020. doi: 10.1109/TIFS.2020.2972129.

A. Mustafa, M. N. U. Islam, and S. Ahmed. Dynamic spectrum sensing under crash and Byzantine failure environments for distributed convergence in cognitive radio networks. *IEEE Access*, 9:23153–23167, 2021.

V. S. S. Nadendla, Y. S. Han, and P. K. Varshney. Distributed inference with M-ary quantized data in the presence of Byzantine attacks. *IEEE Transactions on Signal Processing*, 62(10):2681–2695, 2014. doi: 10.1109/TSP.2014.2314072.

M. Pease, R. Shostak, and L. Lamport. The Byzantine generals problem. *ACM Transactions on Programming Languages and Systems*, 4(3):382–401, 1982.

C. Quan, S. Bulusu, B. Geng, and P. K. Varshney. Ordered transmission-based detection in distributed networks in the presence of Byzantines. *arXiv preprint arXiv:2201.08737*, 2022a.

C. Quan, B. Geng, Y. Han, and P. K. Varshney. Enhanced audit bit based distributed Bayesian detection in the presence of strategic attacks. *IEEE Transactions on Signal and Information Processing Over Networks*, 8:49–62, 2022b.

C. Quan, Y. S. Han, B. Geng, and P. K. Varshney. Reputation and audit bit based distributed detection in the presence of Byzantines. In *2022 56th Asilomar Conference on Signals, Systems, and Computers*, pages 548–552, 2022c. doi: 10.1109/IEEECONF56349.2022.10051853.

C. Quan, N. Sriranga, H. Yang, Y. S. Han, B. Geng, and P. K. Varshney. Efficient ordered-transmission based distributed detection under data falsification attacks. *IEEE Signal Processing Letters*, 30:145–149, 2023. doi: 10.1109/LSP.2023.3244748.

Z. N. Rawas, Q. He, and R. S. Blum. Energy-efficient noncoherent signal detection for networked sensors using ordered transmissions. In *2011 45th Annual Conference on Information Sciences and Systems*, pages 1–5. IEEE, 2011.

N. Sriranga, K. G. Nagananda, R. S. Blum, A. Saucan, and P. K. Varshney. Energy-efficient decision fusion for distributed detection in wireless sensor networks. In *2018 21st International Conference on Information Fusion (FUSION)*, pages 1541–1547. IEEE, 2018.

A. Vempaty, B. Kailkhura, and P. K. Varshney. *Secure Networked Inference with Unreliable Data Sources*. Springer, 2018.

J. Wu, T. Song, Y. Yu, C. Wang, and J. Hu. Sequential cooperative spectrum sensing in the presence of dynamic Byzantine attack for mobile networks. *PLoS One*, 13(7):e0199546, 2018.

L. Yang, H. Zhu, Z. Zhu, X. Luo, and H. Qian. Distributed ordering transmissions for latency-sensitive estimation in wireless sensor networks. In *2019 IEEE 90th Vehicular Technology Conference (VTC2019-Fall)*, pages 1–5, 2019. doi: 10.1109/VTCFall.2019.8891536.

Z. Yang, A. Gang, and W. U. Bajwa. Adversary-resilient distributed and decentralized statistical inference and machine learning: An overview of recent advances under the Byzantine threat model. *IEEE Signal Processing Magazine*, 37(3):146–159, 2020.

L. Zhang, Q. Wu, G. Ding, S. Feng, and J. Wang. Performance analysis of probabilistic soft SSDF attack in cooperative spectrum sensing. *EURASIP Journal on Advances in Signal Processing*, 2014(1):1–9, 2014.

L. Zhang, G. Nie, G. Ding, Q. Wu, Z. Zhang, and Z. Han. Byzantine attacker identification in collaborative spectrum sensing: A robust defense framework. *IEEE Transactions on Mobile Computing*, 18(9):1992–2004, 2018.

Part IV

Applications in Smart Environments

10

Internet of Musical Things for Smart Cities

Paolo Casari[1,2] and Luca Turchet[1]

[1]*Department of Information Engineering and Computer Science, University of Trento, Trento, Italy*
[2]*CNIT, Parma, Italy*

10.1 Introduction

Smart cities are broadly defined as environments that exploit information and communication technologies (ICTs) in order to improve the management of common resources, while adding to the well-being both of its citizens and of the users of city services at large [Hammons and Myers, 2019]. A founding element of smart city operation is data collection [Bui et al., 2012; Khan et al., 2017; Bastos et al., 2022], and automatic processing, e.g., in cloud-based facilities [Krämer, 2014; Alam, 2021]. Smart governance systems then provide real-time and on-demand decision-making and adaptation of city services to optimize the fruition and utility of such services. The vision of smart cities is deeply intertwined with that of the Internet of Things (IoT). The field of the IoT relates to the extension of the Internet into the physical realm, by means of everyday physical objects that are spatially distributed and augmented using ICT. IoT devices provide the data for the operation of a smart city with the needed resolution and sampling rates. The IoT intersects deeply with the concept of smart city. However, scarce attention has been devoted thus far by researchers in considering the concept of smart city under the lenses of one of the activities most widespread in the society and in cities: that of music.

In recent years, the IoT paradigm has been investigated under the lenses of musical creativity, leading to the emergent field of the Internet of Musical Things (IoMusT) [Turchet et al., 2018]. Nowadays, the area is receiving increasing attention by the industrial, academic, and artistic research communities, as testified by a growing number of publications, products, and musical performances. Specifically, the IoMusT relates to the networks of "Musical Things,"

Wireless Sensor Networks in Smart Environments: Enabling Digitalization from Fundamentals to Advanced Solutions, First Edition. Edited by Domenico Ciuonzo and Pierluigi Salvo Rossi.

which are intelligent and connected objects serving a musical purpose. Thanks to their connectivity features, Musical Things are able to interact with each other and cooperate to reach common musical goals. The IoMusT technological infrastructure enables ecosystems of interoperable devices that connect musical stakeholders with each other providing novel interaction possibilities for different kinds of musical activities. These activities include performance [Bevilacqua et al., 2021], composition [Clester and Freeman, 2021], pedagogy [Alexandraki et al., 2023], and music therapy [Timoney et al., 2020], both in co-located and remote settings. As a consequence, manifold are the musical stakeholders who can be beneficiaries of IoMusT technologies, including performers, composers, students, teachers, conductors, studio producers, live sound engineers, and audience members.

We believe that a smart city should not just be a place where ICT is used for creating greater efficiencies in a city's services and resources [Hammons and Myers, 2019], where data is connected and harvested from smart instruments and sensors, and then used to create smart services for citizens. Rather, in this chapter, we propose the concept of "smart musical cities," which specifically targets musical stakeholders: *we push the vision that current- and next-generation connectivity technology should turn smart cities into creative hubs, that enable different actors of artistic efforts to seamlessly work together, both synchronously and asynchronously.* Moreover, a smart city should open and promote the fruition and the very perception of art environments to the public at large, making it possible to interact with them in ways that are creative per se.

For example, with reference to Figure 10.1, smart musical cities will enable musicians to perform together while being distributed across a possibly widespread geographical area. The presence of local computing facilities integrated as core resources of access and transport networks will provide a low-latency and highly reliable resource to perform basic (e.g., mixing) and more advanced audio streaming functions (e.g., error concealment, or superimposition of a live and a recorded audio/video stream). In the same vein, smart musical cities should enable musical performances themselves to be distributed: different bands may perform at different venues (such as arenas, bars, and in the streets) and should allow distributed audiences (possibly located all over a prescribed geographical boundary) not only to interact with the band in ways conceived to enhance the musical experience but also among themselves (synced cheering, choral singing, rhythmical sounds such as foot stomping and hand clapping).

A smart musical city should enable the fruition of the very city districts and environments in an artistically and historically engaging fashion. For example, tourists should be able to roam a city while hearing reproduced voices of past personalities, important events (such as the construction of key civilian infrastructure, the promulgation of a landmark law, past disasters, battles, as well as voice soundscapes),

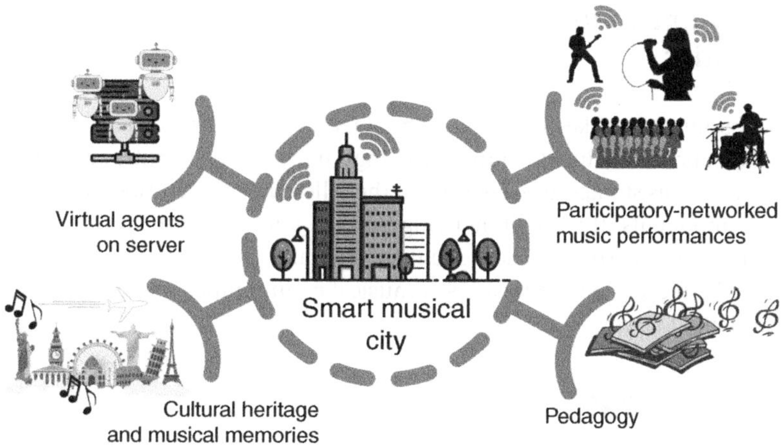

Figure 10.1 Concept of a smart musical city along with the example services considered in Section 10.3.

leading to a form of "4D" tourism, where the time dimension is provided by sound experiences. Engaging descriptions of the surroundings should be available and consumable on demand, possibly made to engage passing people by means of the above techniques. Sentiment analysis for people passing in certain environments or districts may be associated with the best music to foster similar sentiments. In turn, recommender systems for music listening and may tap on the location where a person is moving and suggest tunes and works of art related to that location, its general sentiment, as well as past experiences from other users.

Generative music approaches relying on artificial intelligence (as is the case, e.g., for the Holon app [Holmqvist et al., 2023]) can very well complement all of the above smart interactions, e.g., by inspiring to urban morphology and land use to change recommendations or morph the style a specific audio performance is played back to the user. Notably, the information on land user coverage can be taken from open databases (including, e.g., OpenStreetMap, OpenTreeMap, as well as local, higher-resolution city-level or regional databases), making a perfect case for how the availability, connection, and interaction of data from different sources can improve city services and cultural experiences alike.

Similar applications of real-time musical interactions mediated by fast and reliable network connectivity include, e.g., music teaching and pedagogy in general, as well as musical therapy (e.g., for mental of physical rehabilitation). The remainder of this chapter will present enabling technologies for IoMusT interactions in smart cities (Section 10.2) and explain our vision for IoMusT applications such as those discussed in this introduction (Section 10.3). Finally, Section 10.4 draws concluding remarks.

10.2 Key-Enabling Technologies for IoMusT in Smart Musical Cities

There exist a number of key enabling technologies for smart musical city applications. These primarily include Musical Things, networking infrastructures and protocols, as well as storage and datasets. In the following, we describe such technological components. It is worth noticing that the joint use of the technologies mentioned below is currently leading to novel digital ecosystems of humans and machines, which are expected to support musical communities in their activities in unprecedented ways.

10.2.1 Musical Things

A Musical Thing can take the form of any smart device utilized to control, generate, or track responses to music content. To date, the IoMusT research community has developed various types of Musical Things, both in industrial and academic settings [Keller et al., 2019; Yaseen et al., 2022]. Relevant examples in this space are the so-called smart musical instruments [Turchet, 2019], which are musical devices enhanced with intelligent features capable of conferring the instrument with context-awareness and proactivity abilities. Another example is represented by the musical haptic wearables [Turchet et al., 2021], which are wireless devices conceived to support musical communication leveraging the sense of touch. Furthermore, virtual and augmented reality head-mounted displays can be used for musical applications in networked, shared environments [Loveridge, 2020].

The development of Musical Things is rooted in embedded systems that are dedicated to low-latency audio processing tasks and are equipped with connectivity capabilities [McPherson and Zappi, 2015; Turchet and Fischione, 2021]. Thanks to such operating systems it is possible to build dedicated low-latency applications leveraging embedded music information retrieval methods [Stefani et al., 2022; Pelinski et al., 2023]. The extracted information can then be repurposed in many ways, especially in real-time, to create different kinds of musical applications (e.g., for the control of stage lights or smoke machines). Another relevant technology supporting the development of Musical Things applications is Web Audio [Marasco and Allison, 2019]. The Web Audio API is a proposed standard of the World Wide Web Consortium [Buffa et al., 2022], which makes it possible to generate, analyze, and process audio streams directly in the browser.

From a different perspective, researchers have devoted their attention to the development of different frameworks aiming at the efficient interconnection of Musical Things [Fraietta et al., 2019; Matuszewski, 2020; Dannenberg, 2022; Vieira et al., 2022; Turchet and Antoniazzi, 2023], as well as of geographically displaced musicians via the so-called network music performance systems

[Carôt and Werner, 2008; Cáceres and Chafe, 2010; Drioli et al., 2013; Rottondi et al., 2016; Carôt et al., 2020b; Turchet and Fischione, 2021]. In particular, recent years have witnessed an increasing use of 5G technologies in musical settings [Carôt et al., 2020a; Cheli and Giordano, 2022; Dürre et al., 2022; Turchet and Casari, 2024a].

10.2.2 5G-and-Beyond Networks

One foundational aspect characterizes musical interactions through the IoMusT: a participating user should receive audio stimuli fast enough to both mask any delays due to networking and data transfer and enable a smooth and seamless interaction. Therefore, the most important aspect of IoMusT exchanges is reliable and fast network connectivity. For instance, perceptual studies have proven that the maximum end-to-end delay bearable by performers playing together over the Internet is between 20 and 30 ms [Rottondi et al., 2016] in the absence of synchronization cues, e.g., from a conductor. For higher delay values, the performers become unable to maintain a stable tempo and tend to slow down progressively. The reliability of the connection is also extremely important, as such stringent delay values leave very little space for the retransmission of lost packets. Even if audio streams include replicas of sent audio data,[1] excessive losses cause gaps in the audio stream that cannot be easily compensated for. Some algorithms exist that apply signal processing or machine learning to automatically conceal audio losses [Fink and Zölzer, 2014; Verma et al., 2020]. Yet, a burst of packet errors still implies perceivable interruptions in the audio data.

The above discussion generalizes to all IoMusT interactions and explains why fifth-generation (5G) cellular connectivity receives so much expectations from the IoMusT community. With respect to preceding 4G technology, 5G has several desirable features. Besides adding reliability via more effective channel coding, 5G adds an option to operate scheduling at a higher rate by increasing the numerology of the system [Vihriälä et al., 2016]. Higher numerologies correspond to allocating shorter slots that have a larger bandwidth available. Using a sufficiently high numerology, e.g., $\mu = 1$ to 3, it is possible to tune the 5G scheduler to allocate resources with millisecond granularity, and greatly reduce the baseline scheduling latency of a 4G network, which is on the order of 10 ms.

Further 5G improvements leading to the development of ultra-reliable low-latency communications (URLLC) for multiuser scenarios with periodic and non-sporadic traffic are expected to further improve the feasibility of 5G cellular

1 Forward error correction (FEC) is usually not an option, as packetized audio usually comes in very packets of a few hundred bytes, and complex but effective error-correcting codes would need longer codewords and imply a non-negligible processing time. In turn, the latter would seize part of the latency budget.

technology for the IoMusT. In addition, advancements on grant-free communications should provide an extra means to improve the IoMusT experience, e.g., by supplementing lost data or by enabling the early recovery of exceedingly delayed packets. While URLLC is not fully deployed to date and plans indicate it will not start being commercially available at large before 2025 in the EU [EU Parliament, 2021], such a technology holds great promise of extra-fast data delivery [Yang et al., 2021; Yun et al., 2022] and vanishing errors [Weerasinghe et al., 2020], supporting the fast transmission and high-reliability requirements of the IoMusT.

Due to the stringent latency requirements, the presence of computing capability as close as possible to the IoMusT devices is of paramount importance. Consider a typical IoMusT scenario with a group of networked music performers: in the baseline case, each performer would need to send its audio stream to every other performer, such that their IoMusT system can collect, resynchronize, and mix audio locally for the performer's fruition. In this scenario, the amount of audio flows increases as the square of the number of performers with two undesirable consequences: that each flow may incur errors and become a source of quality loss for the entire mixed output stream; and that separate flows may have to contend for the same network access resources. The latter case is likely in densely populated areas where two performers may be located within the coverage area of the 5G base station (gNB), and potentially affects an entire group of performers if a whole band or orchestra should interact wirelessly on the same stage, typically covered by the same gNB.

The 5G ecosystems provide a first solution to this square-law multiplicity of flows by inherently providing Multi-access Edge Computing (MEC) capabilities embedded in the 5G operator's network [ETSI Group Specification, 2022]. MEC functionalities may reside in the operator's network and be colocated or very close to the edge of the network. For example, a single MEC host may serve a restricted number of geographically close gNBs. This enables a much more efficient interactions from an IoMusT perspective: each performer may send its own data stream to the MEC server, that will host mixing/audio processing functions and jitter buffers itself. The audio streams do not need to reach every performer any longer, but rather concentrate on the MEC server: thus, for n performers, the networked music performance only requires n upstream and n downstream flows, leading to a much more efficient resource usage. Having an architecture based on an MEC server hosting mixing and processing functions also enables more effective transmission error concealment, which can occur along with jitter buffer management, before preparing a final mixed stream to redistribute to the interacting performers. Similarly, a MEC-based architecture can support complex audio computations such as those involved in spatial audio algorithms [Martusciello et al., 2023].

In the 5G ecosystem, network function virtualization (NFV) is the mainstream approach to deploying network functions in the operator's compute sites. Given the current uptake of fully 5G deployments (cf. standalone, SA, versus non-standalone, NSA 5G networks [Turchet and Casari, 2023]), it is reasonable to assume that local computation capabilities will be available at possibly multiple sites over a given geographical or metropolitan area. These sites may host the virtual network functions (VNFs) that enable IoMusT interactions. Relevant examples in the context of a smart musical city include supporting distributed performers as described above, providing streamable contents for musical education, enabling the interaction with location-specific sounds that recall specific events or historical periods, and run the intelligence that detects the state of a patient and matches relevant features thereof with the surroundings, environment, time-of-day, and similar characteristics of therapeutic contexts.

10.2.3 Datasets and Storage

Datasets for the IoMusT require a separate discussion. As is in the nature of smart cities, data from different instruments and sensors should be converged to utilizers that can extract values from them. Data continuously collected via Musical Things from multiple users in manifold times, spaces, and activities can be leveraged to create unprecedented context-aware systems, as well as proactive applications based on the knowledge of the tracked context (such as recommender systems to be used in a variety of musical activities). In general, as it should have appeared clear from the above discussion, several services of the IoMusT are data-driven (e.g., as related to recommendations on location-dependent soundscapes) or AI/ML-based (e.g., error concealment in networked music performance, or generative music tools for pedagogy and immersive interactions with a given environment). These services require significant amounts of data to generate reliable outcomes or to train their underlying models.

Yet, one of the main issues with datasets for the IoMusT is that there exist few if any. For example, despite an extremely rich global tradition in music performances, musical instruments are traditionally not networked. The very IoMusT concept has been conceptualized only recently [Turchet et al., 2018, 2023]: thus, very little data is available about, e.g., utilization patterns, user preferences, styles, and contextual information related to smart musical instrument and Musical Things usage. Similarly, recommender systems would require a sufficiently large amount of data to at least be able to profile user needs/tastes and provide relevant recommendations. While general musical user profiling is commonly performed, e.g., by popular apps and web sites, the fact that the musical recommendation panorama has not significantly progressed beyond user-directed recommendations also hints at the absence of workable and actionable datasets. For example, a quick search for the "music generation" task on paperswithcode.com [Papers With Code, 2023] reveals 23 datasets. Comparatively, datasets related to large language models, generation of text, image segmentation, and recognition can count on more than 1000 datasets. Therefore, additional work is needed to favor a data-driven approach that keeps services tailored to fruition by different users, and up to date with the most recent utilization trends, in the spirit of a smart musical city [Hammons and Myers, 2019].

A complementary problem is that of making the datasets available to the services and users. This entails methods facilitating the storage of such data, their interoperability across heterogeneous devices and services, as well as guarantee of fast accessibility in networked settings.

10.3 Smart Musical City Concept and Services

Some researchers have recently started to investigate the musical flavor of the smart city concept. However, only a handful of studies exist to date on such a topic. In the project Sonic City, the authors developed a system that enables users to create electronic music in real time by walking through and interacting with the urban environment [Gaye et al., 2003]. Other authors proposed to use sonification for the interpretation of smart musical city data and the production of novel musical experiences [Sarmento et al., 2020; Roddy and Bridges, 2021]. Nevertheless, the themes and perspective addressed thus far in the literature are rather different from the vision proposed in the present work.

Section 10.3.1 present and discuss different services that a smart musical city could offer to citizens and visitors alike. Section 10.3.1 discusses the interaction between musicians and virtual agents assisting search, play, and composition tasks; Section 10.3.2 presents participatory music performances as a novel way for players and audiences to interact, mediated by high-reliability and low-latency wireless networks; Section 10.3.3 discusses how to exploit music to foster immersive visiting experiences both in touristic and in everyday contexts; Section 10.3.4 outlines how pedagogy can be enhanced by musical applications.

10.3.1 Interaction Between Musicians and Virtual Agents on Server

The IoMusT technological infrastructure has the potential to enable novel forms of interactions of musicians with musical content mediated by virtual agents running on remote servers. An example of such applications is reported in Turchet et al. [2020]. Authors developed a system comprising a smart guitar and a cloud-based server hosting a music repository. The purpose of the system was that of supporting the guitar player in querying the repository by a novel modality: instead of performing textual queries using the name of the artist or the title of the piece (like it occurs with conventional music streaming services such as Spotify), the musician plays an excerpt with given musical features (e.g., tempo, chords, mood). The intelligence embedded in the instrument retrieves such features and uses them to perform a content-based query. The server then returns to the player pieces of music matching those requested musical features. Such a study highlighted the need for progressing the computation capabilities of embedded audio systems as well as the need for reducing the network latency, using for instance 5G and a MEC server.

Along the same lines, it is possible to envision other types of services based on fast access to a server using Musical Things and a dedicated ultra-reliable low latency communication. These include for instance novel forms of recommendation systems to support music teaching and learning, and in particular self-learning. In the latter case, a virtual agent running on a remote server could determine in real-time the errors performed by a learner and provide recommendations in a timely manner. Moreover, it is possible to devise novel forms of tools assisting composition processes, where virtual agents hosted on a server could generate music excerpts based on inputs from composers. The same concept could also apply to other kinds of musical activities such as automatic mixing performed in real time. In general, for these scenarios to occur and succeed, it is necessary to apply advanced techniques for big data analytics, as well as it is important to create interfaces that ultimately can provide users with a satisfying user experience.

10.3.2 Participatory Networked Music Performances

The IoMusT infrastructure allows to imagine novel kinds of artistic forms positioned at the confluence of networked music performance (NMPs) and participatory art, in particular technology-mediated audience participation (TMAP). These new art forms can be performed in the context of a smart musical city.

NMP systems are hardware and software solutions conceived to allow musicians to play at a distance thanks to a communication link. Their end goal is to reproduce the same conditions as acoustic-instrumental on-site performances. To achieve this goal different issues must be overcome, namely latency, jitter, and the credible rendering of the acoustic scene shared between musicians. The main issue among these is represented by the latency introduced by the acquisition, packetization, and transmission of audio data through the network. A related issue is packet jitter (i.e., the latency variation between consecutive packets carrying audio data), which needs to be kept constant and as low as possible. Several studies have shown that to guarantee performative conditions as realistic as possible, the delay of the content produced at one end and perceived at the other end shall not exceed 20–30 ms, which correspond to the time taken by sound waves propagating in air to cover a distance of 8–10 m [Rottondi et al., 2016]. Such distance is normally assumed to be the maximum tolerance threshold for the physical displacement among players in a room to ensure a stable interplay. Furthermore, to achieve a high degree of realism in an NMP application, it is essential to recreate for all musicians involved a perception of sharing the same space, as recently shown in Tomasetti and Turchet [2023]. This can be achieved by means of spatial audio algorithms which enable a three-dimensional localization of audio sources [Paterson and Lee, 2021], as well as by means of room acoustic modeling techniques [Savioja and Svensson, 2015], which can simulate the type of room in which musicians virtually play (e.g., a concert hall or a rehearsal room).

In a different vein, TMAP systems [Hödl et al., 2017] capitalize on information and communication technologies with the aim of democratizing access to music making and increasing the active engagement of audiences in live music performances. These systems disrupt the traditional unidirectional chain of musical communication in which the musical messages are exchanged sequentially from composers to performers to listeners.

The IoMusT makes the combination of NMP and TMAP possible, enabling novel forms of artistic expression where displaced audiences can interact with displaced musicians, and actually contribute to the performance. For instance, a concert can be performed by different bands playing at different venues of a city, and audiences both colocated with the bands and displaced (e.g., in bars, or at their homes) could actively participate to the end result of the performance.

10.3.3 Cultural Heritage

Recently, several applications have been supporting tourism and local content creators by providing platforms where creators could store voiced description of a place, stories about it as well as music, and tourists could roam around a city, be localized, and play such previously recorded audio streams [del Carmen Rodríguez-Hernández et al., 2012]. For instance, such apps as Citytalks[2] and Autio[3] follow the above paradigm. Other apps such as izi.Travel[4] and Vox City[5] also provide an interface to consume location-dependent audio content, but focus on travel guides. In the context of a smart musical city, services for leisure and tourism, such as the ones above, are extremely relevant. Yet, the cultural heritage of a given location is rich in non-voice sounds. From a musical perspective, smart musical city and landscape environments may be associated to music, sounds, and voices related to famous events [Cheng and Shen, 2016]. Being able to revamp

2 https://www.citytalks.eu.
3 https://autio.com.
4 https://izi.travel.
5 https://voxcity.com.

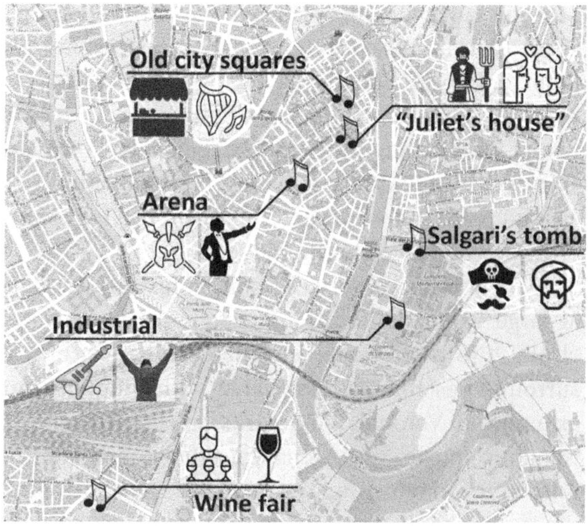

Figure 10.2 Plan of the city center of Verona, Italy, with landmarks associated to location-specific musical tunes and sounds. Source: OpenStreetMap/CC BY SA 2.0.

these sounds may greatly enrich the interaction between people and sites of cultural interest [Braunhofer et al., 2013].

Take as an example the city of Verona, Italy, as shown in the simplified map of Figure 10.2, and imagine a person wandering through the city while being able to access location-dependent sounds from an app or similar facility. Verona is ripe with Roman age masterpieces, including some of the old city walls and the Arena (former gladiator amphitheater, now a world-famous opera venue): the person could approach the Arena, listen to sounds of gladiator clashes as well as famous opera arias. Verona is also the setting of Shakespeare's "Romeo and Juliet," and a fourteen-century house-tower of medieval origins has been converted into a museum known as "Juliet's house," showing typical furnishing, clothing, and tools for different activities of the time. The visitor to the neighborhood of the house may listen to settings of medieval markets in the nearby streets (famous surrounding squares hosted several markets and landmarks of the political and civilian life of the medieval town) and could listen to popular music tunes, both those famous at the time and later associated with the same historical period. Passing by Verona's monumental cemetery may turn up pirate songs as the person approaches the tomb of Emilio Salgari, known for his novels about the fictional late ninetieth-century pirate Sandokan. Verona's fair area is home to the most well-known Italian wine exhibition, "Vinitaly," which may trigger sounds of toasting and bottle opening to people passing by. These are just a few examples of the way location-related sound and music suggestions powered by the IoMusT can make the experience of many places much more immersive as well as contribute to cultural dissemination.

In the same vein, people could pass by different places not just get suggestions about typical musical tunes for those places, but also a chance to interact with them, e.g., by singing or playing over these musics [Cramer, 2016]. For example, Verona's Arena nearabouts may be associated with famous opera arias, over which casual singers may superimpose their own voice, and other amateurs or professionals may provide extra voicing, harmonization, or similar vocal additions. Players and performers may add their own playing to the aria, and transform its style or empower a given harmonic section.

A similar musical service a smart musical city could offer to citizens and people at large would be location-dependent generative musical suggestions [Krause et al., 2016]. For example, different districts of a city or different areas of its surroundings may be associated with different music moods. Continuing the example of Verona's landmarks considered above, Juliet's house may trigger romantic music, the Arena epic tunes, and the medieval squares may pass a relaxed mood matching the many bars and surrounding meeting places. Notably, the system could accumulate data over time by recording the preferences of the listeners that decide to proceed with the suggested tunes or request different musical motifs or moods. In the long run, the system may learn and provide a "musical mapping" of a city's neighborhoods such that, e.g., places with some strong ethnical characterization propose typical music for the local population and favor cultural interchange.

Finally, a sufficiently fine tracking of the association between musical tastes and locations may produce a strong mapping of typical musical memories, and help people passing by revamp both their own memories and collective memories of past important events. This service would be different from serving users with particularly famous tunes that most people would like to listen at a given place: rather, it relates to tracking what a user visited in the past and which musical works, tunes, or styles were associated with the experience at the visited locations. Then, future visits to those same locations may rekindle memories more effectively by proposing the same music, following the well-known pattern for which synesthetic experience and associated emotions are more effective at strengthening or recalling memories [Jäncke, 2008; Rothen et al., 2018].

10.3.4 Pedagogy

While it has been noted that the panorama of musical pedagogy remains often confined to tools such as metronomes and tuners, leaving it to expert teachers to address instrument-specific skills such as attack sharpness [Acquilino and Scavone, 2022], a smart musical city offers a geographically bounded area where music pedagogy can be taken to a much more interactive and connected level than in traditional group or one-to-one classes. For example, the MusiCoLab system provides an environment for collaborative music learning, based not just on lectures and practice, but also on a number of tools to empower interactive practices [Alexandraki et al., 2023]. Such tools include collaboration engines with musical data and metadata, score following and annotation, a music generation engine that includes re-harmonization and blending, all supported by a conferencing server to enable personal communications between the teacher and the pupil(s). As the authors note, the system suffers from the well-known issues that affect any networked music performance, whereby the tolerance to the latency of the interactions is one order of magnitude less than in speech. If network transport delays are one of the main contributions to end-to-end delays [Turchet and Casari, 2024a], the relatively limited geographical extension of smart cities and the availability of slicing in the 5G ecosystem [Turchet and Casari, 2024b] would enable controlled transport latency, and thus a workable experience in musical pedagogy. The use of artificial intelligence complements the above features by providing teachers and students with automatic algorithms to assess rhythm, intonation, and level of learning or correctness of execution for teacher assignments [Wei et al., 2022]. As noted above, training such AIs will be the only means to make them sufficiently reliable tools for the best fruition by educators and pupils. The smart musical city and its inherent possibilities to coalesce data, readings, and feedback from several interactions across different environments, constitutes an ideal mediator to collect data required by musical pedagogy services.

Smart musical cities as interconnected social conglomerates are also the perfect place to implement social pedagogy practices, which has been shown to greatly help address social inequality by considering upbringing as a responsibility of the whole social fabric rather than of specific individuals. From a musical perspective, several studies suggest that partaking in a joint activity such as musical performance can help develop not only the learning capabilities of children (including, e.g., disadvantaged, looked-after, and adopted children) but also create deeper bonds between them and their responsible/caregiver adults [Humphrey, 2020]. In the same vein, playful learning in primary schools has been associated with an increase of meaningful engagement when associated to music [Byrne, 2021]. These examples should convince that smart musical cities are not just the drive of new IoMusT-based musical pedagogy services but also vehicles of pedagogical innovation per se.

10.4 Conclusions

This chapter aimed to offer our perspective on how smart cities should integrate with one of the main activities that is actually performed in cities: musical and sound-related interactions. We deem the Internet of Musical Things to be the technological driver of such future integration, supported by the continuously increasing reliability of low-latency transmissions, e.g., from 5G-and-beyond technology. We described several types of interactions and applications that smart cities can enjoy with the support of the IoMusT, including networked performances both by humans and by a solo player interacting with virtual agents, cultural heritage promotion, location-based recommendation systems, memory revamping as well as music pedagogy.

In particular, we discussed that, like for other smart city services, musical services also need data to train machine learning models, as well as to tune services to specific user profiles. IoMusT devices can be the driver of such new data collection streams and provide relevant measurement of music and musical instruments utilization habits and style, along with commonplace musical preferences based on searches and listening history.

Notably, significant support for such developments exists from the scientific community, both through technical groups such as the IEEE Communications Society Internet of Sounds Emerging Technologies Initiative,[6] and via periodic meetings that are taking momentum, such as the International Symposium on the Internet of Sounds, which will reach its 5th edition in 2024.

Bibliography

A. Acquilino and G. Scavone. Current state and future directions of technologies for music instrument pedagogy. *Frontiers in Psychology*, 13, 2022. doi: 10.3389/fpsyg.2022.835609.

T. Alam. Cloud-based IoT applications and their roles in smart cities. *Smart Cities*, 4(3):1196–1219, 2021. doi: 10.3390/smartcities4030064.

C. Alexandraki, N. Mimidis, Y. Viglis, A. Nousias, D. Milios, and K. Tsioutas. Collaborative playalong practices in online music lessons: The MusiCoLab Toolset. In *Proceedings of the 4th International Symposium on the Internet of Sounds*, pages 1–10, 2023.

D. Bastos, A. Fernández-Caballero, A. Pereira, and N. P. Rocha. Smart city applications to promote citizen participation in city management and governance: A systematic review. *Informatics*, 9(4):89, 2022. doi: 10.3390/informatics9040089.

F. Bevilacqua, B. Matuszewski, G. Paine, and N. Schnell. On designing, composing and performing networked collective interactions. *Organised Sound*, 26(3):333–339, 2021.

6 https://ios.committees.comsoc.org/.

M. Braunhofer, M. Kaminskas, and F. Ricci. Location-aware music recommendation. *International Journal of Multimedia Information Retrieval*, 2:31–44, 2013. doi: 10.1007/s13735-012-0032-2.

M. Buffa, S. Ren, O. Campbell, T. Burns, S. Yi, J. Kleimola, and O. Larkin. Web audio modules 2.0: An open web audio plugin standard. In *Companion Proceedings of the Web Conference 2022*, pages 364–369, 2022.

N. Bui, A. P. Castellani, P. Casari, and M. Zorzi. The internet of energy: A web-enabled smart grid system. *IEEE Network*, 26(4):39–45, 2012. doi: 10.1109/MNET.2012.6246751.

R. Byrne. *An exploration of playful music pedagogy within selected Irish primary school contexts*. PhD thesis, Dublin City University, 2021.

J. P. Cáceres and C. Chafe. JackTrip: Under the hood of an engine for network audio. *Journal of New Music Research*, 39(3):183–187, 2010.

M. del Carmen Rodríguez-Hernández, S. Ilarri, R. Trillo-Lado, and R. Hermoso. Location-aware recommendation systems: Where we are and where we recommend to go. In *Proceedings of LocalRec'15*, 2012.

A. Carôt and C. Werner. Distributed network music workshop with soundjack. In *Proceedings of the 25th Tonmeistertagung*, Leipzig, Germany, 2008.

A. Carôt, M. Dohler, S. Saunders, F. Sardis, R. Cornock, and N. Uniyal. The world's first interactive 5G music concert: Professional quality networked music over a commodity network infrastructure. In *Proceedings of the Sound and Music Computing Conference*, pages 407–412, June 2020a.

A. Carôt, C. Hoene, H. Busse, and C. Kuhr. Results of the fast-music project—five contributions to the domain of distributed music. *IEEE Access*, 8:47925–47951, 2020b.

F. Cheli and S. Giordano. Service parameters identification for adaptive networked music performance. In *2022 Global Information Infrastructure and Networking Symposium*, pages 94–98. IEEE, 2022.

Z. Cheng and J. Shen. On effective location-aware music recommendation. *ACM Transactions on Information Systems (TOIS)*, 34(2):1–32, Apr 2016. doi: 10.1145/2846092.

I. Clester and J. Freeman. Composing the network with streams. In *Proceedings of the Audio Mostly Conference*, pages 196–199. 2021.

H. Cramer. Local sounds & singing along in cars. In *Proceedings of ACM MobileHCI*, page 1055–1058, 2016. doi: 10.1145/2957265.2964197.

R. B. Dannenberg. Scalable and easy-to-use NIME networking. In *International Conference on New Interfaces for Musical Expression*. PubPub, 2022.

C. Drioli, C. Allocchio, and N. Buso. Networked performances and natural interaction via LOLA: Low latency high quality A/V streaming system. In *International Conference on Information Technologies for Performing Arts, Media Access, and Entertainment*, pages 240–250. Springer, 2013.

J. Dürre, N. Werner, S. Hämäläinen, O. Lindfors, J. Koistinen, M. Saarenmaa, and R. Hupke. In-depth latency and reliability analysis of a networked music performance over public 5G infrastructure. In *Audio Engineering Society Convention 153*. Audio Engineering Society, 2022. URL https://aes2.org/publications/elibrary-page/?id=21950.

ETSI Group Specification. Multi-access Edge Computing (MEC): Framework and Reference Architecture. GS MEC 003, ETSI, Mar 2022.

EU Parliament. 5G technology in Europe: Ultra-reliable low latency communication, 2021. URL https://map.sciencemediahub.eu/5gm=6/1388.88067/659.5,p=62. last visited: Dec. 2023.

M. Fink and U. Zölzer. Low-delay error concealment with low computational overhead for audio over IP applications. In *Proceedings of the International Conference on Digital Audio Effects*, pages 309–316, 2014.

A. Fraietta, O. Bown, S. Ferguson, S. Gillespie, and L. Bray. Rapid composition for networked devices: HappyBrackets. *Computer Music Journal*, 43(2):89–108, 2019.

L. Gaye, R. Mazé, and L. E. Holmquist. Sonic city: The urban environment as a musical interface. In *Proceedings of the International Conference on New Interfaces for Musical Expression*, volume 3, pages 109–115, 2003.

R. Hammons and J. Myers. Smart cities. *IEEE Internet of Things Magazine*, 2(2):8–9, 2019. doi: 10.1109/MIOT.2019.8892761.

O. Hödl, G. Fitzpatrick, and F. Kayali. Design implications for technology-mediated audience participation in live music. In *Proceedings of the Sound and Music Computing Conference*, pages 28–34, 2017.

O. Holmqvist, M. Meddings, and H. David. Holon app, 2023. URL https://www.holonic.systems/. Last accessed: Dec. 2023.

R. Humphrey. Social pedagogy, music making and adopted children. *International Journal of Social Pedagogy*, 9(1):7, 2020. doi: 10.14324/111.444.ijsp.2020.v9.x.007.

L. Jäncke. Music, memory and emotion. *Journal of Biology*, 7(21):1–5, 2008. doi: 10.1186/jbiol82.

D. Keller, C. Gomes, and L. Aliel. The handy metaphor: Bimanual, touchless interaction for the Internet of Musical Things. *Journal of New Music Research*, 48(4):385–396, 2019.

M. A. Khan, S. Sargento, and M. Luis. Data collection from smart-city sensors through large-scale urban vehicular networks. In *Proceedings of IEEE VTC-Fall*, pages 1–6, 2017. doi: 10.1109/VTCFall.2017.8288308.

M. Krämer. Controlling the processing of smart city data in the cloud with domain-specific languages. In *Proceedings of IEEE/ACM UCC*, pages 824–829, 2014. doi: 10.1109/UCC.2014.134.

A. E. Krause, A. C. North, and L. Y. Hewitt. The role of location in everyday experiences of music. *Psychology of Popular Media Culture*, 5(3):232–257, Apr 2016. doi: 10.1037/ppm0000059.

B. Loveridge. Networked music performance in virtual reality: Current perspectives. *Journal of Network Music and Arts*, 2(1):2, 2020.

A. T. Marasco and J. Allison. Connecting web audio to cyber-hacked instruments in performance. In *Proceedings of the Web Audio Conference*, 2019.

F. Martusciello, C. Centofanti, C. Rinaldi, and A. Marotta. Edge-enabled spatial audio service: Implementation and performance analysis on a MEC 5G infrastructure. In *Proceedings of the 4th International Symposium on the Internet of Sounds*, pages 1–8, 2023. doi: 10.1109/IEEECONF59510.2023.10335480.

B. Matuszewski. A web-based framework for distributed music system research and creation. *Journal of the Audio Engineering Society*, 68(10):717–726, 2020.

A. McPherson and V. Zappi. An environment for Submillisecond-Latency audio and sensor processing on BeagleBone black. In *Audio Engineering Society Convention 138*. Audio Engineering Society, 2015. URL http://www.aes.org/e-lib/browse.cfm?elib=17755.

Papers With Code. "music generation" datasets, 2023. URL https://paperswithcode.com/datasets?task=music-generation. last visited: Dec. 2023.

J. Paterson and H. Lee. *3D Audio*. Routledge, 2021.

T. Pelinski, R. Diaz, A. L. B. Temprano, and A. McPherson. Pipeline for recording datasets and running neural networks on the Bela embedded hardware platform. In *Proceedings of the International Conference on New Interfaces for Musical Expression*, 2023.

S. Roddy and B. Bridges. The design of a smart city sonification system using a conceptual blending and musical framework, web audio and deep learning techniques. In *International Conference on Auditory Display 25-28 June 2021*, pages 105–110, 2021.

N. Rothen, A. K. Seth, and J. Ward. Synesthesia improves sensory memory, when perceptual awareness is high. *Vision Research*, 153:1–6, 2018. doi: 10.1016/j.visres.2018.09.002.

C. Rottondi, C. Chafe, C. Allocchio, and A. Sarti. An overview on networked music performance technologies. *IEEE Access*, 4:8823–8843, 2016.

P. Sarmento, O. Holmqvist, and M. Barthet. Musical smart city: Perspectives on ubiquitous sonification. 2020.

L. Savioja and U. P. Svensson. Overview of geometrical room acoustic modeling techniques. *The Journal of the Acoustical Society of America*, 138(2):708–730, 2015.

D. Stefani, S. Peroni, and L. Turchet. A comparison of deep learning inference engines for embedded real-time audio classification. In *Proceedings of the Digital Audio Effects Conference*, 2022.

J. Timoney, A. Yaseen, and D. Mcevoy. The potential role of Internet of Musical Things in therapeutic applications. In *Proceedings of the 10th Workshop on Ubiquitous Music (UbiMus 2020)*. g-ubimus, 2020.

M. Tomasetti and L. Turchet. Playing with others using headphones: Musicians prefer binaural audio with head tracking over stereo. *IEEE Transactions on Human-Machine Systems*, 53(3):501–511, 2023.

L. Turchet. Smart musical instruments: Vision, design principles, and future directions. *IEEE Access*, 7:8944–8963, 2019.

L. Turchet and F. Antoniazzi. Semantic web of musical things: Achieving interoperability in the Internet of Musical Things. *Journal of Web Semantics*, 75:100758, 2023.

L. Turchet and P. Casari. Assessing a private 5G SA and a public 5G NSA architecture for networked music performances. In *Proceedings of the 4th International Symposium on the Internet of Sounds*, pages 1–6, 2023.

L. Turchet and P. Casari. Latency and reliability analysis of a 5G-enabled Internet of Musical Things system. *IEEE Internet of Things Journal*, 11(1):1228–1240, 2024a. doi: 10.1109/JIOT.2023.3288818.

L. Turchet and P. Casari. On the impact of 5G slicing on an Internet of Musical Things system. *IEEE Internet of Things Journal*, 11(19):32079–32088, 2024b.

L. Turchet and C. Fischione. Elk Audio OS: An open source operating system for the Internet of Musical Things. *ACM Transactions on Internet of Things*, 2(2):1–18, 2021.

L. Turchet, C. Fischione, G. Essl, D. Keller, and M. Barthet. Internet of Musical Things: Vision and challenges. *IEEE Access*, 6:61994–62017, 2018.

L. Turchet, J. Pauwels, C. Fischione, and G. Fazekas. Cloud-smart musical instrument interactions: Querying a large music collection with a smart guitar. *ACM Transactions on Internet of Things*, 1(3):1–29, 2020.

L. Turchet, T. West, and M. M. Wanderley. Touching the audience: Musical haptic wearables for augmented and participatory live music performances. *Personal and Ubiquitous Computing*, 25(4):749–769, 2021.

L. Turchet, M. Lagrange, Rottondi C., G. Fazekas, N. Peters, J. Østergaard, F. Font, T. Bäckström, and C. Fischione. The internet of sounds: Convergent trends, insights and future directions. *IEEE Internet of Things Journal*, 10(13):11264–11292, 2023.

P. Verma, A. I. Mezza, C. Chafe, and C. Rottondi. A deep learning approach for low-latency packet loss concealment of audio signals in networked music performance applications. In *2020 27th Conference of Open Innovations Association (FRUCT)*, pages 268–275. IEEE, 2020.

R. Vieira, D. C. Muchaluat-Saade, and F. L. Schiavoni. Sunflower: An interactive artistic environment based on IoMusT concepts. In *ACM International Conference on Interactive Media Experiences*, pages 245–248, 2022.

J. Vihriälä, A. A. Zaidi, V. Venkatasubramanian, N. He, E. Tiirola, J. Medbo, E. Lähetkangas, K. Werner, K. Pajukoski, A. Cedergren, and R. Baldemair. Numerology and frame structure for 5G radio access. In *2016 IEEE 27th Annual*

International Symposium on Personal, Indoor, and Mobile Radio Communications (PIMRC), pages 1–5, 2016. doi: 10.1109/PIMRC.2016.7794610.

T. N. Weerasinghe, I. A. M. Balapuwaduge, and F. Y. Li. Priority-based initial access for URLLC traffic in massive IoT networks: Schemes and performance analysis. *Computer Networks*, 178:107360, 2020. doi: 10.1016/j.comnet.2020.107360.

J. Wei, M. Karuppiah, and A. Prathik. College music education and teaching based on AI techniques. *Computers and Electrical Engineering*, 100:107851, 2022. doi: 10.1016/j.compeleceng.2022.107851.

P. Yang, X. Xi, T. Q. S. Quek, J. Chen, X. Cao, and D. Wu. Ran slicing for massive IoT and bursty URLLC service multiplexing: Analysis and optimization. *IEEE Internet of Things Journal*, 8(18):14258–14275, 2021. doi: 10.1109/JIOT.2021.3068518.

A. Yaseen, S. Chakraborty, and J. Timoney. A cooperative and interactive gesture-based drumming interface with application to the Internet of Musical Things. In *International Conference on Human-Computer Interaction*, pages 85–92. Springer, 2022.

J. Yun, Y. Goh, W. Yoo, and J.-M. Chung. 5G multi-RAT URLLC and eMBB dynamic task offloading with MEC resource allocation using distributed deep reinforcement learning. *IEEE Internet of Things Journal*, 9(20):20733–20749, 2022. doi: 10.1109/JIOT.2022.3177425.

11

Robust Target Tracking in Sensor Networks with Measurement Outliers#

Hongwei Wang[1], Hongbin Li[2], and Jun Fang[1]

[1] *National Key Laboratory of Wireless Communications, University of Electronic Science and Technology of China, Chengdu, China*
[2] *Department of Electrical and Computer Engineering, Stevens Institute of Technology, Hoboken, NJ, USA*

11.1 Introduction

Advances in wireless communication and micro-electro-mechanical system (MEMS) technologies make it possible to deploy a sensor network (SN) for surveillance, patrolling, and target tracking [Souza et al., 2016]. Generally, target-tracking methods in SNs can be divided into two main categories, i.e. the centralized schemes and the decentralized ones, as illustrated in Figure 11.1. In centralized approaches, a fusion center is deployed to collect all measurements from the entire sensor nodes of the SN and process these data to obtain the information (including the position and the velocity) of the target of interest. The fusion center can employ the measurement-augmented approach [Lee, 2008; Ge et al., 2016] and implement the well-known Kalman filter (KF) and its nonlinear variants (e.g. the cubature information filter (CIF) [Pakki et al., 2011]) as centralized tracking solutions.

In centralized solutions, readings of each sensor node are required to be transmitted to the data fusion center, which may lead to substantial communication overhead, thereby limiting scalability. In addition, the fusion center is solely responsible for data processing. Therefore, a centralized target-tracking scheme over SNs is heavily reliant on the reliability of the fusion center and may malfunction if the fusion center fails. Moreover, the knowledge of the measurement model at each node should be transmitted to the fusion center, which introduces additional challenges, especially for heterogeneous SNs.

#The work of H. Wang was supported by the National Natural Science Foundation of China under Grants no. 62103083. The work of H. Li was supported in part by the National Science Foundation under Grants CCF-2316865, ECCS-2212940, and ECCS-2332534.

Wireless Sensor Networks in Smart Environments: Enabling Digitalization from Fundamentals to Advanced Solutions, First Edition. Edited by Domenico Ciuonzo and Pierluigi Salvo Rossi.

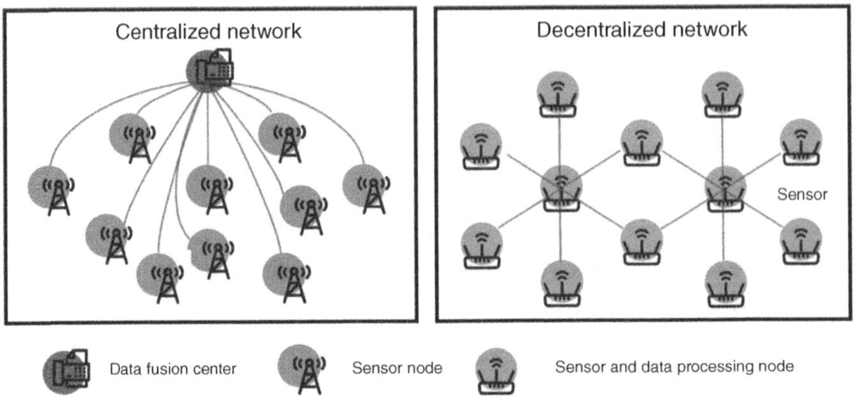

Figure 11.1 Centralized network and decentralized network.

To address these challenges, decentralized target-tracking solutions are gaining interest. In these approaches, each node within the SN tracks the target's state using its own observations, as well as those from its neighbors through local communication. The decentralized method enhances scalability and reliability since it does not require the detailed topology of SNs, which is advantageous for time-variant SNs. Different from centralized solutions, which consolidate all network observations for jointly processing, decentralized solutions may experience some performance degradation and increased sensitivity to outliers. Therefore, a proper information exchange strategy is needed for decentralized solutions to closely match the centralized solutions' performance. Meanwhile, additional procedures should be devised to ensure robustness to outliers in decentralized solutions.

Decentralized solutions require strategies for data exchange among neighboring nodes to avoid over-utilization or under-utilization of related measurements. To this end, several consensus schemes have been proposed, e.g. [Olfati-Saber and Shamma, 2005; Battistelli et al., 2014a, 2015; Chen et al., 2017]. Conventional decentralized target-tracking methods usually rely on the Gaussian assumptions for measurement noises. In practice, however, such a Gaussian assumption may be invalid due to the presence of outliers, especially in scenarios where massive low-cost sensors are deployed. Several solutions have been developed to address outliers in SNs. Specifically, to approximate the underlying non-Gaussian noises caused by outliers, a distributed algorithm based on the Gaussian mixture model is proposed in Li and Jia [2012]. An interactive multiple model (IMM) scheme is utilized to devise robust solutions in Battistelli et al. [2015] and Tian et al. [2015]. In addition, since outliers lead to a heavy-tailed measurement noise, Student's t distribution, which has a heavy tail, is employed to model the measurement noise for outlier-contaminated measurements, resulting in a centralized target-tracking algorithm in Zhu et al. [2013], and a decentralized solution in Dong et al. [2018].

In this chapter, we employ an outlier-detection strategy [Wang et al., 2018] and derive both centralized robust target-tracking and decentralized methods. Specifically, we introduce an outlier-detection measurement model by combining the original measurement model with an outlier indicator using a beta-Bernoulli prior. The mean-field variational Bayesian (VB) inference method is then utilized to jointly estimate the state of the target and the outlier indicators. In the decentralized solution, each node implements the VB inference to estimate both the state and the outlier indicator, and the hybrid consensus scheme is employed to ensure consistency across nodes in SNs. Numerical simulations are conducted to illustrate the effectiveness of the proposed algorithms.

11.2 Problem Formulation

Consider a target-tracking problem in an SN, as illustrated in Figure 11.2. The target is governed by a discrete-time process

$$\boldsymbol{x}_t = \boldsymbol{f}(\boldsymbol{x}_{t-1}) + \boldsymbol{v}_t, \tag{11.1}$$

in which $\boldsymbol{x}_t \in \mathbb{R}^n$ denotes the state of a target, $\boldsymbol{f}(\cdot)$ is the state transition function, and $\boldsymbol{v}_t \sim \mathbb{N}(0, \boldsymbol{Q}_t)$ denotes the process noise. $\boldsymbol{x}_0 \sim \mathbb{N}(\hat{\boldsymbol{x}}_{0|0}, \boldsymbol{P}_{0|0})$ is the initial value of the state. In practice, $\boldsymbol{f}(\cdot)$ is obtained from either some prior knowledge of target motion [Li and Jilkov, 2003] or a data-driven approach [Aftab and Mihaylova, 2020]. The considered SN has N nodes including both sensor nodes and

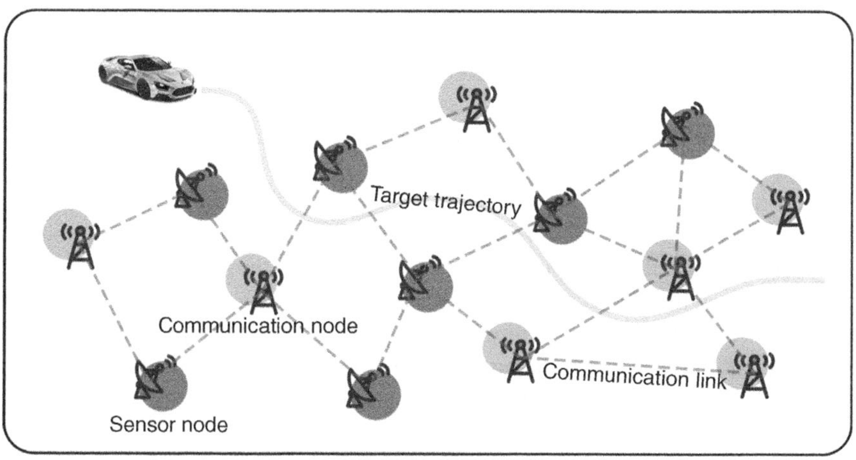

Figure 11.2 Target tracking in an SN.

communication nodes, and its topology can be described by an undirected graph $\mathcal{G} = (\mathcal{E}, \mathcal{D})$, where $\mathcal{D} = \{1, \dots, N\}$ is the vertex set, and $\mathcal{E} \subset \{\{i,j\} | i,j \in \mathcal{D}, i \neq j\}$ is the edge set. The vertex set can be represented by $\mathcal{D} = \mathcal{S} \cup \mathcal{C}$, where $\mathcal{S} = \{1, \dots, S\}$ is the sensor nodes set, and $\mathcal{C} = \mathcal{D} \backslash \mathcal{S}$ denotes the communication nodes set. Generally, the SN deploys the communication nodes to enhance the connectivity. We assume that there exists a path between every pair of vertices, i.e. the considered SN is connected. For convenience, the set containing the sth node and its neighbors is denoted by $\mathcal{N}_s = \{j \in \mathcal{D} | \{s,j\} \in \mathcal{E}\} \cup \{s\}$.

These sensor nodes have the capabilities to make measurements associated with the state vector of the target. Specifically, for the sth ($s \in \mathcal{S}$) sensor, its observation is given by

$$\boldsymbol{y}_{t,s} = \boldsymbol{h}_s(\boldsymbol{x}_t) + \boldsymbol{w}_{t,s}, \tag{11.2}$$

where $\boldsymbol{y}_{t,s} \in \mathbb{R}^{m_s}$ is the recorded reading, $\boldsymbol{h}_s(\cdot)$ is the measurement mapping, and $\boldsymbol{w}_{t,s} \in \mathbb{R}^{m_s}$ denotes the measurement noise. For the measurement noises of different sensor nodes, we have the following assumptions: (i) $\boldsymbol{w}_{t,s_1}$ and $\boldsymbol{w}_{t,s_2}$ for $s_1, s_2 \in \mathcal{S}$ and $s_1 \neq s_2$ are mutually independent; (ii) $\boldsymbol{w}_{t,s}$ for $s \in \mathcal{S}$ is independent of both $\boldsymbol{x}_0$ and process noise.

Nominally, each measurement noise $\boldsymbol{w}_{t,s}$ follows a Gaussian distribution, say, $\boldsymbol{w}_{t,s} \sim \mathbb{N}(0, \boldsymbol{R}_{t,s})$. Nevertheless, the Gaussian property becomes invalid when measurements are contaminated by outliers. To identify outliers in measurements, a binary variable $z_{t,s}$ is introduced to indicate whether $\boldsymbol{y}_{t,s}$ is an outlier or not. Specifically, $z_{t,s}$ is set to 1 if $y_{t,s}$ is a reliable observation, while $z_{t,s}$ is assigned to 0 when $\boldsymbol{y}_{t,s}$ is an outlier. With such an indicator, we introduce the following outlier-detection model for $\boldsymbol{y}_{t,s}$:

$$p(\boldsymbol{y}_{t,s} | \boldsymbol{x}_t, z_{t,s}) \propto \left(\mathbb{N}(\boldsymbol{y}_{t,s}; \boldsymbol{h}_s(\boldsymbol{x}_t), \boldsymbol{R}_{t,s})\right)^{z_{t,s}}, \tag{11.3}$$

where $z_{t,s}$ follows the following beta-Bernoulli distribution:

$$p(z_{t,s} | \pi_{t,s}) = \pi_{t,s}^{z_{t,s}} (1 - \pi_{t,s})^{(1-z_{t,s})}. \tag{11.4}$$

$$p(\pi_{t,s}) \propto \pi_{t,s}^{e_{0,s}-1} (1 - \pi_{t,s})^{f_{0,s}-1}. \tag{11.5}$$

It should be noticed that $\pi_{t,s}$ is a stochastic parameter characterized by $e_{0,s}$ and $f_{0,s}$. From (11.3), we know that $p(\boldsymbol{y}_{t,s} | \boldsymbol{x}_t, z_{t,s})$ is the same as the signal model in (11.2) when $z_{t,s} = 1$, while it is a constant when $z_{t,s} = 0$. In the latter case, $\boldsymbol{y}_{t,s}$ is marked as an outlier. This is because the value of the likelihood function is a constant that is independent of the state. In the outlier-detection model, one can adjust the value $e_{0,s}$ and $f_{0,s}$ to control the probability that $\boldsymbol{y}_{t,s}$ is an outlier, i.e.

$$\Pr\left(\boldsymbol{y}_{t,s} \text{ is an outlier}\right) = \frac{f_{0,s}}{e_{0,s} + f_{0,s}}. \tag{11.6}$$

The objective is to develop solutions to track the target (i.e. estimate the states of the target), as well as to detect outliers that may be encountered at each sensor node of the SN. Both the centralized and the decentralized tracking schemes are considered. Specifically, for the decentralized tracking approach, we integrate the consensus strategy with the outlier-detection technique.

11.2.1 Cubature Information Filter

In this section, we briefly summarize a typical Gaussian approximation filter, e.g. the cubature information filter (CIF). For convenience, we consider the state-space model (i.e. (11.1) and (11.2)) with only one sensor and omit s for convenience. In addition, given a Gaussian distribution $\mathbb{N}(\boldsymbol{\mu}, \boldsymbol{\Sigma})$, its information matrix (or precision matrix) and information vector are, respectively, defined as $\boldsymbol{\Gamma} = \boldsymbol{\Sigma}^{-1}$ and $\boldsymbol{\gamma} = \boldsymbol{\Gamma}\boldsymbol{\mu}$. The procedure of the CIF is described as follows.

Initialization: Denote $\boldsymbol{I} \in \mathbb{R}^{n\times n}$ being an identify matrix and $\boldsymbol{\Delta} = [\boldsymbol{I} - \boldsymbol{I}]$. The basic cubature point set $\{\boldsymbol{\eta}_i, \omega_i\}$ for $i = 1, \dots, 2n$ can then be computed, where $\omega_i = 1/(2n)$ and $\boldsymbol{\eta}_i = \sqrt{n}\boldsymbol{\Delta}_i$ with $\boldsymbol{\Delta}_i$ being the ith column of $\boldsymbol{\Delta}$.

Prediction: The transformed sigma points are generated as

$$\boldsymbol{P}_{t-1|t-1} = \boldsymbol{S}_{t-1|t-1}\boldsymbol{S}^{T}_{t-1|t-1}, \tag{11.7}$$

$$\boldsymbol{\eta}_{i,t-1} = \boldsymbol{S}_{t-1|t-1}\boldsymbol{\eta}_i + \hat{\boldsymbol{x}}_{t-1|t-1}, \tag{11.8}$$

where $\mathbb{N}(\hat{\boldsymbol{x}}_{t-1|t-1}, \boldsymbol{P}_{t-1|t-1})$ is the posterior density at the $(t-1)$th time instant. Then, the predicted state and its associated covariance are calculated by

$$\hat{\boldsymbol{x}}_{t|t-1} = \sum_{i=1}^{2n} \omega_i \boldsymbol{\psi}_{i,t-1}, \tag{11.9}$$

$$\boldsymbol{P}_{t|t-1} = \sum_{i=1}^{2n} \omega_i (\boldsymbol{\psi}_{i,t-1} - \hat{\boldsymbol{x}}_{t|t-1})(\boldsymbol{\psi}_{i,t-1} - \hat{\boldsymbol{x}}_{t|t-1})^T + \boldsymbol{Q}_{t-1}, \tag{11.10}$$

where $\boldsymbol{\psi}_{i,t-1} = f(\boldsymbol{\eta}_{i,t-1})$. The corresponding information format is then given by

$$\boldsymbol{\Gamma}_{t|t-1} = \boldsymbol{P}^{-1}_{t|t-1}, \boldsymbol{\gamma}_{t|t-1} = \boldsymbol{\Gamma}_{t|t-1}\hat{\boldsymbol{x}}_{t|t-1}. \tag{11.11}$$

Filtering: Applying the statistical linear error propagation methodology [Lee, 2008], the pseudo-measurement matrix is defined as

$$\boldsymbol{H}_t = \boldsymbol{P}^{-1}_{t|t-1}\boldsymbol{P}_{xy}, \tag{11.12}$$

where $\boldsymbol{P}_{xy}$ is calculated by

$$\boldsymbol{P}_{xy} = \sum_{i=1}^{2n} \omega_i \left(\boldsymbol{\zeta}_{i,t} - \hat{\boldsymbol{x}}_{t|t-1}\right)\left(\boldsymbol{\varrho}_{i,t} - \hat{\boldsymbol{y}}_t\right)^T, \tag{11.13}$$

in which $\zeta_{i,t} = \boldsymbol{S}_{t|t-1}\boldsymbol{\eta}_i + \hat{\boldsymbol{x}}_{t|t-1}$ with $\boldsymbol{P}_{t|t-1} = \boldsymbol{S}_{t|t-1}\boldsymbol{S}_{t|t-1}^T$, $\boldsymbol{\varrho}_{i,t} = \boldsymbol{h}(\zeta_{i,t})$, and $\hat{\boldsymbol{y}}_t = \sum_{i=1}^{2n} \omega_i \boldsymbol{\varrho}_{i,t}$. Given $\boldsymbol{H}_t$, one can compute the following correlation information terms:

$$\tilde{\boldsymbol{y}}_t = \left(\boldsymbol{y}_t - \hat{\boldsymbol{y}}_t + \boldsymbol{H}_t\hat{\boldsymbol{x}}_{t|t-1}\right). \tag{11.14}$$

$$\boldsymbol{I}_t = \boldsymbol{H}_t\boldsymbol{R}_t^{-1}\boldsymbol{H}_t^T. \tag{11.15}$$

$$\boldsymbol{i}_t = \boldsymbol{H}_t\boldsymbol{R}_t^{-1}\tilde{\boldsymbol{y}}_t. \tag{11.16}$$

Finally, the filtered state can be updated. Specifically, its information format is computed by

$$\boldsymbol{\Gamma}_{t|t} = \boldsymbol{\Gamma}_{t|t-1} + \boldsymbol{I}_t,\ \boldsymbol{\gamma}_{t|t} = \boldsymbol{\gamma}_{t|t-1} + \boldsymbol{i}_t, \tag{11.17}$$

and its mean and variance are recovered as

$$\boldsymbol{P}_{t|t} = \boldsymbol{\Gamma}_{t|t}^{-1},\ \hat{\boldsymbol{x}}_{t|t} = \boldsymbol{P}_{t|t}\boldsymbol{\gamma}_{t|t}. \tag{11.18}$$

11.3 Centralized Robust Target Tracking

For centralized processing-based tracking, measurements from sensor nodes are collected at the fusion center, where they are utilized to track the target. Considering the fact that measurements from different sensor nodes are mutual independent, the likelihood function of the measurements can be formulated as the product of the likelihood functions from all sensor nodes. Therefore, given the latent variables $\boldsymbol{\Theta}_t \triangleq \{\boldsymbol{x}_t, \mathcal{Z}_t, \boldsymbol{\pi}_t\}$, we have

$$p(\mathcal{Y}_t|\boldsymbol{\Theta}_t) = \prod_{s\in\mathcal{S}} p(\boldsymbol{y}_{t,s}|\boldsymbol{x}_t, z_{t,s})p(z_{t,s}|\pi_{t,s})p(\pi_{t,s}), \tag{11.19}$$

where $\mathcal{Y}_t \triangleq \{\boldsymbol{y}_{t,1}, \dots, \boldsymbol{y}_{t,S}\}$, $\boldsymbol{\pi}_t \triangleq \{\pi_{t,1}, \dots, \pi_{t,S}\}$, and $\mathcal{Z}_t \triangleq \{z_{t,1}, \dots, z_{t,S}\}$. The posterior distribution with respect to $\boldsymbol{\Theta}_t$ conditioned on $\mathcal{Y}_{1:t}$ is given by

$$p(\boldsymbol{\Theta}_t|\mathcal{Y}_{1:t}) = \frac{p(\boldsymbol{\Theta}_t, \mathcal{Y}_{1:t})}{p(\mathcal{Y}_{1:t})}. \tag{11.20}$$

Directly calculating $p(\boldsymbol{\Theta}_t|\mathcal{Y}_{1:t})$ is challenging because $p(\mathcal{Y}_{1:t})$ is in general computationally intractable. To deal with this difficulty, we employ the variational Bayesian (VB) approach [Tzikas et al., 2008]. Specifically, a mean-field variational distribution $q(\boldsymbol{\Theta}_t) = q(\boldsymbol{x}_t)q(\mathcal{Z}_t)q(\boldsymbol{\pi}_t)$ is utilized as an approximation for the posterior distribution $p(\boldsymbol{\Theta}_t|\mathcal{Y}_{1:t})$, and the variational distribution is obtained via solving the following optimization problem:

$$\{q(\boldsymbol{x}_t), q(\mathcal{Z}_t), q(\boldsymbol{\pi}_t)\} = \arg\min_{q(\boldsymbol{\Theta}_t)} \text{KLD}\left(q(\boldsymbol{\Theta}_t)\|p(\boldsymbol{\Theta}_t|\mathcal{Y}_{1:t})\right), \tag{11.21}$$

where $\mathrm{KLD}(p_1 \| p_2)$ is the Kullback–Leibler divergence (KLD) between two distributions p_1 and p_2. The solution to (11.21) is given by

$$q(\boldsymbol{x}_t) \propto \exp\left(\langle \ln p(\mathcal{Y}_t, \boldsymbol{\Theta}_t | \mathcal{Y}_{1:t-1}) \rangle_{q(\mathcal{Z}_t)q(\boldsymbol{\pi}_t)}\right), \tag{11.22}$$

$$q(\mathcal{Z}_t) \propto \exp\left(\langle \ln p(\mathcal{Y}_t, \boldsymbol{\Theta}_t | \mathcal{Y}_{1:t-1}) \rangle_{q(\boldsymbol{x}_t)q(\boldsymbol{\pi}_t)}\right), \tag{11.23}$$

$$q(\boldsymbol{\pi}_t) \propto \exp\left(\langle \ln p(\mathcal{Y}_t, \boldsymbol{\Theta}_t | \mathcal{Y}_{1:t-1}) \rangle_{q(\boldsymbol{x}_t)q(\mathcal{Z}_t)}\right), \tag{11.24}$$

where $\langle a \rangle_b$ represents the expectation of a over the distribution of b, and $p(\mathcal{Y}_t, \boldsymbol{\Theta}_t | \mathcal{Y}_{1:t-1})$ is decomposed as

$$p(\mathcal{Y}_t, \boldsymbol{\Theta}_t | \mathcal{Y}_{1:t-1}) = p(\boldsymbol{x}_t | \mathcal{Y}_{1:t-1}) p(\mathcal{Y}_t | \boldsymbol{x}_t, \mathcal{Z}_t) p(\mathcal{Z}_t | \boldsymbol{\pi}_t) p(\boldsymbol{\pi}_t), \tag{11.25}$$

where $p(\boldsymbol{x}_t | \mathcal{Y}_{1:t-1})$ is the predictive density function, which is generally approximated by $\mathbb{N}(\hat{\boldsymbol{x}}_{t|t-1}, \boldsymbol{P}_{t|t-1})$ via (11.9) and (11.10). It is apparent that the update rules in the VB inference, as given in (11.22)–(11.24), are coupled. The alternating updating strategy is a prevalent method used to sequentially update variational distributions, wherein each distribution is updated individually while the others remain fixed.

In (11.22), keeping the terms that only relate to $\boldsymbol{x}_t$ leads to

$$q(\boldsymbol{x}_t) \propto \exp\left(-\frac{1}{2} \|\boldsymbol{x}_t - \hat{\boldsymbol{x}}_{t|t-1}\|^2_{\boldsymbol{P}^{-1}_{t|t-1}} - \sum_{s \in \mathcal{S}} \frac{\langle z_{t,s} \rangle}{2} \|\boldsymbol{y}_{t,s} - \boldsymbol{h}_s(\boldsymbol{x}_t)\|^2_{\boldsymbol{R}^{-1}_{t,s}} \right), \tag{11.26}$$

where $\|\boldsymbol{x}\|^2_{\boldsymbol{A}}$ represents $\boldsymbol{x}^T \boldsymbol{A} \boldsymbol{x}$, and $\langle z_{t,s} \rangle$ denotes the mean of $z_{t,s}$. It appears that $q(\boldsymbol{x}_t)$ can be obtained via a Kalman filtering algorithm, particularly in its information format. Specifically, $q(\boldsymbol{x}_t)$ can be approximated by $\mathbb{N}(\hat{\boldsymbol{x}}_{t|t}, \boldsymbol{P}_{t|t})$, where the associated two parameters are updated as

$$\boldsymbol{I}_{t,s} = \langle z_{t,s} \rangle \boldsymbol{H}_{t,s} \boldsymbol{R}^{-1}_{t,s} \boldsymbol{H}_{t,s}, \ \boldsymbol{i}_{t,s} = \langle z_{t,s} \rangle \boldsymbol{H}_{t,s} \boldsymbol{R}^{-1}_{t,s} \tilde{\boldsymbol{y}}_{t,s}, \tag{11.27}$$

$$\boldsymbol{\Gamma}_{t|t} = \boldsymbol{\Gamma}_{t|t-1} + \sum_{s \in \mathcal{S}} \boldsymbol{I}_{t,s}, \ \boldsymbol{\gamma}_{t|t} = \boldsymbol{\gamma}_{t|t-1} + \sum_{s \in \mathcal{S}} \boldsymbol{i}_{t,s}, \tag{11.28}$$

$$\boldsymbol{P}_{t|t} = \boldsymbol{\Gamma}^{-1}_{t|t}, \ \hat{\boldsymbol{x}}_{t|t} = \boldsymbol{P}_{t|t} \boldsymbol{\gamma}_{t|t}, \tag{11.29}$$

in which $\boldsymbol{H}_{t,s}$ is obtained via (11.12) using $\boldsymbol{h}_s(\boldsymbol{x}_t)$, $\tilde{\boldsymbol{y}}_{t,s}$ is updated via (11.14) based on $\boldsymbol{y}_{t,s}$, and $\boldsymbol{\Gamma}_{t|t-1}$ and $\boldsymbol{\gamma}_{t|t-1}$ are obtained via (11.11).

For $\mathcal{Z}_t$, we have $p(\mathcal{Z}_t) = \prod_{s \in \mathcal{S}} p(z_{t,s})$. This is because the elements in $\mathcal{Z}_t$ are mutually independent. Consequently, we can update $q(z_{t,s})$ in a separate manner. Take $q(z_{t,s})$ as an example. Dropping off the terms in (11.23) that do not depend on $q(z_{t,s})$ results in

$$\begin{aligned} q(z_{t,s}) &\propto \exp \left\langle \ln p(\boldsymbol{y}_{t,s} | \boldsymbol{x}_t, z_{t,s}) + \ln p(z_{t,s} | \pi_{t,s}) \right\rangle_{q(\boldsymbol{x}_t)q(\pi_{t,s})} \\ &\propto \exp\left\{ -0.5 z_{t,s} \, \mathrm{tr}(\boldsymbol{D}_{t,s} \boldsymbol{R}^{-1}_{t,s}) + z_{t,s} \zeta_1 + (1 - z_{t,s}) \zeta_2 \right\}, \end{aligned} \tag{11.30}$$

where

$$\boldsymbol{D}_{t,s} = \int \left(\boldsymbol{y}_{t,s} - \boldsymbol{h}_s(\boldsymbol{x}_t)\right)\left(\boldsymbol{y}_{t,s} - \boldsymbol{h}_s(\boldsymbol{x}_t)\right)^T q(\boldsymbol{x}_t)d\boldsymbol{x}_t, \tag{11.31}$$

$$\zeta_1 \triangleq \langle \ln \pi_{t,s} \rangle_{q(\pi_{t,s})} = \Psi(e_{t,s}) - \Psi(e_{t,s} + f_{t,s}), \tag{11.32}$$

$$\zeta_2 \triangleq \langle \ln(1 - \pi_{t,s}) \rangle_{q(\pi_{t,s})} = \Psi(f_{t,s}) - \Psi(e_{t,s} + f_{t,s}), \tag{11.33}$$

in which $\Psi(\cdot)$ is the digamma function. From (11.30), it can be seen that $z_{t,s}$ follows a Bernoulli distribution, and hence its expectation is given by

$$\langle z_{t,s} \rangle_{q(z_{t,s})} = \frac{e^{\zeta_1 - 0.5\mathrm{tr}(\boldsymbol{D}_{t,s}\boldsymbol{R}_{t,s}^{-1})}}{e^{\zeta_1 - 0.5\mathrm{tr}(\boldsymbol{D}_{t,s}\boldsymbol{R}_{t,s}^{-1})} + e^{\zeta_2}}. \tag{11.34}$$

Analogously, owing to the independence, $q(\boldsymbol{\pi}_t)$ can also be formulated as $\prod_{s \in \mathcal{S}} q(\pi_{t,s})$. Therefore, the update for $q(\pi_{t,s})$ is given as

$$q(\pi_{t,s}) \propto \exp\left((e_{t,s} - 1)\ln \pi_{t,s} + (f_{t,s} - 1)\ln(1 - \pi_{t,s})\right), \tag{11.35}$$

which indicates that $q(\pi_{t,s}) \sim \mathrm{Beta}(e_{t,s}, f_{t,s})$ with

$$e_{t,s} = e_{t,s}^0 + \langle z_{t,s} \rangle_{q(z_{t,s})}. \tag{11.36}$$

$$f_{t,s} = f_{t,s}^0 + 1 - \langle z_{t,s} \rangle_{q(z_{t,s})}. \tag{11.37}$$

For clarity, we provide a summary of the centralized robust CIF (cRCIF) for target tracking with K VB iterations in Algorithm 11.1.

Algorithm 11.1 The cRCIF for target tracking

Require: $\mathcal{Y}_{1:T}, \hat{\boldsymbol{x}}_{0|0}, P_{0|0}, Q_{1:T}, R_{1:T}$.
for $t = 1 : T$ **do**
  Calculate $\{\hat{\boldsymbol{x}}_{t|t-1}, P_{t|t-1}\}$ via {(11.9),(11.10)};
  Calculate $\{\boldsymbol{\gamma}_{t|t-1}, \boldsymbol{\Gamma}_{t|t-1}\}$ via (11.11);
  Initialize $k = 0$, $e_{t,s}^k, f_{t,s}^k$ and $\langle z_{t,s}^k \rangle = 1$ for $s \in \mathcal{S}$;
  for $k = 1 : K$ **do**
    Compute $\{\boldsymbol{I}_{t,s}^k, i_{t,s}^k\}$ via (11.27) with $\langle z_{t,s}^{k-1} \rangle$;
    Compute $\{\boldsymbol{\Gamma}_{t|t}^k, \boldsymbol{\gamma}_{t|t}^k\}$ via (11.28);
    Compute $\{P_{t|t}^k, \hat{\boldsymbol{x}}_{t|t}^k\}$ via (11.29);
    Compute $\langle z_{t,s}^k \rangle$ via (11.34), $e_{t,s}^k$ via (11.36), and $f_{t,s}^k$ via (11.37) for $s \in \mathcal{S}$;
  end for
  $\hat{\boldsymbol{x}}_{t|t} = \hat{\boldsymbol{x}}_{t|t}^K$, $P_{t|t} = P_{t|t}^K$.
end for

11.4 Decentralized Robust Target Tracking

Although the centralized tracking scheme can fully utilize the information in measurements and provide a performance benchmark, it has high communication overhead and relatively low reliability as the fusion center potentially represents a single point of failure. To deal with these challenges, we derive a decentralized robust target-tracking algorithm, which integrates the consensus averaging with outlier detection to achieve performance comparable to that of the centralized tracking scheme. We start by briefly introducing the consensus strategy, followed by developing a decentralized solution by considering both the outlier detection and the consensus strategy.

11.4.1 Consensus Strategy

For the sake of clarity, we introduce the following operators:

$$\bigoplus_a \left(\eta_a \odot p_a(x)\right) \triangleq \frac{\prod_a (p_a(x))^{\eta_a}}{\int \prod_a (p_a(x))^{\eta_a} dx}, \tag{11.38}$$

$$p_a(x) \oplus p_b(x) \triangleq \frac{p_a(x)p_b(x)}{\int p_a(x)p_b(x)dx}, \tag{11.39}$$

in which $p_a(x)$ and $p_b(x)$ are probability density functions (PDFs) with respect to x, and $\eta_a > 0$ is a weighting factor.

In the considered networked system, assume that the PDF of each node is $p_n(x)$, $n \in \mathcal{D}$, and the consensus PDF is defined by the Kullback–Leibler average of these PDFs of each node, i.e.

$$\overline{p}(x) = \arg\inf_{p(x)} \sum_{n\in\mathcal{D}} \frac{1}{N} \text{ KLD } \left(p(x) \| p_n(x)\right). \tag{11.40}$$

Theoretically, $\overline{p}(x)$ is given by

$$\overline{p}(x) = \bigoplus_n \left(\frac{1}{N} \odot p_n(x)\right). \tag{11.41}$$

Apparently, calculating $\overline{p}(x)$, which requires the PDFs of all nodes, is intractable since it is generally impossible for a node to acquire the PDFs of other nodes. Therefore, we turn to compute $\overline{p}(x)$ at each node distributively in an iterative manner, i.e.

$$\lim_{l\to\infty} p_n^l(x) = \overline{p}(x), n \in \mathcal{D}, \tag{11.42}$$

where l is the iteration index. Specifically, the update of any node in each iteration only relates to its own PDF and its neighbors' PDFs. For the $(l+1)$th iteration, the consensus density at the nth node is given by

$$p_n^{l+1}(x) = \bigoplus_{j\in\mathcal{N}_s} \left(\pi_{n,j} \odot p_j^l(x)\right), \tag{11.43}$$

where $\pi_{n,j}$ is the consensus weight. We here employ the well-known Metropolis weights,

$$\pi_{n,j} = \begin{cases} \dfrac{1}{\max\{|\mathcal{N}_n|, |\mathcal{N}_j|\}}, & n \in \mathcal{D}, j \in \mathcal{N}_n, j \neq n, \\ 1 - \sum\limits_{j \in \mathcal{N}_i, n \neq j} \pi_{n,j}, & j = n, \\ 0, & \text{others.} \end{cases} \tag{11.44}$$

Other consensus weights, e.g. the constant weight, can also be employed.

In addition, if $p_j^l(x)$ follows a Gaussian distribution represented by its information parameters γ_j^l and $\mathbf{\Gamma}_j^l$, then $p_n^{l+1}(x)$ in (11.43) is also a Gaussian random variable with its information parameters given by

$$\mathbf{\Gamma}_n^{l+1} = \sum_{j \in \mathcal{N}_n} \pi_{n,j} \mathbf{\Gamma}_j^l. \tag{11.45}$$

$$\gamma_n^{l+1} = \sum_{j \in \mathcal{N}_n} \pi_{n,j} \gamma_j^l. \tag{11.46}$$

In the tracking problem, each node has two kinds of information, i.e. the prior information from the predicted density, and the innovation from the likelihood density. Therefore, the consensus on the posterior density at each node involves three steps: consensus on prior, consensus on likelihood, and fusing the consensus results of the priors and likelihoods. In the following, we provide details of these three steps.

11.4.2 Consensus on Prior

Due to the fact that the local prior distribution is independent of the outlier detection procedure, achieving the consensus on the prior for the nth node can be directly attained by the following L iterations:

$$p_{t|t-1,n}^l(\boldsymbol{x}_t) = \bigoplus_{j \in \mathcal{D}} \left(\pi_{n,j} \odot p_{t|t-1,j}^{l-1}(\boldsymbol{x}_t) \right), \tag{11.47}$$

where $\pi_{n,j}$ is the weight given by (11.44), and $l = 1, \dots, L$ represents the index of the consensus step. It should be noticed that in (11.47), the predicted density $p_{t,n}(\boldsymbol{x}_t | \mathcal{Y}_{1:t-1})$ is utilized to initialize $p_{t|t-1,n}^0(\boldsymbol{x}_t)$. Since $p_{t,n}(\boldsymbol{x}_t | \mathcal{Y}_{1:t-1})$ follows a Gaussian, (11.47) can be calculated in a closed form. Specifically, the $p_{t|t-1,n}^l(\boldsymbol{x}_t) \sim \mathbb{N}((\mathbf{\Gamma}_{t|t-1,n}^l)^{-1} \gamma_{t|t-1,n}^l, (\mathbf{\Gamma}_{t|t-1,n}^l)^{-1})$, where

$$\mathbf{\Gamma}_{t|t-1,n}^l = \sum_{j \in \mathcal{D}} \pi_{n,j} \mathbf{\Gamma}_{t|t-1,j}^{l-1}. \tag{11.48}$$

$$\gamma_{t|t-1,n}^l = \sum_{j \in \mathcal{D}} \pi_{n,j} \gamma_{t|t-1,j}^{l-1}. \tag{11.49}$$

In (11.48) and (11.49), $\gamma^0_{t|t-1,j}$ and $\Gamma^0_{t|t-1,j}$ are, respectively, initialized by $\gamma_{t|t-1,j}$ and $\Gamma_{t|t-1,j}$.

11.4.3 Consensus on Likelihood

Similarly, achieving consensus on likelihood can be accomplished through the following L-step iterations:

$$r^l_{t,n}(\boldsymbol{x}_t) = \bigoplus_{j\in D}\left(\pi_{n,j} \odot r^{l-1}_{t,j}(\boldsymbol{x}_t)\right), \tag{11.50}$$

where $r^0_{t,n}(\boldsymbol{x}_t)$ is, depending on whether the node is a sensor node or a communication node, determined by

$$r^0_{t,n}(\boldsymbol{x}_t) = \begin{cases} p(\boldsymbol{y}_{t,n}|\boldsymbol{x}_t), & n \in S, \\ \text{constant}, & n \in C. \end{cases} \tag{11.51}$$

It is apparent that $p(\boldsymbol{y}_{t,n}|\boldsymbol{x}_t)$ is not a Gaussian distribution because it depends on the binary indicator variable $z_{t,n}$. As a consequence, achieving the consensus on likelihood (11.50) generally lacks a closed-form solution. Fortunately, given the indicator $z_{t,n}$, the likelihood function $p(\boldsymbol{y}_{t,n}|\boldsymbol{x}_t, z_{t,n})$ is a Gaussian distribution. The consensus on the likelihood step and the VB iteration is coupled because the update of $z_{t,n}$ is within the VB iteration (see details in (11.30)). It should be noticed that each sensor node performs outlier detection locally via a similar procedure in the centralized robust tracking scheme. We therefore omit the details of estimating $z_{t,n}$.

With the updated $z^k_{t,n}$, $p(\boldsymbol{y}_{t,n}|\boldsymbol{x}_t, z^k_{t,n})$, i.e. the likelihood of the nth ($n \in S$) sensor node, is given by

$$p(\boldsymbol{y}_{t,n}|\boldsymbol{x}_t, z^k_{t,n}) \propto \exp\left(-\frac{1}{2}(\boldsymbol{x}_t^T \boldsymbol{I}^k_{t,n}\boldsymbol{x}_t - 2\boldsymbol{x}_t^T \boldsymbol{i}^k_{t,n})\right), \tag{11.52}$$

where $\boldsymbol{I}_{t,n}$ and $\boldsymbol{i}_{t,n}$ are, respectively, given by

$$\boldsymbol{I}^k_{t,n} = \langle z^k_{t,n}\rangle \boldsymbol{H}_{t,n}\boldsymbol{R}^{-1}_{t,n}\boldsymbol{H}_{t,n}, \tag{11.53}$$

$$\boldsymbol{i}^k_{t,n} = \langle z^k_{t,n}\rangle \boldsymbol{H}_{t,n}\boldsymbol{R}^{-1}_{t,n}\tilde{\boldsymbol{y}}_{t,n}, \tag{11.54}$$

in which $\boldsymbol{H}_{t,n}$ and $\tilde{\boldsymbol{y}}_{t,n}$ are, respectively, can be found in (11.12) and (11.14). Since a communication node does not provide measurements, its likelihood is in fact a constant, meaning that at the kth VB iteration for the nth ($n \in C$) node, we have

$$\boldsymbol{I}^k_{t,n} = 0,\ \boldsymbol{i}^k_{t,n} = 0. \tag{11.55}$$

After obtaining the information terms through (11.53)–(11.55), consensus on likelihood for the nth node can be achieved via the following L iterations:

$$\boldsymbol{I}^{k,l}_{t,n} = \sum_{j\in N_n} \kappa_{n,j}\boldsymbol{I}^{k,l-1}_{t,j}, \tag{11.56}$$

$$\boldsymbol{i}_{t,n}^{k,l} = \sum_{j \in N_n} \kappa_{n,j} \boldsymbol{i}_{t,j}^{k,l-1}, \tag{11.57}$$

with the following initialization:

$$\boldsymbol{I}_{t,n}^{k,0} = \boldsymbol{I}_{t,n}^{k}, \ \boldsymbol{i}_{t,n}^{k,0} = \boldsymbol{i}_{t,n}^{k}.$$

11.4.4 Fusing the Consensus Results

With the results of the consensus on prior and likelihood, the fusion result for the nth node is given by

$$p_{t,n}(\boldsymbol{x}_t) = p_{t|t-1,n}^{L}(\boldsymbol{x}_t) \oplus \left(\delta_{t,n} \odot r_{t,n}^{L}(\boldsymbol{x}_t)\right), \tag{11.58}$$

where $p_{t|t-1,n}^{L}(\boldsymbol{x}_t)$ is from (11.47), $r_{t,n}^{L}(\boldsymbol{x}_t)$ is from (11.50), and $\delta_{t,n}$ is a parameter to prevent the overweighting of innovations. In the conventional Kalman filter, the fusion rule is similar to the one in (11.58) except that $\delta_{t,n}$ is set to 1. Therefore, $\delta_{t,n}$ should be theoretically set to $|\mathcal{N}|$ such that $\pi_{t,n}^{L}\delta_{t,n} = 1$, where $\pi_{t,s}^{L} = 1/|\mathcal{N}|$ when $L \to \infty$. In practice, however, L is relatively small because of the power supply constraint and limited communication capacity of each node. The choice of $\delta_{t,n} = |\mathcal{N}|$ will cause performance degradation due to these practice issues. Details can be found in Battistelli et al. [2014b]. A feasible distributed approach to calculate $\delta_{t,n}$ is

$$\delta_{t,n} = \begin{cases} 1, & \theta_{t,n}^{L} = 0, \\ 1/\theta_{t,n}^{L}, & \text{else}, \end{cases} \tag{11.59}$$

where $\theta_{t,n}^{L}$ is iteratively computed via

$$\theta_{t,n}^{l} = \sum_{j \in \mathcal{N}_n} \pi_{n,j} \theta_{t,j}^{l-1}, \tag{11.60}$$

with the initialization value of $\theta_{t,n}^{0}$ being

$$\theta_{t,n}^{0} = \begin{cases} 1, & \text{for } n \in \mathcal{S}, \\ 0, & \text{for } n \in \mathcal{C}, \end{cases}$$

Substituting $\delta_{t,n}$ in (11.59) into (11.58) will solve $p_{t,n}(\boldsymbol{x}_t)$, which, clearly, is a Gaussian random variable with the parameters being given by

$$\boldsymbol{P}_{t|t,n}^{k} = (\boldsymbol{\Gamma}_{t|t,n}^{k})^{-1}, \tag{11.61}$$

$$\boldsymbol{x}_{t|t,n}^{k} = \boldsymbol{P}_{t|t,n}^{k} \boldsymbol{\gamma}_{t|t,n}^{k}, \tag{11.62}$$

where $\boldsymbol{\Gamma}_{t|t,n}^{k}$ and $\boldsymbol{\gamma}_{t|t,n}^{k}$ are, respectively, computed by

$$\boldsymbol{\Gamma}_{t|t,n}^{k} = \boldsymbol{\Gamma}_{t|t-1,n}^{k,L} + \delta_{t,n} \boldsymbol{I}_{t,n}^{k,L}. \tag{11.63}$$

$$\boldsymbol{\gamma}_{t|t,n}^{k} = \boldsymbol{\gamma}_{t|t-1,n}^{k,L} + \delta_{t,n}\boldsymbol{i}_{t,n}^{k,L}. \tag{11.64}$$

After updating the state via (11.63) and (11.64), the $(k+1)$th step of the VB iteration can proceed. For clarity, the resulting dRCIF for target tracking is summarized in Algorithm 11.2. As can be seen in Algorithm 11.2, the dRCIF comprises three main components: achieving the consensus on prior, achieving the consensus on likelihood, and the VB iterations. The latter two components are coupled, with the consensus on likelihood specifically integrated into the VB iterations. The computational complexity of the VB iterations depends on K, while that of the consensus on prior or likelihood depends on L. Therefore, the computational complexity of the dRCIF is approximately $\mathcal{O}(\zeta_2(L)) + \mathcal{O}(\varsigma_1(K))\mathcal{O}(\varsigma_2(L))$, where $\varsigma_1(\cdot)$ and $\varsigma_2(\cdot)$ are functions relative to their arguments. In addition, during the consensus step (either on prior or on likelihood), the precise matrix (of dimensions $n \times n$) and the information vector (of dimensions $n \times 1$) are shared. Therefore, considering that the precise matrix is symmetric, $K(n^2 + 3n)/2$ real numbers are required to broadcast from a node to its neighbors.

Algorithm 11.2 Decentralized robust CIF (dRCIF) for target tracking

Require: $\mathcal{Y}_{1:T}$, $\hat{\boldsymbol{x}}_{0|0}$, $P_{0|0}$, $Q_{1:T}$, $R_{1:T}$, and L.
for $t = 1 : T$ **do**
  Calculate $\{\hat{\boldsymbol{x}}_{t|t-1,n}, P_{t|t-1,n}\}$ and $\{\boldsymbol{\gamma}_{t|t-1,n}, \boldsymbol{\Gamma}_{t|t-1,n}\}$ via (11.9)–(11.11) for $n \in \mathcal{D}$.
  Calculate $H_{t,n}$ via (11.12) for $n \in \mathcal{S}$.
  for $l = 1 : L$ **do**
    Achieving consensus on prior via (11.48) and (11.49) for $n \in \mathcal{D}$;
  end for
  Initialize $k = 0$, $e_{t,n}^k, f_{t,n}^k$ and $\langle z_{t,n}^k \rangle = 1$ for $n \in \mathcal{S}$.
  for $k = 1 : K$ **do**
    Compute $I_{t,n}^k$ and $i_{t,n}^k$ via (11.27) with $\langle z_{t,n}^{k-1} \rangle$ for $n \in \mathcal{S}$;
    Set $I_{t,n}^k = 0$ and $i_{t,n}^k = 0$ for $n \in \mathcal{C}$;
    for $l = 1 : L$ **do**
      Achieving consensus on likelihood as (11.56) and (11.57) for $n \in \mathcal{D}$;
    end for
    Calculate the parameter $\delta_{t,n}$ for $n \in \mathcal{D}$;
    Update $\boldsymbol{\Gamma}_{t|t,n}^k$ and $\boldsymbol{\gamma}_{t|t,n}^k$ via (11.63) and (11.64) for $n \in \mathcal{D}$;
    Compute $P_{t|t,n}^k$ and $x_{t|t,n}^k$ for $n \in \mathcal{S}$;
    Update $\langle z_{t,n}^k \rangle$ via (11.34), $e_{t,n}^k$ via (11.36), and $f_{t,n}^k$ via (11.37) for $n \in \mathcal{S}$;
  end for
  Output $P_{t|t,n} = (\boldsymbol{\Gamma}_{t|t,n}^K)^{-1}$, $\hat{\boldsymbol{x}}_{t|t,n} = P_{t|t,n}\boldsymbol{\gamma}_{t|t,n}^K$ for $n \in \mathcal{D}$,
end for

11.5 Numerical Examples

In this section, we consider a scenario where a maneuvering target is tracked by a networked sensing system. The networked system, the topology of which is illustrated in Figure 11.3, has 100 nodes consisting of 5 active sensor nodes, 10 passive sensor nodes, and 85 communication nodes.

We consider a coordinated turning model with an unknown turning rate to model the moving target, i.e.

$$\boldsymbol{x}_{t+1} = \begin{pmatrix} 1 & \frac{\sin(\omega_t T_s)}{\omega_t} & 0 & \frac{\cos(\omega_t T_s)-1}{\omega_t} & 0 \\ 0 & \cos(\omega_t T_s) & 0 & -\sin(\omega_t T_s) & 0 \\ 0 & \frac{1-\cos(\omega_t T_s)}{\omega_t} & 1 & \frac{\sin(\omega_t T_s)}{\omega_t} & 0 \\ 0 & \sin(\omega_t T_s) & 0 & \cos(\omega_t T_s) & 0 \\ 0 & 0 & 0 & 0 & 1 \end{pmatrix} \boldsymbol{x}_t + \boldsymbol{v}_t, \tag{11.65}$$

where the state $\boldsymbol{x}_t$ is defined as $[x_t, \dot{x}_t, y_t, \dot{y}_t, \omega_t]^T$, including the position (x_t, y_t), the corresponding velocities $(\dot{x}_t, \dot{y}_t)$, and the turning rate ω_t; T_s is the sampling time which is set to 1s; and $\boldsymbol{v}_t \sim \mathbb{N}(0, \boldsymbol{Q}_t)$, where

$$\boldsymbol{Q}_t = \begin{pmatrix} \phi_1 \overline{\boldsymbol{Q}} & 0 & 0 \\ 0 & \phi_1 \overline{\boldsymbol{Q}} & 0 \\ 0 & 0 & \phi_2 \end{pmatrix}, \overline{\boldsymbol{Q}} = \begin{pmatrix} T_s^3/3 & T_s^2/2 \\ T_s^2/2 & T_s \end{pmatrix}, \tag{11.66}$$

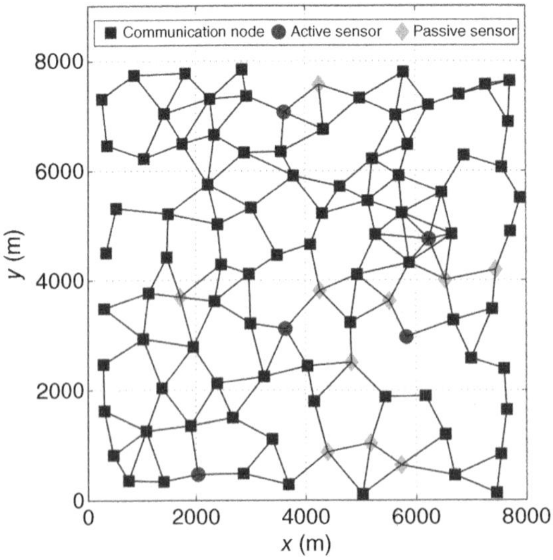

Figure 11.3 The topology of the considered SN.

in which $\phi_1 = 10^{-1}$ and $\phi_2 = 1.75 \times 10^{-4}$. In the simulation, the trajectory of the target is generated by the dynamics (11.65) with the initial state randomly selecting from $\boldsymbol{x}_0 \sim \mathbb{N}(\boldsymbol{\mu}, \boldsymbol{\Sigma})$where

$$\boldsymbol{\mu} = [10^3\ \text{m}, 5 \times 10\ \text{m/s}, 2 \times 10^3\ \text{m}, -5 \times 10\ \text{m/s}, 5.3 \times 10^{-2}\ \text{rad/s}]^T.$$
$$\boldsymbol{\Sigma} = \text{diag}\left([10^4, 10^2, 10^4, 10^2, 3.04 \times 10^{-6}]\right).$$

Denote the location of the sth sensor node by (p_x^s, p_y^s). There are two types of sensors in the networked system, i.e. the active sensors and the passive sensors. The active sensor can provide both the range and bearing measurements, and its measurement mapping is given by

$$\boldsymbol{y}_t^s = \begin{bmatrix} \sqrt{(x_t - p_x^s)^2 + (y_t - p_y^s)^2} \\ \text{atan2}(y_t - p_y^s, x_t - p_x^s) \end{bmatrix} + \boldsymbol{w}_t^s, \tag{11.67}$$

where atan2 denotes the four-quadrant inverse tangent function, and $\boldsymbol{\omega}_t^s$ is the measurement noise. In contrast, the passive sensor only provides the bearing of the target, i.e.

$$\boldsymbol{y}_t^s = \text{atan2}(y_t - p_y^s, x_t - p_x^s) + \boldsymbol{w}_t^s. \tag{11.68}$$

In the simulation, the measurements may be contaminated by outliers, and hence we utilize the Gaussian mixture model for sensor nodes:

$$\boldsymbol{w}_t^s \sim (1 - \lambda)\mathbb{N}(0, \boldsymbol{R}_t^s) + \lambda\mathbb{N}(0, \alpha\boldsymbol{R}_t^s), \tag{11.69}$$

where λ controls the probability of outliers, α are parameters to describe the power of outliers, and $\boldsymbol{R}_t$ is given by

$$\boldsymbol{R}_t = \begin{cases} \text{diag}[10^2, 1.22 \times 10^{-5}], & \text{for the active sensor,} \\ 1.22 \times 10^{-5}, & \text{for the passive sensor.} \end{cases}$$

In addition, the simulation length T is set to 50, and $e_{t,s}^0 = 0.9$ and $f_{t,s}^0 = 0.1$.

To demonstrate the performance of the proposed cRCIF and the dRCIF, we compare with the following existing solutions:

- **The clairvoyant centralized CIF (cCIF-t)**: In this approach, the exact knowledge of the measurement noise model (11.69) is known at the fusion center;
- **The clairvoyant decentralized CIF (dCIF-t)**: In this method, each node is assumed to know the exact measurement noise model (11.69);
- **The robust decentralized CIF based on a student's t distribution (dTCIF) [Dong et al., 2018]**: In this algorithm, we set the parameters as recommended in Dong et al. [2018];
- **The robust decentralized CIF based on the IMM (dIMMCIF) [Battistelli et al., 2015]**: Two models are considered in the dIMMCIF. In the first model, $\boldsymbol{w}_t^s$ is considered as $\mathbb{N}(0, \boldsymbol{R}_t^s)$, while it is set to $\mathbb{N}(0, \alpha\boldsymbol{R}_t^s)$ in the second model.

The probability transition matrix in the dIMMCIF is [0.9, 0.1; 0.9, 0.1]; and the initial weights of these two models are 0.9 and 0.1, respectively.

The root mean-square error (RMSE) and the time-averaged RMSE (TRMSE) of the target position are considered as the performance metrics, and these metrics are obtained by $M = 100$ independent Monte Carlo runs. In the centralized target-tracking scenario, the definition of the RMSE of position is given by

$$RMSE_t = \left(\frac{1}{M} \sum_{m=1}^{M} \left\| \boldsymbol{p}_t^{(m)} - \hat{\boldsymbol{p}}_t^{(m)} \right\|^2 \right)^{1/2}, \tag{11.70}$$

where $\boldsymbol{p}_t^{(m)}$ is the actual position of the target at the mth Monte Carlo run, while $\hat{\boldsymbol{p}}_t^{(m)}$ is the estimated one. For the decentralized tracking scenario, the average RMSE over the entire nodes is considered, i.e.

$$RMSE_t = \left(\frac{1}{NM} \sum_{m=1}^{M} \sum_{n=1}^{N} \left\| \boldsymbol{p}_t^{(m)} - \hat{\boldsymbol{p}}_{t,n}^{(m)} \right\|_2^2 \right)^{1/2}, \tag{11.71}$$

where $\hat{\boldsymbol{p}}_{t,n}^{(m)}$ is the estimated target position of the nth node. The TRMSE of the position is given by

$$TRMSE = \frac{1}{T} \sum_{t=1}^{T} RMSE_t \tag{11.72}$$

First of all, we test how K (i.e. iteration number) of the VB affects the proposed methods. Figure 11.4 illustrates the TRMSEs of the proposed robust tracking methods as a function of K in the scenario that $\lambda = 0.4$ and $\alpha = 100$. The consensus step of the dRCIF is set to $L = 5$, which we will discuss later. The results indicate that our methods achieve a stable estimate after two or three VB iterations. Therefore, we set the default number of the VB iterations in our methods to 3.

Table 11.1 presents the TRMSEs of decentralized algorithms with different consensus steps. Since the dIMMCIF relies solely on consensus on likelihood strategy, it necessitates more steps to achieve a reliable estimate. As expected,

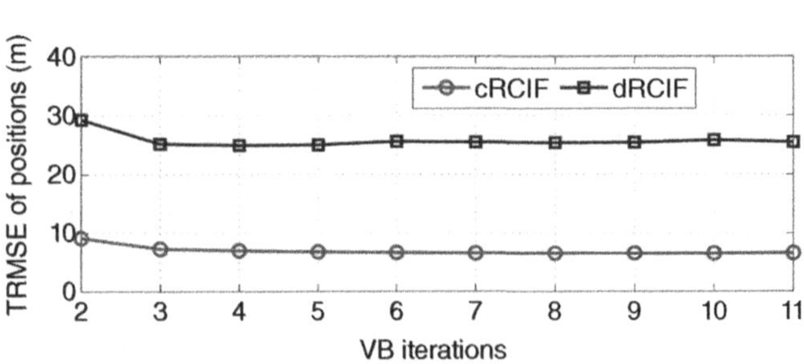

Figure 11.4 TRMSE of the position of the proposed methods when $\lambda = 0.4$ and $\alpha = 100$.

Table 11.1 TRMSE of position for different algorithms with different consensus steps.

	$L = 1$	$L = 2$	$L = 3$	$L = 4$	$L = 5$
dRCIF	29.17	19.07	14.58	12.67	11.51
dTCIF	39.53	28.17	21.44	17.90	16.22
dCIF-t	27.41	17.01	12.26	10.65	9.57
	$L = 6$	$L = 7$	$L = 8$	$L = 9$	$L = 10$
dIMMCIF	39.53	29.0	21.26	20.43	19.49

performance improves with the increased consensus steps across all decentralized fusion algorithms. Notably, our proposed dRCIF exhibits the smallest deviation from the benchmark solution dCIF-t, followed by the dTCIF. Despite the dIMMCIF requiring more steps, it performs the poorest. In the following we set $L = 10$ for the dIMMCIF, while $L = 5$ for the others.

Finally, we investigate the impact of λ and α on the proposed solutions. Figure 11.5 displays the position TRMSEs of various target-tracking algorithms for varying λ from 0.05 to 0.5, with $\alpha = 100$. In Figure 11.6, we present similar results while fixing $\lambda = 0.2$ and varying α. From these figures, it is apparent that the TRMSEs generally increase with λ for all algorithms, while all

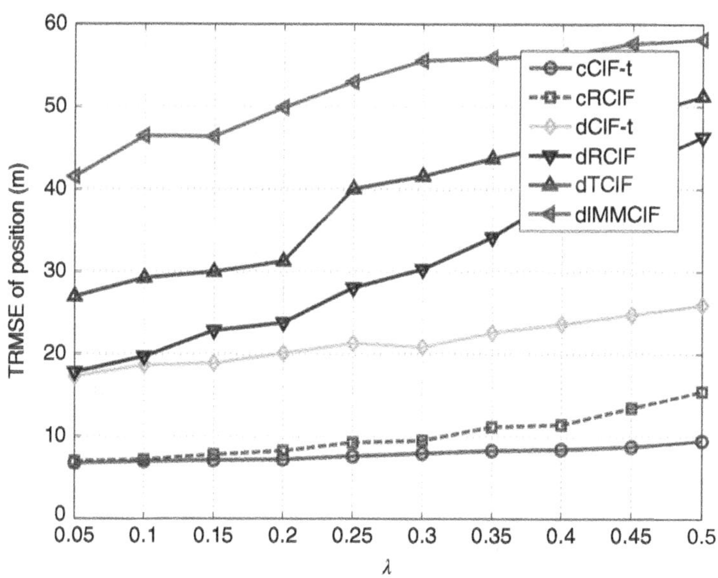

Figure 11.5 TRMSEs of position for the respective methods versus λ when $\alpha = 100$.

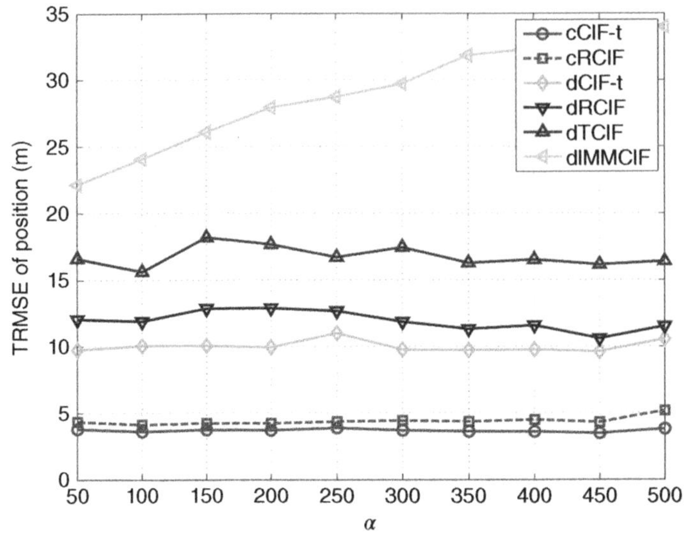

Figure 11.6 TRMSEs of position for the respective methods versus α when $\lambda = 0.2$.

except the dIMMCIF are nearly unaffected by the growth of α. This suggests that these algorithms are responsive to changes in the contamination ratio but are comparatively less affected by the intensity of the contaminating noise.

11.6 Conclusion

In this chapter, we addressed the robust target-tracking problem in networked systems where measurements may be affected by outliers. We expanded the traditional measurement model to incorporate potential outliers using a binary variable that distinguishes between nominal readings and outliers. Additionally, we introduced a beta-Bernoulli prior for the binary indicator, facilitating simultaneous target tracking and indicator estimation within a VB framework. Moreover, we devised a decentralized target-tracking approach by integrating consensus averaging into the VB iteration process. Simulation results demonstrated that our proposed solutions yield superior performance compared to existing robust approaches.

Bibliography

W. Aftab and L. Mihaylova. A learning Gaussian process approach for maneuvering target tracking and smoothing. *IEEE Transactions on Aerospace and Electronic Systems*, 57(1):278–292, 2020.

G. Battistelli, L. Chisci, and C. Fantacci. Parallel consensus on likelihoods and priors for networked nonlinear filtering. *IEEE Signal Processing Letters*, 21(7):787–791, 2014a.

G. Battistelli, L. Chisci, G. Mugnai, A. Farina, and A. Graziano. Consensus-based linear and nonlinear filtering. *IEEE Transactions on Automatic Control*, 60(5):1410–1415, 2014b.

G. Battistelli, L. Chisci, C. Fantacci, A. Farina, and A. Graziano. Consensus-based multiple-model Bayesian filtering for distributed tracking. *IET Radar, Sonar & Navigation*, 9(4):401–410, 2015.

Q. Chen, W. Wang, C. Yin, X. Jin, and J. Zhou. Distributed cubature information filtering based on weighted average consensus. *Neurocomputing*, 243:115–124, 2017.

P. Dong, Z. Jing, H. Leung, K. Shen, and M. Li. Robust consensus nonlinear information filter for distributed sensor networks with measurement outliers. *IEEE Transactions on Cybernetics*, 49(10):3731–3743, 2018.

Q. Ge, T. Shao, Q. Yang, X. Shen, and C. Wen. Multisensor nonlinear fusion methods based on adaptive ensemble fifth-degree iterated cubature information filter for biomechatronics. *IEEE Transactions on Systems, Man, and Cybernetics: Systems*, 46(7):912–925, 2016.

D.-J. Lee. Nonlinear estimation and multiple sensor fusion using unscented information filtering. *IEEE Signal Processing Letters*, 15:861–864, 2008.

W. Li and Y. Jia. Distributed consensus filtering for discrete-time nonlinear systems with non-Gaussian noise. *Signal Processing*, 92(10):2464–2470, 2012.

X. R. Li and V. P. Jilkov. Survey of maneuvering target tracking. Part I. Dynamic models. *IEEE Transactions on Aerospace and Electronic Systems*, 39(4):1333–1364, 2003.

R. Olfati-Saber and J. S. Shamma. Consensus filters for sensor networks and distributed sensor fusion. In *Proceedings of the 44th IEEE Conference on Decision and Control*, pages 6698–6703. IEEE, 2005.

K. Pakki, B. Chandra, D.-W. Gu, and I. Postlethwaite. Cubature information filter and its applications. In *Proceedings of the 2011 American Control Conference*, pages 3609–3614. IEEE, 2011.

É. L. Souza, E. F. Nakamura, and R. W. Pazzi. Target tracking for sensor networks: A survey. *ACM Computing Surveys (CSUR)*, 49(2):1–31, 2016.

Y. Tian, Z. Chen, and F. Yin. Distributed IMM-unscented Kalman filter for speaker tracking in microphone array networks. *IEEE/ACM Transactions on Audio, Speech and Language Processing*, 23(10):1637–1647, 2015.

D. G. Tzikas, A. C. Likas, and N. P. Galatsanos. The variational approximation for Bayesian inference. *IEEE Signal Processing Magazine*, 25(6):131–146, 2008.

H. Wang, H. Li, J. Fang, and H. Wang. Robust Gaussian Kalman filter with outlier detection. *IEEE Signal Processing Letters*, 25(8):1236–1240, 2018.

H. Zhu, H. Leung, and Z. He. A variational Bayesian approach to robust sensor fusion based on Student-t distribution. *Information Sciences*, 221:201–214, 2013.

12

A Federated Prototype-Based Model for IoT Systems: A Study Case for Leakage Detection in a Real Water Distribution Network

Diego P. Sousa[1], José M. B. da Silva Jr[2], Charles C. Cavalcante[1], and Carlo Fischione[3]

[1] *Department of Teleinformatics Engineering, Federal University of Ceara, Fortaleza, Ceara, Brazil*
[2] *Department of Information Technology, Uppsala University, Uppsala, Sweden*
[3] *School of Electrical Engineering and Computer Science and Digital Futures Research Center, KTH Royal Institute of Technology, Stockholm, Sweden*

12.1 Introduction

With the goal of minimizing the loss of resources, the early detection of anomalies is a relevant issue in smart cities (SCs). Among different challenges in monitored environments, the efficient management of water [Kulkarni and Farnham, 2016], energy [Ullah et al., 2017], and air pollution [Bacco et al., 2017] constitute valuable examples of attractive solutions concerning the development of SCs.

In the context of water management for leak prevention, the authors Zaman et al. [2020] have divided the leak management strategies into before pipeline operation (hydraulic model assessment and structural methods) and after pipeline operation (leak assessment, leak prevention, and leak detection). Moreover, the study by Gupta and Kulat [2018] has evidenced that research on detecting leaks in water distribution networks (WDNs) has been continuously conducted for over two decades.

Furthermore, as detailed by Li et al. [2015], the leak detection solutions are classified into hardware solutions (acoustic and non-acoustic) and software solutions (numerical and non-numerical modeling). Regarding the hardware solutions, the works by Senin et al. [2019] and Moubarak et al. [2011] have extensively studied hardware-based strategies for water leakage detection, including ground radar and acoustic solutions, respectively. In regard to software solutions, the authors Chan et al. [2018] have reviewed high-powered solutions grounded on non-numerical modeling, such as support vector machines (SVMs) and convolutional neural networks (CNNs).

Wireless Sensor Networks in Smart Environments: Enabling Digitalization from Fundamentals to Advanced Solutions, First Edition. Edited by Domenico Ciuonzo and Pierluigi Salvo Rossi.

Despite the benefits obtained when using SVMs or CNNs, the study by Villmann et al. [2017] has presented a comprehensive overview of prototype-based models (PBMs) and highlighted their potential for producing more interpretative results compared to nonlinear solutions. To this end, in this work, we explore the principles of PBMs rather than the ones from the more complex, but less interpretative, aforementioned approaches.

In addition to the machine learning strategies context, the recent studies of Soldevila et al. [2017] and Xing and Sela [2019] have confirmed the efficiency of leakage detection modeling based on pressure analysis and machine-learning techniques. Moreover, the authors Wan et al. [2022] have summarized the state-of-the-art on pressure and flow WDN monitoring by smart sensor usage. Among the listed data-driven studies, most of the proposed strategies have adopted the frequency (in observations per minute) in the range from 1 to 15 minutes.

On the concern of data security for SCs solutions, the recent survey of Moubayed et al. [2021] has summarized the state-of-the-art on water leakage detection strategies and suggested potential research opportunities, including federated learning (FL). FL is an emerging machine learning technique enabling multiple devices to train a global model collaboratively without sharing their private data [Verbraeken et al., 2020]. Therefore, FL is a candidate solution to address privacy concerns associated with centralized machine learning.

Inspired by those reported successful cases, our objective is to propose an efficient and low-complexity distributed solution for identifying potential leaks in WDNs in municipal areas while ensuring the privacy of the hydraulic data. Therefore, our solution is a federated modeling for detecting leakage in WDNs by analyzing observed water pressure and flow data using low-complexity learning strategies.

We extend our previous work Sousa et al. [2023], which has analyzed the proposed study case using a centralized learning approach and only processed the observed water pressure measurements. We further extend the previous analysis by including flow measurements and, mainly, proposing a distributed approach.

Furthermore, we propose a solution that is independent of hydraulic modeling and that focuses exclusively on the observed hydraulic measurements. As a software-based approach, our proposition is applicable to leakage detection setups where the measurement of hydraulic data at multiple locations within the WDN is feasible. In this regard, we analyze water pressure and flow measurements obtained from pumps within district-metered areas (DMAs) in Stockholm, Sweden. To assess the effectiveness of our proposed solution, we concentrate on a specific monitored subarea of the WDN.

Moreover, we create realistic and compact sets by using a reduced number of prototypes for generating representative samples for fault detection/classification of a monitored WDN. Specifically, we first train the prototypes to represent the observed hydraulic data into comprehensible subgroups. In the following, we use the trained prototypes to process operational condition predictions. Finally, we compare the performance between the distributed and centralized approaches.

The remainder of this chapter is structured as follows. We present the formulation of prototype-based learning (PBL) in Section 12.2. We briefly introduce the concepts of federated learning in Section 12.3. Following that, we discuss our proposed formulation for federated prototype-based models (FPBMs) in Section 12.4. We describe our case study in Section 12.5. Finally, we discuss our results and conclusion in Sections 12.6 and 12.7, respectively.

12.2 Prototype-Based Learning

PBL is a type of machine learning that constructs models based on representative examples. As discussed in Kohonen [2013], PBMs are also denominated as competitive learning algorithms within the field of artificial neural networks. Specifically, the term "competitive" is related to the main principle of PBMs, which is the competition among the available reference units (known as prototypes) to represent input data partitions.

Algorithms from the PBL literature include supervised learning models, such as the family of learning vector quantization (LVQ) algorithms [Nova and Estévez, 2014] and unsupervised learning models, such as the Kohonen's self-organizing map (SOM) [Miljković, 2017].

Usually, prototypes are created by selecting a subset of instances that are representative of each class or category in the dataset. Then, the prototypes are updated during a training stage. Finally, the prototypes are used to predict the class (in supervised learning) or cluster (in unsupervised learning) of new instances based on their similarity to the prototypes.

The similarity between instances is typically measured using a dissimilarity metric, such as the Euclidean distance. The prototype that is closest to new instances is used to predict the class label of those instances. Therefore, since it is possible to directly compare input data using prototypes, PBMs are known in machine learning theory due to their potential of explicitly representing observations [Biehl et al., 2016].

In the following, we give some necessary definitions for PBL methods. We denote $\mathcal{A}$ as the finite set of specified a priori R reference units, $\mathcal{A} = \{c_1, c_2, \dots, c_r, \dots, c_R\}$, and each reference unit $c_r \in \{c_1, c_2, \dots, c_R\}$ has an associated reference vector $\mathbf{w}_r \in \mathbb{R}^p$, which represents its position (called receptive field center) in the input space. Hence, the set of prototypes is formulated as $\mathbf{W}_{[p\times R]} = [\mathbf{w}_1, \mathbf{w}_2, \dots, \mathbf{w}_R]$, where the rth column of the matrix of prototypes $\mathbf{W}$

denotes the reference vector $\mathbf{w}_r$. Essentially, the problem of building prototypes consists in finding these vectors $\mathbf{w}_r$ that best represent a group of input data $\mathbf{X}$.

The learning structures are designed through a data mapping (projection) from the input space $\chi \in \mathbb{R}^p$ onto the set of R reference units. Thus, the overall training procedure of competitive algorithms is grounded on the competition of the column vectors of a given matrix $\mathbf{W}$. Here, these reference vectors compete to represent data regions of the data matrix $\mathbf{X}$, formulated as $\mathbf{X}_{[p \times N]} = [\mathbf{x}_1, \mathbf{x}_2, \dots, \mathbf{x}_N]$, with N denoting the total number of input samples, where $R \ll N$.

Specifically, the assignment of an input sample $\mathbf{x} \in \mathbf{X}$ to a prototype originates the competition among the reference units in $\mathcal{A}$, in the sense that they "compete" to get an input sample assigned to them. The update rules of PBL algorithms are based on the winner-takes-all (WTA) principle, in which only the winner(s) prototype(s) is(are) updated, and the losers do not forget what they have already learned.

Let us consider a trained matrix of prototypes $\mathbf{W}$. Then, a useful computational geometry concept is the Voronoi region [Boots et al., 2009]. For each reference unit $i \in \mathcal{A}$, a Voronoi region V_i is the convex area of all samples $\mathbf{x} \in \mathbb{R}^p$ for which $\mathbf{w}_i$ is the nearest reference:

$$V_i = \left\{ \mathbf{x} \in \mathbb{R}^p \mid i = \underset{j=1,\dots,R}{\arg\min}\, d(\mathbf{x}, \mathbf{w}_j) \right\}. \tag{12.1}$$

Moreover, we define the Voronoi set $\mathcal{R}_i$ as the data in $\mathbf{X}$ for which i is the nearest reference unit. Therefore, the data space is partitioned as

$$\mathbb{R}^p = V_1 \cup V_2 \cup \cdots \cup V_{R-1} \cup V_R \quad \text{and} \quad V_i \cap V_j = \emptyset \quad \text{for} \quad i \neq j, \tag{12.2}$$

where such a partitioning concept is valid for both unsupervised and supervised paradigms.

In the following, we introduce the training procedures for the unsupervised and supervised learning.

12.2.1 Unsupervised Learning

Since the PBL training procedures are established on iterative learning by means of the reference units' competition, we initiate by defining the concept of iteration. In PBL, we denote as iteration a single stimulus provoked over the set of reference units when we present an input sample $\mathbf{x}$ to this set. Hence, for a given tth iteration, the competition is based on the following decision criterion:

$$c_r(\mathbf{x}(t), \mathbf{w}_r(t)) = \underset{i=1,\dots,R}{\arg\min}\, d(\mathbf{x}(t), \mathbf{w}_i(t)), \tag{12.3}$$

in which $d(\cdot;\cdot)$ denotes a dissimilarity measure specific to the PBL algorithm used, and $c_r(\cdot) : \mathbb{R}^p \times \mathbb{R}^p \to \mathbb{R}$ is the reference unit of the nearest (known as winner) prototype among the R available.

Algorithm 12.1: Winner-takes-all

initialize $\mathbf{W}, E, N, \eta_0$

$i \leftarrow 0$ (ith iteration), $\quad I \leftarrow E \times N$ (number of iterations)

for *each epoch* $e = 1, \ldots, E$ **do**

  $\mathbf{x} \leftarrow$ randomPermutationSamples(N)

  for *each sample* $n = 1, \ldots, N$ **do**

    $i \leftarrow i + 1, \quad \eta(i) = \eta_0(\frac{\eta_I}{\eta_0})^{i/I}$

    $c_r(i) \leftarrow \arg\min_{j=1,\ldots,R} \| \mathbf{X}_{(:,\mathbf{x}(n))} - \mathbf{W}_{(:,j)}(i) \|$

    $\mathbf{W}_{(:,r)}(i+1) \leftarrow \mathbf{W}_{(:,r)}(i) + \eta(i)[\mathbf{X}_{(:,\mathbf{x}(n))} - \mathbf{W}_{(:,r)}(i)]$

  end

end

return W

In general, unsupervised PBL algorithms are extensions of the WTA learning rule [Kohonen, 1990a]:

$$\begin{aligned} \mathbf{w}_r(t+1) &= \mathbf{w}_r(t) + \eta(t)[\mathbf{x}(t) - \mathbf{w}_r(t)], \\ \mathbf{w}_i(t+1) &= \mathbf{w}_i(t), \qquad \text{if } i \neq r, \end{aligned} \tag{12.4}$$

where $0 < \eta(t) < 1$ is the learning rate.

The WTA algorithm is described in Algorithm 12.1. We denote $\mathbf{X}_{(:,n)}$ as the nth column vector of $\mathbf{X}$, E is the number of epochs, I is the number of iterations, and η is the decaying learning rate, where the initial and final rates are η_0 and η_I, respectively.

Some relevant unsupervised PBL algorithms are listed below respecting their chronological order. We first outline the WTA algorithm [Kohonen, 1988], which does not have a neighborhood function to update more than one winner prototype per training iteration. The following two, frequency-sensitive competitive learning (FSCL) [Ahalt et al., 1990] and rival penalized competitive learning (RPCL) [Xu et al., 1993] present improvements to reduce the occurrence of underutilized reference units. Finally, the SOM [Kohonen, 1990b], neural-gas (NG) [Martinetz and Schulten, 1991], and growing NG [Fritzke, 1994] include improvements by proposing neighborhood functions to preserve the existent geometry of the reference units.

12.2.2 Supervised Learning

Let us consider a set of training input–output samples $\{(\mathbf{x}(t), y(t))\}_{t=1}^{N}$, where $\mathbf{x}(t) \in \mathbb{R}^p$ denotes the tth input sample, and $y(t) \in C$ denotes its corresponding class label. Note that $y(t)$ is a categorical variable, which assumes only one out of L values in the finite set $C = \{c_1, c_2, \dots, c_L\}$. For the supervised PBL algorithms, we have $R > L$, i.e. the number of prototypes (R) is higher than the number of classes (L). As a consequence, different prototypes may share the same label.

For a given tth iteration, the class assignment for a new input sample $\mathbf{x}(t)$ is based on the following decision criterion:

$$\hat{y}(t) = \text{Class of } \mathbf{w}_r(t), \quad \text{where} \quad r = \arg\min_{i=1,\dots,R} d(\mathbf{x}(t), \mathbf{w}_i(t)), \tag{12.5}$$

in which $d(\cdot,\cdot)$ denotes a dissimilarity measure specific to the supervised PBL algorithm, and r is the index of the nearest prototype among the R available. Concerning the supervised PBL algorithms, the learning rules are extensions of the LVQ1 classifier [Kohonen, 1990a]:

$$\begin{aligned} \mathbf{w}_r(t+1) &= \begin{cases} \mathbf{w}_r(t) + \eta(t)[\mathbf{x}(t) - \mathbf{w}_r], & \text{if } \hat{y}(t) = y(t), \\ \mathbf{w}_r(t) - \eta(t)[\mathbf{x}(t) - \mathbf{w}_r], & \text{otherwise.} \end{cases} \\ \mathbf{w}_i(t+1) &= \mathbf{w}_i(t), \quad \text{if } i \neq r. \end{aligned} \tag{12.6}$$

Supervised PBL algorithms are known in the literature as LVQ classifiers, where the first contribution (LVQ1) was proposed by Kohonen [1988]. Since the LVQ1 does not have a cost function that ensures convergence, diverse variants have been produced to improve the original proposition, including LVQ2.1 and LVQ3 [Kohonen, 1990a], with improvements to obtain higher convergence speed; RLVQ [Bojer et al., 2001], the pioneer on using a distance learning approach; and GLVQ [Sato and Yamada, 1996], the first one to introduce a cost function. For an in-depth understating of the taxonomy of the LVQ variants concerning their chronological order, we recommend the study by Nova and Estévez [2014].

12.3 Federated Learning

Federated learning is a machine learning technique that allows multiple devices to collaboratively train a common global model without sharing their data with each other or a central server. The goal of FL methods is to enable training of machine learning models across multiple decentralized devices while preserving privacy and reducing communication costs [Li et al., 2020]. Therefore, FL has a number of benefits, including preserving the privacy of users' data, reducing communication costs, and improving model accuracy by leveraging the diversity of data across multiple devices. It is especially useful in scenarios where data cannot be centralized due to privacy or regulatory concerns, such as healthcare [Xu et al., 2021] or finance [Long et al., 2020].

In FL, each kth device from $k = 1, \dots, K$, such as a smartphone or a smart sensor, has its own data that it uses to train the model locally. The local models are then aggregated to create a global model that is more accurate across all classes than the individual models, while preserving the privacy of each device's data. The process of FL involves the following steps at a tth global iteration:

1) **Initialization:** A central server initializes the model and sends it to each participating device $k \in S_t$, in which S_t denotes a random set of m selected clients. The cardinality of S_t is in the range from 1 until K;
2) **Local training:** Each participating device trains the model on its own data using its own resources during E local epochs;
3) **Model aggregation:** The local models are sent back to the central server, where they are aggregated to create a more accurate global model;
4) **Update and repeat:** The updated global model is sent back to each device, and the process is repeated until convergence.

The first FL method proposed in the literature was federated averaging (FedAvg) [McMahan et al., 2017]. Since then, several extensions and contributions to the original FedAvg method have been proposed in the literature, including FL using differential privacy [Wei et al., 2020], fair FL [Li et al., 2019a], and convergence analysis for non-iid data [Li et al., 2019b].

12.4 Federated Prototype-Based Models

FPBMs constitute a class of machine learning algorithms that are used for decentralized training of models [Brinkrolf and Hammer, 2021]. For simplicity, it is a combination of FL and PBL techniques.

As discussed in Section 12.2, PBL is a machine learning paradigm that reduces the redundancy of the data by representing considerable amounts of data into limited quantity of reference units. In FPBMs, it is used to minimize the required hyperparameters to be sent to the central server.

As discussed in Section 12.3, FL is a machine learning technique, where the model is trained on a decentralized network of devices, without requiring the data to be transferred to a central server. Instead, each device trains the model on its own data and sends only the model updates back to the central server. This helps to ensure privacy and security of the data.

On the concern of unsupervised FPBMs, the authors Servetnyk et al. [2020] proposed a distributed clustering method based on a federated approach of the SOM network, and the authors Servetnyk and Fung [2022] proposed a distributed clustering method grounded on the distributed k-means++. Moreover, it is valuable to emphasize that there are solutions that consider FL and clustering methods but without prototypes-based modeling, e.g. the work by Kim et al. [2021] introduced dynamic clustering in FL by proposing a generative adversarial network-based clustering algorithm. Within the context of supervised FPBMs, federated learning vector quantization (FLVQ) is a relatively new machine learning paradigm, where the work by Brinkrolf and Hammer [2021] is the first one to use this terminology, and then the work by Vaquet et al. [2022] applied the FLVQ to deal with the concept of drift between nodes.

To the best of our knowledge, there is limited literature available on the FBPM topic. Overall, the limited literature on FPBMs suggests that they have potential applications in decentralized and privacy-preserving machine learning. However, further research is needed to explore the limitations and scalability of the FPBMs for real-world applications. Therefore, this work concentrates on the problem formulation of FPMBs by comprehensively defining an adequate terminology by combining federated learning and prototype-based methods.

In particular, as it is also used in centralized PBL (see Section 12.2), we denote N as total number of input samples, R as the number of reference units, p as the number of features, E is the number of local epochs, and I is the number of local iterations. We mean by local iteration the unitary local model updating step resulting from the iteration between a single input sample and the local model. Hence, a single local epoch comprises N local iterations, and the total number of local epochs demands $E \times N$ iterations. From the federated hyperparameters, we denote K as the number of clients, N_k as the number of input samples in the kth client, $\mathbf{X}_k$ as the input data matrix of the kth client, $\mathbf{W}$ as the global matrix of prototypes, $\mathbf{W}_k$ as the local matrix of prototypes of the kth client, T as the number of global iterations, α is the ratio of device participation, and η is the learning rate.

FPBMs work by first initializing a set of prototypes $\mathbf{W}(t = 0)$. These reference units are used to represent different data regions. The participating devices then train the model $\mathbf{W}(t)$ on their own data $\mathbf{X}_k$ and update the prototypes accordingly. The updated prototypes $\mathbf{W}_k(t + 1)$ are then sent back to the central server, where they are aggregated to form a global model $\mathbf{W}(t + 1)$. This procedure is summarized in Algorithm 12.2 and minimizes the following cost function $C(t)$ formulated by Servetnyk and Fung [2022]:

$$
\begin{aligned}
C(t) = \min_{\mu_{knr}, \mathbf{w}_r} & \sum_{k \in K} \sum_{n \in N_k} \sum_{r \in R} \frac{1}{2} \mu_{knc} \|\mathbf{x}_{kn} - \mathbf{w}_r(t)\|^2 \\
\text{s.t.} & \sum_{k \in K} \mu_{knr} = 1, k \in K, n \in N_k, r \in R, \mu_{knr} \in [0, 1],
\end{aligned}
\tag{12.7}
$$

which denotes the sum of the squared errors (SSEs) obtained of the samples of all participating clients when using the global model $\mathbf{W}(t)$. Moreover, all the PBMs listed in both Section 12.2.1, such as WTA, and Section 12.2.2, such as LVQ1, can be adopted in the function named ClientUpdate in Algorithm 12.2. To contextualize the local training, we present the ClientUpdate function when adopting the WTA learning scheme in Algorithm 12.3.

Algorithm 12.2: Federated prototype-based models

Server executes:
initialize $\mathbf{W}(t = 0)$
for *each round* $t = 1, \dots, T$ **do**
  $m \leftarrow \max(\alpha(t) \times K, 1)$
  $S_t \leftarrow$ random set of m clients
  for *each client* $k \in S_t$ **do**
    $\mathbf{W}_k(t+1) \leftarrow$ **ClientUpdate**$(\mathbf{X}_k, \mathbf{W}(t))$
  end
  $\mathbf{W}(t+1) \leftarrow \sum_{k=1}^{K} \frac{N_k}{N} \mathbf{W}_k(t+1)$
end
return W

Algorithm 12.3: ClientUpdate based on Winner-Takes-All algorithm

$i \leftarrow 0$ (ith local iteration)
$I_k \leftarrow E \times N_k$ (number of local iterations)
Client k receives $\mathbf{W}(t)$ from Server
if $t = 0$ **then**
  initialize $\mathbf{W}_k(i)$
end
else
  $\mathbf{W}_k(i)$=$\mathbf{W}(t)$
end
for *each local epoch* $e = 1, \dots, E$ **do**
  $\mathbf{x} \leftarrow$ randomPermutationSamples(N_k)
  for *each sample* $n = 1, \dots, N_k$ **do**
    $i \leftarrow i + 1$
    $c_r(i) \leftarrow \arg\min_{j=1,\dots,R} \| \mathbf{X}_{k(:,\mathbf{x}(n))} - \mathbf{W}_{k(:,j)}(i) \|$
    $\mathbf{W}_{k(:,r)}(i+1) \leftarrow \mathbf{W}_{k(:,r)}(i) + \eta(i)[\mathbf{X}_{k(:,\mathbf{x}(n))} - \mathbf{W}_{k(:,r)}(i)]$
  end
end
return $\mathbf{W}_k(I_k)$

Furthermore, we illustrate the entire training process of FPBMs in Figure 12.1. Note that we represent full device participation by using continuous lines connecting the devices to the demultiplexer and multiplexer operators located at the input and output of the Server, respectively. However, we can delineate partial participation scenario if we substitute some of those continuous lines to dotted ones. Fundamentally, the dotted lines would denote the inactive devices.

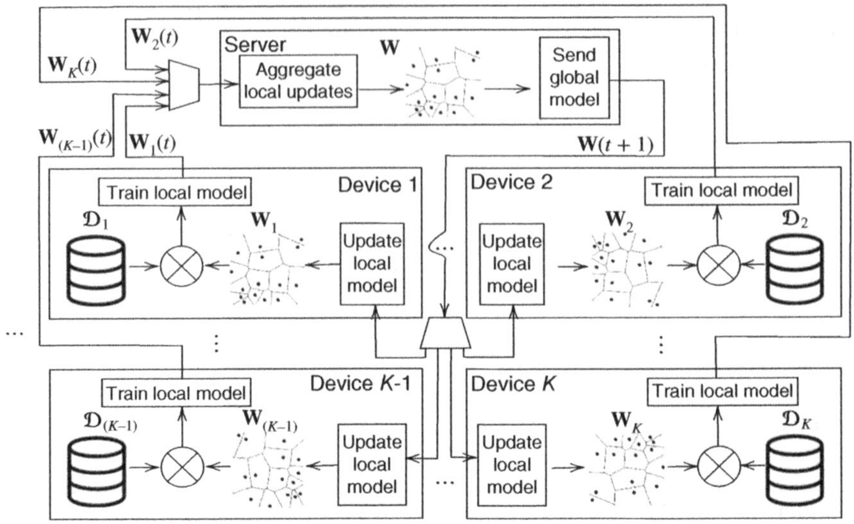

Figure 12.1 Federated learning scenario with K devices and a server. Federated learning scenario with K devices and a server. Note in this representation, we assume full device participation during the training of the global model. Source: The authors.

In the partial participation scenario, we only consider the active devices at a given tth global iteration instead of the K existing. In particular, the active devices are the ones that belong to the set S_t. Hence, only those devices are updated and participate in the training of the global model along the tth global iteration.

12.5 Case Study: Water Distribution Network in Stockholm

In this case study, we consider a set of water pressure and flow observations from the WDN of the municipality of Stockholm, Sweden. The water pressure and flow observations represent the pumping stations of the WDN operating in normal and faulty (presence of leakage) working conditions, where these conditions are distinguished through a maintenance report. In particular, the adopted dataset was gently provided by the water and wastewater company of Stockholm (SVOA, *Stockholm Vatten och Avfall*), and the company has been continuously storing new observations.

12.5.1 Dataset Description

In details, the shared data includes water measurements collected from January 2018 to March 2019. Particularly, we analyze a DMA of the WDN that corresponds to a residential area that has a total of four pumping stations. Moreover, there

Table 12.1 Number of observed days per working condition

	Year		
	2018	**2019**	**Total**
Normal	324	70	394
Leakage	34	20	54
Total	358	90	448

Source: Sousa et al. [2023]/IWA Publishing.

are no tanks or reservoirs in the monitored area and the population of the area is 70,250. As a consequence that leaks in WDNs constitute anomalous working conditions, it is valuable to empathize that most of the observations are categorized as normal working conditions. This imbalance is shown in Table 12.1.

Due to a privacy agreement with SVOA, we neither describe the network architecture nor reveal sensitive information of the DMA. Hence, we generically label these pumping stations as A, H, K, and S.

In the hydraulic dataset, the flow and pressure data are stored for entire days of acquisition with a one minute sampling frequency. Considering the raw database, we denote by *sample* a flow and a pressure time series concatenated, where both signals are stored as a vector of 1440 components for each station and each day. Moreover, there are 7 days with excessive missing values in the time series during the 15-month mentioned period. For simplicity, we remove these non-representative samples from this analysis. Therefore, there are 448 available observed days to build the predictive model, and the total number of samples is denoted by $N = 448$.

Let $\mathbf{p}_{n,i} = [p_{1,n,i}, \dots, p_{1440,n,i}]^T$ and $\mathbf{f}_{n,i} = [f_{1,n,i}, \dots, f_{1440,n,i}]^T$ denote the 1440 pressure and flow measurements from pump $i \in \{A, H, K, S\}$ during the nth observation day for $n = 1, \dots, N$. Then, the sample that represents the nth day is denoted by $\mathbf{S}_n = [\mathbf{p}_{n,A}, \mathbf{f}_{n,A}; \mathbf{p}_{n,H}, \mathbf{f}_{n,H}; \mathbf{p}_{n,K}, \mathbf{f}_{n,K}; \mathbf{p}_{n,S}, \mathbf{f}_{n,S}]$, which has 2880 rows and 4 columns.

Figures 12.2 and 12.3 illustrate the stored pressure and flow time series at each pump station separated by each working condition. From these figures, we note that there is a notable correlation between the normal and faulty (presence of leakage) operation conditions on the selected stations. Therefore, for effective data analysis, the hydraulic data must be treated from raw data to extracted data, where the observations are mapped to a set of representative feature vectors.

Furthermore, a major characteristic of samples collected from distinct sources, such as distinct pumping stations, is that we are supposed to obtain different data distributions at each source. Accordingly, Figures 12.4 and 12.5 illustrate the occurrence of this phenomenon. Consequently, for a successful data analysis, the hydraulic data must be treated in a distributed manner.

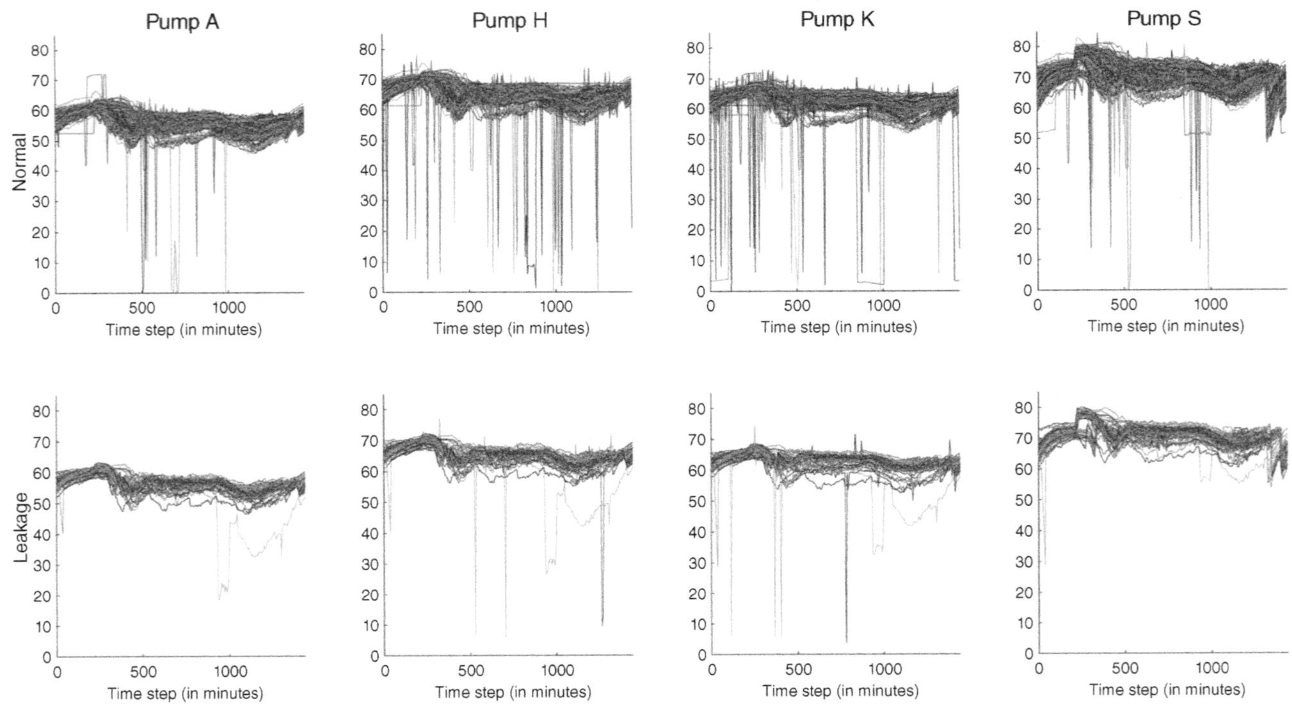

Figure 12.2 Pressure time series. Stored daily water pressure time series for each pump and working conditions (in meters of water column). Source: Sousa et al. [2023]/IWA Publishing.

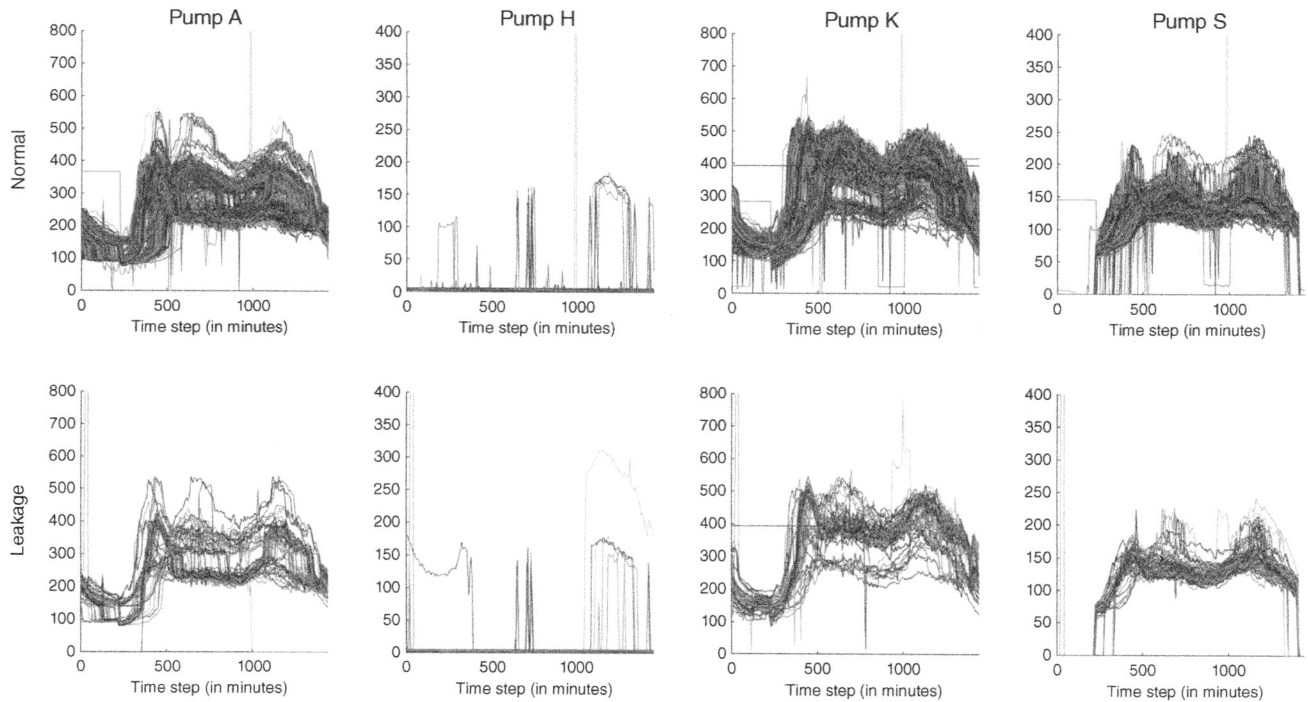

Figure 12.3 Flow time series. Stored daily water flow time series for each pump and working conditions (in m^3/h). Source: The authors.

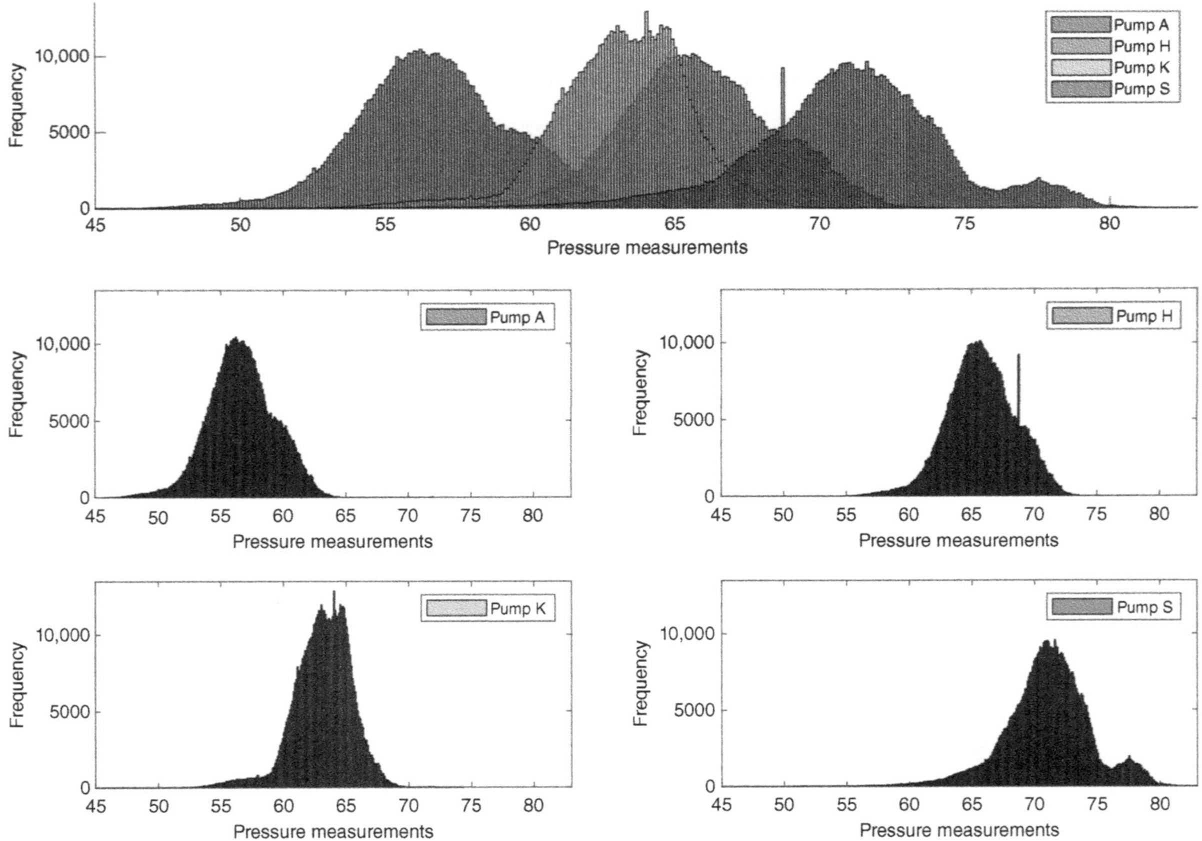

Figure 12.4 Histograms of water pressure measurements. Histograms of the stored water pressure measurements (in meters of water column) at each pump. Source: The authors.

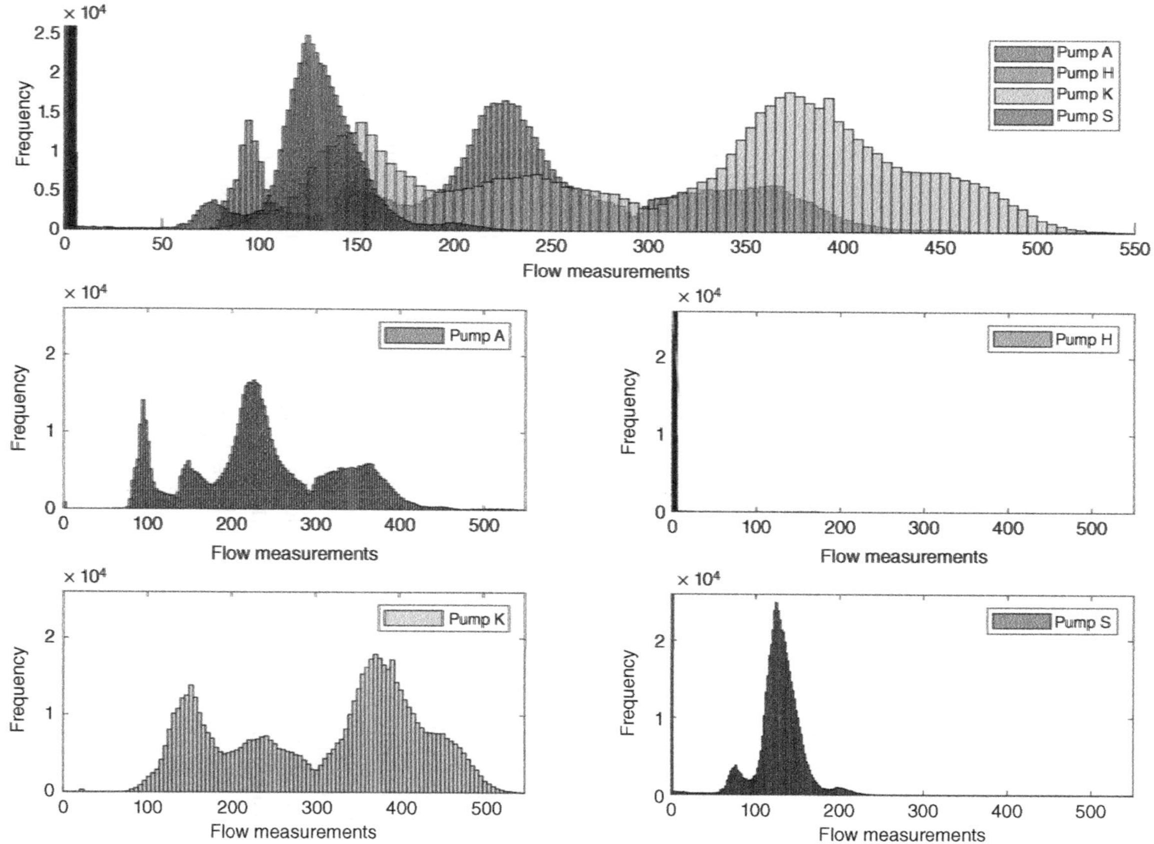

Figure 12.5 Histograms of water flow measurements. Histograms of the stored water flow measurements (in m^3/h) at each pump. Source: The authors.

12.5.2 Feature Extraction

To generate suitable feature vectors that represent the proposed engineering application, we apply canonical discriminant function on the original signals aiming to obtain linear combinations of the interval variables, known as *canonical variables*, that summarize between-class variation. Further explanation of canonical analysis is given in Rencher [1992].

Thus, the procedure for building the feature vectors comprises the following eight steps:

1) Define the acquisition period and the sampling rate.
2) Read the hydraulic, pressure or flow, signals from a selected station.
3) Separate the samples according to their corresponding labels, such as normal and leakage.
4) Calculate the within-group $\mathbf{W}_\mathbf{g}$ and between-group $\mathbf{B}_\mathbf{g}$ scatter matrices (see definition in Rencher [1992]).
5) Find the eigenvector $\mathbf{v}_{1_{[1\times\rho]}}$ associated to the largest eigenvalue of the matrix $\mathbf{W}_\mathbf{g}^{-1}\mathbf{B}_\mathbf{g}$, in which $\rho = 1440$ denotes the number of components of the raw hydraulic signal.
6) Obtain the projected data $\hat{p}$ by applying the inner product between $\mathbf{v}_1$ and the raw data matrix $\mathbf{X}$:

$$\hat{p}_{[1\times N]} = \mathbf{v}_{1_{[1\times\rho]}}\mathbf{X}_{[\rho\times N]}; \tag{12.8}$$

7) Repeat the steps 2–6 for the remaining pump stations.
8) Finally, concatenate every projected data for pumps A, H, K, and S:

$$\mathbf{D} = \left[\hat{p}_A^T|\hat{p}_H^T|\hat{p}_K^T|\hat{p}_S^T\right]. \tag{12.9}$$

In summary, the treated dataset $\mathbf{D}$ is comprised of 448 four-dimensional labeled feature vectors, in which the attribute values represent the canonical values obtained from the most representative canonical function. We execute this procedure to each hydraulic feature, and then we concatenate both treated pressure dataset $\mathbf{D}_p = [\mathbf{D}_p(A)|\mathbf{D}_p(H)|\mathbf{D}_p(K)|\mathbf{D}_p(S)]$ and treated flow dataset $\mathbf{D}_f = [\mathbf{D}_f(A)|\mathbf{D}_f(H)|\mathbf{D}_f(K)|\mathbf{D}_f(S)]$ to generate the treated hydraulic dataset $\mathbf{H}$ as:

$$\mathbf{H}_{[448\times 8]} = \left[\mathbf{D}_f|\mathbf{D}_p\right]. \tag{12.10}$$

12.5.3 Dataset Settings

For the federated framework, the hydraulic features are distributed among the pumping stations, while the centralized framework considers all the features. We hypothesize that the federated framework can further improve the recognition rates, including scenarios in which each pump has only part of the available information.

In this work, we specify the decentralized and centralized datasets as follows. We define each pumping station of the WDN as a device. Moreover, we set the number of features p as two, denoting the canonical values of the pressure and flow of each device. Therefore, we set the number of devices K as four, the number of samples of the kth device N_k as 448 for every device, and the total number of samples as 1792.

Hence, the decentralized and centralized datasets are built as follows. We denote the decentralized datasets as $\mathcal{D}_i$ for $i \in \{A, H, K, S\}$, and the centralized dataset as $\mathcal{D}_C$. According to the treated hydraulic dataset $\mathbf{H}$, we define the decentralized and centralized datasets as

$$\begin{aligned}
&\mathcal{D}_{A_{[448\times2]}} = [\mathbf{D}_f(A)|\mathbf{D}_p(A)], \quad \mathcal{D}_{H_{[448\times2]}} = [\mathbf{D}_f(H)|\mathbf{D}_p(H)],\\
&\mathcal{D}_{K_{[448\times2]}} = [\mathbf{D}_f(K)|\mathbf{D}_p(K)], \quad \mathcal{D}_{S_{[448\times2]}} = [\mathbf{D}_f(S)|\mathbf{D}_p(S)],\\
&\text{and}\, \mathcal{D}_{C_{[1792\times2]}} = [\mathcal{D}_A; \mathcal{D}_H; \mathcal{D}_K; \mathcal{D}_S].
\end{aligned} \tag{12.11}$$

12.6 Results and Discussions

In this section, we evaluate the proposed federated machine learning framework to analyze the real dataset for leakage detection, whose operational conditions are represented as N (normal) and L (leakage). Moreover, we compare the performance of the proposed FPBM algorithm with the results obtained when applying the conventional centralized learning paradigm. Specifically, our proposed FPBM is an extension of the PBM WTA algorithm. Therefore, we represent the standard PBM as centralized WTA and our proposed FPBM as federated WTA.

12.6.1 Numerical Results

We execute 100 independent runs of both centralized and federated clustering algorithms. For each run, three steps of the proposed methodology are carried out: (i) canonical discriminant analysis of the training set for each pumping station, (ii) training of the matrix of prototypes, and (iii) evaluation of the clustering procedure. At the end of each run, the purity rate of each learning framework is obtained. Specifically, we label each cluster according to the class that appears most frequently within it. Consequently, the purity rate denotes the ratio of the correctly matched class and cluster labels to the total of data samples.

We examine our empirical results considering full device participation. It means that all pumping stations participate in the aggregation step. We consider a maximum of $T = 800$ global iterations and then return the global solution $\mathbf{W}(T)$. In addition, we set the total number of local epochs E as 10, the total number of prototypes R as $\frac{\sqrt{N_k}}{2} \approx 10$, and the learning rate of the kth pumping station as $\eta_k(i) = \frac{2}{i \times N_k}$, where $i = 1, \ldots, I_k$ denotes the ith iteration of the WTA along the local training. Note that $I_k = E \times N_k$ denotes the total number of WTA iterations of the kth pumping station on an entire single global iteration.

For all federated experiments, we start by randomly selecting five N samples and five L samples from the kth pumping station to build the initial kth local model, $\mathbf{W}^k(t = 0)$. For the centralized framework, we randomly select those N and L samples from the centralized dataset to build the initial model, $\mathbf{W}(t = 0)$.

12.6.2 Validation of the Canonical Discrimination Function

We begin our analysis by validating the canonical discriminant function, which is the technique we use to extract relevant information from the raw flow and pressure time series. We consider all pressure and flow samples of each pump to generate the treated hydraulic dataset **H**. For this scenario, we obtain the class separation as shown in Figure 12.6. In this figure, the N and L samples are

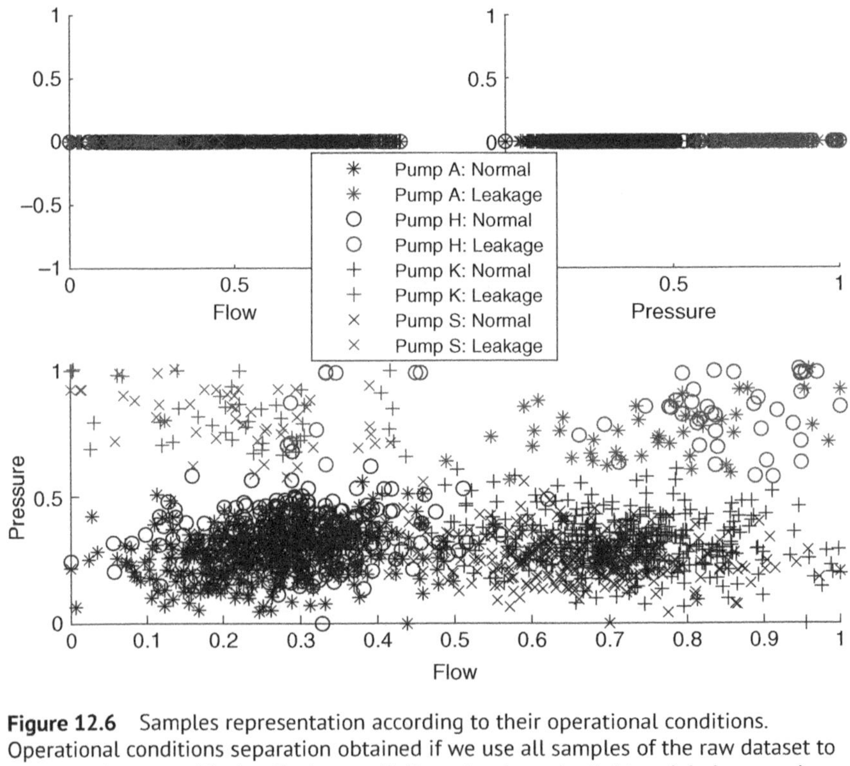

Figure 12.6 Samples representation according to their operational conditions. Operational conditions separation obtained if we use all samples of the raw dataset to generate the treated hydraulic dataset **H**. Note that here the light and dark gray colors represent the normal and leakage samples, respectively, while the four symbols represent the pump to which the sample belongs to. Source: The authors.

represented by the light and dark gray colors. As observed in the scatter plot, the proposed strategy is able to generate explainable separations between the contrasting operational conditions.

12.6.3 Minimization of the Cost Function

Then, we proceed with our analysis by validating the FED-WTA algorithm by the means of their capability to minimize the cost function formulated in Eq. (12.7). The cost functions of the two frameworks along the training for the 100 independent runs are shown in Figure 12.7. From this figure, we note that the centralized WTA is the option that generated the largest dispersion of the MSE among the independent runs, i.e. it is the least reliable. Meanwhile, the federated WTA consistently provides lower values of MSE along the training. Specifically, the centralized

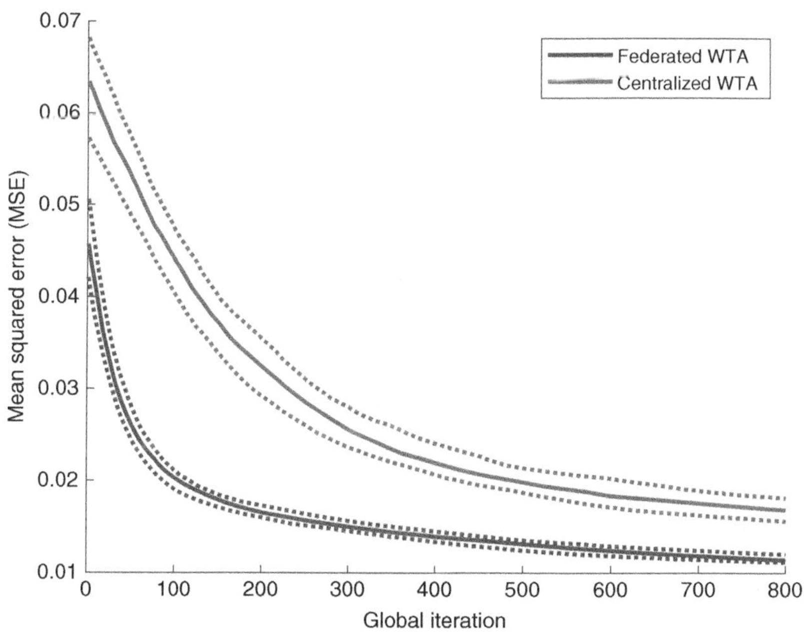

Figure 12.7 MSE obtained along the global iterations. The cost functions obtained from the federated and centralized frameworks along the training for 100 independent runs. The light and dark gray curves represent the federated WTA and centralized WTA cost functions, respectively. In addition, the continuous curves denote the median values of MSE along the training of each framework. Moreover, the dotted curves illustrate the superior and inferior boundaries of the second quartile values of MSE, respectively. Source: The authors.

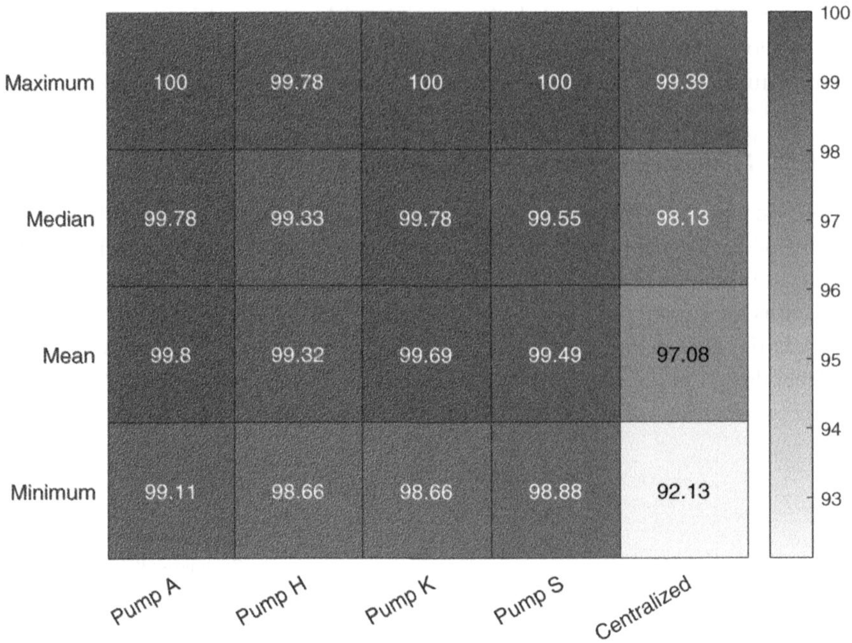

Figure 12.8 Purity rate performance. Purity rate performance obtained by the evaluated clustering paradigms. Source: The authors.

WTA generated a mean variance of 1.4592×10^{-4}, while the FED-WTA generated a mean variance of 3.4618×10^{-5}.

12.6.4 Analysis of the Clustering Performance

The statistical performance of each clustering method is shown in Figure 12.8. A closer look at these metrics reveals a small increase over the maximum purity rates in comparison with the centralized approach, in which every pumping station presents higher performance when using the global model. Specifically, the maximum purity rates improved 0.6% in pumps A, K, and S, and 0.4% in pump H. For the minimum purity rates, we observe a considerable increase in the federated framework, in which it improved 7.6% in pump A, 7.3% in pump S, and 7.1% in pumps H and K. In addition, we verify that for the federated scenario, there is a substantial difference of performance among the participating pumps, where pump A is the one with the best statistical results, e.g. 99.78% and 99.11% regarding the median and minimum purity rates, respectively, while pump H is the one with the worst performance, e.g. 99.33% and 98.66% regarding the median and minimum purity rates, respectively.

12.6.5 Analysis of the Voronoi Regions

A major characteristic of PBM, such as the centralized WTA, is that we are able to check the Voronoi regions in order to have a notion on how the classes are distributed among the prototypes. Therefore, to verify the samples separation obtained when using the global model at the *T*th iteration, we illustrate the Voronoi regions for a given running of the WTA frameworks in Figures 12.9 and 12.10. From these figures, we observe that the federated framework generated better classes separation than the centralized one. Therefore, we consider the federated approach as the preferable learning framework due to the higher cluster purity rates obtained from the proposed federated solution method. Moreover, the FED-WTA provided a solution that preserves the privacy of the pumping stations.

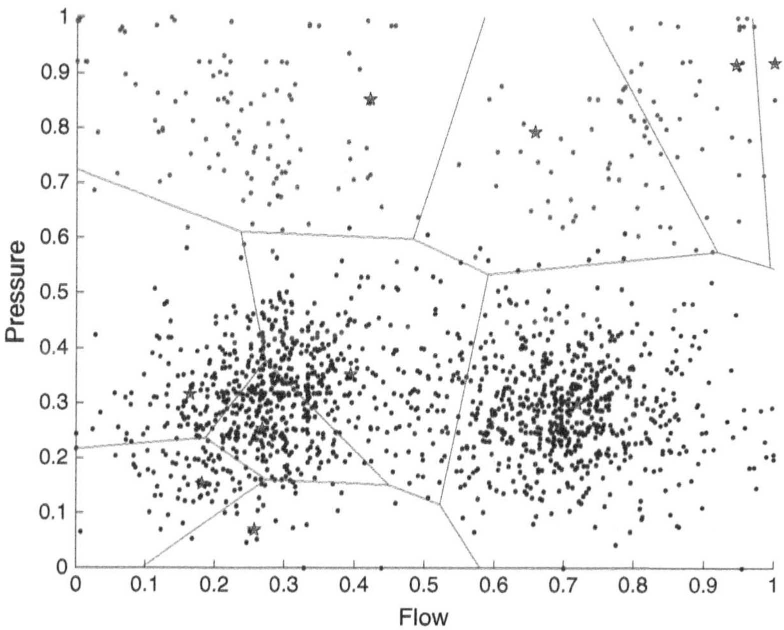

Figure 12.9 Centralized Voronoi cells. Voronoi cells generated when using the centralized WTA at a given run. Source: The authors.

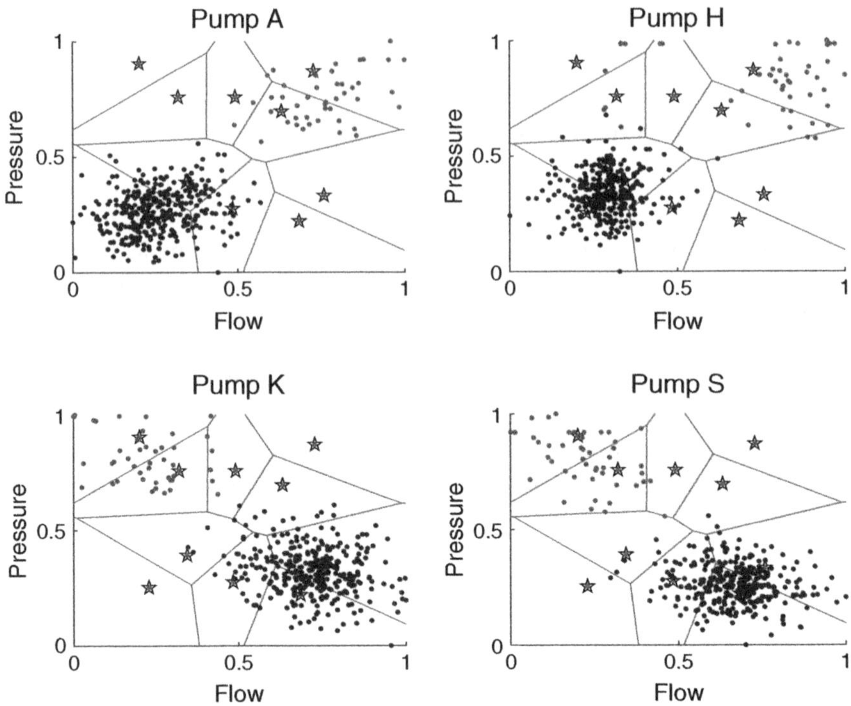

Figure 12.10 Federated Voronoi cells. Voronoi cells generated when using the proposed federated solution at a given run. Source: The authors.

12.7 Conclusions

In this work, we proposed a distributed algorithmic solution for water leakage detection in WDNs through the analysis of observed hydraulic data by means of emerging machine learning strategies. To evaluate our solution, we considered real world data from water pressure and flow measurements from pumps in a residential DMA of the WDN of Stockholm, Sweden.

We proposed a low complexity machine-learning framework for leakage detection. Specifically, our methodology used techniques from prototype-based learning and federated learning paradigms. For the numerical experiments of our proposed solution, we used real dataset from a DMA in Stockholm, Sweden.

The numerical findings showed the viability and potential benefits of combining PBL and FL. Specifically, we obtained better purity results for all pumping stations in the federated learning than the centralized learning scheme. Although our analysis are constrained to FED-WTA, we hope that our insights can inspire future work established on other FPBMs.

For those interested in exploring further, there are several key papers that provide valuable insights into related areas. For a comprehensive understanding of the concept of explainability regarding machine learning solutions, we recommend the studies by Verma et al. [2020], Zhou et al. [2021], and Burkart and Huber [2021]. Lastly, for those looking into challenges and solutions of WDNs operations in the machine learning context, the work by Artelt et al. [2024] introduces a promising Python toolbox for supporting the modeling of complex scenarios to handle diverse operational conditions of WDNs.

Acknowledgments

The work of Diego Perdigão Sousa was partially supported by the Coordenação de Aperfeiçoamento de Pessoal de Nível Superior (CAPES) – Finance Code 001, Grant No. 88887.155782/2017-00, and in part by the Conselho Nacional de Desenvolvimento Científico e Tecnológico (CNPq), Proc. No. 172998/2023-9. The work of José Mairton B. da Silva Jr. was partially supported by the European Union's Horizon Europe Research and Innovation Programme under the Marie Skłodowska-Curie Project FLASH (Grant Agreement No. 101067652), and in part by the Ericsson Research Foundation, the Hans Werthén Foundation, and the Mistra InfraMaint Programme. The work of Charles Casimiro Cavalcante was partially supported by the CNPq, Proc. No. 308512/2023-5, and in part by the Fundação Cearense de Apoio ao Desenvolvimento Científico e Tecnológico (FUNCAP), Grant No. PS-0186-00103.01.00/21. The work of Carlo Fischione was partially supported by the SSF project SAICOM and the VR project WIDCOMP. The authors also thank Stockholm Vatten och Avfall company, Stockholm, Sweden, for providing the data used in this study.

Bibliography

S. C. Ahalt, A. K. Krishnamurthy, P. Chen, and D. E. Melton. Competitive learning algorithms for vector quantization. *Neural Networks*, 3(3):277–290, 1990. ISSN 0893-6080.

A. Artelt, M. S. Kyriakou, S. G. Vrachimis, D. G. Eliades, B. Hammer, and M. M. Polycarpou. A toolbox for supporting research on AI in water distribution networks. *arXiv preprint arXiv:2406.02078*, 2024.

M. Bacco, F. Delmastro, E. Ferro, and A. Gotta. Environmental monitoring for smart cities. *IEEE Sensors Journal*, 17(23):7767–7774, 2017. doi: 10.1109/JSEN.2017.2722819.

M. Biehl, B. Hammer, and T. Villmann. Prototype-based models in machine learning. *WIREs Cognitive Science*, 7(2):92–111, 2016. doi: 10.1002/wcs.1378.

T. Bojer, B. Hammer, D. Schunk, and K. T. Von Toschanowitz. Relevance determination in learning vector quantization. In *European Symposium on Artificial Neural Networks*, pages 271–276, 2001.

B. Boots, K. Sugihara, S. N. Chiu, and A. Okabe. *Spatial Tessellations: Concepts and Applications of Voronoi Diagrams*. John Wiley & Sons, 2009.

J. Brinkrolf and B. Hammer. Federated learning vector quantization. In *ESANN European Symposium on Artificial Neural Networks, Computational Intelligence and Machine Learning*, 2021.

N. Burkart and M. F. Huber. A survey on the explainability of supervised machine learning. *Journal of Artificial Intelligence Research*, 70:245–317, 2021.

T. K. Chan, C. S. Chin, and X. Zhong. Review of current technologies and proposed intelligent methodologies for water distributed network leakage detection. *IEEE Access*, 6:78846–78867, 2018.

B. Fritzke. A growing neural gas network learns topologies. *Advances in Neural Information Processing Systems*, 7, 1994.

A. Gupta and K. D. Kulat. A selective literature review on leak management techniques for water distribution system. *Water Resources Management*, 32(10):3247–3269, 2018.

Y. Kim, E. Al Hakim, J. Haraldson, H. Eriksson, J. M. B. da Silva Jr., and C. Fischione. Dynamic clustering in federated learning. In *IEEE International Conference on Communications*, pages 1–6, 2021. doi: 10.1109/ICC42927.2021.9500877.

T. Kohonen. An introduction to neural computing. *Neural Networks*, 1(1):3–16, 1988.

T. Kohonen. Improved versions of learning vector quantization. In *IEEE IJCNN International Joint Conference on Neural Networks*, volume 1, pages 545–550, 1990a.

T. Kohonen. The self-organizing map. *Proceedings of the IEEE*, 78(9):1464–1480, 1990b. doi: 10.1109/5.58325.

T. Kohonen. Essentials of the self-organizing map. *Neural Networks*, 37:52–65, 2013. ISSN 0893-6080. doi: 10.1016/j.neunet.2012.09.018. Twenty-fifth Anniversay Commemorative Issue.

P. Kulkarni and T. Farnham. Smart city wireless connectivity considerations and cost analysis: Lessons learnt from smart water case studies. *IEEE Access*, 4:660–672, 2016.

R. Li, H. Huang, K. Xin, and T. Tao. A review of methods for burst/leakage detection and location in water distribution systems. *Water Science and Technology: Water Supply*, 15(3):429–441, 2015.

T. Li, M. Sanjabi, A. Beirami, and V. Smith. Fair resource allocation in federated learning. *arXiv preprint arXiv:1905.10497*, 2019a.

X. Li, K. Huang, W. Yang, S. Wang, and Z. Zhang. On the convergence of FedAvg on non-IID data. *arXiv preprint arXiv:1907.02189*, 2019b.

T. Li, A. K. Sahu, A. Talwalkar, and V. Smith. Federated learning: Challenges, methods, and future directions. *IEEE Signal Processing Magazine*, 37(3):50–60, 2020. doi: 10.1109/MSP.2020.2975749.

G. Long, Y. Tan, J. Jiang, and C. Zhang. Federated learning for open banking. In Q. Yang, L. Fan, and H. Yu, editors, *Federated Learning: Privacy and Incentive*, pages 240–254. Springer International Publishing, Cham, 2020.

T. Martinetz and K. Schulten. A "neural-gas" network learns topologies. In *International Conference on Artificial Neural Networks*, pages 397–402, 1991.

B. McMahan, E. Moore, D. Ramage, S. Hampson, and B. A. y Arcas. Communication-efficient learning of deep networks from decentralized data. In A. Singh and J. Zhu, editors, *Proceedings of the 20th International Conference on Artificial Intelligence and Statistics*, volume 54, pages 1273–1282. PMLR, 2017. URL http://proceedings.mlr.press/v54/mcmahan17a/mcmahan17a.pdf.

D. Miljković. Brief review of self-organizing maps. In *40th International Convention on Information and Communication Technology, Electronics and Microelectronics (MIPRO)*, pages 1061–1066. IEEE, 2017.

P. M. Moubarak, P. Ben-Tzvi, and M. E. Zaghloul. A self-calibrating mathematical model for the direct piezoelectric effect of a new MEMS tilt sensor. *IEEE Sensors Journal*, 12(5):1033–1042, 2011.

A. Moubayed, M. Sharif, M. Luccini, S. Primak, and A. Shami. Water leak detection survey: Challenges & research opportunities using data fusion & federated learning. *IEEE Access*, 9:40595–40611, 2021.

D. Nova and P. A. Estévez. A review of learning vector quantization classifiers. *Neural Computing and Applications*, 25(3-4):511–524, 2014.

A. C. Rencher. Interpretation of canonical discriminant functions, canonical variates, and principal components. *The American Statistician*, 46(3):217–225, 1992.

A. Sato and K. Yamada. Generalized learning vector quantization. *Advances in Neural Information Processing Systems*, pages 423–429, 1996.

S. F. Senin, M. S. Jaafar, and R. Hamid. Locating underground water pipe leakages via interpretation of ground penetrating radar signals. *International Journal of Engineering & Technology*, 8(2):72–77, 2019.

M. Servetnyk and C. C. Fung. Distributed dual averaging based data clustering. *IEEE Transactions on Big Data*, 9(1):372–379, 2022.

M. Servetnyk, C. C. Fung, and Z. Han. Unsupervised federated learning for unbalanced data. In *GLOBECOM 2020 - 2020 IEEE Global Communications Conference*, pages 1–6, 2020. doi: 10.1109/GLOBECOM42002.2020.9348203.

A. Soldevila, R. M. Fernandez-Canti, J. Blesa, S. Tornil-Sin, and V. Puig. Leak localization in water distribution networks using Bayesian classifiers. *Journal of Process Control*, 55:1–9, 2017.

D. P. Sousa, R. Du, J. M. B. da Silva Jr., C. C. Cavalcante, and C. Fischione. Leakage detection in water distribution networks using machine-learning strategies. *Water Supply*, 23(3):1115–1126, Feb 2023. ISSN 1606-9749. doi: 10.2166/ws.2023.054.

R. Ullah, Y. Faheem, and B.-S. Kim. Energy and congestion-aware routing metric for smart grid AMI networks in smart city. *IEEE Access*, 5:13799–13810, 2017.

V. Vaquet, F. Hinder, J. Brinkrolf, P. Menz, U. Seiffert, and B. Hammer. Federated learning vector quantization for dealing with drift between nodes. In *ESANN European Symposium on Artificial Neural Networks, Computational Intelligence and Machine Learning*, 2022.

J. Verbraeken, M. Wolting, J. Katzy, J. Kloppenburg, T. Verbelen, and J. S. Rellermeyer. A survey on distributed machine learning. *ACM Computing Surveys*, 53(2):1–33, Mar 2020. ISSN 0360-0300. doi: 10.1145/3377454.

S. Verma, J. Dickerson, and K. Hines. Counterfactual explanations for machine learning: A review. *arXiv preprint arXiv:2010.10596*, 2:1, 2020.

T. Villmann, A. Bohnsack, and M. Kaden. Can learning vector quantization be an alternative to SVM and deep learning? Recent trends and advanced variants of learning vector quantization for classification learning. *Journal of Artificial Intelligence and Soft Computing Research*, 7(1):65–81, 2017.

X. Wan, P. K. Kuhanestani, R. Farmani, and E. Keedwell. Literature review of data analytics for leak detection in water distribution networks: A focus on pressure and flow smart sensors. *Journal of Water Resources Planning and Management*, 148(10):03122002, 2022.

K. Wei, J. Li, M. Ding, C. Ma, H. H. Yang, F. Farokhi, S. Jin, T. Q. S. Quek, and H. V. Poor. Federated learning with differential privacy: Algorithms and performance analysis. *IEEE Transactions on Information Forensics and Security*, 15:3454–3469, 2020.

L. Xing and L. Sela. Unsteady pressure patterns discovery from high-frequency sensing in water distribution systems. *Water Research*, 158:291–300, 2019.

L. Xu, A. Krzyzak, and E. Oja. Rival penalized competitive learning for clustering analysis, RBF net, and curve detection. *IEEE Transactions on Neural networks*, 4(4):636–649, 1993. doi: 10.1109/72.238318.

J. Xu, B. S. Glicksberg, C. Su, P. Walker, J. Bian, and F. Wang. Federated learning for healthcare informatics. *Journal of Healthcare Informatics Research*, 5:1–19, 2021.

D. Zaman, M. K. Tiwari, A. K. Gupta, and D. Sen. A review of leakage detection strategies for pressurised pipeline in steady-state. *Engineering Failure Analysis*, 109:104264, 2020.

J. Zhou, A. H. Gandomi, F. Chen, and A. Holzinger. Evaluating the quality of machine learning explanations: A survey on methods and metrics. *Electronics*, 10(5):593, 2021.

13

Multi-Agent Inverse Learning for Sensor Networks: Identifying Coordination in UAV Networks*

Luke Snow and Vikram Krishnamurthy

Department of Electrical and Computer Engineering, Cornell University, Ithaca, NY, USA

13.1 Introduction

In strategic environments, autonomous systems such as unmanned aerial vehicles (UAVs) are becoming ubiquitous for reconnaissance, surveillance, and combative purposes. Often, such autonomous systems are deployed in groups, e.g. UAV swarms, in order to collect information more efficiently or multiply the combative force. Furthermore, these multi-agent intelligent systems typically have sophisticated sensors and communication capabilities, which allow them to respond in real time to an adversary's probe, e.g. radar tracking signals. This results in a strategic interaction between the multi-agent system and the adversary; the study of this interaction at the physical layer, for instance analyzing electromagnetic suppression techniques, is typically referred to as "electronic warfare."

We consider a multi-agent strategic interaction scenario in which a radar is tracking a network of UAVs. We take the perspective of the radar and ask how can we detect *coordination* in the UAV network? Such coordination detection would not only allow us to understand the functionality of the network, but when combined with estimates for the UAV objectives would allow us to *predict future network behavior*. Thus, the second question we ask is: if the network is coordinating, how can we reconstruct individual objective functions, which induce the observed aggregate behavior?

We study this problem at a higher level of abstraction than traditional electronic warfare investigations; this allows us to formulate the "coordination" problem as a general linearly constrained multi-objective optimization. Then, the problem of

* This research was funded by NSF Grant CCF-2312198 and Army Research Office Grant W911NF-21-1-0093.

Wireless Sensor Networks in Smart Environments: Enabling Digitalization from Fundamentals to Advanced Solutions, First Edition. Edited by Domenico Ciuonzo and Pierluigi Salvo Rossi.

detecting coordination and reconstructing feasible objective functions becomes that of *inverse multi-objective optimization*. We present several tools from microeconomic theory, which allow us to accomplish this inverse learning problem efficiently. While this microeconomic interpretation is conceptualized at a higher level of abstraction than traditional electronic warfare procedures, we also present how this framework arises naturally from physical-layer considerations, such as radar waveform modulation and multi-target filtering algorithms.

This chapter is organized as follows. Section 13.2 presents the mathematical details of (forward and inverse) multi-objective optimization and presents the microeconomic tools, which can be used to accomplish general inverse multi-objective optimization. Then, Section 13.3 presents the UAV network coordination detection procedure. First, the radar – UAV network interaction dynamics are specified, then it is shown how the microeconomic interpretation arises from filtering-level tracking considerations. Finally, in Section 13.4, we present the application of the microeconomic tools from Section 13.2 to the coordination detection problem.

13.2 Multi-Objective Optimization and Revealed Preferences

In order to characterize conditions under which coordination can be detected by an outside observer, one much precisely define what is meant by coordination in the first place. Notions of coordination have appeared in, e.g. Chen et al. [2020], Quintero et al. [2010], and Wise and Rysdyk [2006]. We utilize a well-motivated and widely used framework to define coordination, known as multi-objective optimization. In this section, we present the mathematical details of multi-objective optimization and *inverse* multi-objective optimization and give a microeconomic result, allowing us to achieve the latter efficiently. The application of these frameworks to the UAV coordination detection problem will be detailed in Sections 13.3 and 13.4.

13.2.1 Multi-Objective Optimization

Here, we outline what distinguishes multi-objective optimization from single-objective optimization and provide the resultant generalized notion of a solution concept.

13.2.1.1 Multi-Objective Problem

We consider a system composed of multiple autonomous agents. Each agent has an individual utility function that captures their objective and aims to act in a way that maximizes their utility function. In order to capture a notion of coordination, it is assumed that there is a joint constraint on the actions taken, such that both the set of all actions, which can be taken by a particular agent and the resultant utility achieved by this agent, are dependent on the actions taken by *all* of the agents. This coupling forces the set of all agents to jointly consider the actions taken in order to achieve individual objectives.

13.2.1.2 Multi-Objective Solution Concept

The reader may realize that this is also the setting of game theory, where a standard investigation is that of non-cooperative agents acting solely in self-interest. The classical solution concept in non-cooperative game theory is that of Nash Equilibrium, where no agent can gain in their utility by unilaterally deviating (changing their action). We distinguish this from the *cooperative* solution concept in multi-objective optimization that of *Pareto-optimality*. Pareto optimality occurs when no agent can gain in their utility by unilaterally deviating (changing their action) *without simultaneously decreasing the utility of another agent*. So, an individual agent could feasibly change their action to increase their utility, but this would come at the expense of decreasing another agent's utility. Thus, a Pareto-optimal solution captures a notion of coordination since the agents do not act in complete self-interest but act in order to maximize the entire set of utility functions.

13.2.2 Inverse Multi-Objective Optimization

13.2.2.1 Inverse Multi-Objective Problem

Now that the multi-objective problem has been conceptualized, one may ask: given a dataset of actions, how can it be determined if the group is behaving in a Pareto-optimal manner? This general problem is denoted as inverse multi-objective optimization and originated from the recovery of decision process structures in microeconomic group behavior analysis [Chiappori and Ekeland, 2009]. More specifically, in inverse multi-objective optimization, we aim to determine *if there exist* individual utility functions for which the actions are multi-objective optimal. If so, we aim to reconstruct such utility functions in order to better understand or predict the system dynamics. A key framework for accomplishing this will be that of microeconomic revealed preferences.

13.2.2.2 Revealed Preferences

The microeconomics literature contains the most well-developed formulations of such inverse multi-objective optimization, nominally "Group Revealed Preferences." The Revealed Preferences paradigm dates back to seminal work [Afriat, 1967], where utility maximization behavior is detected from consumer budget-expenditure data. The Group Revealed Preferences [Cherchye et al., 2011] formulation extends these works to the multi-agent scenario, giving necessary and sufficient conditions for group behavior to be consistent with multi-objective optimization. Furthermore, a methodology is provided for reconstructing feasible utility functions under which the observed behavior is multi-objective optimal. This allows for inference of multi-agent group motives or prediction of future behavior.

13.2.3 Outline

In the rest of this section, we make the above concepts more mathematically precise: we first outline the mathematics of multi-objective optimization, then provide the relevant framework for inverse multi-objective optimization, given by the micro-economic Group Revealed Preference formulation. In Sections 13.3 and 13.4, we utilize these mathematical tools in the UAV coordination detection problem.

13.2.4 Multi-Objective Optimization

In this section, we introduce the multi-objective optimization problem we will consider, then present its solution concept of Pareto-optimality, and discuss how Pareto-optimal solutions can be obtained.

13.2.4.1 Multi-Objective Problem

We consider $M \in \mathbb{N}$ agents. We denote $\beta \in \mathbb{R}^n$ a general joint action taken by all agents. For example β can represent a vector containing distinct actions taken by each agent, or it can represent a single action that has been agreed upon by the set of agents. Each agent $i \in [M] := \{1, \dots, M\}$ has a *utility function* $f^i : \mathbb{R}^n \to \mathbb{R}$, representing agent i's utility gained from the joint action taken.

This setting is sufficiently general to capture standard game-theoretic notions. For instance, in non-cooperative Game Theory, the joint action β can represent the set of distinct actions taken by each agent. Then solution concepts such as Nash Equilibria, where no agent has an incentive to unilaterally deviate from its action, can be studied.

Our focus in this setting will instead be on a notion of multi-agent *coordination*, given by a particular linearly constrained multi-objective optimization:

Linearly Constrained Multi-Objective Optimization

$$\begin{aligned} &\arg\max_{\beta} \{f^1(\beta), \dots, f^M(\beta)\} \\ &s.t.\ \beta \in X_c := \{\gamma \in \mathbb{R}^n : \alpha'\gamma \leq 1\}, \end{aligned} \tag{13.1}$$

Equation (13.1) encodes the idea that the agents must cooperate such that joint action β maximizes over all objective functions f^i, provided β is in a linear constraint set $\alpha'\beta \leq 1$ formed by *constraint vector* α. The linear constraint $\alpha'\beta$[1] is bounded by 1 without losing generality (see Section I-A of Krishnamurthy et al. [2020]).

The astute reader may at this point ask what precisely is meant by the maximization in (13.1). Indeed, it turns out that we need to introduce a generalized notion of optimality in order for this maximization to be well-posed.

13.2.4.2 Multi-Objective Solution Concept: Pareto Optimality

In single-objective optimization, the goal is to find a feasible argument, which maximizes the objective, in that the objective evaluated at this argument is greater than or equal to the objective evaluated at any other point in the feasible set. A naive generalization of this to the multi-objective setting might be to find an argument, which maximizes all objectives. However, unless there are very tight restrictions on the objective function structures (e.g. all the same function or all one-dimensional and monotone), there will seldom exist an argument β, which simultaneously maximizes all objectives. Thus, there will be tradeoffs between objectives for varying argument β. The general solution concept for the multi-objective optimization problem (13.1) that captures these tradeoffs is instead that of *Pareto optimality*:

Definition 13.1 ***(Pareto Optimality)*** For fixed $\{\{f^i(\cdot)\}_{i=1}^M, \alpha\}$ and a vector $\beta \in X_c = \{\gamma \in \mathbb{R}^n : \alpha'\gamma \leq 1\}$, let

$$Z^t(\beta) = \{\gamma \in X_c : f^i(\gamma) \geq f^i(\beta) \forall i \in [M]\}$$
$$Y^t(\beta) = \{\gamma \in X_c : \exists k \in [M] : f^k(\gamma) > f^k(\beta)\}.$$

The vector β is said to be *Pareto-optimal* if

$$Z^t(\beta) \cap Y^t(\beta) = \emptyset, \tag{13.2}$$

where $\emptyset$ denotes the empty set.

In words, a vector β is Pareto-optimal if there does not exist another vector γ in the feasible set X_c, which increases the value of some objective $f^i(\cdot)$ without simultaneously decreasing the value of some other objective $f^j(\cdot)$, $i, j \in [M]$.

This is a well-motivated and nontrivial conception of cooperative optimality in multi-agent systems [Marden et al., 2014; Ràdulescu et al., 2020]. It captures the idea that even if a single agent may gain by deviating from the Pareto-optimal joint action, it does not do so since that gain would come at the expense of another agent.

1 For vector x, we let x' represent the transpose of x.

From another perspective, if a joint action is not yet Pareto-optimal, it means that it can be altered such that no agents' utility decreases, and at least one agent's utility increases. Such an alteration may have to be undertaken by a certain agent who gains nothing by changing their action but does so in order to increase the utility of a different agent. Thus, achieving the Pareto-optimum conceptually corresponds to all agents simultaneously acting for the best of the entire group.

In general, there will be a set of Pareto-optimal solutions, some benefiting certain individual agents more than others, but all maximizing the utilities of the entire group in the above-described sense.

Definition 13.2 ***(Pareto Frontier)*** The set of *all* Pareto-optimal solutions to the problem (13.1) is known as the *Pareto-frontier* and is denoted

$$X_{PF}(\{f^i\}_{i=1}^M, \alpha_t) := \{\beta \in X_c : (13.2) \text{ is satisfied}\}. \tag{13.3}$$

Now, we say that β solves (13.1) if and only if β is Pareto-optimal, i.e.

$$\beta \in \{\arg\max_{\gamma} \{f^1(\gamma), \dots, f^M(\gamma)\}\, s.t.\, \gamma \in X_c\} \iff \beta \in X_{PF}(\{f^i\}_{i=1}^M, \alpha).$$

13.2.4.3 Computing Pareto Optimal Solutions

We have discussed the multi-objective optimization problem and its solution concept of Pareto-optimality. The question remains: given joint-action constraints and individual utility functions, how can one (or the multi-agent group itself) actually compute Pareto-optimal solutions? Here, we show how Pareto-optimal solutions can be obtained by simply maximizing linear combinations of objective functions $\{f^i\}_{i=1}^M$ subject to the linear constraint $\alpha'\beta \leq 1$.

Before presenting this result, we need to introduce some notation. Let $\mu = (\mu^1, \dots, \mu^M)' \in \mathbb{R}^M_{\geq 0}$ be a set of real-valued weights on the non-negative unit simplex $\mathcal{W}_M$, defined as

$$\mathcal{W}_M := \{\mu \in \mathbb{R}^M_{\geq 0} : \mathbf{1}'\mu = 1\}. \tag{13.4}$$

Also let

$$\mathcal{W}_M^+ := \{\mu \in \mathbb{R}^M_+ : \mathbf{1}'\mu = 1\} \subset \mathcal{W}_M, \tag{13.5}$$

be the set of strictly positive weights. Let us denote

$$S(\mu) := \left\{ \beta : \beta \in \arg\max_{\gamma} \sum_{i=1}^M \mu^i f^i(\gamma) \quad \text{s.t.} \quad \alpha'\gamma \leq 1 \right\}$$

i.e. $S(\mu)$ is the set of all vectors β maximizing a linear combination of objective functions $\{f^i\}$ with weights μ, such that the linear constraint $\alpha'\beta \leq 1$ is satisfied.

Then, we have the following relation [Miettinen, 2012]:

$$\bigcup_{\mu \in \mathcal{W}_M^+} S(\mu) \subseteq X_{PF}(\{f^i\}_{i=1}^M, \alpha_t) \subseteq \bigcup_{\mu \in \mathcal{W}_M} S(\mu), \tag{13.6}$$

where the second inclusion is an equality if the objective functions are concave. Relation (13.6) implies that if we solve

$$\arg\max_{\gamma} \sum_{i=1}^{M} \mu^i f^i(\gamma)\, s.t.\, \alpha'\gamma \leq 1, \tag{13.7}$$

with weights μ strictly positive, then this solution is guaranteed to be Pareto-optimal. Furthermore, provided the objective functions f^i are concave, *all* Pareto-optimal solutions can be produced by solving (13.7) with weights varying over the non-negative simplex $\mathcal{W}_M$. In particular, this is useful since (13.7) is a *constrained single-objective optimization*, which can be computed efficiently in most cases if the utility functions are concave.

13.2.5 Inverse Multi-Objective Optimization

In this section, we make the concept of inverse multi-objective optimization mathematically precise and introduce a key theorem enabling us to achieve it in a general microeconomic framework.

13.2.5.1 Inverse Multi-Objective Problem

The inverse multi-objective optimization problem can be stated conceptually as follows. Given constrained outputs (actions) of an observed multi-agent system, does there exist a set of utility functions under which the observed outputs are multi-objective optimal? Can these utility functions be reconstructed? At first, a mathematical instantiation of this statement might be: Given (α, β), does there exist a set of utility functions $\{f^i\}_{i=1}^M$ and weights $\mu \in \mathcal{W}_M$ such that

$$\beta \in \left\{ \arg\max_{\gamma} \sum_{i=1}^{M} \mu^i f^i(\gamma) \quad \text{s.t.} \quad \alpha'\gamma \leq 1 \right\}. \tag{13.8}$$

If there exists such a set of weights μ and utility functions $\{f^i\}_{i=1}^M$, then we say that the data (α, β) is *rationalized* by these weights and utility functions. However, for a single data-point (α, β), there will *always* exist sets $\{f^i\}$ and μ, which rationalize it. To see this, take μ in the corner of the simplex, such that $\mu^i = 1$ for some i and $\mu^j = 0\, \forall j \neq 1$. Then, (13.8) reduces to $\beta \in \arg\max_\gamma f^i(\gamma)$ s.t. $\alpha'\gamma \leq 1$, and obviously one can find some f^i for which this is true. Thus, the inverse multi-objective optimization with a single data-point is trivial.

To make the problem nontrivial, we consider multiple data-points indexed by time, i.e. suppose that we observe the dataset $\boldsymbol{\beta} := \{\alpha_t, \beta_t\}$ of constraint vectors α_t

and system outputs β_t indexed over discrete-time $t \in [T] := \{1, \dots, T\}$. Then, this extended inverse multi-objective optimization problem can be stated as follows:

Inverse Multi-Objective Optimization

Given a time-indexed dataset $\boldsymbol{\beta} := \{\alpha_t, \beta_t\}$, do there exist utility functions $\{f^i\}_{i=1}^M$ such that

$$\beta_t \in \left\{ \arg\max_{\gamma} \sum_{i=1}^{M} \mu^i f^i(\gamma) \quad \text{s.t.} \quad \alpha_t' \gamma \leq 1 \right\} \quad \forall t \in [T],$$

for some weights μ in the simplex $\mathcal{W}_M$? If so, how can one reconstruct these utility functions?

The above problem is distinct from the (trivial) single data-point problem explained above, since here, the utility functions $\{f^i\}_{i=1}^M$ must rationalize the data $\{\alpha_t, \beta_t\}$ *for all* $t \in [T]$ simultaneously. One can easily see how this distinction makes the problem nontrivial since the set of utility functions, which rationalize the data-set for some fixed time-point may not rationalize the data for another time-point. In this sense, the inverse multi-objective optimization problem tests whether a multi-agent system behaves optimally (in the Pareto-sense) at each time point and is also consistent in behaving optimally (w.r.t. the same utility functions) over all tested time-points.

Figure 13.1 provides an illustration of the procedure for inverse multi-objective optimization, in relation to the generative process of multi-objective optimization.

Next, we discuss a microeconomic solution to a specific form of this problem.

13.2.5.2 Group Revealed Preferences

The microeconomic field of Revealed Preferences aims to detect utility maximization behavior among observed consumers. We present here the form of multi-objective optimization considered in this literature, which is a special case of the general multi-objective problem (13.1). Suppose that we have the dataset of constraints and system responses $\boldsymbol{\beta} = \{\alpha_t, \{\beta_t^i\}_{i=1}^M, t \in [T]\}$. Here, β_t^i corresponds to the action taken by agent i. We say that the dataset satisfies "collective rationality" if it solves the following multi-objective optimization problem:

Microeconomic Collective Rationality

$\exists\, \mu \in \mathcal{W}_M, \{U^i\}_{i=1}^M, U^i : \mathbb{R}^N \to \mathbb{R}$ concave and monotone increasing, such that:

$$\{\beta_t^i\}_{i=1}^M \in \left\{ \arg\max_{\{\gamma^i\}_{i=1}^M} \sum_{i=1}^{M} \mu^i U^i(\gamma^i) \quad \text{s.t.} \quad \alpha_t' \left(\sum_{i=1}^{M} \gamma^i \right) \leq 1 \right\} \quad \forall t. \tag{13.9}$$

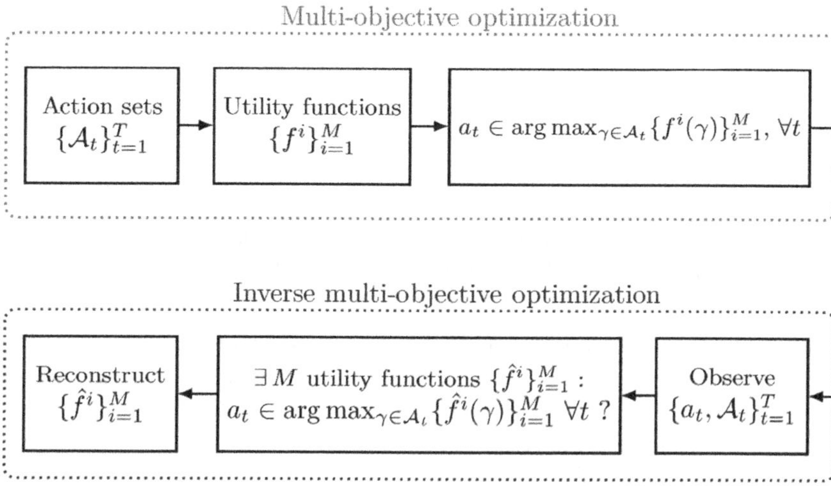

Figure 13.1 Forward and inverse multi-objective optimization. The (forward) multi-objective optimization problem consists of a set of feasible actions and a utility function for each agent. The optimization problem is to find an action a_t that maximizes over the set of utility functions. The *inverse* multi-objective optimization problem is to observe the actions taken and first determine if there exist individual utility functions making the actions Pareto-optimal. Then, if so, these "rationalizing" utility functions should be reconstructed.

Notice that this form of "collective rationality" can be obtained as a special case of the more general form (13.1), where each agent's utility function is only explicitly dependent on its own action. However, (13.9) still optimizes over joint actions in the same sense as (13.1) since the linear constraint limits the sum of individual actions.

The inverse multi-objective problem in this specialized case then is analogous to the general problem in Figure 13.1: we ask if there exist utility functions such that (13.9) holds for all t. In Cherchye et al. [2011], a necessary and sufficient condition is derived for the dataset $\boldsymbol{\beta}$ to be consistent with this notion of multi-objective optimization.

Theorem 13.1 Let $\boldsymbol{\beta} = \{\alpha_t, \{\beta_t^i\}_{i=1}^M, t \in [T]\}$ be a set of observations. The following are equivalent:

1) there exists a set of M concave and continuous objective functions $U^1, \dots, U^m$, weights $\mu \in \mathcal{W}_M^+$ and constraint p^* such that $\forall t \in [T]$:

$$\{\beta_t^i\}_{i=1}^M \in \left\{ \arg\max_{\{\gamma^i\}_{i=1}^M} \sum_{i=1}^M \mu^i U^i(\gamma^i) \quad \text{s.t.} \quad \alpha_t' \left(\sum_{i=1}^M \gamma^i \right) \leq p^* \right\}, \tag{13.10}$$

2) there exist numbers $u_j^i \in \mathbb{R}, \lambda_j^i > 0$ such that for all $s, t \in [T], i \in [M]$:

$$u_s^i - u_t^i - \lambda_t^i \alpha_t' [\beta_s^i - \beta_t^i] \leq 0. \tag{13.11}$$

Proof: See Proposition 1 of Cherchye et al. [2011].

Furthermore, if the above conditions hold, then specific utility functions that "rationalize" the dataset can be reconstructed in the following way.

Corollary 13.1 Given constants $u_t^i, \lambda_t^i, t \in [T], i \in [M]$, which make (13.11) feasible, explicit monotone and continuous objective functions that "rationalize" the dataset
$\{\alpha_t, \beta_t^i, t \in [T], i \in [M]\}$ are given by

$$U^i(\cdot) = \min_{t \in [T]} \left[u_t^i + \lambda_t^i \alpha_t' [\cdot - \beta_t^i] \right], \tag{13.12}$$

i.e. (13.10) is satisfied with objective functions (13.12).

Proof: See Lemma 1 of Snow et al. [2022].

These results give us a principled and efficient way of performing inverse multi-objective optimization, by testing the feasibility of a linear program. We can first test whether the data is consistent with "collective rationality," i.e. whether the group is behaving "intelligently" by consistently optimizing a set of utility functions, then we can reconstruct individual utility functions that rationalize the dataset. This gives us a mechanism for inferring the underlying distribution of objectives in the group or for predicting future group behavior.

In this section, we have first presented the general forward and inverse multi-objective optimization problems, then revealed a specific form of multi-objective optimization that can be tested efficiently by solving a particular linear program. Next, we present the setting in which we will apply these results: detecting UAV coordination. We first outline the UAV-tracking dynamics and interaction model, then show how this can be mapped to the setting presented in this first section, allowing for efficient testing of UAV "coordination."

13.3 Multi-Objective Optimization in UAV Networks

In this section, we consider the specific instantiation of a radar – UAV network tracking scenario. The multi-objective optimization framework presented in

Section 13.2 will allow us to precisely define coordination in the UAV network and efficiently detect such coordination on the radar's end. In this section, we

- present the radar – UAV network interaction dynamics,
- provide the definition of UAV network coordination,
- outline several motivational target-tracking frameworks, which give rise to the above notion.

13.3.1 Interaction Dynamics

Here, we provide the general interaction dynamics between a UAV (target) network and a radar (us). For now, let us define, at time $t \in \mathbb{N}$, the radar's tracking signal as α_t and target i's maneuver as β_t^i. Figure 13.2 displays the high-level interaction dynamics: The radar probes the target network and obtains measurements of the network maneuvers. We will momentarily give explicit motivation for how these variables can be interpreted in a physical-layer multi-target tracking scenario. We consider inverse multi-objective optimization; we aim to detect whether the target network coordinates in a specific sense (corresponding to our previous notion of multi-objective optimization).

At an implementation level, we aim to detect whether the targets jointly adjust their maneuvers such that their overall utility is maximized (in the Pareto-optimal sense), subject to a constraint on their *detectability* by the radar. We will also momentarily provide a definition and motivation for such a notion of detectability.

We consider two timescales for the interaction: the fast time scale $k = 1, 2, \ldots$ represents the scale at which the target state and measurement dynamics occur,

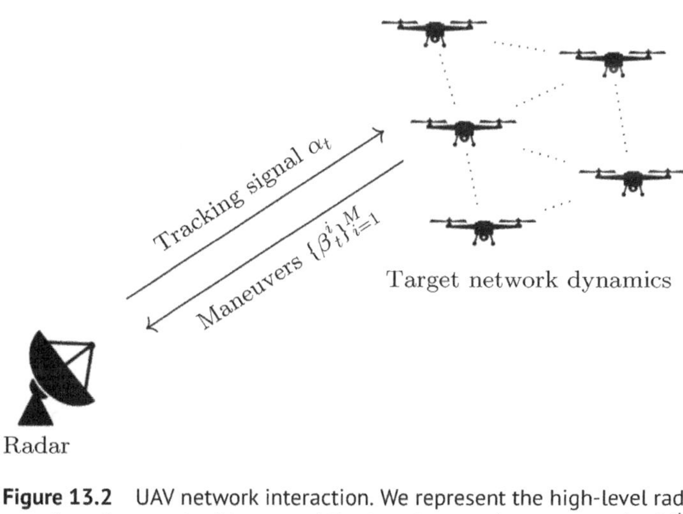

Figure 13.2 UAV network interaction. We represent the high-level radar tracking waveform (parameters) by α_t, and the target network maneuvers by $\{\beta_t^i\}_{i\in[M]}$.

and the slow timescale $t = 1, 2, \ldots$ represents the scale at which the radar probes (tracking signals) and UAV maneuvers $\{\beta_t^i\}_{i=1}^M$ occur.

Definition 13.3 ***(Radar – Multi-Target Interaction)*** The radar – UAV network interaction has the following dynamics:

$$\begin{aligned} \text{radar emission}: &\quad \alpha_t \in \mathbb{R}_+^N \\ \text{UAV i maneuver}: &\quad \beta_t^i \in \mathbb{R}_+^N \\ \text{UAV i state}: &\quad x_k^i \in \mathbb{R}^q,\ x_{k+1}^i \sim p_{\beta_t^i}(x|x_k^i) \\ \text{radar observation}: &\quad y_k^i \in \mathbb{R}^p,\ y_k^i \sim p_{\alpha_t}(y|x_k^i) \\ \text{radar tracker}: &\quad \pi_k^i = \mathcal{T}(\pi_{k-1}^i, y_k^i), \end{aligned} \tag{13.13}$$

where π_k^i is radar i's target state posterior, and $\mathcal{T}$ is a general Bayesian tracker. For a fixed t in the slow timescale, α_t abstractly represents the radar's signal output, which parameterizes its measurement kernel, and β_t^i represents target i's maneuver (radial acceleration, etc.), which parameterizes the state update kernel. These interaction dynamics are illustrated in Figure 13.3. Taking the point of view of the radar, we aim to detect if the targets are *coordinating*.

We next present precisely what is meant by coordination and motivate how the mathematical definition can be derived from practical multi-target filtering algorithms.

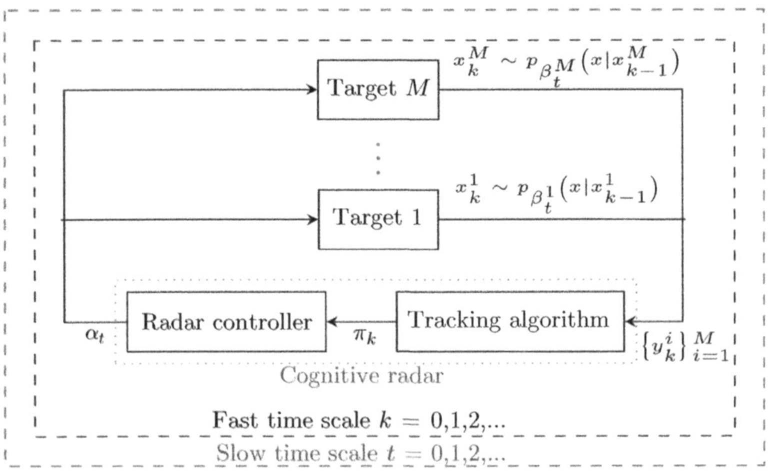

Figure 13.3 UAV network interaction dynamics. The interaction occurs at two timescales. The slow timescale, indexed by $t \in \mathbb{N}$, is the scale at which radar waveform signal parameters α_t and target maneuvers $\{\beta_t^i\}_{i=1}^M$ are adjusted. For a fixed radar tracking waveform and set of target maneuvers, the radar obtains a sequence of target measurements $\{y_k^i\}_{i\in[M]}$, indexed on the fast timescale by $k \in \mathbb{N}$. From these measurements, the radar implements a multi-target filtering algorithm to track the states $\{x_k^i\}_{i\in[M]}$ and thus can recover $\{\beta_t^i\}$.

13.3.2 UAV Network Coordination: Constrained Spectral Optimization

Here, we present a correspondence between the spectral UAV network dynamics and a constrained multi-objective optimization problem, thereby defining what is meant by coordination and showing how it arises from the interaction dynamics (13.13).

13.3.2.1 UAV Network Coordination

In formulating our problem, it is necessary to define rigorously what we mean by UAV coordination. Examples of such coordination definitions have been proposed and studied in works Snow et al. [2022], Snow and Krishnamurthy [2023], and Shi et al. [2017]. We consider the following coordination specification. Each UAV has an individual utility function f^i, which maps from its state dynamics β_t^i, parametrizing the state transition kernel in (13.13), to a real-valued utility, i.e.

$$f^i : \mathbb{R}^N \to \mathbb{R},$$

Such utility functions can capture the UAVs' flight objectives by quantifying a reward profile for flight dynamics. The UAVs then should act to maximize their individual utility functions at each point in time in order to achieve their flight objective. However, such individual maximization would decouple the UAV dynamics such that they act independently of each other's trajectories. A notion of coordination would need to capture a certain coupling or codependency between these trajectories.

We propose to quantify this coupling through a constraint on the radar's average measurement precision. This captures the idea that the UAVs aim to obtain some flight objective while jointly acting such that the entire network remains hidden to a certain degree from the radar. This induces a coupling between UAV trajectories; the UAVs must adjust their individual sequential state dynamics such that the entire network satisfies a certain undetectability constraint.

This coordination formulation can be summarized informally as

$$\begin{aligned} &\text{maximize}\,(f^1, \dots, f^M), \text{ such that} \\ &\text{average radar measurement precision} \leq \text{bound}. \end{aligned} \tag{13.14}$$

The "maximize $(f^1, \dots, f^M)$" can be interpreted in the framework of Pareto optimality, as introduced in Section 13.2. The radar measurement precision bound can be derived from standard multi-target tracking algorithms, as we show in Section 13.3.2.2.

This leads us to our formal definition of coordination in a UAV network, given as follows:

Definition 13.4 ***(Coordinating UAV Network)*** Considering the interaction dynamics (13.13), we define a coordinating UAV network to be a network of M UAVs, each with individual concave, continuous, and monotone increasing[2] objective functions $f^i : \mathbb{R}^N \to \mathbb{R}, i \in [M]$, which produces output signals $\{\beta_t^i\}_{i=1}^M$ on the slow timescale in accordance with

$$\begin{aligned} \{\beta_t^i\}_{i=1}^M \in \arg\max_{\{\beta^i\}_{i=1}^M} & \{f^1(\beta^1), \dots, f^M(\beta^M)\} \\ \text{s.t.} \quad & \alpha_t' \left(\sum_{i=1}^M \beta^i \right) \leq 1. \end{aligned} \tag{13.15}$$

Note that (13.15) is a special case of the general multi-objective optimization problem (13.1), in which the objective functions do not share a common argument but the arguments are jointly constrained. Thus, a coordinating UAV network controls its joint state dynamics (through e.g. controlling a certain formation) such that they are *Pareto optimal* (Definition 13.1) with respect to each objective function, the tracking signal from the radar, and a constraint on the UAV network's detectability.

It is quite straightforward to interpret the individual utility functions f^i of the targets as encoding flight objectives, but one may well ask how the linear constraint in (13.15) corresponds to a bound on the radar's average measurement precision, as suggested in the informal definition (13.14). We next provide an example of multi-target state dynamics and several resultant radar tracking algorithms, which naturally give rise to this constraint. The purpose is to shed light on how the abstract constrained multi-objective optimization (13.15) can be recovered from practical filtering-level tracking dynamics.

13.3.2.2 Multi-Target Spectral Dynamics

Here, we specify a concrete example of the abstract dynamics (13.13). Linear Gaussian dynamics for a target's kinematics [Li and Jilkov, 2003] and linear Gaussian measurements at each radar are widely assumed as a useful approximation [Bar-Shalom et al., 2004]. Thus, we will consider the following linear Gaussian state dynamics and measurements over the *fast time scale* $k \in \mathbb{N}$, with a particular $t \in \mathbb{N}$ fixed:

$$\begin{aligned} x_{k+1}^i &= A^i x_k^i + w_k^i, \; x_0^i \sim \pi_0^i, \\ y_k^i &= C^i x_k^i + v_k^i, \; i \in [m], \end{aligned} \tag{13.16}$$

---

2 This objective function structure is known as "locally non-satiated" in the microeconomics literature and is not necessarily restrictive when considering target objectives, see Krishnamurthy et al. [2020].

where $x_k^i, w_k^i \in \mathbb{R}^q$ are the target i state and noise vectors, respectively, and $A^i \in \mathbb{R}^{q\times q}$ is the state update matrix for target i. $y_k^i \in \mathbb{R}^p$ is the radar's measurement of target i, $C^i \in \mathbb{R}^{p\times q}$ is the measurement transformation, and $v_k^i \in \mathbb{R}^p$ is the measurement noise. The constraints and subsequent radar responses will be indexed over the *slow time scale* $t \in \mathbb{N}$. Abstractly, these will parameterize the state and noise covariance matrices:

$$w_k \sim \mathcal{N}(0, Q_t(\beta_t^i)), \quad v_k^i \sim \mathcal{N}(0, R_t(\alpha_t)). \tag{13.17}$$

In this spectral interpretation, β_t^i represents the vector of eigenvalues of state-noise covariance matrix Q_t, and α_t represents the vector of eigenvalues of the inverse measurement noise covariance matrix R_t^{-1}. Thus, given this interpretation, we can view modulations of α_t and β_t^i as corresponding to increased/decreased measurement precision on the part of the radar. This will be made precise subsequently when we discuss filtering details. First, we briefly illustrate how such noise covariance matrices can be parameterized in the first place.

Waveform Design for Measurement Covariance Modulation To give a precise structure to the radar dynamics, this section provides examples of how the observation noise covariance $R_t(\alpha_t)$ in (13.17) can depend on the radar waveform. Further details on maximum likelihood estimation involving the radar ambiguity function can be found in Van Trees [2004] and Kershaw and Evans [1994]. The waveform specifications involve the following terms:

- c denotes the speed of light (in free space),
- ω_c denotes the carrier frequency,
- θ is an adjustable parameter in the waveform,
- η is the signal to noise ratio at the radar,
- $j = \sqrt{-1}$ is the unit imaginary number,
- $s(t)$ is the complex envelope of the waveform,
- α is the vector of eigenvalues of R^{-1}.

We now provide three example waveforms and their resulting observation noise covariance matrices $R(\alpha)$:

1) Triangular Pulse – Continuous Wave

$$s(t) = \begin{cases} \sqrt{\frac{3}{2\theta}}\left(1 - \frac{|t|}{\theta}\right) & -\theta < t < \theta \\ 0 & \text{otherwise} \end{cases}$$

$$R(\alpha) = \begin{bmatrix} \frac{c^2\theta^2}{12\eta} & 0 \\ 0 & \frac{5c^2}{2\omega_c^2\theta^2\eta} \end{bmatrix},$$

2) Gaussian Pulse – Continuous Wave

$$s(t) = \left(\frac{1}{\pi\theta^2}\right)^{1/4} \exp\left(\frac{-t^2}{2\theta^2}\right)$$

$$R(\alpha) = \begin{bmatrix} \frac{c^2\theta^2}{s\eta} & 0 \\ 0 & \frac{c^2}{2\omega_c^2\theta^2\eta} \end{bmatrix},$$

3) Gaussian Pulse – Linear Frequency Modulation Chirp

$$s(t) = \left(\frac{1}{\pi\theta_1^2}\right)^{1/4} \exp\left(-\left(\frac{1}{2\theta_1^2} - j\theta_2\right)t^2\right)$$

$$R(\alpha) = \begin{bmatrix} \frac{c^2\theta_1^2}{2\eta} & \frac{-c^2\theta_2\theta_1^2}{\omega_c\eta} \\ \frac{-c^2\theta_2\theta_1^2}{\omega_c\eta} & \frac{c^2}{\omega_c^2\eta}\left(\frac{1}{2\theta_1^2} + 2\theta_2^2\theta_1^2\right) \end{bmatrix}.$$

The key idea is that by adapting the waveform parameters, the radar can modulate the covariance matrix $R(\alpha)$. This modulation can be viewed at a higher level as an adaptation of the eigenvalues of $R(\alpha)$. We treat α as the vector of eigenvalues of $R^{-1}(\alpha)$, so that increasing α increases the measurement precision. Such an increase directly corresponds to, or is enacted by, changes to the physical-layer waveform parameterization, as illustrated above.

Next, given the above Linear Gaussian specification of the multi-target dynamics (13.16), we present two multi-target filtering examples. The goal is to illustrate how the spectral interpretation of α_t and β_t^i in (13.17) gives rise within these algorithms to the linear constraint $\alpha_t(\sum_{i=1}^{M} \beta_t^i) \leq 1$ in (13.15). Recall that this linear constraint should correspond to a physical-layer bound on the radar's average measurement precision.

13.3.3 Multi-Target Filtering

The goal of this section is to present several multi-target tracking schemes, a simple decoupled Kalman filter and a more complex joint probabilistic data association filter (JPDAF) and show how the high-level coordination framework (13.19) can be recovered from each. *These serve as illustrative examples of how to map complex multi-target tracking algorithms to the constrained multi-objective optimization* (13.15). One should be able to extend these mappings to other target tracking schemes.

13.3.3.1 Decoupled Kalman Filtering

A simple interpretation of the multi-target tracking procedure is a standard decoupled Kalman filter, whereby after measurements are associated to each target, a standard Kalman filter is applied to track each target state separately. This procedure is idealized but allows for a nice exposition of the connection between filtering precision and the constraint in (13.9).

Filter Dynamics Consider the linear Gaussian dynamics (13.16) and (13.17). Based on observations $y_1^i, \dots, y_k^i$ associated to target i, the tracking functionality in the radar computes the target i state posterior

$$\pi_k^i = \mathcal{N}(\hat{x}_k^i, \Sigma_k^i),$$

where $\hat{x}_k^i$ is the conditional mean state estimate, and Σ_k^i is the covariance, computed by the classical Kalman filter:

$$\begin{aligned}
\Sigma_{k+1|k}^i &= A^i \Sigma_k^i (A^i)' + Q_t(\beta_t^i) \\
K_{k+1}^i &= C^i \Sigma_{k+1|k}^i (C^i)' + R_t(\alpha_t) \\
\hat{x}_{k+1}^i &= A^i \hat{x}^i + \Sigma_{k+1|k}^i (C^i)' (K_{k+1|k}^i)^{-1} (y_{k+1}^i - C^i A^i \hat{x}_k^i) \\
\Sigma_{k+1}^i &= \Sigma_{k+1|k}^i - \Sigma_{k+1|k}^i (C^i)' (K_{k+1}^i)^{-1} C^i \Sigma_{k+1|k}^i .
\end{aligned}$$

Under the assumption that the model parameters in (13.16) satisfy $[A^i, C^i]$ is detectable and $[A^i, \sqrt{Q_t(\beta_t^i)}]$ is stabilizable, the asymptotic predicted covariance $\Sigma_{k+1|k}^i$ as $k \to \infty$ is the unique non-negative definite solution of the *algebraic Riccati equation* (ARE):

$$\begin{aligned}
&\mathcal{A}(\alpha_t, \beta_t^i, \Sigma) := \\
&\quad -\Sigma + A^i(\Sigma - \Sigma(C^i)'[C^i\Sigma(C^i)' + R_t(\alpha_t)]^{-1}C^i\Sigma)(A^i)' + Q_t(\beta_t^i) = 0.
\end{aligned} \tag{13.18}$$

Let $\Sigma_t^*(\alpha_t, \beta_t^i)$ denote the solution of the ARE and $\Sigma_t^{*-1}(\alpha_t, \beta_t^i)$ be its inverse, representing the asymptotic measurement *precision* obtained by the radar.

Extracting a Revealed Preference Bound By Lemma 3 of Krishnamurthy et al. [2020], we can represent a limit $\overline{\Sigma}^{-1}$ on the radar's precision of target i measurement, $\Sigma_t^{*-1}(\alpha_t, \beta_t^i)$ as the simple linear inequality $\alpha_t' \beta_t^i \leq 1$, i.e.

$$\alpha_t' \beta_t^i \leq 1 \iff \Sigma_t^{*-1}(\alpha_t, \beta_t^i) \leq \overline{\Sigma}^{-1}$$

where the constant 1 bound is taken without loss of generality. The key idea behind this equivalence is to show that the asymptotic precision $\Sigma_n^{*-1}(\cdot, \beta_t^i)$ is monotone increasing in the first argument α_t using the information Kalman filter formulation. Then, we can represent a constraint on the radar's average precision over measurements of all targets as

$$\alpha_t' \left(\sum_{i=1}^{M} \beta_t^i \right) \leq 1. \tag{13.19}$$

Thus, we recover a direct correspondence between the radar's average measurement precision and the linear inequality constraint in (13.9). Thus, again, "collective rationality" (13.9) on the part of the UAV network can directly be interpreted as the high-level constrained multi-objective optimization (13.14).

The recovery of this linear constraint (13.19) from the decoupled Kalman filter gives a clear correspondence between the filtering dynamics and the high-level objective constraint (13.14). However, this decoupled Kalman filtering scheme is idealized and simplified; next, we outline a more sophisticated multi-target tracking algorithm, which is widely used in practice [Fortmann et al., 1980; Rezatofighi et al., 2015], and show the same recovery of the linear constraint (13.19).

13.3.3.2 Joint Probabilistic Data Association Filter

The JPDAF operates under the regime, where n measurements $y_k^j, j \in [n]$ (13.24) of m targets are obtained, and it is not known which measurements correspond to which target. See Bar-Shalom and Li [1995] for clarification of any details.

Filter Dynamics Define the empirical validation matrix $\Omega = [\omega_{jt}, j \in [n], t \in \{0, \dots, m\}$, with $\omega_{jt} = 1$ if measurement j is in the *validation gate* of target t, and 0 otherwise. It is common to let the $t = 0$ index correspond to "none of the targets."

Now, we construct an object θ known as the "joint association event," as

$$\theta = \bigcap_{j=1}^{m} \theta_{jt_j},$$

where

- θ_{jt} represents the event that measurement j originated from target t;
- t_j is the index of the target which measurement j is associated with the event under consideration.

So, θ can represent any possible set of associations between measurements and targets.

Then, we can form the *event matrix*

$$\hat{\Omega}(\theta) = [\hat{\omega}_{jt}],$$

where

$$\hat{\omega}_{jt} = \begin{cases} 1, & \theta_{jt} \in \theta \\ 0, & \text{else} \end{cases},$$

$\hat{\Omega}(\theta)$ is thus the indicator matrix of measurement–target associations in event θ.

We say an event θ is a *feasible association event* if

1) a measurement is associated to only one source,

$$\sum_{t=0}^{m} \hat{\omega}_{jt}(\theta) = 1, \quad \forall j \in [n], \tag{13.20}$$

2) at most one measurement originates from each target,

$$\delta_t(\theta) := \sum_{j=1}^{n} \hat{\omega}_{jt}(\theta) \leq 1, \quad \forall t \in [m]. \tag{13.21}$$

Denote by Θ the set of all feasible events.

The binary variable $\delta_t(\theta)$ is known as the "target detection indicator" since it indicates whether, in event θ, a measurement j has been associated to target t. We may also define a "measurement association indicator"

$$\tau_j(\theta) := \sum_{t=1}^{m} \hat{\omega}_{jt}(\theta), \tag{13.22}$$

which indicates whether a particular measurement j is associated with a target t. Note the difference between (13.22) and (13.20); the latter sums from 0 to include the possibility of a measurement being assigned to "no target," i.e. clutter, while the former sums from 1, indicating whether the measurement has been assigned to an actual target.

Using these definitions, we can write the number of false (unassociated) measurements in event θ as

$$\phi(\theta) := \sum_{j=1}^{n} [1 - \tau_j(\theta)]. \tag{13.23}$$

Using these preliminary concepts, the JPDAF can be formulated by first deriving the posterior probability of joint-association events given the measured data, then incorporating this into a standard filtering scheme akin to the Kalman filter. The filtering can be done in an uncoupled or coupled manner; the former assumes that target measurements are independently distributed, and the latter is capable of correlations in target state estimation errors.

Uncoupled Filtering Now given a particular feasible joint-association event $\theta_k \in \Theta$, and letting δ_t, τ_j, and ϕ be shorthand for (13.21), (13.22), and (13.23), respectively, evaluated at θ_k, Bar-Shalom and Li [1995] derives the posterior probability $P(\theta_k | \{y_k^j\}_{j=1}^n)$, under the uncoupled assumption, as

$$P(\theta_k | \{y_k^j\}_{j=1}^n) \propto \frac{\phi!}{m_k!} \mu_F(\phi) V^{-\phi} \prod_j [f_{tj}(y_k^j)]^{\tau_j} \prod_t (P_D^t)^{\delta_t} (1 - P_D^t)^{1-\delta_t}, \tag{13.24}$$

where P_D^t is the detection probability of target t, $m_k = n - \phi$, and

$$f_{tj}(y_k^j) = \mathcal{N}(y_k^j; \hat{y}_{k|k-1}^{t_j}, S_k^{t_j}),$$

with $\hat{y}_{k|k-1}^{t_j}$ being the predicted measurement for target t_j in the previous iteration of the filter, and $S_k^{t_j}$ being the associated innovation covariance matrix. $\mu_F(\phi)$ is the probability mass function governing the number of false measurements ϕ, and such measurements not associated with a target are assumed uniformly distributed in the surveillance region of volume V.

Given, this posterior probability the uncoupled filter proceeds by separately filtering each target state independently. For brevity, we do not introduce this filtering process, but do so for the more sophisticated and robust *coupled* filter.

Coupled Filtering Given a particular feasible joint-association event $\theta_k \in \Theta$, and letting δ_t, τ_j, and ϕ be shorthand for (13.21), (13.22), and (13.23), respectively, evaluated at θ_k, Bar-Shalom and Li [1995] derives the posterior probability $P(\theta_k|\{y_k^j\}_{j=1}^n)$ as

$$P(\theta_k P(\theta_k|\{y_k^j\}_{j=1}^n) \propto \frac{\phi!}{m_k!}\mu_F(\phi)V^{-\phi}f_{t_{j_1},t_{j_2},\ldots}(y_k^j, j : \tau_j = 1)\prod_t (P_D^t)^{\delta_t}(1-P_D^t)^{1-\delta_t}, \tag{13.25}$$

where here $f_{t_{j_1},t_{j_2},\ldots}$ is the joint pdf of the measurements of the targets under consideration, and t_{j_i} is the target in which $y_k^{j_i}$ is associated in event θ_k. Now, we introduce the Joint Probabilistic Data Association Coupled Filter (JPDACF) state estimation and covariance update.

We form the stacked state vector of predicted states, and associated covariance, as

$$\hat{x}_{k|k-1} = \begin{bmatrix} \hat{x}_{k|k-1}^1 \\ \vdots \\ \hat{x}_{k|k-1}^m \end{bmatrix},$$

$$P_{k|k-1} = \begin{bmatrix} P_{k|k-1}^{1\,1} & \cdots & P_{k|k-1}^{1\,m} \\ \vdots & & \vdots \\ P_{k|k-1}^{m\,1} & \cdots & P_{k|k-1}^{m\,m} \end{bmatrix},$$

where $P^{t_1\,t_2}$ is the cross-covariance between targets t_1 and t_2. The coupled filtering is done as follows:

$$\hat{x}_{k|k} = \hat{x}_{k|k-1} + W_k \sum_\theta P(\theta|\{y_k^j\}_{j=1}^n)[\mathbf{y}_k(\theta) - \hat{y}_{k|k-1}],$$

where

$$\mathbf{y}_k(\theta) = \begin{bmatrix} y_k^{j_1(\theta)} \\ \vdots \\ y_k^{j_m(\theta)} \end{bmatrix},$$

and $j_i(\theta)$ is the measurement associated with target i in event θ. The filter gain W_k is given by

$$W_k = P_{k|k-1}\hat{C}_k'\left[\hat{C}_k P_{k|k-1}\hat{C}_k' + \hat{R}_k\right]^{-1},$$

where

$$\hat{C}_k = \text{diag}\left[\delta_1(\theta)C_k^1, \dots, \delta_m(\theta)C_k^m\right]$$
$$\hat{R}_k = \text{diag}\left[R_k^1, \dots, R_k^m\right],$$

are the block diagonal measurement and noise covariance matrices, respectively. The binary detection indicator variables $\delta_i(\theta)$ accounts for the possibility of a measurement not being associated to target i. The predicted stacked measurement vector is

$$\hat{y}_{k|k-1} = \hat{C}_k\hat{x}_{k|k-1} = \hat{C}_k\hat{A}_{k-1}\hat{x}_{k-1},$$

with $\hat{A}_{k-1} = \text{diag}[A_{k-1}^1, \dots, A_{k-1}^m]$ being the block diagonal state update matrix.

The covariance of the updated state is given as

$$P_{k|k} = P_{k|k-1} + [1 - \psi_0]W_k\hat{S}_kW_k' + \tilde{P}_k, \tag{13.26}$$

where $\hat{S}_k = \hat{C}_kP_{k|k-1}\hat{C}_k' + \hat{R}_k$ is the innovation covariance,

$$\psi_{jt} := \sum_{\theta:\theta_{jt}\in\theta} P(\theta|\{y_k^j\}_{j=1}^n),$$

and $\psi_0 := \sum_{j=1}^m \psi_{j0}$ is the probability that no measurements arise from targets. $\tilde{P}_k$ is the spread of the innovation terms:

$$\tilde{P}_k := W_k\tilde{S}_kW_k'.$$

with

$$\tilde{S}_k = \begin{bmatrix} \sum_{j=1}^{m_k}\psi_{j1}\left[y_k^1 - \hat{x}_{k|k-1}^1\right]\cdot\left[y_k^1 - \hat{x}_{k|k-1}^1\right]' - v_{1,k}v_{1,k}' \\ \vdots \\ \sum_{j=1}^{m_k}\psi_{jm}\left[y_k^m - \hat{x}_{k|k-1}^m\right]\cdot\left[y_k^m - \hat{x}_{k|k-1}^m\right]' - v_{m,k}v_{m,k}' \end{bmatrix}.$$

and

$$v_{i,k} = \sum_{j=1}^{m_k}\psi_{ji}\left[y_k^i - \hat{x}_{k|k-1}^i\right].$$

Extracting a Revealed Preference Bound The crucial observation is that, as in the Kalman filter algebraic Riccati equation (13.18), the covariance (13.26) is monotone decreasing in α_t, since this corresponds to increasing $\hat{R}_k$ for fixed k. Thus, the asymptotic measurement precision (inverse of asymptotic predicted covariance) is monotone *non-decreasing* in α_t, and by the same reasoning as Lemma 3 of Krishnamurthy et al. [2020], we may derive the equivalence

$$\alpha_t{}'\left(\sum_{i=1}^{M}\beta_t^i\right) \leq 1 \iff \lim_{k\to\infty} P_{k|k}^{-1}(\alpha_t, \{\beta_t^i\}) \leq \hat{P}^{-1},$$

Thus, we again have that the constraint $\alpha_t{}'\left(\sum_{i=1}^{M}\beta_t^i\right) \leq 1$ is a natural representation for a bound on the average measurement precision.

13.4 Detection of Coordination

In Section 13.3.3, we showed how a notion of coordination, corresponding to linearly constrained multi-objective optimization, arises naturally from several standard multi-target filtering algorithms. In this section, we illustrate how to detect coordination in UAV networks using the microeconomic revealed preference tools in Section 13.2.5. We first consider deterministic detection, which is a straightforward application of the results in Section 13.2.5, then extend this to optimal statistical detection when UAV maneuvers are observed in noise.

13.4.1 Deterministic Coordination Detection

We take $\beta_t^i > 0\,\forall t \in [T], i \in [M]$, i.e. each UAV always has a non-zero process noise. Then by Lemma 1 in Snow and Krishnamurthy [2023], (13.15) is equivalent to

$$\{\beta_t^i\}_{i=1}^M \in \arg\max_{\{\beta^i\}_{i=1}^M} \sum_{i=1}^{M} \mu^i f^i(\beta^i) \quad \text{s.t.} \quad \alpha_t'\left(\sum_{i=1}^{M}\beta^i\right) \leq 1, \tag{13.27}$$

for any $\mu \in \mathcal{W}_M^+$.

Recall that we are interested in the *inverse* multi-objective optimization problem. The equivalence between (13.27) and (13.15) allows us to directly utilize the microeconomic result Theorem 13.1, such that detecting coordination is equivalent to solving the linear program (13.11). Furthermore, we can reconstruct feasible utility functions that rationalize the dataset as (13.12). This procedure for detection of coordination and utility function reconstruction is illustrated in Algorithm 13.1.

Algorithm 13.1 Detecting coordination

1: Record the time-indexed dataset of radar waveforms and UAV network responses $\boldsymbol{\beta} = \{\alpha_t, \{\beta_t^i\}_{i=1}^M, t \in [T]\}$.
2: **if** $\exists u_j^i, \lambda_j^i : \forall s, t \in [T], i \in [M] : u_s^i - u_t^i - \lambda_t^i \alpha_t'[\beta_s^i - \beta_t^i] \leq 0$ **then**
3: Declare coordination present
4: Reconstruct feasible utility functions U^i as$U^i(\cdot) = \min_{t \in [T]} [u_t^i + \lambda_t^i \alpha_t'[\cdot - \beta_t^i]$
5: $\Rightarrow \exists \mu \in \mathcal{W}_M : \{\beta_t^i\}_{i=1}^M \in \arg\max_{\{\gamma^i\}_{i=1}^M} \sum_{i=1}^M \mu_i U^i(\gamma^i) : \alpha_t' \left(\sum_{i=1}^M \gamma^i\right) \leq 1$
6: **end if**

13.4.1.1 Numerical Example

Here, we provide a numerical example for the deterministic coordination detection and utility reconstruction procedure outlined in Algorithm 13.1. We consider $M = 3$ targets and acquire data over $T = 10$ time-steps. The radar waveform measurement covariance eigenvalue vector α_t and each target maneuver vector β_t^i are taken to be two-dimensional. The three targets are taken to have the following simple utility functions:

$$\begin{aligned} f^1(\beta) &= \det(Q(\beta))^2 = \beta(1)^2\beta(2)^2 \\ f^2(\beta) &= \sqrt{\beta(1)\beta(2)} \\ f^3(\beta) &= \beta(1)\sqrt{\beta(2)}. \end{aligned} \tag{13.28}$$

We then generate the vectors α_t, β_t^i, with $\mu^1 = 0.4, \mu^2 = 0.4, \mu^3 = 0.3$, as follows:

- $\alpha_t \sim U[0.1, 1.1]^2$
- $\{\beta_t^i\}_{i=1}^M \in \arg\max_{\{\gamma^i\}_{i=1}^M} \sum_{i=1}^3 \mu^i f^i(\gamma^i) \quad \text{s.t.}\ \alpha_t'(\sum_{i=1}^3 \gamma^i) \leq 1$

Thus, the target responses $\{\beta_t^i\}$ satisfy our notion of coordination (multi-objective optimization). Then, implementing Algorithm 13.1, we confirm that the linear program (13.11) has a feasible solution, indicating the presence of multi-objective optimization, and we may reconstruct feasible utility functions. Reconstructed utility functions are illustrated in Figure 13.4. Notice that the reconstructed utility functions match the relative profiles of the true utility functions, and do so while being concave.

13.4.2 Statistical Detection of Coordination

Recall that thus far we have considered only deterministic UAV i dynamics β_t^i. We now consider the case when these measured responses are corrupted by noise.

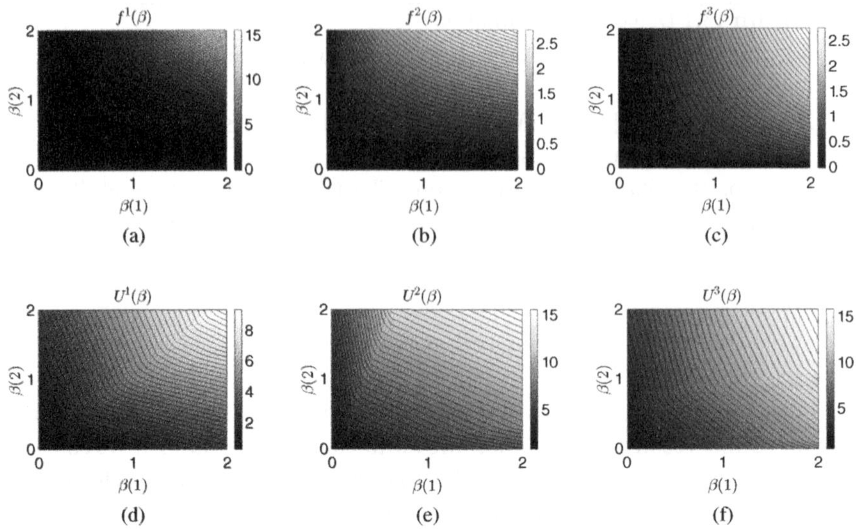

Figure 13.4 $f^i(\beta)$ is the true objective function of the ith radar, inducing the responses $\{\beta_t^i\}_{t=1}^{10}$. $U^i(\beta)$ is the reconstructed objective function for radar i, computed using the dataset $\boldsymbol{\beta} = \{\alpha_t, \{\beta_t^i\}_{i=1}^M, t \in [T]\}$ and (13.12). (a) $f^1(\beta) = \det(Q(\beta))$, (b) $f^2(\beta) = \mathrm{Tr}(Q(\beta))$, (c) $f^3(\beta) = \sqrt{\beta(1)}\beta(2)$, (d) $\min_{t\in[10]} [u_t^1 + \lambda_t^1 \alpha_t'[\cdot - \beta_t^1]$, (e) $\min_{t\in[10]} [u_t^2 + \lambda_t^2 \alpha_t'[\cdot - \beta_t^2]$, and (f) $\min_{t\in[10]} [u_t^3 + \lambda_t^3 \alpha_t'[\cdot - \beta_t^3]$.

We introduce a statistical detector for determining whether these *noisy* responses are consistent with multi-objective optimization, with theoretical guarantees on Type-I error.

Let $\overline{\boldsymbol{\beta}}$ denote the dataset when the radar responses are observed in noise:

$$\overline{\boldsymbol{\beta}} = \{\alpha_t, \tilde{\beta}_t^i, t \in [T], i \in [M]\}, \tag{13.29}$$

where $\tilde{\beta}_t^i = \beta_t^i + \epsilon_t^i$, and ϵ_t^i are independent random variables generated according to distributions Λ_t^i. We propose a statistical detector to optimally determine if the responses are consistent with Pareto optimality (13.1). Define

H_0: null hypothesis that the dataset (13.29) arises from the optimization problem (13.15) for all $t \in [T]$.

H_1: alternative hypothesis that the dataset (13.29) does not arise from the optimization problem (13.15) for all $t \in [T]$.

There are two possible sources of error:

- **Type-I error**: Reject H_0 when H_0 is valid.
- **Type-II error**: Accept H_0 when H_0 is invalid.

We formulate the following test statistic $\Phi^*(\overline{\beta})$, as a function of $\overline{\beta}$, to be used in the detector:

$$\Phi^*(\overline{\beta}) = \max_i \hat{\Phi}^i(\overline{\beta}), \tag{13.30}$$

where $\hat{\Phi}^i(\overline{\beta})$ is the solution to

$$\min \Phi^i \; : \; \exists\, u_t^i > 0, \lambda_t^i > 0 \; : \; u_s^i - u_t^i - \lambda_t^i \alpha_t'(\overline{\beta}_s^i - \overline{\beta}_t^i) - \lambda_t^i \Phi^i \leq 0. \tag{13.31}$$

Form the random variable Ψ as

$$\Psi = \max_{i,\, t \neq s} [\alpha_t'(\epsilon_t^i - \epsilon_s^i)]. \tag{13.32}$$

Then, we propose the following statistical detector (with $\gamma \in (0, 1)$):

$$\int_{\Phi^*(\overline{\beta})}^{\infty} f_\Psi(\psi) d\psi \begin{cases} \geq \gamma \Rightarrow H_0 \\ < \gamma \Rightarrow H_1 \end{cases}, \tag{13.33}$$

where $f_\Psi(\cdot)$ is the probability density function of Ψ. Let F_Ψ be the cdf of Ψ and $\overline{F}_\Psi$ be the complementary cdf of Ψ. Then, we have the following guarantees:

Theorem 13.2 Consider the noisy dataset (13.29), and suppose (13.31) has a feasible solution. Then

1) The following null hypothesis implication holds:

$$H_0 \subseteq \bigcap_{i \in [M]} \{\hat{\Phi}^i(\overline{\beta}) \leq \Psi^i\}, \tag{13.34}$$

2) The probability of Type-I error (false alarm) is

$$\mathbb{P}_{\Phi^*(\overline{\beta})}(H_1 | H_0) = \mathbb{P}(\overline{F}_\Psi(\Phi^*(\overline{\beta})) \leq \gamma \,|\, H_0) \leq \gamma,$$

3) The optimizer $\Phi^*(\overline{\beta})$ yields the smallest Type-I error bound:

$$\mathbb{P}_{\overline{\Phi}(\overline{\beta})}(H_1 | H_0) \geq \mathbb{P}_{\Phi^*(\overline{\beta})}(H_1 | H_0) \quad \forall \overline{\Phi}(\overline{\beta}) \in [\Phi^*(\overline{\beta}), \Psi].$$

Proof: See Snow and Krishnamurthy [2023].

The motivation for this detector is that it allows one to quantify a strict upper bound on the probability of Type-I error; the specific choice of threshold γ is left to the designer and may vary depending on application criteria.

In practice, one would likely not have access to the true density function $f_\Psi(\cdot)$. However, it is typical to assume some structure on the additive noise process $\{\Lambda_t^i, t \in [T]\}_{i \in [M]}$ such as Gaussianity. Thus, under such an assumption, one can compute an approximation $\hat{F}_\Psi(\cdot)$ of the cumulative distribution function $F_\Psi(\cdot)$,

Algorithm 13.2 Detecting multi-objective optimization

for l=1:L **do**
  for i=1:M **do**
    simulate $\epsilon_l^i = [\epsilon_1^i, \dots, \epsilon_N^i]^{(l)}, \quad \epsilon_t^i \sim \Lambda_t^i$
  end for
  Compute $\Psi^l := \max_i \{\max_{t \neq s}[\alpha_t(\epsilon_t^i - \epsilon_s^i)]\}$
end for
Compute $\hat{F}_\Psi(\cdot)$ from $\{\Psi^l\}_{l=1}^L$
Record radar network response $\bar{\beta}$ to the probe α_t
Solve (13.30) for $\Phi^*(\bar{\beta})$
Save $\mathcal{P} := \{\hat{u}_t^i, \hat{\lambda}_t^i, t \in [T], i \in [M]\}$ such that

$$\hat{u}_s^i - \hat{u}_t^i - \hat{\lambda}_t^i \alpha_t'(\bar{\beta}_s^i - \bar{\beta}_t^i) - \hat{\lambda}_t^i \hat{\Phi}^i(\bar{\beta}) \leq 0 \; \forall i \in [M]$$

Implement detector (13.33) as

$$1 - \hat{F}_\Psi(\Phi^*(\bar{\beta})) \begin{cases} > \gamma \Rightarrow H_0 \\ \leq \gamma \Rightarrow H_1 \end{cases} \tag{13.35}$$

then implement the statistical detector using this. Algorithm 13.2 provides such an implementation of the statistical detector (13.33).

This section presented techniques for both deterministic and statistical detection of coordination in UAV networks. These techniques exploit the microeconomic revealed preference results in Section 13.2.5 and the abstract correspondence between UAV dynamics and linearly constrained multi-objective optimization in Sections 13.3.2 and 13.3.3.

13.5 Conclusion

We have investigated the mathematical properties of multi-objective optimization and inverse multi-objective optimization and presented a microeconomic technique for performing the latter. We have demonstrated how this can be applied in a UAV network coordination detection scheme, by utilizing radar tracking signals. This methodology is more abstract than traditional electronic warfare procedures and thus allows for a concise encapsulation of the above stated problem and algorithmic solution. We also show how this abstract formulation can be recovered by several specific multi-target filtering algorithms and specifications of radar waveform design. However, the application of the presented methodology is not limited to these cases and can find use in a variety of inverse multi-objective optimization settings.

Bibliography

S. N. Afriat. The construction of utility functions from expenditure data. *International Economic Review*, 8(1):67–77, 1967.

Y. Bar-Shalom and X.-R. Li. *Multitarget-Multisensor Tracking: Principles and Techniques*, volume 19. YBS Publishing, Storrs, CT, 1995.

Y. Bar-Shalom, X. R. Li, and T. Kirubarajan. *Estimation with Applications to Tracking and Navigation: Theory Algorithms and Software*. John Wiley & Sons, 2004.

W. Chen, J. Liu, H. Guo, and N. Kato. Toward robust and intelligent drone swarm: Challenges and future directions. *IEEE Network*, 34(4):278–283, 2020.

L. Cherchye, B. De Rock, and F. Vermeulen. The revealed preference approach to collective consumption behaviour: Testing and sharing rule recovery. *The Review of Economic Studies*, 78(1):176–198, 2011.

P.-A. Chiappori and I. Ekeland. The microeconomics of efficient group behavior: Identification 1. *Econometrica*, 77(3):763–799, 2009.

T. E. Fortmann, Y. Bar-Shalom, and M. Scheffe. Multi-target tracking using joint probabilistic data association. In *1980 19th IEEE Conference on Decision and Control Including the Symposium on Adaptive Processes*, pages 807–812. IEEE, 1980.

D. J. Kershaw and R. J. Evans. Optimal waveform selection for tracking systems. *IEEE Transactions on Information Theory*, 40(5):1536–1550, 1994.

V. Krishnamurthy, D. Angley, R. Evans, and B. Moran. Identifying cognitive radars-inverse reinforcement learning using revealed preferences. *IEEE Transactions on Signal Processing*, 68:4529–4542, 2020.

X. R. Li and V. P. Jilkov. Survey of maneuvering target tracking. Part I. Dynamic models. *IEEE Transactions on Aerospace and Electronic Systems*, 39(4):1333–1364, 2003.

J. R. Marden, H. P. Young, and L. Y. Pao. Achieving Pareto optimality through distributed learning. *SIAM Journal on Control and Optimization*, 52(5):2753–2770, 2014.

K. Miettinen. *Nonlinear Multiobjective Optimization*, volume 12. Springer Science & Business Media, 2012.

S. A. P. Quintero, F. Papi, D. J. Klein, L. Chisci, and J. P. Hespanha. Optimal UAV coordination for target tracking using dynamic programming. In *49th IEEE Conference on Decision and Control (CDC)*, pages 4541–4546. IEEE, 2010.

R. Ràdulescu, P. Mannion, D. M. Roijers, and A. Nowé. Multi-objective multi-agent decision making: a utility-based analysis and survey. *Autonomous Agents and Multi-Agent Systems*, 34(1):10, 2020.

S. H. Rezatofighi, A. Milan, Z. Zhang, Q. Shi, A. Dick, and I. Reid. Joint probabilistic data association revisited. In *Proceedings of the IEEE International Conference on Computer Vision*, pages 3047–3055, 2015.

C. Shi, S. Salous, F. Wang, and J. Zhou. Power allocation for target detection in radar networks based on low probability of intercept: A cooperative game theoretical strategy. *Radio Science*, 52 (8):1030–1045, 2017.

L. Snow and V. Krishnamurthy. Statistical detection of coordination in a cognitive radar network through inverse multi-objective optimization. In *IEEE International Conference on Information Fusion*, 2023.

L. Snow, V. Krishnamurthy, and B. M. Sadler. Identifying coordination in a cognitive radar network-a multi-objective inverse reinforcement learning approach. In *International Conference on Acoustics, Speech, and Signal Processing*, 2022.

H. L. Van Trees. *Detection, Estimation, and Modulation Theory, Part I: Detection, Estimation, and Linear Modulation Theory*. John Wiley & Sons, 2004.

R. Wise and R. Rysdyk. UAV coordination for autonomous target tracking. In *AIAA Guidance, Navigation, and Control Conference and Exhibit*, page 6453, 2006.

14

Immersive IoT Technologies for Smart Environments

Subhas C. Mukhopadhyay[1], *Anindya Nag*[2,3], *and Nagender K. Suryadevara*[4]

[1] *School of Engineering, Macquarie University, Sydney, New South Wales, Australia*
[2] *Faculty of Electrical and Computer Engineering, Technische Universität Dresden, Dresden, Germany*
[3] *Centre for Tactile Internet with Human-in-the-Loop (CeTI), Technische Universität Dresden, Dresden, Germany*
[4] *School of Computer and Information Sciences, University of Hyderabad, Hyderabad, India*

14.1 Introduction

The process in the field of science and technology has allowed researchers to develop smart environments. This not only improves the quality of life but also assists researchers in studying and analyzing human behavior. Prior to the popularization of sensing systems in earlier times, there would be a considerable amount of time and energy required to complete a task. This was tackled in the late 1980s [Wieder and Neppl, 1992] when single-crystal silicon sensors [Nag et al., 2015a, 2015b; Xu et al., 2019] were popularized for different applications. These sensors have been formed with the conventional micro-electromechanical systems (MEMS) [Nag et al., 2015c] technique, where there is an easy possibility to customize the size and shape of the prototypes. These silicon-based sensors have been used for various applications [Afsarimanesh et al., 2017; Alahi et al., 2017]. Over time, scientists have also designed and developed sensors with a flexible nature [Nag et al., 2017a; Han et al., 2019a]. These sensors have been formed with a wide range of nanomaterials [He et al., 2022; Afsarimanesh et al., 2022; Nag et al., 2022a] and polymers [Nag et al., 2018a, 2019a] using different kinds of printing techniques [Khan et al., 2014]. Some of the common nanomaterials like carbon nanotubes (CNTs) [Han et al., 2019b; Gao et al., 2021; Nag et al., 2021a], graphene [Nag et al., 2021b, 2022b, 2023], graphite [Nag et al., 2018b, 2018c], and other metallic nanomaterials [Liu et al., 2021; Rafiee et al., 2021] have been used as per the requirement of the sensors. Each of these nanomaterials has been fused with polymers and other nanomaterials

Wireless Sensor Networks in Smart Environments: Enabling Digitalization from Fundamentals to Advanced Solutions, First Edition. Edited by Domenico Ciuonzo and Pierluigi Salvo Rossi.

to form the resultant prototypes. Similar to the nanofillers, different kinds of polymers like polydimethylsiloxane (PDMS) [Nag et al., 2016a, 2018d, 2019b], polyethylene terephthalate (PET) [Wang et al., 2010; Nag et al., 2016b; Emamian et al., 2017], polyimide (PI) [Nag et al., 2017b; Nag and Mukhopadhyay, 2018; Han et al., 2019c], and others were used. These raw materials have been processed by different printing methods, including 3D printing [He et al., 2020a; Kalkal et al., 2021], laser ablation [He et al., 2020b; Alheshibri et al., 2022], and others. These MEMS and printing techniques have been very effective in developing sensors that have been used for smart environments.

The use of sensors for smart environments has been very effective in terms of robustness and performance. The sensors have been deployed in various locations [Mukhopadhyay, 2014; Zafeirelli and Kavroudakis, 2024] in a controlled environment to study human activities on a daily basis. This not only assists in determining human nature behavior but also protects patients and elderly people during emergencies [Leelaarporn et al., 2021]. Since the inception of smart environments [Ramírez-Moreno et al., 2021; Zorkany et al., 2022], researchers have tried to develop these scenarios with the assistance of actuators [Mukhopadhyay et al., 2021] and wireless communication protocols [Mukhopadhyay, 2022]. The smart environments have been able to utilize reliable networking infrastructure for data transmission between the different units forming the particular smart services [Kanellopoulos et al., 2023]. In specific cases, more than one wireless protocol is being connected due to their individualistic dissimilar scales [Kanellopoulos et al., 2023]. For example, smart home services have been known to use certain personal area networks (PANs) like ZigBee (IEEE 802.1.4) [Kelly et al., 2013], LoRa [Gupta and Van Zyl, 2021], and Bluetooth (IEEE 802.15.1) [James et al., 2022] in smaller environments, while using WiMAX (IEEE 802.16) [Al-Azzawi et al., 2020] in larger smart grid areas. All of these protocols have been able to trans-receive the data in asynchronous and synchronous manner. They have been capable of dealing with smart city applications in terms of data traffic and delay. Each of these protocols has been tested with a tougher quality of service to improve their bandwidth and decrease delay. Table 14.1 [Jawhar et al., 2018] shows some of the networking architectures and protocols considered for smart city environments.

14.2 State-of-the-Art

Where is All The Water is a multi-disciplinary research collaboration project involving four Australian universities, namely Macquarie University, University of New South Wales, Australian National University, and the University of Sydney. The project was managed by New South Wales Smart Sensing Networks (NSSN) and financially supported by the Department of Planning, Industry

Table 14.1 Comparison of the networking architectures used for smart environments.

76541Protocol	Char.	Physical layer specs	Data link layer specs	Data rate	Transmission range	Smart city app.
IEEE 802.15.4 (ZigBee)	Energy saving, very short-range	2.4 GHz Band, DSSS	Carrier sense multiple access with collision avoidance (CSMA/CA)	20–250 kbps	10–20 m	Smart buildings, smart grid, smart water
IEEE 802.15.1 (Bluetooth)	Cable replacement	2.4 GHz Band, FHSS/FSK	Master/Slave, time division duplexing (TDD)	1 Mbps	10–100 m	Smart buildings, smart grid, smart water
IEEE 802.11a	Data networking, local area network	5 GHz, OFDM	CSMA/CA, distributed coordination function (DCF)/point coordination function (PCF)	6, 9, 12, 18, 24, 36, 48, 54 Mbps	120 m outdoors	All
IEEE 802.11b	Data networking, local area network	2.4 GHz Band, DSSS	CSMA/CA, DCF/PCF	1, 2, 5.5, 11 Mbps	140 m outdoors	All
IEEE 802.11g	Data networking, local area network	2.4 GHz Band, DSSS, OFDM	CSMA/CA, DFS/PFS	6, 9, 12, 18, 24, 36, 48, 54 Mbps	140 m outdoors	All
IEEE 802.11n	Data networking, local area network	2.4 GHz and 5 GHz Band, DSSS, OFDM	CSMA/CA, DFS/PFS	15, 30, 45, 60, 90, 120, 135, 150 Mbps	250 m outdoors	All
IEEE 802.16 (WiMAX)	Metropolitan area network	2–66 GHz Band, OFDMA	TDD, frequency division duplexing (FDD)	2–75 Mbps	Up to 35 miles	Smart grid, smart water, pipeline monitoring
Satellite	Wide area network	1.53–31 GHz	Frequency division multiple access (FDMA), time division multiple access (TDMA)	10 Mbps (upload), 1 Gbps (download)	Satellites cover 100s of km to the entire earth	Pipeline monitoring, unmanned aerial vehicles (UAVs)

Source: Jawhar et al. [2018]/Springer Nature/CC BY 4.0.

and Environment, Government of New South Wales [Andersen et al., 2025]. The project was completed in 2022, and it has shown for the first time that low-cost sensor networks, in combination with other technologies, can respond to the Australian infrastructure problem of great distances and low population [Shearan et al., 2022]. It has demonstrated that local gravity measurements – once commercially made available – can be a technology that can help us to map the underground and assist with quantifying recharge to the groundwater. Along with those technologies, satellites have an important role to play in the management of resources. As an outcome of the project, large uncertainties are identified in water accounting, with the interaction between surface and groundwater being one of the major sources of uncertainty. Thus, innovative solutions to quantify the fluxes between above and below ground remain a priority.

Around New South Wales, Australia, there are a total of 4673 sites with telemetry-based monitoring systems, but only 396 recorded rainfall. Effectively, it is extremely difficult to know the exact amount of water received by the region, and then there are issues such as labor-intensive data collection, difficulty in sensing underground parameters, limitations in modeling, gaps in the collected data, as well as unaccounted differences. The completed project investigated the design and development of low-cost sensors for real-time measurement of parameters of interest, installed at the right places, along with local gravity sensing for underground parameters, satellite-based measurement, fusion of different types of sensor data, and finally, analysis of data to predict water quantity to reduce uncertainty. The details of the activities with different organizations, along with the division of spatial resolution to justify the research activities, are shown in Figure 14.1. For monitoring underground water movement with high spatial resolution, low-cost sensors for the measurement of soil moisture and soil temperature have been developed at Macquarie University, as shown in Figure 14.2. Figure 14.3 shows the process and installation of the low-cost environmental sensor node. The sensor can measure parameters up to a depth of 50 cm from the ground surface, along with many other environmental parameters. The measured data will be uploaded into the cloud through the Internet of Things (IoT) and will be used to calculate evapotranspiration in real time. The knowledge of real-time values of evapotranspiration can provide significant information on the missing water from the water system.

A proper understanding of water loss in the water system needs the fusion of different types of sensors. The reported sensors will produce measured data, a combination of which will create an accurate and better picture of the status of the quantification of water in our water system. There is a need for significant research and development to develop low-cost, high-performance, selective sensors for the detection of different parameters for the qualification of water to safeguard human lives from consuming contaminated waters [Akhter et al., 2020, 2021a, 2021b;

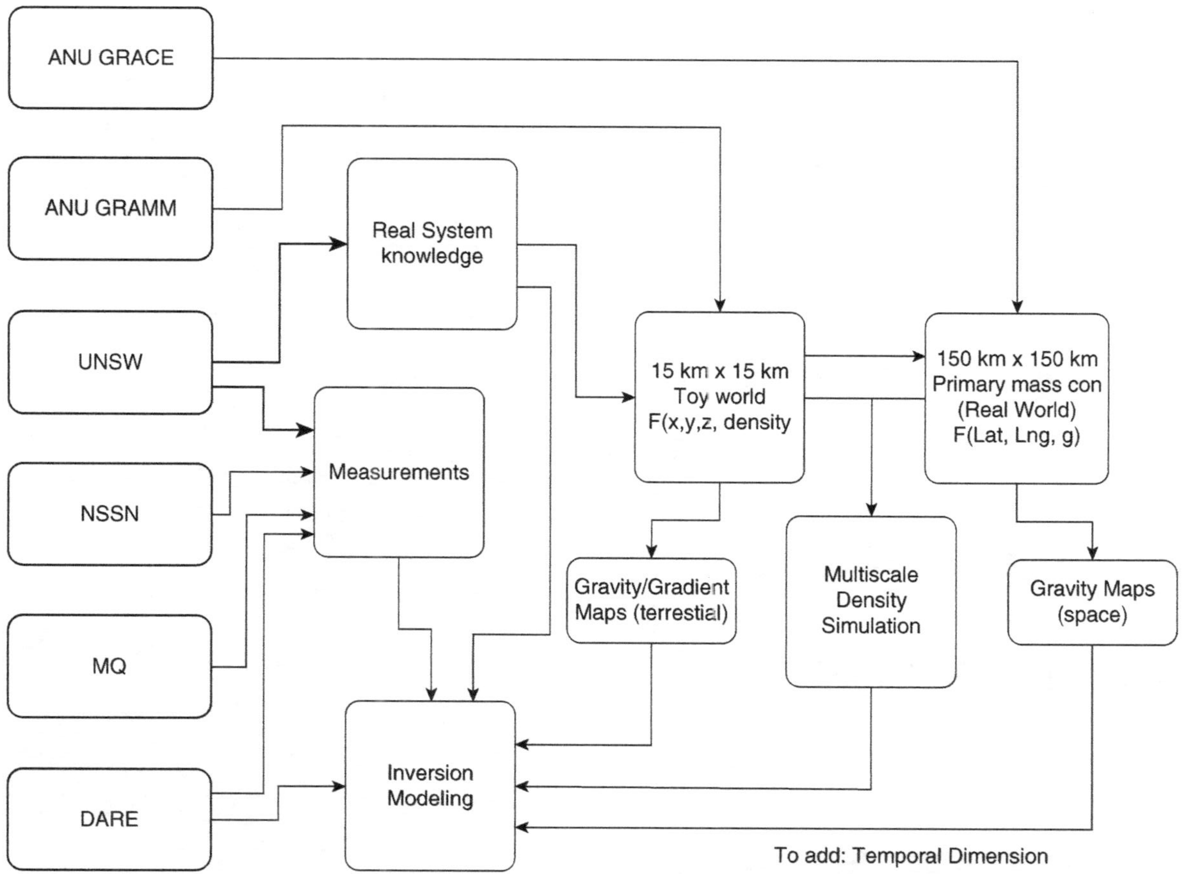

Figure 14.1 Representation of the steps followed in the project: Where is All The Water. Source: Adapted from XXX [2021].

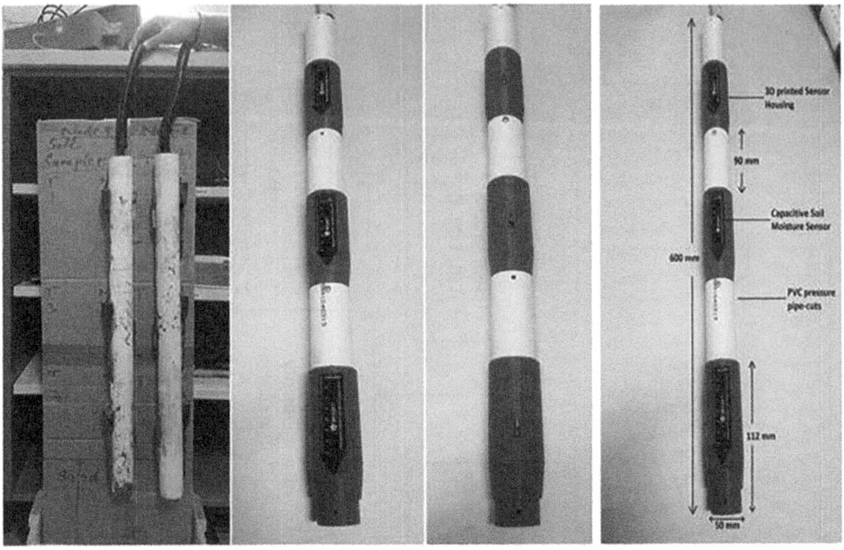

Figure 14.2 Development of a low-cost sensor for monitoring underground soil moisture and soil temperature. Source: Afridi et al. [2023]/IEEE.

Figure 14.3 Installation of low-cost environmental sensor node.

Shearan et al., 2021; Areekath et al., 2022; Hui et al., 2022; Wang et al., 2023]. The forward roadmap for the future, which shows the combination of knowledge of hydrology, modeling, physical low-cost sensors, remote sensing, and data analytics, will be extremely useful in determining parameters for the governance of water systems.

14.3 Immersive Technologies

Immersive technologies are playing an important role in promoting the Internet of Things (IoT) and smart environments. They help create interactive, engaging, and effective experiences within these environments. Here are some of the immersive technologies commonly associated with smart IoT environments (Figure 14.4):

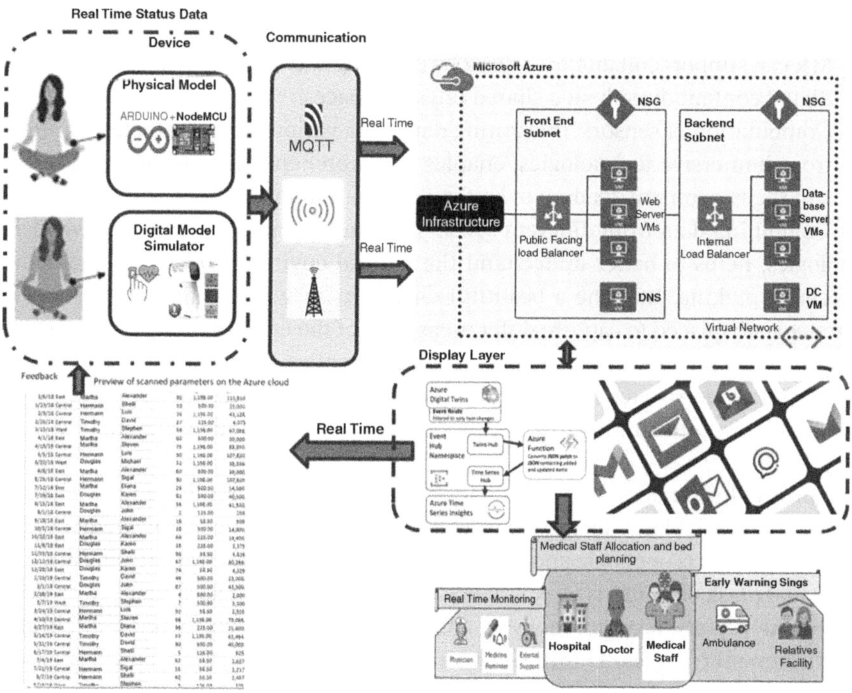

Figure 14.4 Digital twin – device layer, digital model, data computation and display (DDDD) architecture for a use case on health monitoring.

14.3.1 Augmented Reality (AR)/Virtual Reality (VR) and Mixed Reality (MR)

- Augmented reality (AR) can be used to provide technicians with timely information and relevant data covering physical equipment, making maintenance and repair easier.
- AR can help users navigate virtual environments by enhancing directions, points of interest, and other relevant information in their field of view. Training and simulation: virtual reality is used for immersive learning situations, allowing users to simulate situations in a real-world environment. This is especially useful for training users in a smart environment without exposing them to real risks. Remote monitoring and control: virtual reality can enable remote monitoring and control of devices and systems, providing environmental indicators.
- Mixed reality (MR) combines the elements of AR and VR, allowing users to interact with the physical and virtual worlds at the same time. This can improve the user experience in a smart environment.
- MR can support collaborative work by allowing users to share and interact with digital content directly in a shared physical space.
- Combination of sensors: integrating data from various sensors, including those from immersive technologies, enables a comprehensive understanding of the smart environment, leading to better decision-making.

Combining data from different sensors, including those from immersive technologies, helps to better understand the natural environment, leading to better decision-making. Imagine a beautiful environment where different sensors are strategically placed to capture different aspects of the environment. These sensors can include traditional IoT devices, cameras, motion sensors, temperature sensors, and immersive technologies such as AR and VR devices. Each sensor collects specific data, contributing to a comprehensive view of the environment.

Example: IoT devices can collect data on temperature, humidity, and other environmental variables. Camera: captures visual information, detects movement, and identifies objects or people and provides user interaction and real-time insight into a virtual mask or immersive experience. By combining the data streams from these different sensors, a comprehensive and comprehensive picture of the intelligent environment emerges. This universal understanding allows the following:

The integration of data helps to understand the context of the environment. For example, temperature data from IoT sensors can be combined with visual data from cameras to understand the impact of temperature changes on human behavior or equipment performance. Immersive technology contributes to virtual reality monitoring, providing dynamic overlays or virtual displays that update as the physical environment changes. This quick response is useful for quick decision-making. Linked datasets can be analyzed using advanced analytics and machine learning algorithms. This helps predict situations, anomalies, or potential problems before they become serious, helping to make better decisions. Integrating data from immersive technologies, such as augmented reality and virtual reality, with traditional security systems can provide a comprehensive security system. For example, combining visual data with information from motion sensors can improve the accuracy of threat detection. In the case of smart spaces designed for human interaction, understanding user behavior through immersive technology can lead to a personalized and user-friendly environment. The integration of data from various sensors, including immersive technology, helps decision-makers better understand the environment. This comprehensive data collection plays an important role in making the right decisions, optimizing processes, and creating a safe and efficient smart environment.

14.3.2 Smart Environments

A Digital Twin (DT) is a ***virtual representation*** of a physical object that enables direct data exchange between its physical counterparts. In healthcare, DT provides revolutionary solutions, solving problems such as early detection of health problems or monitoring chronic diseases in time. For example, a cardiac patient's DT can be analyzed to predict an upcoming cardiac event, thus providing prompt intervention. In the future, it is expected that intervention will affect personalized treatment and intervention. Digital technologies and services have been shown to be beneficial to both health professionals and patients as they support data collection, clinical communication, disease management, and other related services [Xu et al., 2019]. In addition, DT can solve the gaps in the current health care system, such as delays in obtaining patient data immediately in emergencies or in remote areas, providing potential benefits of immediate diagnostic information and immediate medical assistance, as described in Nag et al. [2015a]. This capability helps deploy a variety of critical applications on the Internet of Things (IoT), as shown in the study by Nag et al. [2015b, 2015c]. The digital transformation process that currently affects many sectors, including health, began with the launch of the Industry 4.0 project in 2013 [Alahi et al., 2017]. This process is based on advanced technologies such as IoT, cloud and edge computing, AI, and big data analysis [Afsarimanesh et al., 2017]. The DT system, based on the mentioned technologies, makes a digital transformation of any system and is often used by industrial and engineering companies. Over the past decade, DT technology has been widely adopted for healthcare applications. One of the best DT tools is Healthcare DT (HDT) [Nag et al., 2017a; Han et al., 2019a].

Device layer: this layer represents the controller, NodeMCU ESP8266, and wearable sensors. The aforementioned device layer plays a very important role in data collection for the IoT Hub using the MQTT protocol. MQTT is a message protocol designed specifically for networks with limited bandwidth, high latency, and inconsistent connectivity, which is often encountered in the IoT environment. Communication immediately plays an important role in the DT process.

Digital model: virtual reality tends to provide users with medication reminders and emergency notifications while monitoring the physiological state of the patient directly using the wearable device. This involves building a history of the participant, including sensor examples and activities to identify activities, such as medication and health-related activities. These types can predict future behavior and enable thinking and forecasting.

Data computation-cloud layer: comprises components essential for D2C telemetry and data integration at the cloud layer.

Communication: the actual data exchange between the front layer, layer, and cloud layer is possible through the communication layer. The protocols applied to it are HTTPS and MQTT.

14.4 Immersive IoT Technologies

The combination of immersive IoT and mobile edge computing (MEC) provides a powerful collaboration that can open new opportunities in different environments, delivering immersive content with low latency, improved efficiency, and optimism. MEC is a network architecture concept that brings processing and security capabilities closer to the edge of the network and closer to the data source.

Benefits:

- By processing data closer to the source, MEC reduces latency, which is critical for applications. Bandwidth efficiency: offloading processing operations to the edge reduces the need to transfer large amounts of raw data to central cloud servers.
- Distributing computing resources at the edge enables better scalability and better performance for a large number of connected devices.

The synergy between immersive IoT and MEC:

- **Timing**: MEC supports timing, which is important for an immersive experience where low latency is essential to avoid delays.
- **Reduced network load**: by localizing data at the edge, immersive IoT applications can reduce the load on network infrastructure, thereby reducing compression and improving overall performance. Keeping sensitive data close to the

edge can improve the privacy and security of immersive applications because sensitive information is not transmitted over long distances, thereby improving privacy and security.

Use cases:

- Immersive IoT applications for smart city planning, navigation, and infrastructure, as well as MEC ensure low response.
- AR/VR tools are used for training, management, and monitoring in industrial environments, and MEC supports real-time analysis and decision-making.
- MEC can support AR-assisted surgery or remote patient monitoring, increasing the power of immersive IoT solutions for healthcare.

Given the inherent complexities of wireless communication and computation technologies, decision-making and resource management in such dynamic environments have become increasingly challenging. Making correct offloading decisions is crucial for system efficiency, as incorrect choices can negatively impact performance. Machine learning approaches offer a solution by learning from data and enabling more efficient decision-making in offloading scenarios. This is very useful when we deal with complex problems that require considering multiple factors. In contrast to classical methods, machine learning methods excel in dynamic systems where decisions need to be made quickly and efficiently. Classical methods may struggle with non-polynomial optimization problems, which are impractical to solve in highly dynamic real-time systems. Therefore, in today's highly dynamic and time-critical smart systems, employing machine learning methods is a more intelligent and beneficial approach. One of the important tasks of **immersive IoT technologies** is the "Increasing system lifetime by optimal and accurate offloading of compute-intensive tasks."

The purpose of this case study is to automate the decision-making capabilities of a system that will select a mobile device for task execution based on certain parameters. Computation offloading is of two types: fine-grained (dividing the task into multiple components) and coarse-grained (treating each task as a whole). Here, we are providing a solution for coarse-grained offloading only, which tells us we should execute the task locally or remotely as a whole. Most of these tasks or data-offloading decision problems are dataset-dependent, and your problem formulation should be highly corelated to the dataset. A common method followed by everyone to determine the task size or task load is calculating the total CPU cycle needed for the mobile device to complete that task (this can be achieved with the help of task execution time). This value might not be very accurate as the devices that we are using can be from different generations. In other words, they can have different fabrication processes and distinct transistor sizes, so they might use individual instruction sets, and each of these instructions might take random

CPU cycles to complete. Therefore, instead of generalizing the CPU cycle count, we should focus on a more consistent approach. For example, we can benchmark a CPU to find out its capabilities in various task scenarios, and then we can see what percentage of that processor is being used for how much time. Using this method, we can generate real-time accurate task load data for further processing.

14.4.1 System Model

In order to execute the task remotely, we followed the client–server paradigm using JAVA in Android Studio. Here, only one device can act as the local device (Samsung on 5 Pro), and the better device (Realme 5 Pro) can be used as a remote MEC server at a time. As the implementation and problem formulation are dependent on the dataset, if you change any parameters of the research, then we have to collect the dataset again for the new scenario. This selection is done after testing the capabilities of the devices. Both devices will have the {pdf2text} Android application installed within them for dataset creation. Another Android application named {Decision} will be there only in the local device that will give the offloading decision after taking input from the user (Figure 14.5).

We built an application named pdf2text. This Android app extracts text data from PDF files and shows that content in an Android text-view. This will be our main task of choice. We will run many PDF files on both devices locally and remotely to see how they perform, and, in the background, we will run Android Studio

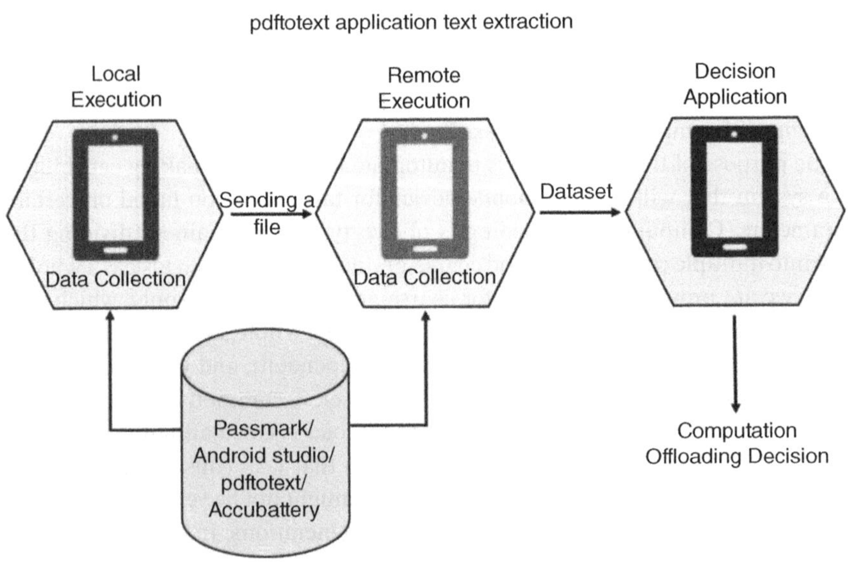

Figure 14.5 System model.

profiling and battery tracing apps to collect a lot of information. After training the ML model, we cannot run this model directly in JAVA, an Android environment. First, this model is downloaded as a pickle file, and with the help of the Flask framework, we convert our model to an application program interface (API) that can be stored in the cloud services, and later on runtime, our mobile app can access this model via API call. In the bottom left corner, battery consumption is shown, and in the bottom right corner, CPU core usage is shown by Accubattery. This instance is taken from a Realme 5 Pro device, which has an octa-core processor. When we ran a file of size 1017 KB, we could see both high-performance cores reached peak usage, but the other efficiency core's usage stayed at a bare minimum.

Local execution model: we are solving a coarse-grained [Nag et al., 2022a] offloading decision problem by trying to automate the device selection, and for that, we need to calculate the cost of execution in the local device. First, we will find out the CPU's capability. The Samsung device performs 1407 million integer operations per second, which is shown in this benchmark. Let us call that *local*, as it is directly proportional to the CPU frequency rate. When we extracted text from a PDF file in the pdf2text application, the mobile device consumed p_l mA (milliampere) energy per second, the task was completed after t_l seconds, and the average CPU usage was given by $s\%$. Then, the task (i) size for the local device will be

$$T_i^l = f_{local} \times s \times t_l,$$

and the total energy consumed by the local device for the task (i) will be

$$E_i^l = p_l \times t_l.$$

For example, text extraction from a PDF file of size 627 KB took 4.677 seconds in the Samsung device, with 31.33% CPU usage and 235.33 mA/s battery consumption.

$$\textit{Task size} = T^l = \frac{1407 \times 31.33 \times 4.677}{100} = 2061.68 \text{ (million ops)}.$$

$$\textit{Energy consumed} = E^l = 235.33 \times 4.677 = 1100.63 \text{ mA}.$$

In our experiments, we saw that the Samsung device uses 110 mA when all the background tasks are turned off except Bluetooth connectivity, and while transferring any file, it rises to 160–200 mA.

14.5 Network and Remote Execution Model

The computation offloading vastly depends on the communication mechanism and distance between the devices. As our proposal is to make an energy-efficient algorithm, we used Bluetooth. For remote execution, we have two components: data offloading cost and execution cost. We have calculated the capability of the Realme device, which is 21,519 million operations per second. Let us call that f_{rem}.

To send a PDF file to a remote device, the local device consumes pc_{om} mA energy per second, and communication time is t_{com}. The remote mobile device consumes p_r mA (milliampere) energy per second, on average, for execution; the task is completed after t_r seconds, and the average CPU usage is given by $s\%$.

Then, the task (i) size for the remote device will be

$$T_i^r = f_{rem} \times s \times t_r,$$

and the total energy consumed by the local device for the task (i) would be

$$E_i^r = (p_{com} \times t_{com}) + (p_r \times t_r).$$

Bluetooth transfer energy consumption is an overhead in remote execution = sender side + receiver side = 65 mA + 40 mA = 105 mA. Let us take the same example: to send the file from local to a remote device at a distance 40 ft, it took 12.5 seconds, and text extraction from a PDF file of size 627 KB took 1.077 seconds in the Realme device, with 19.66% CPU usage and 221 mA/s battery consumption took place.

$$\textit{Task size } (T^l) = \frac{21{,}519 \times 19.66 \times 1.077}{100} = 4556.39 \text{ (million ops).}$$

$$\textit{Energy consumed } (E^r) = (105 \times 12.5) + (221 \times 1.077) = 1550.51 \text{ mA.}$$

14.5.1 Decision-Making Procedure

In order to provide a decision, the decision model requires a composite set of inputs, which we call a system state given by $S = \{F, Pg, Tl, Tr, D, C\}$, where F is the file size, Pg means the pages, Tl denotes the task size for the local device, Tr means the task size for the remote device, D is the distance between devices, and C is communication time. These are the same parameters we used to generate our dataset. We observe the composite state S and then calculate the immediate costs of executing the task locally and offloading it. Based on these costs, we make the decision ri, which can be either executing task (i) in the Samsung device ($ri = 0$) or offloading it to the Realme device ($ri = 1$). If we take the same example of the subsection local execution model and remote execution model, the local and remote energy costs are 1100.63 and 1550.51 mA, respectively. Therefore, we select the local device for this task.

14.5.2 Data Collection

One of the major challenges in solving this decision problem was the lack of a dataset. After completing the problem formulation, we decided to generate two types of data: (i) computation cost and energy cost and (ii) network data. Finally, they were merged together to generate a novel dataset of [2] of 6600 rows (Figures 14.6 and 14.7).

We ran numerous PDF files of different types on the pdf2text Android application and collected data using the Android Studio profiler and the Accubattery application. Major features were PDF file size, the page count, RAM consumption, local and remote CPU usage, local and remote task size (composite data), local and remote execution time, local and remote energy consumption, etc. Even though the energy cost computation ideas have been taken from the old literature in this

	Task size (KB)	Pages	Local cpu use (%)	local ram use (MB)	local Exec. time(s)	ener loc/s (mA)	remote cpu use (%)	remote ram use (KB)	remote Exectime (s)	Exec ener rem/s (mA)	Distance (ft)	Comm. time (s)	Com. ener (mA)	Task size local (ops)	Task size rem (ops)
0	1	1	15	44	0.504	160	2.5	65	0.308	175	1	5.5	105	106.36	165.69
1	1	1	15	44	0.504	160	2.5	65	0.308	175	2	5.5	105	106.36	165.69
2	1	1	15	44	0.504	160	2.5	65	0.308	175	3	5.8	105	106.36	165.69
3	1	1	15	44	0.504	160	2.5	65	0.308	175	4	5.9	105	106.36	165.69
4	1	1	15	44	0.504	160	2.5	65	0.308	175	5	5.9	105	106.36	165.69
......	......	......	......	......	......	......	......	......	......	......	......	......	......	......	......
6595	46394	564	32.38	111.4	96.932	247.3	16	147.3	20.708	482.81	36	697.7	105	44161	71298
6596	46394	564	32.38	111.4	96.932	247.3	16	147.3	20.708	482.81	37	711.3	105	44161	71298
6597	46394	564	32.38	111.4	96.932	247.3	16	147.3	20.708	482.81	38	745	105	44161	71298
6598	46394	564	32.38	111.4	96.932	247.3	16	147.3	20.708	482.81	39	778.5	105	44161	71298
6599	46394	564	32.38	111.4	96.932	247.3	16	147.3	20.708	482.81	40	840	105	44161	71298

Figure 14.6 Extended dataset.

	Task size	Pages	Task size local	Task size rem	Distance	Comtime
0	1	1	106.36	165.69	1	5.5
1	1	1	106.36	165.69	2	5.5
2	1	1	106.36	165.69	3	5.8
3	1	1	106.36	165.69	4	5.9
4	1	1	106.36	165.69	5	5.9
......	......	......	......	......	......	......
6595	46394	564	44160.92	71298.47	36	697.7
6596	46394	564	44160.92	71298.47	37	711.3
6597	46394	564	44160.92	71298.47	38	745
6598	46394	564	44160.92	71298.47	39	778.5
6599	46394	564	44160.92	71298.47	40	840

Figure 14.7 Final training dataset.

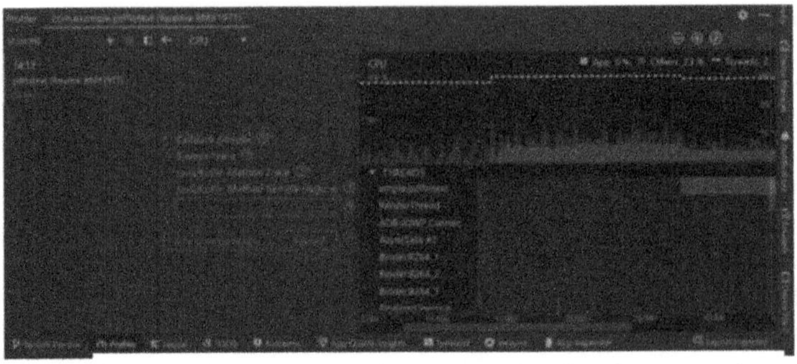

Figure 14.8 Android Studio CPU and execution time profiling.

domain, we have given the idea to use CPU capability for most of the calculation.

$$Tasksize = s \times f \times t \text{ (million-ops)},$$

where s is the average CPU usage, f is the CPU capability to perform integer operations, and t is the execution time (Figures 14.8 and 14.9).

As both of the devices belong to a different generation of silicon fabrication and use a distinct version of OS, it is inevitable that they use a variable amount of RAM while running the app pdf2text. Therefore, while calculating, we have normalized the RAM usage value. While testing the Bluetooth capabilities in an open environment, we saw that both devices were able to send data up to 200 ft (60 m) of distance, but due to packet loss and inconsistent data transfer, we reduced our range to 40 ft. We have sent files of sizes between 1 and 50 MB multiple times while varying the distance between local and remote devices (Figure 14.10).

Similarly, for various files, we saw a uniform network speed drop when the distance between the two devices increased. When we are sending a 5 MB file 40 ft away, it takes 56 seconds, but when we are sending a 40 MB file to the same distance, it takes 840 seconds, which is not 8 times the previous communication time, so it is evident that at long distance, packet loss and signal barriers play a big role.

14.5.3 Optimal Problem Formulation

Using our mathematical model and dataset, we are training an ML/DL model. The objective of that model is to make optimal offloading decisions for each task or application. Let us denote the optimal offloading policy as σ, which minimizes system cost in this way.

$$\sigma^{r} = \operatorname{argmin} \sum_{i=1}^{n} E(S, r_i).$$

Minimizing system cost means each of our decisions will help us conserve energy for the system as we select the mobile device that is optimal for the task.

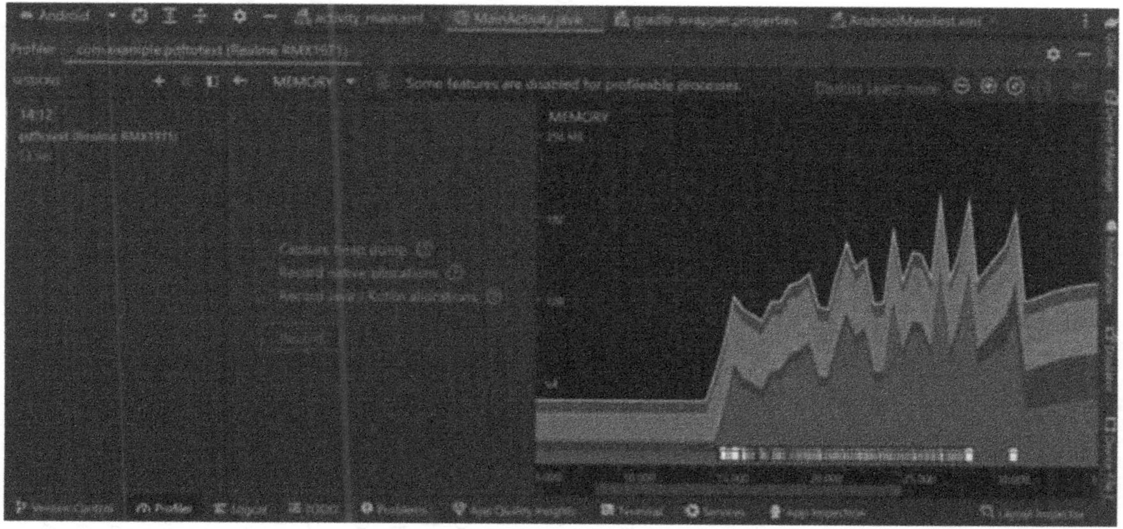

Figure 14.9 Android Studio RAM profiling.

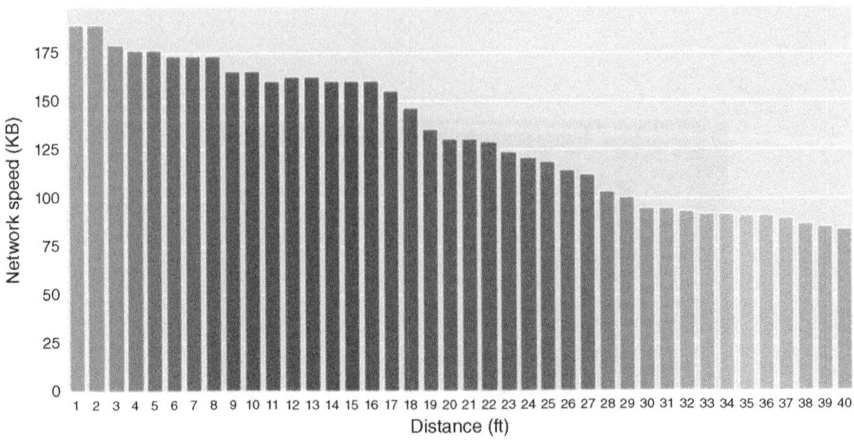

Figure 14.10 Network speed drops with distance.

14.6 Results

After running different ML classification algorithms on our dataset, we found that a 70 : 30 train (4620 rows) and test (1980 rows) split worked best. Out of these five ML algorithms, random forest performs the best with 99.6% accuracy (Table 14.2).

Then, we trained a neural network with different parameters. The best performance we obtained is 96.01% with 60 EPOCH and a 70 : 30 train–test split by using two hidden layers of 128 neurons with ReLU activation function and an output layer with softmax function (Tables 14.3 and 14.4).

Finally, we tested some of the elementary ways to offload the task or data.

Let us now see how decision accuracy affects energy consumption.

1) We try to extract text from a PDF file of size 10,203 KB (671 pages). If it is done locally (Samsung device), then CPU usage is 38%, RAM usage is 90 MB, time taken for completion is 180.272 seconds, and energy consumption will be 229.08 mA/s.

$$\text{Energy used } (E^l) = 180.272 \times 229.08 = 41{,}296 \text{ mA.}$$

In the case of offload (Realme device), CPU usage is 17.07%, RAM usage is 154 MB, time taken for completion is 34.631, and energy consumption will be 504.45 mA. Assuming that these two devices are situated 2 ft apart, the file transfer time is 50.1 seconds, and Bluetooth energy consumption is 105 mA/s.

$$\text{Energy used } (E^r) = 34.631 \times 504.45 + 50.1 \times 105 = 22{,}730 \text{ mA.}$$

Table 14.2 Accuracy results of ML algorithms.

Algorithms	Train/test split	Accuracy (%)
SVM	60:40	67.77
	70:30	67.78
	80:20	67.95
KNN	60:40	98.86
	70:30	99.24
	80:20	99.17
Decision tree	60:40	99.36
	70:30	99.29
	80:20	99.32
Logistic regression	60:40	93.86
	70:30	94.49
	80:20	94.47
Random forest	60:40	99.43
	70:30	99.60
	80:20	99.47

Here, we can see that a single correct decision can save us 41,296 – 22,730 = 18,566 mA energy.

2) Our system battery capacity = (Samsung) 2080 mA h + (Realme) 3159 mA h = 5239 mA h^2, which means our system can provide 5239 milliampere continuous energy for 1 hour or (5239 × 60 × 60) = 18,860,400 mA energy. The random forest algorithm gave us the best decision accuracy (99.6%) out of all the models we tested. Assume that our system gets 800 tasks of 10,203 KB files. It will select 797 (right) remote execution and 3 (wrong) local execution, resulting in energy consumption of: 797 × 22,730 + 3 × 41,296 = 18,239,698, which is less than our system battery capacity, so even after finishing 800 tasks, it leaves us with 620,702 mA system battery capacity.
3) Random offloading had an accuracy of 51.69%. Assume that our system again gets 800 tasks of the same 10,203 KB files, it will select 413 remote execution and 387 local execution, resulting in energy consumption of: 413 × 22,730 + 387 × 41,296 = 25,369,042 mA, which is much higher than our system battery capacity, therefore both the devices will run out of battery before most of the tasks are completed.

Table 14.3 Accuracy of neural network.

Training rounds	Split	Accuracy (%)
10 EPOCH	60 : 40	89.66
	70 : 30	92.27
	80 : 20	92.04
30 EPOCH	60 : 40	93.07
	70 : 30	94.85
	80 : 20	94.09
50 EPOCH	60 : 40	95.42
	70 : 30	95.05
	80 : 20	94.14
60 EPOCH	60 : 40	95.45
	70 : 30	96.01
	80 : 20	94.92
70 EPOCH	60 : 40	95.76
	70 : 30	95.45
	80 : 20	95.30
80 EPOCH	60 : 40	94.81
	70 : 30	95.61
	80 : 20	95.98

Table 14.4 Accuracy of basic algorithms.

Basic algorithms	Accuracy (%)
Random offload	41.9–51.69
Total offload	57.16
No offload	42.83

4) In case the random offloading achieves the best decisions in the first half of the execution of 800 tasks, which is 413 correct decisions, our system will still only be able to run 239 more tasks with the remaining battery. This way, we can prove that our energy-efficient machine learning-based offloading scheme is many times better than other basic approaches.

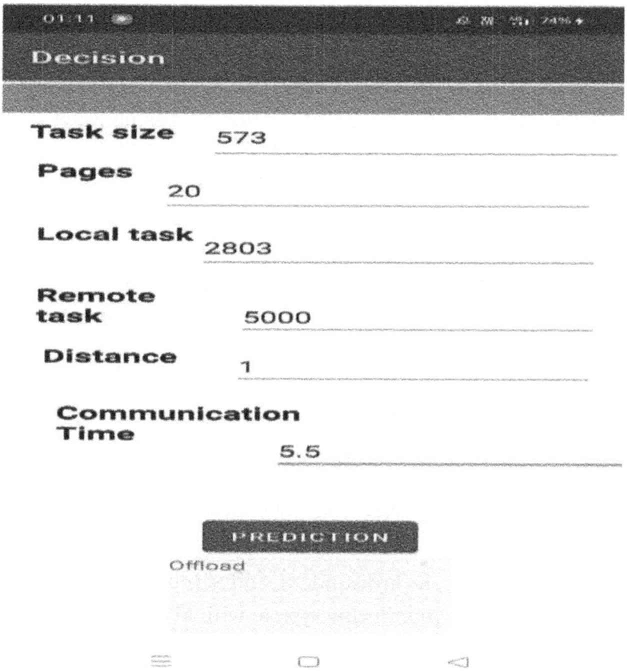

Figure 14.11 Decision Android app.

In Figure 14.11, we can see our second application, which is called Decision. It is taking the same input set that we have given to our ML model. A file of size 573 KB (20 pages), with local and remote task sizes of 2803 million ops and 5000 million ops, respectively, comes to the system. At that time, the distance between the local and remote devices is 1 ft, and they can transfer this file within 5.5 seconds. Then, the accurate decision is to offload the task to the Realme device.

The study focuses on the problem of offloading resource-intensive applications in MEC networks for immersive IoT theme of applications. We generate cost functions associated with local execution and offloading scenarios and present our decision-making idea. To minimize the overall execution cost, we formulate the problem as a binary classification problem and propose an algorithm, EEMOS, which is trained on our original dataset. Our approach aims to minimize system energy consumption. We compare our proposed approach (random forest implementation) with three alternative basic solutions, including total, random, and no offloading policy, and demonstrate that it outperforms them in terms of offloading accuracy by 42.44%, 47.91%, 56.77%, respectively, which improves system lifetime immensely. We did not compare our work with the literature, as our data collection method and problem formulation of task size calculation are novel.

Bibliography

W. A. Afridi, I. Vitoria, K. Jayasundera, S. Mukhopadhyay, and Z. Liu. Development and field installation of smart sensor nodes for quantification of missing water in soil. *IEEE Sensors Journal*, 23 (21):26495–26502, 2023.

N. Afsarimanesh, S. C. Mukhopadhyay, and M. Kruger. Molecularly imprinted polymer-based electrochemical biosensor for bone loss detection. *IEEE Transactions on Biomedical Engineering*, 65 (6):1264–1271, 2017.

N. Afsarimanesh, A. Nag, M. E. e Alahi, S. Sarkar, S. Mukhopadhyay, G. S. Sabet, and M. E. Altinsoy. A critical review of the recent progress on carbon nanotubes-based nanogenerators. *Sensors and Actuators A: Physical*, 344:113743, 2022.

F. Akhter, A. Nag, M. E. E. Alahi, H. Liu, and S. C. Mukhopadhyay. Electrochemical detection of calcium and magnesium in water bodies. *Sensors and Actuators A: Physical*, 305:111949, 2020.

F. Akhter, H. Siddiquei, M. E. E. Alahi, and S. C. Mukhopadhyay. An IoT-enabled portable sensing system with MWCNTs/PDMS sensor for nitrate detection in water. *Measurement*, 178:109424, 2021a.

F. Akhter, H. R. Siddiquei, M. E. E. Alahi, K. Jayasundera, and S. Mukhopadhyay. An IoT-enabled portable water quality monitoring system with MWCNT/PDMS multifunctional sensor for agricultural applications. *IEEE Internet of Things Journal*, 9 (16):14307–14316, 2021b.

M. E. E. Alahi, L. Xie, S. Mukhopadhyay, and L. Burkitt. A temperature compensated smart nitrate-sensor for agricultural industry. *IEEE Transactions on Industrial Electronics*, 64 (9):7333–7341, 2017.

F. F. Al-Azzawi, Z. F. Al-Azzawi, S. Shandal, and F. A. Abid. Modulation and RS-CC rate specifications in WiMAX IEEE 802.16 Standard with MATLAB Simulink model. In *IOP Conference Series: Materials Science and Engineering*, volume 881, no. 1, page 012109. IOP Publishing, 2020.

M. Alheshibri, K. Elsayed, S. A. Haladu, S. M. Magami, A. Al Baroot, İ. Ercan, F. Ercan, A. A. Manda, E. Çevik, T. S. Kayed, and A. A. Alsanea. Synthesis of Ag nanoparticles-decorated on CNTs/TiO_2 nanocomposite as efficient photocatalysts via nanosecond pulsed laser ablation. *Optics & Laser Technology*, 155:108443, 2022.

M. S. Andersen, J. Close, T. Hu, S. Legge, D. McCallum, S. Mukhopadhyay, P. Runcie, H. Rutlidge, L. Tamsitt, B. Shearan, P. Tregoning, and R. W. Vervoort. Where is All the Water? https://www.nssn.org.au/where-is-all-the-water, 2021 [accessed 4-February-2025].

M. S. Andersen, J. Close, T. Hu, S. Legge, D. McCallum, S. Mukhopadhyay, P. Runcie, H. Rutlidge, L. Tamsitt, B. Shearan, P. Tregoning, and R. W. Vervoort. Where is All the Water? NSW Government report prepared by the NSW Smart Sensing Network. https://www.nssn.org.au/where-is-all-the-water, 2025. [accessed 4-February-2025].

L. Areekath, G. Lodha, S. Kumar Sahana, B. George, L. Philip, and S. C. Mukhopadhyay. Feasibility of a planar coil-based inductive-capacitive water level sensor with a quality-detection feature: an experimental study. *Sensors*, 22 (15):5508, 2022.

S. Emamian, B. B. Narakathu, A. A. Chlaihawi, B. J. Bazuin, and M. Z. Atashbar. Screen printing of flexible piezoelectric based device on polyethylene terephthalate (PET) and paper for touch and force sensing applications. *Sensors and Actuators A: Physical*, 263:639–647, 2017.

J. Gao, S. He, A. Nag, and J. W. C. Wong. A review of the use of carbon nanotubes and graphene-based sensors for the detection of aflatoxin M1 compounds in milk. *Sensors*, 21 (11):3602, 2021.

G. Gupta and R. Van Zyl. Energy harvested end nodes and performance improvement of LoRa networks. *International Journal on Smart Sensing and Intelligent Systems*, 14 (1):1–15, 2021.

T. Han, A. Nag, N. Afsarimanesh, S. C. Mukhopadhyay, S. Kundu, and Y. Xu. Laser-assisted printed flexible sensors: a review. *Sensors*, 19 (6):1462, 2019a.

T. Han, A. Nag, S. C. Mukhopadhyay, and Y. Xu. Carbon nanotubes and its gas-sensing applications: a review. *Sensors and Actuators A: Physical*, 291:107–143, 2019b.

T. Han, A. Nag, R. B. Simorangkir, N. Afsarimanesh, H. Liu, S. C. Mukhopadhyay, Y. Xu, M. Zhadobov, and R. Sauleau. Multifunctional flexible sensor based on laser-induced graphene. *Sensors*, 19 (16):3477, 2019c.

S. He, S. Feng, A. Nag, N. Afsarimanesh, T. Han, and S. C. Mukhopadhyay. Recent progress in 3D printed mold-based sensors. *Sensors*, 20 (3):703, 2020a.

S. He, S. Feng, A. Nag, N. Afsarimanesh, M. E. E. Alahi, S. Li, S. C. Mukhopadhyay, and J. W. C. Wong. IoT-based laser-inscribed sensors for detection of sulfate in water bodies. *IEEE Access*, 8:228879–228890, 2020b.

S. He, Y. Zhang, J. Gao, A. Nag, and A. Rahaman. Integration of different graphene nanostructures with PDMS to form wearable sensors. *Nanomaterials*, 12 (6):950, 2022.

Y. Hui, Z. Huang, M. E. E. Alahi, A. Nag, S. Feng, and S. C. Mukhopadhyay. Recent advancements in electrochemical biosensors for monitoring the water quality. *Biosensors*, 12 (7):551, 2022.

A. James, A. Seth, S. C. Mukhopadhyay, A. James, A. Seth, and S. C. Mukhopadhyay. Bluetooth based IoT system. *IoT System Design: Project Based Approach*:137–166, 2022.

I. Jawhar, N. Mohamed, and J. Al-Jaroodi. Networking architectures and protocols for smart city systems. *Journal of Internet Services and Applications*, 9:1–16, 2018.

A. Kalkal, S. Kumar, P. Kumar, R. Pradhan, M. Willander, G. Packirisamy, S. Kumar, and B. D. Malhotra. Recent advances in 3D printing technologies for wearable (bio) sensors. *Additive Manufacturing*, 46:102088, 2021.

D. Kanellopoulos, V. K. Sharma, T. Panagiotakopoulos, and A. Kameas. Networking architectures and protocols for IoT applications in smart cities: recent developments and perspectives. *Electronics*, 12 (11):2490, 2023.

S. D. T. Kelly, N. K. Suryadevara, and S. C. Mukhopadhyay. Towards the implementation of IoT for environmental condition monitoring in homes. *IEEE Sensors Journal*, 13 (10):3846–3853, 2013.

S. Khan, L. Lorenzelli, and R. S. Dahiya. Technologies for printing sensors and electronics over large flexible substrates: a review. *IEEE Sensors Journal*, 15 (6):3164–3185, 2014.

P. Leelaarporn, P. Wachiraphan, T. Kaewlee, T. Udsa, R. Chaisaen, T. Choksatchawathi, R. Laosirirat, P. Lakhan, P. Natnithikarat, K. Thanontip, and W. Chen. Sensor-driven achieving of smart living: a review. *IEEE Sensors Journal*, 21 (9):10369–10391, 2021.

L. Liu, Y. Jiang, J. Jiang, J. Zhou, Z. Xu, and Y. Li. Flexible and transparent silver nanowires integrated with a graphene layer-doping PEDOT: PSS film for detection of hydrogen sulfide. *ACS Applied Electronic Materials*, 3 (10):4579–4586, 2021.

S. C. Mukhopadhyay. Wearable sensors for human activity monitoring: A review. *IEEE Sensors Journal*, 15 (3):1321–1330, 2014.

S. Mukhopadhyay. An enhancement to channel access mechanism for the IEEE 802.15. 3C MillimeterWave (5G) Standard to support stringent QoS requirements of IoT. In *IoT as a Service: 7th EAI International Conference, IoTaaS 2021, Sydney, Australia, December 13–14, 2021, Proceedings*, volume 421, page 19. Springer Nature, 2022.

S. C. Mukhopadhyay, N. K. Suryadevara, and A. Nag. Wearable sensors and systems in the IoT. *Sensors*, 21:7880, 2021.

A. Nag and S. C. Mukhopadhyay. Fabrication and implementation of printed sensors for taste sensing applications. *Sensors and Actuators A: Physical*, 269:53–61, 2018.

A. Nag, A. I. Zia, X. Li, S. C. Mukhopadhyay, and J. Kosel. Novel sensing approach for LPG leakage detection: part I—operating mechanism and preliminary results. *IEEE Sensors Journal*, 16 (4):996–1003, 2015a.

A. Nag, A. I. Zia, X. Li, S. C. Mukhopadhyay, and J. Kosel. Novel sensing approach for LPG leakage detection—part II: effects of particle size, composition, and coating layer thickness. *IEEE Sensors Journal*, 16 (4):1088–1094, 2015b.

A. Nag, A. I. Zia, S. Mukhopadhyay, and J. Kosel. Performance enhancement of electronic sensor through mask-less lithography. In *2015 9th International Conference on Sensing Technology (ICST)*, pages 374–379. IEEE, 2015c.

A. Nag, S. C. Mukhopadhyay, and J. Kosel. Flexible carbon nanotube nanocomposite sensor for multiple physiological parameter monitoring. *Sensors and Actuators A: Physical*, 251:148–155, 2016a.

A. Nag, S. C. Mukhopadhyay, and J. Kosel. Tactile sensing from laser-ablated metallised PET films. *IEEE Sensors Journal*, 17 (1):7–13, 2016b.

A. Nag, S. C. Mukhopadhyay, and J. Kosel. Wearable flexible sensors: a review. *IEEE Sensors Journal*, 17 (13):3949–3960, 2017a.

A. Nag, S. C. Mukhopadhyay, and J. Kosel. Sensing system for salinity testing using laser-induced graphene sensors. *Sensors and Actuators A: Physical*, 264:107–116, 2017b.

A. Nag, R. B. Simorangkir, E. Valentin, T. Björninen, L. Ukkonen, R. M. Hashmi, and S. C. Mukhopadhyay. A transparent strain sensor based on PDMS-embedded conductive fabric for wearable sensing applications. *IEEE Access*, 6:71020–71027, 2018a.

A. Nag, S. Feng, S. Mukhopadhyay, J. Kosel, and D. Inglis. 3D printed mould-based graphite/PDMS sensor for low-force applications. *Sensors and Actuators A: Physical*, 280:525–534, 2018b.

A. Nag, N. Afasrimanesh, S. Feng, and S. C. Mukhopadhyay. Strain induced graphite/PDMS sensors for biomedical applications. *Sensors and Actuators A: Physical*, 271:257–269, 2018c.

A. Nag, B. Menzies, and S. C. Mukhopadhyay. Performance analysis of flexible printed sensors for robotic arm applications. *Sensors and Actuators A: Physical*, 276:226–236, 2018d.

A. Nag, M. E. E. Alahi, S. Feng, and S. C. Mukhopadhyay. IoT-based sensing system for phosphate detection using Graphite/PDMS sensors. *Sensors and Actuators A: Physical*, 286:43–50, 2019a.

A. Nag, S. C. Mukhopadhyay, and J. Kosel. *Printed Flexible Sensors*. Springer, 2019b.

A. Nag, M. Alahi, E. Eshrat, S. C. Mukhopadhyay, and Z. Liu. Multi-walled carbon nanotubes-based sensors for strain sensing applications. *Sensors*, 21 (4):1261, 2021a.

A. Nag, M. E. E. Alahi, and S. C. Mukhopadhyay. Recent progress in the fabrication of graphene fibers and their composites for applications of monitoring human activities. *Applied Materials Today*, 22:100953, 2021b.

A. Nag, N. Afsarimanesh, S. Nuthalapati, and M. E. Altinsoy. Novel surfactant-induced MWCNTs/PDMS-based nanocomposites for tactile sensing applications. *Materials*, 15 (13):4504, 2022a.

A. Nag, R. B. Simorangkir, D. R. Gawade, S. Nuthalapati, J. L. Buckley, B. O'Flynn, M. E. Altinsoy, and S. C. Mukhopadhyay. Graphene-based wearable temperature sensors: a review. *Materials & Design*, 221:110971, 2022b.

A. Nag, A. Chakraborthy, M. Vega, S. Nuthalapati, H. Harija, M. S. Özer, and M. E. Altınsoy. Ultra-low-cost graphene/fabric-based sensors: part I-Characterisation and Preliminary results. *IEEE Sensors Journal*, 23 (23):28640–28648, 2023.

Z. Rafiee, H. Roshan, and M. H. Sheikhi. Low concentration ethanol sensor based on graphene/ZnO nanowires. *Ceramics International*, 47 (4): 5311–5317, 2021.

M. A. Ramírez-Moreno, S. Keshtkar, D. A. Padilla-Reyes, E. Ramos-López, M. García-Martínez, M. C. Hernández-Luna, A. E. Mogro, J. Mahlknecht, J. I. Huertas, R. E. Peimbert-García, and R. A. Ramírez-Mendoza. Sensors for sustainable smart cities: a review. *Applied Sciences*, 11 (17):8198, 2021.

B. Shearan, F. Akhter, and S. C. Mukhopadhyay. Development of an IoT-enabled portable sulphur sensor: a tutorial paper. *IEEE Sensors Journal*, 22 (11):10075–10088, 2021.

B. Shearan, S. Mukhopadhyay, P. Tregoning, S. Legge, J. Close, H. Rutlidge, J. Simmons, R. Scalzo, G. Francis, M. Isaacs, and L. Tamsitt. Where is All the Water? In *2022 22nd International Symposium on Electrical Apparatus and Technologies (SIELA)*, pages 1–7. IEEE, 2022.

L. Wang, J. Luo, J. Yin, H. Zhang, J. Wu, X. Shi, E. Crew, Z. Xu, Q. Rendeng, S. Lu, and M. Poliks. Flexible chemiresistor sensors: thin film assemblies of nanoparticles on a polyethylene terephthalate substrate. *Journal of Materials Chemistry*, 20 (5):907–915, 2010.

X. Wang, Y. Wang, H. Leung, S. Mukhopadhyay, Y. Cui, and D. Bai. An enhanced measurement for inorganics in water based on a novel planar three-electrode sensor. *IEEE Sensors Journal*, 23 (13):14988–14996, 2023.

A. W. Wieder and F. Neppl. CMOS technology trends and economics. *IEEE Micro*, 12 (4):10–19, 1992.

Y. Xu, X. Hu, S. Kundu, A. Nag, N. Afsarimanesh, S. Sapra, S. C. Mukhopadhyay, and T. Han. Silicon-based sensors for biomedical applications: a review. *Sensors*, 19 (13):2908, 2019.

S. Zafeirelli and D. Kavroudakis. Comparison of outlier detection approaches in a Smart Cities sensor data context. *International Journal on Smart Sensing and Intelligent Systems*, 17 (1):1–18, 2024.

M. Zorkany, E. Abd-Elrahman, and G. A. Fahmy. Real time IoT mobile anchor nodes outdoor localisation mechanism. *International Journal on Smart Sensing and Intelligent Systems*, 15 (1):1–14, 2022.

15

Deployment of IoT in Smart Environments: Challenges and Experiences

Waltenegus Dargie[1], Michel Rottleuthner[2], Thomas C. Schmidt[2], and Matthias Wählisch[1]

[1]*Faculty of Computer Science, Institute of Systems Architecture, Chair of Distributed and Networked Systems, TU Dresden, Dresden, Germany*
[2]*Faculty of Computer Science, HAW Hamburg, Hamburg, Germany*

15.1 Introduction

The Internet of Things (IoT) enables a seamless interaction with a wide range of everyday objects via the Internet. Earlier developments in the field primarily focused on stand-alone applications, such as home and industry automation [Mandula et al., 2015; Wollschlaeger et al., 2017], supply-chain management [Ben-Daya et al., 2019], and precision agriculture [Ahmed et al., 2018; Khanna and Kaur, 2019]. The common role of IoT technologies in these applications consists of (i) embedding a wide range of sensors and actuators in homes, machines, consumer electronics, moving vehicles, robots, etc. and (ii) establishing low-power wireless networks to interconnect the sensors and the objects of interest with the global Internet. As of today, the IoT has become an integral part of complex and distributed systems involving many stakeholders and many applications.

One of the most thrilling applications in the IoT is the implementation of smart environments. This task involves policymakers and urban planners, environment protection agencies, and environment researchers – this diversity leads to complexity, which begins with the various understandings of the term "smart environment" among stakeholders. For sociologists, a smart environment is primarily one that empowers communities, provides great transparency as regards the significance and implementation of high-level policies, promotes equity and diversity, enables the well-being of inhabitants [Toli and Murtagh, 2020], and fosters resilience [Petersen et al., 2015]. For economists, it is one which warrants economic vitality, prosperity, and easy commerce [Albino et al., 2015; Mouratidis and Poortinga, 2020]. For environmentalists, it is one which ensures the protection

Wireless Sensor Networks in Smart Environments: Enabling Digitalization from Fundamentals to Advanced Solutions, First Edition. Edited by Domenico Ciuonzo and Pierluigi Salvo Rossi.

and sustainability of physical environments and a high quality of life [Aguilar and San Román, 2019]. Most importantly, actual people live in, interact with, and are assisted and affected by "smart environments." Any of these stakeholders has different expectations and interacts differently with the IoT. The data abstraction they are interested in likewise differs significantly.

The technical implementations of the IoT face the challenges of designing a robust, distributed system, which meets a wide range of heterogeneous and varying requirements. Sensors, meters, appliances, and other data sources have to be sampled, and the data have to be compressed, aggregated, communicated, and stored. In most cases, a single sensor node integrates multiple sensors, each of which monitors a particular physical parameter. When the fundamental frequencies of the underlying physical phenomena are known, the sampling rates of the corresponding sensors can be determined by taking the Nyquist theorem into consideration. In reality, however, the distributions and rates of change of most parameters vary both in time and in space. As an illustration, consider the monitoring of the six principal air pollutants defined by the US Environmental Protection Agency (EPA) under the Clean Air Act [American Lung Association, 2020] (see Table 15.1). The EPA standard specifies the exposure limits, duration, and frequency for each pollutant but does not provide technical details as to their implementation. This includes the absence of a statement on the minimum sampling rates, accuracy, and temporal and spatial resolutions with which the parameters should be sampled. These issues can be partially addressed if individual users take charge of monitoring the parameters for personal ends. For example, they can employ wearable sensors or sensors integrated with personal devices (e.g. smartwatches, smartphones) whose location also implies personal location. Since the EPA standard implies that data and high-level events – e.g., the frequency of crossing important thresholds – have to be stored for years, individuals, in addition to sensing, should also provide data storage and background applications that process, aggregate, analyze, and act on the sensed data. If, however, (i) individuals should have to rely on publicly available infrastructure or data to keep track of their exposure history to the principal pollutants or (ii) if the pollutants have to be regulated, then the quality with which the parameters are sensed, aggregated, shared, interpreted, and stored, among others, is of profound importance.

High quality data – having high spatio-temporal resolutions – can be obtained by developing highly distributed and scalable sensing systems. This can be achieved by using low-power and affordable sensors and intelligent sensor nodes. But these systems will be limited in many respects, such as their computational power, communication bandwidth, and energy. The purpose of this chapter is to share our experience with developing IoT systems for smart environments and to highlight both the opportunities and the challenges facing such systems.

Table 15.1 The six principal air pollutants identified by the US Environment Protection Agency and the corresponding exposure limits.

Pollutant	Primary (p)/ secondary (s)	Averaging time	Level	Form
CO	p	8 hours	9 ppm	Not to be exceeded more than once per year
		1 hour	35 ppm	
Pb	p and s	Rolling 3 month average	0.15 μg/m^3	Not to be exceeded
NO_2	p	1 hour	100 ppb	98th percentile of 1-hour daily maximum concentrations, averaged over 3 years
	p and s	1 year	53 ppb	Annual mean
O_3	p and s	8 hours	0.070 ppm	Annual fourth-highest daily maximum 8-hour concentration, averaged over 3 years
$PM_{2.5}$	p and s	1 year	12.0 μg/m^3	Annual mean
		1 year	15.0 μg/m^3	Averaged over 3 years
	p and s	24 hours	35.0 μg/m^3	98th percentile, averaged over 3 years
PM_{10}	p and s	24 hours	150 μg/m^3	Not to be exceeded more than once per year on average over 3 years
SO_2	p	1 hour	75 ppb	99th percentile of 1-hour daily maximum concentrations, averaged over 3 years
	s	3 hours	0.5 ppm	Not to be exceeded more than once per year

CO: carbon monoxide; Pb: lead; NO_2: nitrogen dioxide; O_3: ozone; PM: particle pollution; SO_2: sulfur dioxide.
Source: Adapted from NAAQS Table, https://www.epa.gov/criteria-air-pollutants/naaqs-table, last accessed on 15 January, 2024.

The remainder of this chapter is organized as follows. In Section 15.2, we present two use cases and discuss opportunities and challenges in developing low-power and energy-efficient sensing systems. In Section 15.3, we identify important technical requirements to develop sustainable and scalable IoT. In Section 15.4, we discuss the latest developments in system support considering operating systems, communication protocols, and infrastructure. Finally, in Section 15.5, we identify and discuss some open issues and make concluding remarks.

15.2 Application Scenarios and Use Cases

15.2.1 Water Quality Monitoring

Water quality monitoring is one of the most important assignments of a smart environment. The task involves various subtasks, including monitoring (i) contaminants, (ii) pollutant discharge, (iii) impairment, (iv) drought, and (v) water use [Messer et al., 2014; Altenburger et al., 2019; Mishra et al., 2021]. Each of these subtasks involves several parameters to be monitored. Indeed, water quality monitoring goes beyond monitoring actual water bodies to include air and land. Research reveals that drought affects the concentration of pathogens and chemicals in water [Wolfram et al., 2021]; similarly, the deposition of some chemicals – such as calcium, magnesium, potassium, sodium, ammonium, nitrate, chloride, sulfate, and mercury – in the atmosphere affects water quality [Messer et al., 2014]. Most of these tasks require advanced and costly instrumentation and careful laboratory analysis. Considering the high price of sensors and the vast regions to be monitored, achieving high spatio-temporal sensing is a formidable challenge. At present, a combination of different approaches – wired sensing systems, remote sensing, wireless in situ monitoring, unmanned surface vehicles, and demand-based mobile sensing using special-purpose boats and crews – is employed in different settings to ensure safe and sustainable use of water, though a substantial body of research work confirms that this is far from adequate and that the quality of water is deteriorating worldwide [Lemm et al., 2021; Salehi, 2022; Mishra, 2023].

In collaboration with researchers at the Knight School of Computing and Information Sciences, Florida International University, we are developing low-cost and low-power IoT sensor nodes for water quality monitoring. Currently, the nodes are based on the Zolertia platform[1] and integrate (i) two different radios (one operating at 2.4 GHz and the other in the sub-gigahertz frequency bands), (ii) a 3D gyroscope, (iii) a 3D accelerometer, (iv) water and air temperature sensors, (v) a pH sensor, and (vi) a total suspended solids (TSS) sensor.

15.2.1.1 Challenges of Autonomous Mobile Sensing

Our prototype partially addresses the questions of cost and resolution in water quality monitoring. However, some challenges still remain to establish resilient and scalable IoT sensing systems. One of these challenges is ensuring that the nodes operate reliably in extreme weather conditions (excessive heat, high precipitation, and heavy rainfall) and without inhibiting or being affected by surrounding aquatic existence. Another challenge is establishing reliable wireless links and

1 https://docs.contiki-ng.org/en/release-v4.8/_api/group__zoul.html.

maintaining network topology and network connectivity in the presence of a significant amount of motion on the surface of the water on which the nodes are deployed. To surmount the first challenge, we seal the sensor nodes and other sensitive components inside floating waterproof boxes. This decision, however, introduces its own challenges:

- In the absence of ventilation system, how can we manage the heat dissipation produced by the various electronic components (the radio chips, the microcontroller, the power bank, etc.)?
- To what extent do the waterproof boxes affect the radiation and reception of electromagnetic waves?
- Even though the power banks are fitted with solar panels to harvest solar energy, confounding them inside the waterproof boxes prevents sunshine from reaching them. How can we provide adequate protection to the power banks and still harvest energy?

Some of these concerns are partially addressed at design time. For instance, the second issue is partially addressed by employing waterproof marine antennas, which are compatible with the RF front-ends of the two radios. Nevertheless, the extent to which the antennas compensate for the waterproof boxes can only be determined by carrying out experiments. The other concerns likewise require experimental investigations. For these reasons, we carried out several experiments with different configurations and involving different settings. In the first round of experiments, we put the sensor nodes in open boxes (refer to Figure 15.1), enabling the heat dissipation to exit from the boxes naturally and the electromagnetic signals to propagate relatively freely. In the second round of experiments,

Figure 15.1 Deployment of sensor nodes in open boxes on the surface of two different water bodies.

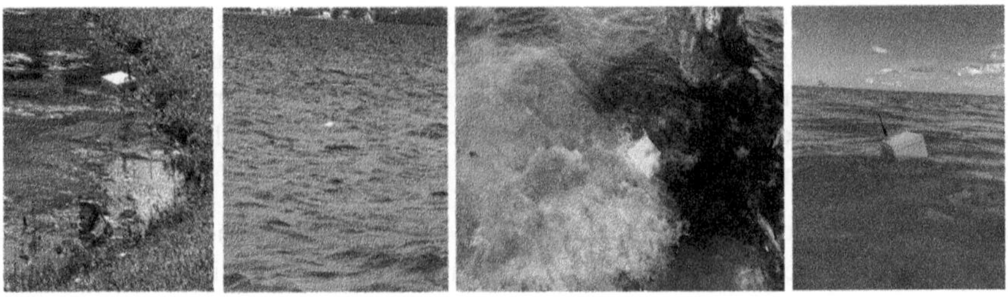

Figure 15.2 Deployment of waterproof sensor nodes on the surface of different water bodies.

we sealed the sensor nodes in waterproof boxes (refer to Figure 15.2). In all the experiments, the nodes self-organized to establish multi-hop wireless sensor networks.

In both types of experiments, the effect of heat dissipation originating from the electronic components was negligible, but the effect of excessive external heat was not. In order to distinguish between these, we performed experiments in different weather conditions in Miami, Florida, between the beginning of June and the end of August 2024. When the experiments were conducted early in the morning or late in the afternoon, or when it was cloudy, there was no appreciable difference in performance (packet delivery ratio, link quality fluctuation, and network fragmentation) between the nodes in the open boxes and the sealed boxes. But during the daytime, when the sun was shining with full power, the performance of the networks established by the nodes in the open boxes deteriorated considerably, packet loss exceeding on average 30% and some nodes failing to perform properly altogether, thus causing the network to fragment. The performance of the networks consistently improved when the nodes were sealed in the waterproof boxes in all weather conditions. We suspect that the boxes protected the nodes not only from the water but also from the extreme temperature and high relative humidity. Moreover, the gain of the waterproof marine antennas was high enough so as to make the effect of the boxes on the propagation and reception of the electromagnetic signal negligible. But all the networks were significantly affected by the motion of the water – motion not only exacerbated link quality fluctuation but also continuously changed the topology of the networks so that unicast communication often resulted in high packet transmission delay, high packet loss, and high retransmission cost. Figure 15.3 displays the change in the received signal strength indicator (RSSI) values of received packets and the 3D linear acceleration the receiving node experienced as a result of the movement of the water on the surface of which the node was deployed.

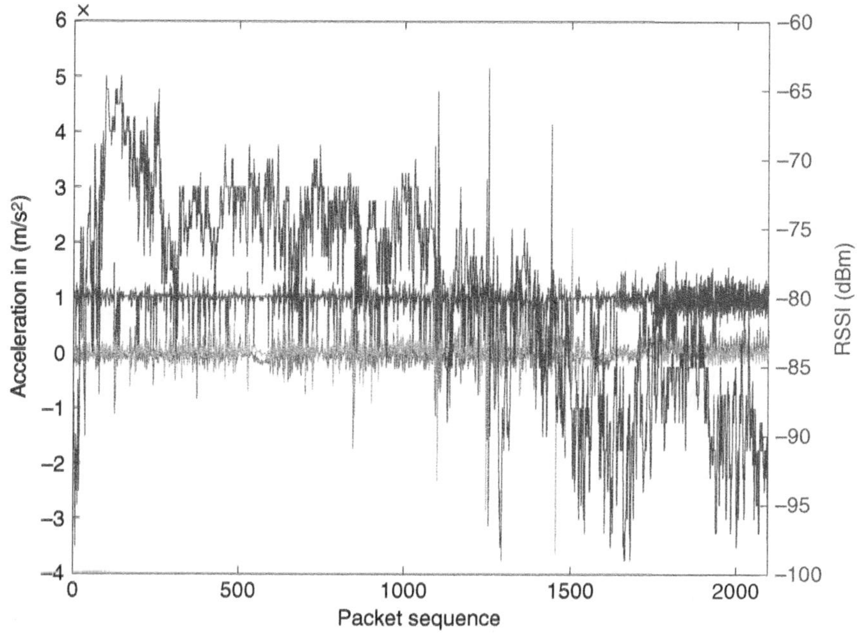

Figure 15.3 Relationship between the change in the RSSI of received packets and the change in the linear accelerations (along the three spatial dimensions) of a receiving node.

15.2.1.2 System Architecture and Implementation

We extend the 6LoWPAN architecture [Mulligan, 2007] to establish and maintain a fully connected network. The extended system aims to deal with mobility and link quality fluctuation and consists of four components, namely, the link quality estimation component, the transmission power adaptation component, the mobility model, and the packet transmission scheduler (refer to Figure 15.4). The mobility model establishes a relationship between the past, the present, and the future change in the RSSI of received packets using the Kalman Filter. This relationship is translated into multiple thresholds of the transmission power. The link quality estimation component evaluates link quality metrics (RSSI, packet reception ratio, and link quality indicator [LQI]) and decides to either delay transmission or adjust the transmission power. The transmission power adaptation component manages the transmission power of the underlying radio. Finally, the packet transmission scheduler manages the rate at which the user datagram protocol (UDP) layer sends out packets for transmission. The reason we place the scheduler at this layer is in order not to overwhelm the routing and the media access control (MAC) layers, since these two layers also temporarily store packets from their neighbors, and

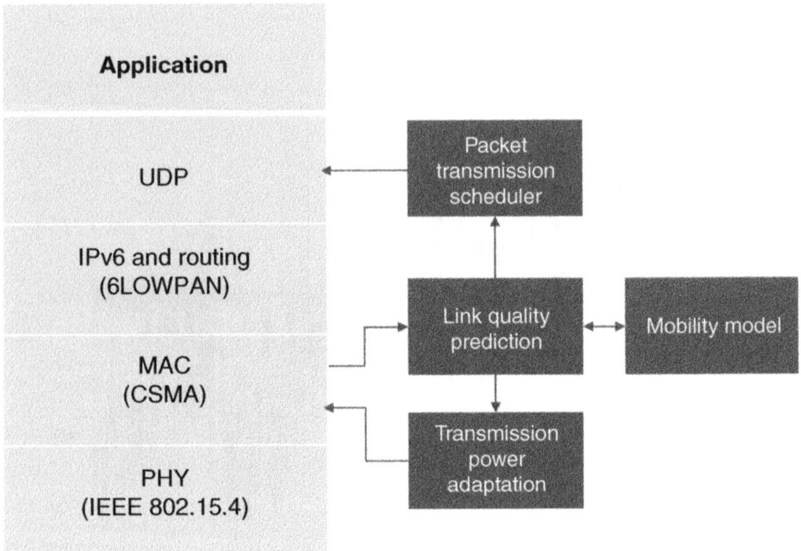

Figure 15.4 The architecture and implementation of the IoT sensing system for water quality monitoring.

their queues can easily overflow, leading to a considerable amount of packets being dropped.

15.2.1.3 Deployment Results and Lessons Learned

We deployed wireless sensor networks on the surface of four water bodies – on a lake at FIU's Main Campus, North Biscayne Bay, South Beach, and Crandon Beach – to study how their performance was affected by the movement of water. The water bodies experience different levels of motion. Similarly, each node's experience of motion was different from that of its peers even though they were all deployed on the same water body and the distance of separation between them was ca. 50 m. This resulted in a disparity of performance even within a single network. Figure 15.5 displays the change in the RSSI of received packets for two neighbor nodes and two different radios (CC1200 and CC2538). When using the CC1200, the nodes were transmitting at 2 Hz rate; and when using the CC2538 radio, at 10 Hz rate. The water at Crandon Beach – to which the figure refers – was moving appreciably (the experiments were conducted on August 28, 2023, at the time when the entire region was experiencing Hurricane Idalia, a powerful and destructive Category 4 hurricane[2]).

---

2 https://en.wikipedia.org/wiki/Hurricane_Idalia.

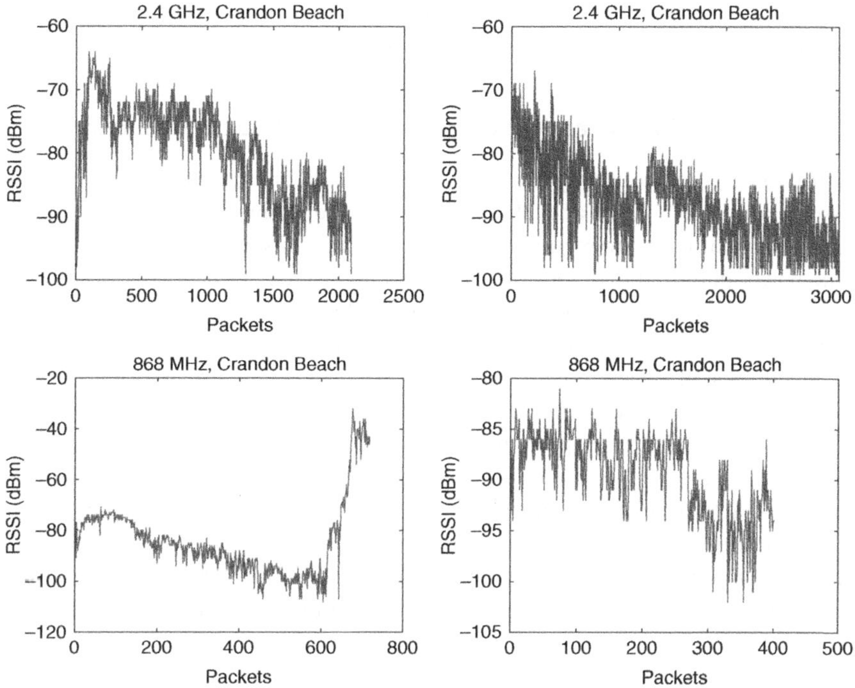

Figure 15.5 Link quality fluctuation of two neighbor nodes deployed on the surface of water at Crandon Beach, Miami, Florida.

Since the sensor nodes were unfettered, the movement of water produced two types of RSSI fluctuations, short-term fluctuations, which resembled fast fading, and long-term fluctuations, which resembled slow fading. Apparently, these fluctuations correspond to the localized and translational motions the two nodes underwent. If we consider the CC2538 radio, the long-term link quality fluctuation experienced an overall change of about 30 dBm from an initial state of connectivity (−70 dBm) to a state of disconnection (−100 dBm). When the water was modestly moved, a transmission power of 0 dBm was sufficient to establish a connection between the two nodes. The radio's output power can be programmed up to a maximum value of 7 dBm. If one adjusts the output power by 1 dBm, then seven stages of adaptation are possible to compensate for the change resulting from the movement of water. A steady link quality can be achieved if one divides the overall change in the RSSI values (30 dBm) into seven non-overlapping groups. This means that for an average change of ca. 4 dBm, a corresponding adjustment of 1 dBm in the output power enables the nodes to maintain a steady link quality until the connection can no longer be maintained. This is how the mobility model enables link quality adaptation.

15.2.2 Mobile Urban Sensing: Energy-Neutral Air Quality Monitoring

Maintaining good air quality is critical for the health of citizens [American Lung Association, 2020; Chen and Hoek, 2020]. People living in densely populated urban environments are regularly exposed to dangerous air pollution emitted by cars, power plants, and factories [American Lung Association, 2020]. Monitoring the air quality at different locations in the city allows for early detection of critical pollution levels and helps city planners to set up countermeasures. However, a small number of measurement stations at fixed locations provide only limited coverage of urban spaces. Bicycles, pedestrians, and car passengers move freely within the city and only punctually pass by such monitored locations. As a result, the collected data leaves out many blind spots and paints an incomplete picture of actual pollution exposure. The monitoring coverage also cannot be improved arbitrarily by just increasing deployment density, as that would result in unreasonable cost and maintenance overhead. Mobile sensing, on the other hand, promises better monitoring coverage without deploying many additional sensors.

In this deployment case, we demonstrate how existing mobile urban infrastructure entities and autonomously operating sensing devices can be capitalized to enhance sensing coverage. Figure 15.6 shows this deployment case: a self-sustainable sensing system mounted on a public transport bus to measure emission exposure in moving traffic. The bus is an emission-free electric vehicle operated in the citywide public transportation network in Hamburg, Germany. By virtue of the emission-free operation, our sensor node is able to observe pollution

Figure 15.6 ECO box prototype deployment on an electric bus.

levels within ongoing traffic, without being affected by the emissions of the carrier vehicle itself.

15.2.2.1 Challenges of Autonomous Mobile Sensing

This setting combines a unique set of challenges of perpetual, maintenance-free operation in smart environments. Due to regulations by the bus operator, no infrastructure of the vehicle can be used, such as power or network uplink. Hence, noninvasive mounting of the sensor node without any modifications of the bus must be ensured. Regularly swapping batteries is also infeasible in this scenario as the sensor is only accessible in the vehicle maintenance facility, and labor cost would significantly increase operational expenses. The sensor node must therefore operate self-sustainably, solely powered via energy harvested from its solar panel. Due to the mobility, the sensor has to locate itself via GPS so that measurements can be linked to specific geographical coordinates. Time synchronization is needed to accurately time stamp all collected data.

Energy harvesting is able to successfully utilize various energy sources in very different application domains, such as wearables and smart buildings [Ma et al., 2020], it promises to be an appropriate solution. However, a problem with measuring particulate matter (PM) concentration is that it requires complex sensors of relatively high-power consumption. GPS is also expensive in power consumption, which challenges energy-neutral operation at typically obtainable energy harvesting power levels [Sudevalayam and Kulkarni, 2011]. With the absence of a local network uplink, the system relies on Low Power Wide Area Network (LPWAN) technology with very limited data rates in the order of less than 1 kbit up to a few kbits per second.

The mobile setting introduces significant dynamics in external conditions. Energy available via solar energy harvesting depends on many factors, such as slow seasonal variations and short-term weather conditions. For mobile scenarios, this is further aggravated as orientation and obstacles that cast shadows change frequently over time. Unpredictable operation of the bus that serves different routes makes it even harder to predict energy availability. A reliable prediction cannot be expected under these conditions. Instead, a self-adaptive feedback mechanism is needed that can quickly adjust the system consumption to the changing energy availability [Rottleuthner et al., 2021]. The node does not only transmit measurement data but also statistics on the consumption of individual actions it performs to allow for collecting metrics related to the power management.

15.2.2.2 System Architecture and Implementation

The sensing system (see Figure 15.7) consists of multiple subsystems for power management, sensing, processing, and communication. A low-power 32-bit

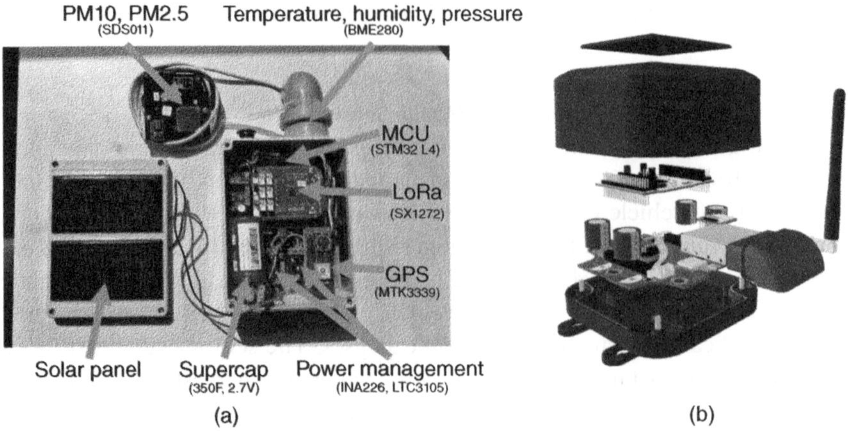

Figure 15.7 Components used in the ECO box prototype (a) and a visualization of the fully integrated design (b). The design is open source and targeted for building energy-neutral sensing systems for long-term outdoor deployments.

microcontroller unit (MCU) runs firmware based on the open-source operating system RIOT [Baccelli et al., 2018]. It controls all other subsystems, executes the adaptive power management, and collects data from external sensors for further processing. Power generated by the top-mounted solar panels is used to charge a super-capacitor. The power management module is responsible for self-measuring harvested and used energy by monitoring the super cap charging and the system power consumption. An SDS011 sensor measures particulate matter concentrations of the air (PM10 and PM2.5), with a resolution of 0.3 $\mu g/m^3$ and a maximum relative error of 15% $\pm$ 10 $\mu g/m^3$. With the BME280 sensor, temperature, humidity, and pressure are measured. An MTK3339 module is used for GPS positioning and time synchronization. The SX1272 radio module uses the LoRa modulation to transmit data over long distances to be received by community administered gateways of The Things Network (TTN).

15.2.2.3 Deployment Results and Lessons Learned

In our mobile sensing deployment, we collected data across an area of roughly 220 km^2 between August 2019 and April 2024. The gathered data provides insights into three main directions: urban pollution, network coverage, and self-sufficient dynamic power management of mobile sensing devices. Very different environmental conditions were encountered during the deployment with temperatures ranging from −8 to 52°C due to the outside placement in winter temperatures and direct exposure to sunlight during summer. Temperature swings within a single day reached up to 40°K.

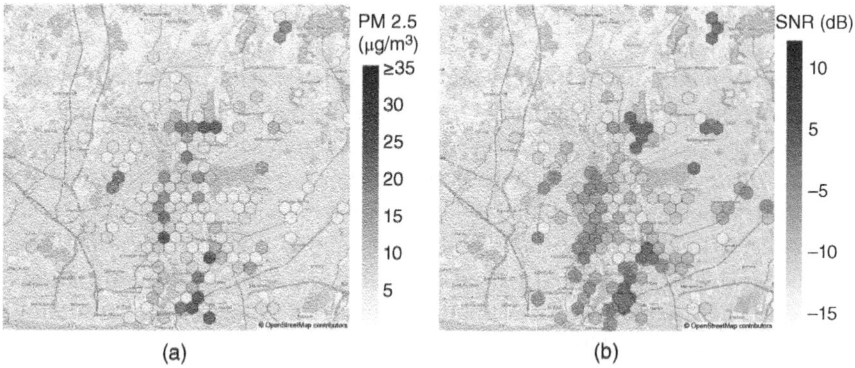

Figure 15.8 Mobile sensing deployment in the city of Hamburg, Germany: mean particulate matter pollution (PM2.5, a). Average signal-to-noise ratio of received data transmissions with dots marking reported gateway locations (b).

We focus our results on particulate matter (PM) measurements of PM2.5, as it was previously shown to increase mortality rate more significantly than PM10 [Chen and Hoek, 2020]. Figure 15.8a shows a heat map of the average PM2.5 concentration, which indicates several hotspots where average pollution levels overshoot primary and secondary concentration levels as defined by the NAAQS (see Table 15.1). The mean value across the whole area is 5.2 μg/m^3, with the 98th percentile at 25 μg/m^3. More polluted hotspots were measured between 20 μg/m^3 and 100 μg/m^3, with one extreme outlier at 700 μg/m^3.

Next to pollution data, the collected data also enables a better understanding of urban LoRa network coverage via geolocated signal-to-noise (SNR) measurements of received transmissions. An inherent property of LoRaWAN infrastructure is that each transmission may be received by multiple gateways. We therefore average the SNR value of the weakest reception of each transmission to determine a conservative coverage map. Figure 15.8b shows the average of the minimum SNR for transmissions originated from each location sector. Circles mark the locations of the gateways that forwarded our measurements and provided location information. This data allows inferring insights about transmission distances and the quality of location data provided by gateway operators. During the deployment 27 different gateways forwarded data from our sensor. Six of those gateways did not expose any location via metadata reported by The Things Network and one incorrectly reported a location in Hong Kong. Only five gateways reported coordinates that plausibly originated from an actual GPS measurement. We therefore conclude that the gateway location information is not very reliable. Under these conditions, we expect that a hypothetical localization based only on LoRaWAN, to be only of limited use for very rough estimates. The average distance of successful transmissions was 780 m, the 90th percentile is at 2.2 km, with the highest achieved distance recorded at 18.5 km.

The power management statistics show that during normal operation, the overall energy consumed for obtaining a GPS fix varied largely with a relative standard deviation of 75%. During normal operation, the system sent around 250 packets per day. However, on average, only 38 packets per day successfully reached one of the distributed gateways, which indicates problems with the network coverage and transmission reliability. Caching data for delayed transmission at locations with better coverage or alternatively using higher spreading factors for transmissions could help to improve this. Long-term persistent storage on the sensor node could provide additional insights by considering all the data that could not be successfully transmitted.

The bus was regularly parked for charging, where repeated measurements at the same location did not contribute to better monitoring coverage. A low-energy movement detection sensor based on an inertial measurement unit (IMU) or a vibration detector could be a useful extension to further improve obtained coverage for a given amount of harvested energy. In other cases, the bus was parked in a garage for longer periods of time to do maintenance. Neither enough sunlight nor GPS reception or network connectivity was available there, which depleted the energy significantly faster than usual, thereby increasing bootstrapping time.

The GPS module of the sensing system relied on a backup battery to keep its volatile state, which is needed for better location acquisition speed. The lifetime was estimated to last throughout the deployment as the backup battery is only used if no power is supplied to the GPS module. In our system, the GPS module is powered via renewable energy harvested from the solar panels. During later development, we determined that the software controlled low-power mode of the GPS module resulted in too high-power consumption during sleep mode. Alternatively, a transistor controlled by the MCU was then employed to cutoff the GPS supply when unused. However, this resulted in a higher load on the backup domain battery, which was not noticed before the sensor box was deployed. After the battery was depleted, the GPS module had to perform a GPS cold start for every location acquisition, which significantly increased the GPS runtime and energy consumption of the GPS. For better performance, the backup domain should also be powered by the renewable energy harvesting source.

Some of the changes above could have been retrofitted during deployment via over-the-air (OTA) firmware updates. OTA updates over LoRaWAN, however, are impractical due to very limited downlink bandwidth.

The backend infrastructure caused problems several times during the deployment. On some occasions, data was not correctly received from the TTN server due to sporadic message queuing telemetry transport (MQTT) session timeouts and server maintenance on our side. The impact of our own outages could have been reduced by setting up multiple distributed systems as simultaneous backup subscribers for the data published via MQTT by the TTN server. The most critical interruption was, however, caused by a breaking change in the TTN stack, which required intervention on the management interface and a manual migration of the sensor end device to the new TTN server instance. After the proposed migration process, our sensor took multiple weeks before it joined the network server again. A control interface in the sensor firmware to schedule a node rejoin procedure remotely would have helped to speed up the migration. The new version of the community-driven TTN stack (v3) was only slowly adopted by the community and shows severely reduced coverage, as can be seen in Figure 15.9, which compares both versions as presented by a third-party coverage map.

15.3 Requirements Analysis

The two use cases presented in Section 15.2 demonstrate the different ways of developing adaptive, scalable, and low-power sensing systems, which can achieve high spatio-temporal sensing. The presented system architectures – one high-level, with a focus on networking aspects, and the other, low-level, with a focus on the organization of hardware components, are intended to portray complementary features. At the same time, the use cases serve to identify shared features between heterogeneous platforms and applications. The usefulness of low-power devices and networks greatly depends on how well the data they generate represent underlying realities. Individual sensors or sensor nodes may only be marginally reliable, but the IoT as a whole has to be reliable and transparent as regards the quality of data it generates. One key aspect of useful IoT deployments is, therefore, to communicate and store data as reliably as possible. This, in turn, requires highly resilient networks, which operate in extreme conditions and rough environments. In such environments, individual nodes may unexpectedly exhaust energy, wireless links may experience intense cross-technology interference, local heat dissipation may render some of the nodes temporarily dysfunctional, and so on. These conditions may force networks to fragment and individual nodes to be disconnected from their parent networks for a long time. Self-adaptive features, such as decentralized ad hoc and mesh networking, enable nodes to surmount some of these challenges. Second, in the presence of outage and bad coverage, local data storage on IoT end devices is critical. However, the data may be voluminous and heterogeneous (metadata, performance indicators, logs, and other device telemetry, which may require long-term persistence), requiring advanced compression and data storage. As persistent flash storage becomes more and more affordable, it will play a significant role in alleviating some of these concerns. Third, reliance on services and infrastructure accessible via proprietary

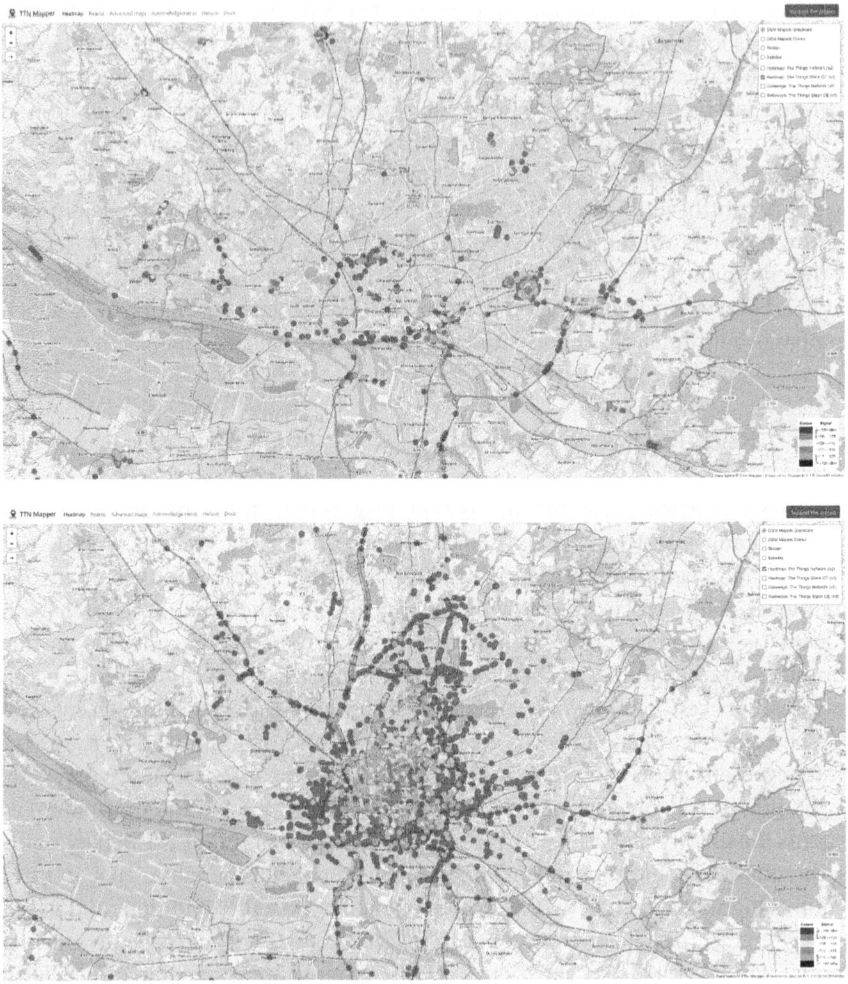

Figure 15.9 Coverage map maintained by the third-party project *TTN Mapper*, comparing the discontinued *The Things Network v2* to *The Things Stack Community Edition v3*. Source: https://ttnmapper.org/heatmap.

cloud services is not long-lasting, as it depends on the success and goodwill of manufacturers and operators. Open standards, unrestricted access rights, and vendor-agnostic interoperability, on the other hand, ensure the desired flexibility of operation, data access, continuity, and seamless integration and adaptation. Fourth, as reliance on IoT services becomes ubiquitous, scalability, security, and longevity become essential features. A sustainable growth of IoT deployments is only feasible insofar as their maintenance requirements are low. Similarly, IoT devices typically operate with exhaustible batteries and, in most cases, replacing or recharging batteries is time consuming and expensive. In some deployments, it is even impractical. On the bright side, the energy efficiency of embedded systems has steadily improved and enabled IoT devices to harvest energy from their surrounding in a much more feasible way today than it was a decade ago. Even so, work still remains to harvest energy reliably and achieve energy-neutral operation.

15.4 System Support

Most low-power wireless IoT devices require specialized software support due to their constrained capabilities. In addition, efficient and scalable use of the data they generate can be possible through systematic data description, aggregation, and abstraction.

15.4.1 IoT Operating Systems

With embedded hardware getting more sophisticated, it becomes increasingly more complex to build and maintain hardware-specific software from scratch. A multitude of open-source operating systems are available for small embedded systems, which address this issue and simplify building modular, reusable software. Projects such as Contiki [Dunkels et al., 2004], TinyOS [Levis et al., 2005], FreeRTOS [Amazon Web Services, 2020], Zephyr [Zephyr Project, 2020], RIOT [Baccelli et al., 2018], Tock [Levy et al., 2017], and Apache Mynewt [Apache Software Foundation, 2020] significantly reduce development time and ease building – and building upon – maintainable software. Development in this domain has already been going on for decades [Levis, 2012], and there are projects, which undergo evolutionary redesigns, such as Contiki-NG [Oikonomou et al., 2022].

An additional benefit of bespoke platform-agnostic, open-source software frameworks is the broad support for hardware platforms of various manufacturers. It helps to minimize the risk of vendor lock-ins and grants flexibility in selecting the best suitable components for a targeted task. With sufficient hardware abstraction layers, developers of embedded IoT applications are relieved from the intricacies of low-level hardware specifics. Low-level communication buses, hardware-accelerated crypto operations [Boeckmann et al., 2022], timers, and other MCU peripherals are therefore easily accessible via hardware-agnostic application programming interfaces (APIs). To verify that different hardware-specific implementations interact with the physical world according to their specification, continuous integration tests can be conducted on automated infrastructure for Hardware-in-the-Loop testing [Weiss et al., 2021].

Many IoT OSs also ship a big number of natively supported sensor and actuator drivers, which enable the running software to bridge the gap between virtual logic and our physical world. In order to distribute this information, the network stack supplies network access and primitives, such as UDP sockets and protocol implementations for RESTFUL communication patterns (CoAP [Shelby et al., 2014]). A versatile set of link technologies is supported by modern IoT OSs to offer suitable connectivity for, often orthogonal, application requirements, e.g. to prioritize power, range, or bandwidth. Common examples are WiFi, local mesh, and ad hoc networking (e.g. via IEEE 802.15.4 [IEEE 802.15 Working Group, 2016] and Bluetooth), long-range communication via LoRa and other sub-GHz technologies.

This provides means for interoperation and coordination between multiple networked endpoints across the Internet. Via the Internet Protocol (IP), already billions of machines are connected, and with the adoption of IPv6, it is ready for massive long-term growth. Being limited by strict memory and power constraints of low-power and lossy IoT networks, demands for a light-weight adaptation layer like 6LoWPAN do not interfere with the technical requirements of IPv6 (e.g. MTU size).

15.4.2 Smart City Infrastructure

No smart environment can be conceived of without extending the concept to cities. Poorly managed cities exacerbate pollution, increase impervious area, and transform the configuration, composition, and context of land covers, thus having both direct and indirect contributions to the quality of the environments surrounding them [Yu et al., 2013]. Smart cities, on the other hand, enable efficient use of resources, accountability, transparency, and decision based on scientific evidence. Not surprisingly, their realization also requires involvement on many levels: laws and regulations on the basis of which they are established; availability and accessibility of data, inhabitants committed to the realization of the vision, and technical solutions to actually make it happen. Despite these challenges, the concept of smart cities is currently being transformed from a theoretical concept and experimental pilot studies to real-world operation.

International initiatives and national standardization bodies reflect this with their work on defining and unifying reusable smart city architectures. The *European Innovation Partnership on Smart Cities and Communities* (EIP-SCC) is supported by the European Commission and defines a reference architecture for implementing urban platforms for smart cities, with an interoperable, vendor-agnostic approach based on open standards [Heuser et al., 2017]. This initiative was also picked up by national standardization, such as DIN SPEC 91357, which extends the EIP-SCC architecture with additional city- and community-specific capabilities. Standardization in this domain is still an open topic, as can be seen with the soon-to-be-released DIN SPEC 91387, which aims to define a unified architecture specifically for urban digital twins and will cover usage scenarios on a technical, as well as the user and decision-maker perspective [Schubbe et al., 2023].

Similar trends can also be seen on the more local legislation level. For example, the city of Hamburg passed a new progressive law for transparency in October 2012. The law obliges all state agencies to proactively provide access to public data for all citizens and interested individuals, as long as it does not interfere with personal privacy rights [Herr et al., 2017]. It is meant to give people better insight and control of government actions, to foster democratic processes, and it warrants direct and immediate access to information.

While standardization and legislation are regularly being updated, system implementations based on previously described reference architectures are already deployed and used in practice. The urban data platform of the city of Hamburg, which follows definitions by the EIP-SCC and DIN SPEC 91357, operates as an integrated system-of-systems. It provides open- and restricted-access data of public authorities and third parties in the form of geospatial data, documents, and live sensor data. Many categories are covered, such as education, science, health, environment, culture, urban infrastructure and development, and traffic. Based on this data, it runs cloud services for analytics, mobile app integrations, and end-user web applications to interact with the system and access information. Its web services provide OGC compatible REST-APIs (e.g. the SensorThings API [Liang et al., 2021]) and data models for unified access and integration using HTTP and MQTT.

In order to identify relevant data in such smart environments, it is also crucial to be able to find available data and understand its meaning and structure. Therefore, metadata describing the published data can be accessed via a central online service called *MetaVer*, where currently 8 of 16 German states index their public metadata [Lukas et al., 2023].

A publicly available cockpit displays metadata and statistics about the urban data platform usage.[3] As of December 2023, it states in its yearly statistics that it has over 300,000 unique monthly visitors, more than 600 million requests were made, and almost 12 million downloads were served. It also integrates more than 6500 sensors, 578 different datasets, and 114 apps. Some examples for information provided by these applications are availability information on electronic vehicle chargers, rental bikes, and parking lots, locations of public outdoor swimming sites with recent water quality measurements, noise level of traffic within the city, and drinking water quality at different locations in the city.

3 https://geoportal-hamburg.de/udp-cockpit.

In addition to just giving access to individual raw information, there are also examples where the combination of different open data sources allows for building new applications. One example for such a project is PrioBike [Krämer, 2022], which uses live traffic data, public map data, and speed measurements of a navigation app or sensor installation, to provide a better biking experience for cyclists, by prioritizing bikes over motorized vehicles via dynamic traffic light control. Another example is a platform called *Mundraub* [Gildhorn, 2023], a community project, which collects locations where edible fruits, nuts, and herbs, can be collected in public spaces. It integrates data from Open Street Map, new findings contributed by the community, and data of the city-wide tree register, which is publicly available via the urban data platform.

15.5 Open Issues and Conclusions

In this chapter, we presented two different use cases – water quality monitoring and urban air quality monitoring – to demonstrate the wide scope and usefulness of low-power and energy-neutral IoT. The deployments clearly demonstrate the potentials of IoT – low-cost, scalability, high spatial, and temporal resolution – but also highlight some formidable challenges, including the need to process, communicate, and store a large amount of data. The use cases also demonstrate several uncertainties arising from the dynamics of the deployment environments, including strong link quality fluctuations due to excessive weather conditions and the mobility of nodes, difficulty of predicting harvestable energy due to sudden changes in the deployment environment (for our case, sudden changes in the driving schedule of the vehicle carrying one of our IoT devices), and frequent disconnection of links and high packet loss due to the movement of water. Resolving these issues is critical to enable reliable and resilient IoT. We expect future IoT devices to leverage more dynamic self-adaption mechanisms, which readjust the hardware configuration for most efficient execution of variable software tasks [Rottleuthner et al., 2022]. Moreover, accurate models of the deployment environments are needed not only to account for the impacts of external factors but also to overcome them. Addressing some of these challenges is the focus of our future research.

Bibliography

M. R. Aguilar and J. San Román. *Smart Polymers and Their Applications*. Woodhead Publishing, 2019.

N. Ahmed, D. De, and I. Hussain. Internet of Things (IoT) for smart precision agriculture and farming in rural areas. *IEEE Internet of Things Journal*, 5(6):4890–4899, 2018.

V. Albino, U. Berardi, and R. M. Dangelico. Smart cities: Definitions, dimensions, performance, and initiatives. *Journal of Urban Technology*, 22(1):3–21, 2015.

R. Altenburger, W. Brack, R. M. Burgess, W. Busch, B. I. Escher, A. Focks, L. M. Hewitt, B. N. Jacobsen, M. L. de Alda, S. Ait-Aissa, and T. Backhaus. Future water quality monitoring: Improving the balance between exposure and toxicity assessments of real-world pollutant mixtures. *Environmental Sciences Europe*, 31(1):1–17, 2019.

Amazon Web Services. FreeRTOS Real-time operating system for microcontrollers. https://www.freertos.org/, last accessed 30-November-2020, 2020.

American Lung Association. State of the air 2020, 2020.

Apache Software Foundation. Apache Mynewt. https://mynewt.apache.org, last accessed 17-July-2020, 2020.

E. Baccelli, C. Gündogan, O. Hahm, P. Kietzmann, M. Lenders, H. Petersen, K. Schleiser, T. C. Schmidt, and M. Wählisch. RIOT: An open source operating system for low-end embedded devices in the IoT. *IEEE Internet of Things Journal*, 5(6):4428–4440, Dec 2018.

M. Ben-Daya, E. Hassini, and Z. Bahroun. Internet of Things and supply chain management: A literature review. *International Journal of Production Research*, 57(15-16):4719–4742, 2019.

L. Boeckmann, P. Kietzmann, L. Lanzieri, T. C. Schmidt, and M. Wählisch. Usable security for an IoT OS: Integrating the zoo of embedded crypto components below a common API. In *Proceedings of Embedded Wireless Systems and Networks (EWSN'22)*, pages 84–95, New York, USA, Oct 2022. ACM. URL https://dl.acm.org/doi/10.5555/3578948.3578956.

J. Chen and G. Hoek. Long-term exposure to PM and all-cause and cause-specific mortality: A systematic review and meta-analysis. *Environment International*, 143:105974, 2020. ISSN 0160-4120.

A. Dunkels, B. Grönvall, and T. Voigt. Contiki-a lightweight and flexible operating system for tiny networked sensors. In *Proceedings of IEEE Local Computer Networks (LCN)*, pages 455–462, Los Alamitos, CA, USA, 2004. IEEE Computer Society.

K. Gildhorn. Mundraub. https://mundraub.org, 2023. [Online; accessed 11-December-2023].

M. Herr, C. Müller, B. Engewald, A. Piesker, and T. Ritter. Abschlussbericht zur Evaluation des Hamburgischen Transparenzgesetzes. Technical report, Institut für Gesetzesfolgenabschätzung und Evaluation, 2017.

L. Heuser, J. Scheer, P. den Hamer, and B. de Lathouwer. Reference Architecture & Design Principles. Technical report, EIP-SCC Work Stream 2, Sept 2017. Version 1.00.

IEEE 802.15 Working Group. IEEE Standard for Low-Rate Wireless Networks. Technical Report IEEE Std 802.15.4™–2015 (Revision of IEEE Std 802.15.4-2011), IEEE, New York, NY, USA, 2016.

A. Khanna and S. Kaur. Evolution of Internet of Things (IoT) and its significant impact in the field of precision agriculture. *Computers and Electronics in Agriculture*, 157:218–231, 2019.

D.s Krämer. PrioBike-HH. https://www.hhva.de/presse/2022/priobike-hh/priobike-hh2, 2022. [Online; accessed 11-December-2023].

J. U. Lemm, M. Venohr, L. Globevnik, K. Stefanidis, Y. Panagopoulos, J. van Gils, L. Posthuma, P. Kristensen, C. K. Feld, J. Mahnkopf, and D. Hering. Multiple stressors determine river ecological status at the European scale: Towards an integrated understanding of river status deterioration. *Global Change Biology*, 27(9):1962–1975, 2021.

P. Levis. Experiences from a decade of TinyOS development. In *Proceedings of USENIX OSDI*, pages 207–220, Berkeley, CA, USA, 2012. USENIX Association.

P. Levis, S. Madden, J. Polastre, R. Szewczyk, K. Whitehouse, A. Woo, D. Gay, J. Hill, M. Welsh, E. Brewer, and D. Culler. TinyOS: An operating system for sensor networks. In W. Weber, J. M. Rabaey, and E. Aarts, editors, *Ambient Intelligence*, pages 115–148. Springer, Berlin Heidelberg, 2005.

A. Levy, B. Campbell, B. Ghena, D. B. Giffin, P. Pannuto, P. Dutta, and P. Levis. Multiprogramming a 64kB computer safely and efficiently. In *Proceedings of the 26th Symposium on Operating Systems Principles*, SOSP '17, page 234–251, New York, NY, USA, Oct 2017. Association for Computing Machinery.

S. Liang, T. Khalafbeigi, H. van Der Schaaf, B. Miles, K. Schleidt, S. Grellet, M. Beaufils, and M. Alzona. OGC SensorThings API Part 1: Sensing Version 1.1, 2021.

B. Lukas, J. Beulke, K.-U. Krause, V. Gorsic, I. Tauber, H. Roos, R. Seidel, and B. Köther. Metadatenverbund metaver. https://metaver.de, 2023. [Online; accessed 11-December-2023].

D. Ma, G. Lan, M. Hassan, W. Hu, and S. K. Das. Sensing, computing, and communications for energy harvesting IoTs: A survey. *IEEE Communications Surveys & Tutorials*, 22(2):1222–1250, 2020.

K. Mandula, R. Parupalli, C. H. A. S. Murty, E. Magesh, and R. Lunagariya. Mobile based home automation using Internet of Things (IoT). In *2015 International Conference on Control, Instrumentation, Communication and Computational Technologies (ICCICCT)*, pages 340–343. IEEE, 2015.

L. C. Messer, J. S. Jagai, K. M. Rappazzo, and D. T. Lobdell. Construction of an environmental quality index for public health research. *Environmental Health*, 13(1):1–22, 2014.

R. K. Mishra. Fresh water availability and its global challenge. *British Journal of Multidisciplinary and Advanced Studies*, 4(3):1–78, 2023.

A. Mishra, A. Alnahit, and B. Campbell. Impact of land uses, drought, flood, wildfire, and cascading events on water quality and microbial communities: A review and analysis. *Journal of Hydrology*, 596:125707, 2021.

K. Mouratidis and W. Poortinga. Built environment, urban vitality and social cohesion: Do vibrant neighborhoods foster strong communities? *Landscape and Urban Planning*, 204:103951, 2020.

G. Mulligan. The 6LoWPAN architecture. In *Proceedings of the 4th Workshop on Embedded Networked Sensors*, pages 78–82, 2007.

G. Oikonomou, S. Duquennoy, A. Elsts, J. Eriksson, Y. Tanaka, and N. Tsiftes. The Contiki-NG open source operating system for next generation IoT devices. *SoftwareX*, 18:101089, 2022. ISSN 2352-7110.

H. Petersen, E. Baccelli, M. Wählisch, T. C. Schmidt, and J. Schiller. The role of the Internet of Things in network resilience. In *Internet of Things. IoT Infrastructures. First International Summit, IoT360 2014, Revised Selected Papers, Part II*, volume 151 of *LNICST*, pages 283–296, Berlin, Heidelberg, 2015. Springer.

M. Rottleuthner, T. C. Schmidt, and M. Wählisch. Sense your power: The ECO approach to energy awareness for IoT devices. *ACM Transactions on Embedded Computing Systems (TECS)*, 20(3):24:1–24:25, Mar 2021. doi: 10.1145/3441643.

M. Rottleuthner, T. C. Schmidt, and M. Wählisch. Dynamic clock reconfiguration for the constrained IoT and its application to energy-efficient networking. In *Proceedings of Embedded Wireless Systems and Networks (EWSN'22)*, pages 168–179, New York, USA, Oct 2022. ACM. URL https://dl.acm.org/doi/10.5555/3578948.3578964.

M. Salehi. Global water shortage and potable water safety; today's concern and tomorrow's crisis. *Environment International*, 158:106936, 2022.

N. Schubbe, M. Boedecker, M. Moshrefzadeh, J. Dietrich, M. Mohl, M. Brink, N. Reinecke, S. Tegtmeyer, and P. Gras. Urbane digitale zwillinge als baukastensystem: Ein konzept aus dem projekt connected urban twins (cut). *ZfV-Zeitschrift für Geodäsie, Geoinformation und Landmanagement*, 148(zfv 1/2023):14–23, 2023.

Z. Shelby, K. Hartke, and C. Bormann. The Constrained Application Protocol (CoAP). RFC 7252, IETF, Jun 2014. URL https://doi.org/10.17487/RFC7252.

S. Sudevalayam and P. Kulkarni. Energy harvesting sensor nodes: Survey and implications. *IEEE Communications Surveys & Tutorials*, 13(3):443–461, Mar 2011.

A. M. Toli and N. Murtagh. The concept of sustainability in smart city definitions. *Frontiers in Built Environment*, 6:77, 2020.

K. Weiss, M. Rottleuthner, T. C. Schmidt, and M. Wählisch. PHiLIP on the HiL: Automated multi-platform OS testing with external reference devices. *ACM Transactions on Embedded Computing Systems (TECS)*, 20(5s):91:1–91:26, Sept 2021. doi: 10.1145/3477040. Selected for presentation at EMSOFT 2021.

J. Wolfram, S. Stehle, S. Bub, L. L. Petschick, and R. Schulz. Water quality and ecological risks in European surface waters–monitoring improves while water quality decreases. *Environment International*, 152:106479, 2021.

M. Wollschlaeger, T. Sauter, and J. Jasperneite. The future of industrial communication: Automation networks in the era of the Internet of Things and Industry 4.0. *IEEE Industrial Electronics Magazine*, 11(1):17–27, 2017.

D. Yu, P. Shi, Y. Liu, and B. Xun. Detecting land use-water quality relationships from the viewpoint of ecological restoration in an urban area. *Ecological Engineering*, 53:205–216, 2013.

Zephyr Project. Zephyr. https://www.zephyrproject.org, last accessed 17-July-2020, 2020.

Index

Wireless Sensor Networks in Smart Environments: Enabling Digitalization from Fundamentals to Advanced Solutions, First Edition. Edited by Domenico Ciuonzo and Pierluigi Salvo Rossi.

f

g

h

i

j

k

l

m

n

o

p

IEEE Press Series on Sensors

Series Editor: Vladimir Lumelsky, Professor Emeritus, Mechanical Engineering, University of Wisconsin-Madison

Sensing phenomena and sensing technology is perhaps the most common thread that connects just about all areas of technology, as well as technology with medical and biological sciences. Until the year 2000, IEEE had no journal or transactions or a society or council devoted to the topic of sensors. It is thus no surprise that the IEEE Sensors Journal launched by the newly-minted IEEE Sensors Council in 2000 (with this Series Editor as founding Editor-in-Chief) turned out to be so successful, both in quantity (from 460 to 10,000 pages a year in the span 2001–2016) and quality (today one of the very top in the field). The very existence of the Journal, its owner, IEEE Sensors Council, and its flagship IEEE SENSORS Conference, have stimulated research efforts in the sensing field around the world. The same philosophy that made this happen is brought to bear with the book series.

- *Magnetic Sensors for Biomedical Applications*
  Hadi Heidari and Vahid Nabaei
- *Smart Sensors for Environmental and Medical Applications*
  Hamida Hallil and Hadi Heidari
- *Whole-Angle MEMS Gyroscopes: Challenges, and Opportunities*
  Doruk Senkal and Andrei M. Shkel
- *Optical Fibre Sensors: Fundamentals for Development of Optimized Devices*
  Ignacio Del Villar and Ignacio R. Matias
- *Pedestrian Inertial Navigation with Self-Contained Aiding*
  Yusheng Wang and Andrei M. Shkel
- *Sensing Technologies for Real Time Monitoring of Water Quality*
  Libu Manjakkal, Leandro Lorenzelli, and Magnus Willander
- *Solid-State Sensors*
  Ambarish Paul, Mitradip Bhattacharjee, and Ravinder Dahiya
- *Biosensors: Nanomaterials, Approaches, and Performance-Enhancement Strategies*
  Baljinder Kaur, Santosh Kumar, and Brajesh Kumar Kaushik
- *Micro Electromechanical Systems (MEMS): Practical Lab Manual*
  Sanket Goel
- *Wireless Sensor Networks in Smart Environments: Enabling Digitalization from Fundamentals to Advanced Solutions*
  Domenico Ciuonzo and Pierluigi Salvo Rossi

Printed and bound by CPI Group (UK) Ltd, Croydon, CR0 4YY
17/09/2026
14964329-0007